Climatology from Satellites

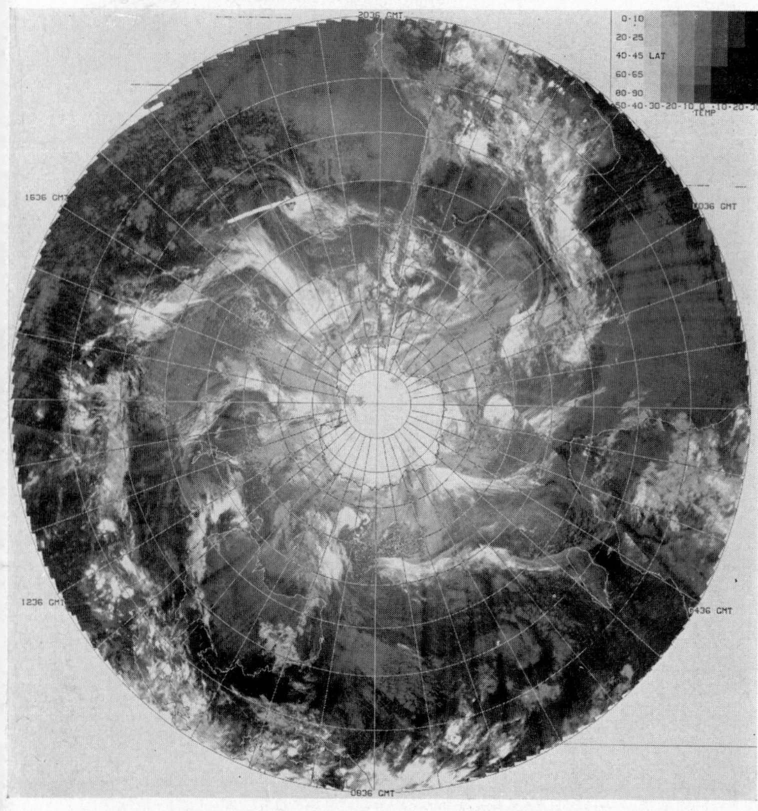

Digitized polar projections of global weather after dark on 24 October 1970, from Itos-1 infra-red evidence. The block diagrams in the upper right-hand corners describe the relationships, at various latitudes, of image tones and surface (or cloud-top) temperatures in °C.
(See p. 43)
(E.S.S.A. photographs)

Climatology from Satellites

E. C. BARRETT

METHUEN & CO LTD
11 New Fetter Lane, London EC4

First published in 1974 by
Methuen & Co Ltd,
11 New Fetter Lane, London EC4

ISBN 0 416 65940 3

Filmset in Photon Times 11 on 12 pt by
Richard Clay (The Chaucer Press) Ltd, Bungay, Suffolk
Printed in Great Britain by
Fletcher & Son Ltd, Norwich

Distributed in the U.S.A. by
HARPER & ROW PUBLISHERS, INC.
BARNES & NOBLE IMPORT DIVISION

Contents

Preface

We all know that feeling of despair caused by just missing a bus. One runs vainly after the desired vehicle, knowing full well that it can only accelerate away despite one's efforts to overtake it. Working and writing in any field involving the scientific exploration of space quickly recalls our experiences as frustrated would-be passengers. New spacecraft, instruments, analytical techniques and absorbing satellite discoveries come and go with ever-increasing speed; it is more than usually difficult to ensure that one's grasp of the situation is up-to-date.

This daunting scenario has led me many times to question the worthwhileness of an attempted synthesis of climatology from satellites. But each time I have been reminded of the increasing need for someone to try to piece together, within a global framework, the welter of fascinating atmospheric facts and discoveries that have emerged from more than ten years of satellite studies. Stock-taking is a useful practice from the research point of view. It is also of value to potential users of a particular range of products. It is essential for students trying to identify contemporary scientific bus-stops, the services and their future destinations.

This book attempts to gather, organize and, where possible, inter-relate many satellite-based findings pertinent to the study of climatology, and to do so in such a way that its topics are reasonably rounded and self-complete. Many of the works to which references are made have not been available hitherto in readily accessible publications. Clearly, some aspects of the atmosphere are more comprehensively surveyed than others, and some readers may detect substantial areas of climatology treated scantily or not at all. Others will dislike the lack of standardization in the units employed by workers whose results are summarized. I believe, however, that my coverage reflects reasonably the present state of the game and that any greater effort to smooth or standardize the contents would misrepresent or otherwise do an injustice to some work while according more than justice to the rest.

In these days of growing preoccupation with remote sensing generally, the approaches and findings of satellite climatology – tentative and fragmentary though they may be – should prove of interest not only to students of the atmosphere but also to many others who may wish to examine the planet Earth from space.

My thanks are due to all who have helped me with this project. In particular, I thank Mr Vincent J. Oliver, Chief, Applications Branch, National Environmental Satellite Center, Washington, for his most helpful comments on the draft manuscript. Several members of his staff, especially Ralph K. Anderson and Frances Parmenter also helped greatly in resolving points of detail in the text. Dr T. H. Vonder Haar of the Department of Atmospheric Science, Colorado State University, gave kind assistance in checking chapters 3 and 6, and Professor S. Gregory, Department of Geography, University of Sheffield, made helpful suggestions with respect to chapters 1 and 14. My thanks are also due to Dr M. G. Hamilton, recently appointed to the staff of the Edgbaston Observatory, Birmingham, for discussions during his period of research at Bristol. Last, but not least, my wife Gillian has contributed much by her assistance with the Indexes, through her typing skills, and, most of all, with her unfailing encouragement. Needless to say, the responsibility for any errors that remain is mine alone.

E. C. Barrett,
Bristol.
May, 1973

Acknowledgments

The author and the publisher would like to thank the following organizations and individuals for supplying material for the plates and granting permission for its use:

The United States Department of Commerce; National Oceanic and Atmospheric Administration; National Aeronautics and Space Administration; the U.S. Air Force; the University of Wisconsin Meteorology Department; the American Meteorological Society; the Stanford Research Institute, California; Prof. V. E. Suomi; Dr J. Kornfield; Dr A. F. Hasler; Dr K. J. Harrison; Prof. T. T. Fujita; T. Izawa; K. Watanabe; John A. Leese; C. S. Novak; Dr Chih-Pei Chang; Lewis J. Allison; E. P. McClain and D. R. Baker.

Especial thanks are due to Leslie A. Watson, Jr, of the Visual Products Support Branch of N.O.A.A. for his invaluable assistance in obtaining many of these illustrations.

Thanks are also due to the following for permission to reproduce copyright diagrams:

American Meteorological Society for figs. 2.11, 2.12, 5.16, 9.20, 9.21, 9.22, 12.3, 12.4, 13.5; ARACON for figs. 4.5, 4.6 (a), (b), (c), 14.5; Edward Arnold Ltd for figs. 2.16, 4.7, 4.8, 5.2, 5.3, 9.7, 9.13, 10.5, 10.7, 12.9; The Association of American Geographers for figs. 14.2, 14.3, 14.6, 14.7, 14.8; The Director of Aerospace Services, U.S.A.F. for figs. 4.9, 12.8, 12.10; Dr R. G. Barry & R. J. Chorley for figs. 1.1, 2.8, 9.15; Prof. J. Bjerknes for fig. 9.9; Collins Publishers for fig. 14.1; John H. Conover for fig. 4.4; Elsevier for fig. 12.11; Prof. W. H. Gray for figs. 10.3 and 10.4; The Controller, Her Majesty's Stationery Office for figs. 2.5, 8.3, 8.4; John

A. Leese for figs. 2.14, 2.15, 5.16; Longman Group Ltd for figs. 2.2, 2.4, 2.17, 5.5, 8.6, 8.7, 8.8, 9.6; Dr B. J. Mason, F.R.S. & Macmillan (Journals) Ltd for fig. 8.15; McGraw-Hill Book Company for figs. 7.9, 8.5, 8.14; Microforms International Marketing Corporation for figs. 5.11, 5.12, 5.13, 5.14, 5.15, 6.3, 6.5, 6.10, 8.9, 8.10, 8.11, 9.2, 9.8, 10.1; National Aeronautics and Space Administration for figs. 2.3, 2.13, 3.1, 3.2, 3.3, 4.2, 4.3, 4.13, 13.1, 13.2; New Zealand Meterological Service for fig. 11.2; Oxford University Press for fig. 2.7; *Petermanns Geographische Mitteilungen* for fig. 11.1; Dr P. K. Rao for figs. 4.10, 4.11; The Royal Geographical Society for figs. 7.10, 7.11, 7.13, 7.14, 11.12, 11.13, 11.14, 14.1; The Royal Meteorological Society for figs. 3.4, 3.5, 4.14, 9.1, 9.16, 10.16, 10.17, 11.7, 11.8, 12.2; Prof. S. C. Serra for fig. 10.11; The University Press of Hawaii for fig. 4.13; World Meteorological Organization for figs. 5.6, 10.2, 12.6.

Common satellite abbreviations employed in the text

(Acronymic terms are italicized: these are almost invariably rendered without stops after each letter. Some of the other abbreviations may be rendered without stops by individual authorities, but practice varies.)

A.P.T.	Automatic Picture Transmission
A.T.S.	Applications Technology Satellite
A.V.C.S.	Advanced Vidicon Camera System
C.D.A.	Command & Data Acquisition Stations
COSPAR	Committee for Space Research
ERTS	Earth Resources Technology Satellite
E.S.S.A.	Environmental Sciences Services Administration
ESSA	Environmental Survey Satellite
F.M.R.	Final Meteorological Radiation Tape Products
GHOST	Global Horizontal Sounding Technique
GOES	Geostationary Operational Environmental Satellite
H.R.I.R.	High Resolution Infra-red Radiometer
I.D.C.S.	Image Dissector Camera System
IRIS	Infra-red Interferometer Spectrometer
ITOS	Improved Tiros Operational Satellite
L.R.I.R.	Low Resolution Infra-red Radiometer
M.R.I.R.	Medium Resolution Infra-red Radiometer
M.S.S.C.C.	Multicolour Spin-scan Cloud camera
MUSE	Monitor of Ultra-violet Solar Energy
N.A.S.A.	National Aeronautics & Space Administration, U.S.A.

N.E.S.C. National Environmental Satellite Center, Washington, D.C., U.S.A.
N.E.S.S. National Environmental Satellite Service, Washington, D.C., U.S.A.
N.M.C. National Meteorological Center, Washington, D.C., U.S.A.
N.O.A.A. National Oceanic & Atmospheric Administration
NOAA National Oceanic & Atmospheric Administration environmental satellite
N.W.R.C. National Weather Records Center, Asheville, Tennessee
OSO Orbiting Solar Observatory satellite
S.C.R. Selective Chopper Radiometer
SIRS Satellite Infra-red Spectrometer
S.S.C.C. Spin-scan Cloud Camera
T.H.I.R. Temperature Humidity Infra-red Radiometer
TIROS Television & Infra-red Observation Satellite
TOS Tiros Operational Satellite
U.S.W.B. United States Weather Bureau
V.C.S. Vidicon Camera System
V.H.R.R. Very High Resolution Radiometer
W.M.O. World Meteorological Organization
W.W.W. World Weather Watch

PART ONE
Introduction

1 The nature and scope of climatology from satellites

Climatology, perhaps more than most fields of scientific inquiry, has been defined variously in the past. Some authorities have referred to it as a science of the atmosphere (e.g. Rumney 1968), while others have stressed its essentially geographical approach and emphasis (e.g. Kendrew 1957). Still other writers have adopted more bipartisan attitudes, e.g. Shaw (1926), for whom climatology was concerned with general weather studies for particular localities. Most agree, however, that climatology is distinct from the closely allied science of meteorology in that it is concerned more with the results of processes at work in the atmosphere and less with their instantaneous operations, while being distinct from geography in that its interest is focused mainly on only a part of the total physical environment of man. Certainly climatology is located where meteorology and geography overlap. Its interest in the atmosphere generally involves periods longer than those with which most meteorologists are preoccupied, while its areal coverage often necessitates an approach along geographical lines.

In order to appreciate the full significance of the satellite contribution to climatology it is necessary to begin by examining in more detail the nature and scope of climatology itself. The major subdivisions may be grouped under two headings: 'topical' and 'scalar' climatology.

Topical subdivisions of climatology

Regional climatology

This involves the description of climates over selected (usually large) areas of the surface of the planet earth. In the past it has depended heavily upon accumulated conventional observations of surface weather elements, such as temperature, humidity, cloudiness, sunshine, rainfall and wind. Traditionally, monthly and annual means have been its chief cornerstones. More recently, comparisons between different climatic stations have been made also by more advanced statistics ranging from medians, modes and standard deviations to appropriate harmonics and eigenvectors. Thus the search continues for better representations of salient features of observed weather through time, nowadays not only at the earth/atmosphere interface but also at supplementary higher levels.

Perhaps the ultimate goal of regional climatology has traditionally been the grouping of areas that are mutually alike within regional climatic classifications. The most widely publicized classification or family of classifications was developed by W. Köppen, whose concepts evolved progressively from 1884 to 1936. Today, however, criticism of his rather unwieldy, inconsistent schemes is rife (Wilcock 1968). Other, contrasting, classifications of popular appeal have included those of Miller (1957) and Thornthwaite (1948). These matters will be discussed more fully in chapter 14.

Synoptic climatology

This younger approach seeks to provide a fresh basis for regional climatology not subject to such abstract statistical approaches. Synoptic climatology involves the study of general weather characteristics through their relationships with patterns of airflow, or 'airflow types' as they have become known (Barry and Chorley 1968). Such analyses depend more on the categorization of pressure map patterns than upon analyses of the more traditional climatic statistics. Thus new problems of classification are encountered, but it has been argued that such approaches are worthwhile since synoptic climatological models 'allow climatological averages to be calculated on a realistic synoptic basis rather than for arbitrary time periods' (Barry 1963). Fig. 1.1 illustrates the range of types that may be invoked in such a study. It depicts average air mass frequencies over north-west Europe in January (after Belasco 1948).

Physical climatology

Rumney (1968) writes that 'physical climatology has developed through the last one hundred years to describe, classify and explain the highly complex and eternally variable relationships among the many qualities of the atmosphere'. Landsberg (1958), at the outset of his survey of this field, observed that 'the field of climatology was being pre-empted by the geographical approach. This was primarily descriptive in nature.' He went on to say that 'the task of physical climatology . . . is to collect and analyse climatological data and *to reveal causes and effects of climate*' (the emphasis is the present author's). Thus, while the physical climatologist may process many of the

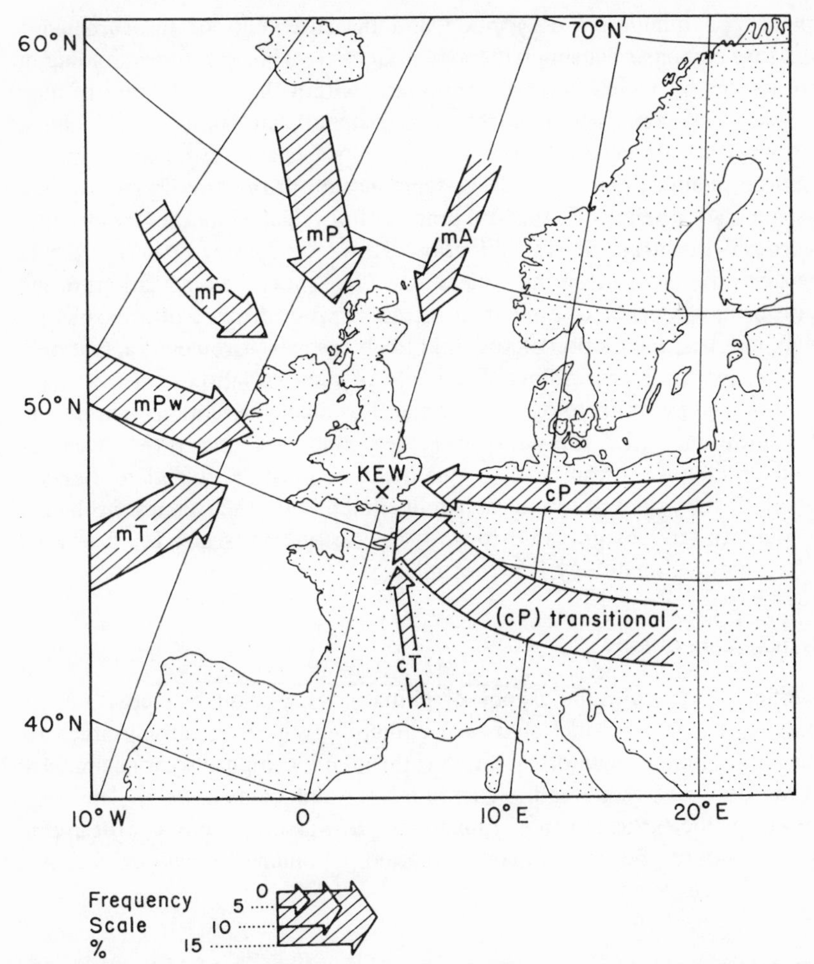

Fig. 1.1 Average air-mass frequencies for Kew (London) in January. Anticyclonic types are included according to their direction of origin. (Based on Belasco 1948; from Barry & Chorley 1968)

data of interest to the regional climatologist, he does this for a different basic purpose. As a result of this much greater emphases are placed on explanation, physical laws and relationships, especially those concerned with the global energy and water balance regimes of the earth and its atmosphere (see e.g. Sellers 1969).

Dynamic climatology

As in most other environmental sciences first nurtured by physical geography, there has been a strong tendency in recent years for less emphasis to be placed on qualitative description and measurement for measurement's sake. More emphasis is being laid today upon increasing our understanding of the processes by which observed variations within the environment of man are caused and come about. Barry (1967) stated that 'the aim of dynamic climatology is ultimately to provide a comprehensive explanation of the general circulation; that is, the large-scale motion of the atmosphere in time and space'. So it seems that this is a branch of physical climatology especially concerned with larger scales of motion. However, Sutton (1965) struck a somewhat different note on the same theme: 'The short-lived deviations of the general circulation must be regarded as an intrinsic feature of atmospheric motion, much as the motion of any fluid tends towards turbulent fluctuations, whilst preserving its mean flow. The study of such variation . . . is the main theme of dynamic climatology.' Since the 'turbulent fluctuations' within the fluid continuum which the atmosphere resembles are of a lower order of magnitude than its mean flow, a qualification must be added to Barry's assertion of the principal aim of dynamic climatology: the trickiest problems it encounters tend to gravitate towards the secondary scale, i.e. that with which synoptic climatology is preoccupied.

Statistical climatology

The different topical subdivisions of climatology obviously merge and/or overlap, and it is clear that statistical climatology is more closely allied to regional climatology than other varieties through its emphasis upon summarizing raw and processed weather data in table, graph or map forms. This may be viewed as the earliest form of climatology, as well as one of the most useful – as the standard products of many climatology branches of national meteorological services testify.

Bioclimatology

Next in our brief survey of systematic branches of climatology, mention must be made of the group of climatologies subsumed under the family heading 'Bioclimatology'. Here attention is focused upon atmospheric factors affecting growth and health within the biosphere. Understandably, an important subsection is concerned directly with man, and his environments, both built and natural. Human comfort and atmospheric pollution figure prominently among its studies. American and Russian manned space flight programmes have contributed significantly to bioclimatological knowledge with special reference to man and other mammals. Other important subsections of this science are agroclimatology, which is concerned with crop production, and medical geography, concerned with patterns of human health, disease and death.

Historical climatology

Here the emphasis is upon the development of climates through time, and many of the climatic irregularities that are smoothed out of data being prepared for other climatological analyses assume great significance in their own right. Studies of climatic change can be based on actual measurements of atmospheric parameters only as far back as the instrumental records extend. Even in the best- locumented regions of north-west Europe this is only to the late eighteenth century. However, other forms of documentary evidence may be invoked with care to give indications of climatic change in historical times (see e.g. Lamb 1967), but this approach yields acceptable results back to only about the eleventh century. Longer-term changes of climate can be inferred best from archaeological, biological and geological evidences, back to the realms of palaeoclimatology, which is concerned with large-scale, long-term variations in global climates through geological time. Studies of climatic change are potentially significant for improved methods of long-term climatic forecasting.

Applied climatology

Evidently all the previously mentioned modes of 'pure' climatology have their own peculiar counterparts within the general scope of 'applied' climatology. In many cases, however, these applied sciences developed rather more slowly, as we may have anticipated remembering that both knowledge and understanding must be possessed before environmental forecasting or engineering become feasible. Historically, the dominant applications of climatological knowledge have been in the earth and biological sciences through accumulated climatic data. We should not underestimate this kind of service still afforded today by climatology to a variety of consumers. However, in recent years, more thought has been given to the development of climatology to permit other potentially valuable applications. Such thought has been stimulated partly by the developing needs of science and industry, but also partly by the continually improving and expanding networks of weather observatories of different kinds.

Various economic studies, many based on benefit/cost analyses (see e.g. Maunder 1970), have demonstrated unequivocally the financial common sense of taking weather and climate much more fully into account in the planning of expensive operations outdoors, and in devising precautions against natural catastrophes. Still other studies have underlined the merits of certain weather modification schemes in support of economic ventures, especially agricultural ventures critically dependent on local climatic factors (see, for example, Smith 1958). At larger scales, climatic considerations are of no small significance in feasibility studies of environmental engineering projects such as those that propose to close the Bering Straits, melt the Arctic ice cap, or reverse the flow of Soviet Russia's major Siberian rivers from northward into the Arctic Ocean to southward into proposed irrigation areas in the more arid heart of Asiatic Russia (see Sawyer 1971). Finally, recent developments in the expanding business of extended-range and long-range forecasting have underlined the fact that 'Meteorology is being confronted with a demand for climatic forecasting for which no adequate scientific basis yet exists' (Lamb 1966). It appears that the problems to be solved in developing an adequate forecasting methodology are climatological rather than meteorological where the forecast period exceeds a few days in length. As more climatological knowledge and understanding become available, so climatology ought to be able to contribute significantly to the solution of long-range forecasting problems.

Scalar subdivisions of climatology

An alternative approach to the subdivision of climatology, popular in some circles, depends upon scalar rather than physical or methodological considerations. The scales of motion systems commonly referred to in the meteorological literature are listed in table 1.1. They apply to such respective features as long waves in the upper tropospheric westerlies, migratory cyclones and anticyclones in middle latitudes, local wind systems and wind gusts and lulls. Clearly these are all parts of a single continuous spectrum of weather systems; in other words, the four groups are by no means discrete. They do provide, however, a useful basis for the subdivision of climatology into the following three sections:

Macroclimatology

This is concerned with features of the climates of substantial parts of the globe, and with the large-scale atmospheric organizations that contribute to them. It embraces considerations of global wind belts identified by mean annual, seasonal and monthly statistical analyses, as well as mean patterns and exchanges of atmospheric energy and moisture. Regional macroclimatology involves characteristic dispositions of long-wave patterns of the atmosphere, and also the general behaviour of synoptic weather perturbations (Barry 1970). As Lee (1966) observed, the geographical distribution of meteorological phenomena is partly controlled by systems larger than themselves. Hence it is proper within macroclimatology not only to consider the large-scale relationships between climatic patterns and earth surface geography, but also the influences of larger atmospheric circulations upon others down to the levels at which the perturbations cease to have significance in the macroscale context. These considerations all apply to the conventional time scale in climatology, namely thirty to thirty-five years. However, in view of the fact that the longer the persistence of a fluctuation of climate, the greater is the area it affects, it should be noted that substantially longer periods of time are properly involved in some studies of variations in the global pattern of climate.

Mesoclimatology

This is seen by some to be concerned with relatively small areas of the earth's surface, between about 10 and 100 km across (cf. 'mesoscale phenomena' in table 1.1). Thus relatively local embroideries upon a larger, more generalized climatic picture are accommodated where the objects of a particular study permit. Urban heat island studies may fall within this category, as well as studies of severe local weather systems such as tornadoes and thunderstorms, whose characteristics not only affect regions of mesoscale, but also are influenced by their dominant weather and topography in turn. Compared with macro- and microclimatology, relatively little work has been done here. As a consequence, this is likely to become one of the big growth areas in climatology as observing networks become denser and climatic techniques and theories get more sophisticated.

Table 1.1
The scales of meteorological motion systems (from Barry 1970)

		Approximate characteristic dimensions			
Motion system		*Horizontal scale (km)*	*Vertical scale (km)*	*Time scale (hours)*	*Total energy**
Macroscale	Planetary waves	5×10^3	10	2×10^2 to 4×10^2	—
	Synoptic perturbations	5×10^2 to 2×10^3	10	10^2	Average depression: 10^{-3}
Mesoscale phenomena		1 to 10^2	1 to 10	1 to 10	Average thunderstorm: 10^{-8}
Microscale phenomena		$< 10^{-1}$	$< 10^{-2}$	10^{-2} to 10^{-1}	Average wind gust: 10^{-17}

* Base 1 = daily solar energy intercepted by the earth.

Microclimatology

Here the climatologist is concerned with fine details of climate, and relatively fine distinctions between quite short periods of time. Microclimatology owes much to the writings of Geiger (1957) and his emphasis on features of the lowest 2 m of the atmosphere contributing to the 'climate near the ground'. According to Huschke (1959), microclimate is 'the fine climatic structure of the air space which extends from the earth's surface to a height where the effects of the immediate character of the underlying surface cannot be distinguished from the general local climate'. Such details are clearly beyond the scope of current meteorological satellite systems.

Climatology from satellites

This title has been chosen carefully to avoid suggesting, implicitly or explicitly, that the advent of space platforms designed specifically for weather observation has led, or is leading, to a new distinctive subdivision of climatology. Certainly a 'satellite climatology' may be conceived *per se*, but only, strictly speaking, in terms of the microclimates within and around the spacecraft themselves. From a technical point of view such a study is quite significant, of course, since many satellite subsystems operate best within readily recognizable ranges of environmental conditions. This fact necessitates the keeping of careful operational checks on satellites through analyses of the so-called 'housekeeping data' they return to earth about themselves. The real importance of climatology from satellites must be that it is able to contribute to several of the areas outlined above. Modern meteorological satellites are fulfilling three most important roles:

1 As observing systems of the earth and its atmosphere. As such, weather satellites have several specially valuable characteristics and abilities. These are discussed below.
2 As highly convenient data collection platforms. Second generation weather satellites are able to collect data from within the atmosphere, or from stations at its base, and to relay those data to specially developed central processing facilities.
3 As communication links between widely spaced ground stations between which large daily exchanges of weather data must take place.

This book is concerned primarily with the first of these three. It deals first with satellites as observing systems; second, it discusses the characteristics of the atmospheric data that satellites return to earth; third, it reviews the early applications of such data to the solution of problems that are essentially climatological in nature. For reasons of both logic and convenience these applications and interpretations of satellite data are organized along geographical lines in chapters 9 to 13, following the more general problems of analysis and interpretation dealt with in chapters 3 to 8. This arrangement reflects the structure of several recent texts on climatology or meteorology. It is chosen here to facilitate comparisons between 'pre-' and 'post-satellite' climatology. Thus the progress that weather satellites have prompted and permitted is underlined.

Although the first specialized meteorological satellite was only orbited early in 1960, satellite data have already contributed much to climatology through analyses of data covering periods ranging from a few days to a few years in length. Initially, a key question seems to be: 'Why are satellites such useful sources of climatological information?' Six reasons may be suggested:

(1) They improve greatly the data coverage of the globe. Surface meteorological stations, unevenly scattered as they are, provide inadequate data from

large parts of the globe even for macroscale climatological purposes. Vetlov (1966) of the Russian Hydrometeorological Service observed that even by then only 10% of the world was adequately provided with conventional meteorological stations for contemporary forecasting programmes. Since the climatological community is largely dependent upon the same data, lack of information has been traditionally the climatologists' chief complaint. Table 1.2 summarizes the immediate objectives of the World Weather Watch (W.W.W.) scheme drawn up in 1967 to improve the network of aerological recording stations. By the time of writing, however, even these fairly modest objectives have not been fully met. Substantial areas in high latitudes, across major ocean surfaces and around the world in low latitudes are still relatively data-remote. Weather satellites have done much already to bring even the most inhospitable and unfrequented areas into view, bringing those obscure regions adequately into the global climatological picture for the very first time. The operational weather satellite systems that have operated since February 1966 have been of particular significance in this respect, first the Essa satellite system and lately Noaa.

Table 1.2
The maximum average spacing between conventional meteorological stations aimed at by the W.W.W. (after W.M.O. 1967)

Type of area	*Level*	*Maximum average spacing (km)*
Continental	Upper air *	1000
Oceanic, with suitably distributed islands	Upper air *	1000
Open oceanic	Upper air *	1500
Oceanic	Surface	1000
Continental	Surface	500

* Observations to be made twice daily, at 0 and 1200 G.M.T.

(2) Their data are internally more homogeneous than, say, cloud or radiation data derived from numerous surface stations. Despite the elaborate precautions taken to standardize the exposure and operation of conventional instruments, national differences still exist notwithstanding the efforts of the W.M.O. to reduce them, local peculiarities are often unavoidable, and each instrument package is unique despite careful calibration. Clear advantages must be inherent in an observing system capable of observing selected variables around most, if not all, of the globe by means of a very few instrument packages. Further advantages must accrue from the associated possibility of operating a single data processing facility for all the resulting circumglobal observations.

(3) Satellite data are spatially continuous across the surface of the globe, whereas most conventional data are point measurements of the atmosphere from selected sample sites. The construction of isopleth maps from satellite data consequently involves less scientific guesswork through interpolation than the construction of comparable maps even for well-documented regions such as North America and western Europe.

(4) Satellite observations are complementary in nature to conventional measurements. Since satellites are remote sensing platforms, investigating the earth/atmosphere system by instruments not in direct contact with that system, it is not surprising that their observations are dissimilar in kind to those made within the atmosphere itself. In some cases satellites observe parameters closely akin to some measured within the atmosphere. In these cases interesting comparisons, cross-corrections and extrapolations are sometimes possible. Where satellite observed parameters are foreign to those observable below, many fresh facts about the atmosphere emerge. The net radiation balance at the top of the earth's atmosphere exemplifies a group of new patterns of key importance in atmospheric science that can now be evaluated from satellite evidence, not just estimated as in pre-satellite days.

(5) Satellite data lend themselves readily to computer processing. Even the photographic data are received as radio signals whose variations can be recorded best as numerical measures of radiation within the visible waveband of the electromagnetic spectrum. Since the volume of new data points acquired each day from a standard Essa or Noaa satellite is probably of the order of $10^9 - 10^{12}$ in magnitude, automatic data processing is obviously essential for any routine programme of satellite data analysis and interpretation.

(6) Satellites can provide a more frequent data coverage than standard meteorological observatories reporting every six, twelve or twenty-four hours. Although low altitude satellites observe given target areas usually only once each day, geosynchronous or geostationary satellites (e.g. the A.T.S. or Applications Technology Satellites) are able to observe their targets at twenty minute intervals by virtue of their apparently fixed positions above chosen points on the equator (see p. 27). Future A.T.S. satellites will interrogate their targets even more frequently than this. Such highly repetitive observation is already bearing fruit in meteorology. Time-lapse cloud films of the tropics from the satellite viewpoint have been prepared by V. E. Suomi and his co-workers in the Space Science and Engineering Center, Wisconsin, U.S.A. (Suomi 1970). These have revealed fresh aspects of the nature and extent of atmospheric mobility evident in the progressive displacements and developments of cloud fields. Some new models of significance to climatology have already emerged from A.T.S. studies, for example the anticyclonic cells and associated 'burst bands' off western Central America described by Fujita (Fujita *et al.* 1969; see p. 257). Doubtless more such models will be suggested in the future.

Undoubtedly there must be enormous scope for climatological studies based wholly or in part upon data from weather satellites, but such satellites will never entirely replace conventional weather stations as atmospheric observatories. In the earliest days of satellite meteorology Neiberger and Wexler (1961) suggested that 'meteorologists who have been following the data received from Tiros I and Tiros II are convinced that weather satellites will have a revolutionary impact on their science'. To suggest that any innovation is 'revolutionary', however, is bound to elicit strong reactions from more cautious men, and many atmospheric scientists would go no further than describing the advent of satellites as an *evolutionary* step, but one of a series continuing to enable mankind to examine ever more closely the atmosphere around us. Thus it is pertinent to note that 'those who believed that meteorological satellites had heralded the coming of the millennium in atmospheric studies were over-optimistic'; although, nevertheless, 'these satellites have proved, and will continue to prove, to be highly important tools for work in both meteorology and climatology' (Barrett 1970). Obviously there must be some limitations to purely satellite based inquiries; this fact is underlined by the simultaneous attempts of the W.M.O. through its World Weather Watch programme to expand and improve the conventional observing network as well as the global satellite system. So plans are well advanced for other evolutionary devices also, such as free flying, constant level balloons (see p. 215), instrumented weather buoys and automatic land stations (see Barrett 1967) among many others. Why should these be necessary? The chief answers to this question should be implicit in the following discussion concerned with the other side of the satellite coin. Many practical problems are associated with satellite operations and data usages. From the climatological point of view the chief problems include:

1 Satellite data are already very voluminous.
2 It is not always easy to see how the data may be analysed best.
3 The resolutions of satellite data are not always optimal for climatological uses.
4 Evaluating the data is often complicated by degradations of satellite–ground complex subsystems, and/or extraneous factors.
5 The best representations of satellite derived climatological facts are not always immediately obvious.

Let us consider each of these problems in more detail.

Volume of data

The enormous volumes of satellite data already available are almost an equal embarrassment to both meteorologist and climatologist alike. Much current research is still seeking the most appropriate way to utilize such data operationally. There are three main stages in this research:

(*a*) The selection of specific satellite indications of key phenomena amenable to the remote sensing approach.
(*b*) The reduction of the appropriate satellite data to manageable quantities.
(*c*) The development of automatic procedures for processing and analysing the relevent data within operational frameworks.

Stages (*a*) and (*b*) are generally simpler to complete than stage (*c*) if research is carried out in that direction. Working in reverse, it is not unduly difficult to develop automatic data processing procedures, but the results are often of relatively little operational use.

The most obvious features of interest figuring in satellite data are cloud forms and distributions. The most widely used reduction form of satellite recorded cloudiness is the nephanalysis (see p. 52). This is a simplified representation of photographed cloud fields. It depicts the dominant cloud types and distributions, as well as other features of interest to the short-term weather forecaster, by now internationally accepted symbols. Nephanalyses have been used widely in short-term forecasting by various national meteorological services for several years. Although satisfactory procedures have not been developed yet to produce them automatically it should not prove impossible to devise such a technique.

The associated problems of data selection, reduction and automatic processing for climatological rather than meteorological purposes are more acute on account of the preoccupation of that science with longer periods of time. Perhaps, indeed, the development of climatological research programmes involving satellite data has been slower than that of their meteorological counterparts largely because both the volume and variety of newly available satellite data have caught unprepared a discipline only just adding several new facets to its more traditional statistical and regional approaches. Much careful thought must be given to managing the rich new funds of data.

Analysis of data

Problems of data interpretation are, meanwhile, at least as difficult as they are interesting. Various satellite observed patterns are subtly different from casual expectations; put differently, few satellite observations relating to the earth/atmosphere system correspond simply to simultaneous conventional observations. Careful comparative analyses are necessary in order that the satellite observations may be understood and interpreted in terms of more traditional types of measurements made within the atmosphere. One example must suffice, to be enlarged on later. It has been shown that linear bands of cloud photographed from weather satellites may indicate variously the directions of wind flows through the cloud layers, or the thermal wind directions through those layers, or some directions intermediate between those of wind flows at the cloud tops and the cloud bases. Most satellite data demand entirely new procedures for interpretation and analysis.

Resolutions of data

The resolutions of meteorological satellite data cannot be optimized for all potential users simultaneously. Hence, for example, while a photo resolution of some 3–5 km is generally deemed satisfactory by the weather forecaster, a finer resolution would be welcomed by many research workers, climatologists among them. It is not by chance that most of what may be loosely designated as 'climatology from satellites' relates to scales of global wind belts or regional macroclimates rather than those of local (topo) climates or microclimates (see Barry 1970). The contents of the present book illustrate and reflect the range of scales across which useful work has been possible thus far.

Degradations in satellite systems

Degradations occur in all satellite sensory systems with the passage of time, due mainly to the wear of component parts and the present impossibility of replacing them. The installation of duplicate sets of instruments in many satellites has alleviated this problem without solving it. Even the relatively modern Itos I satellite was given a lifetime of only six months minimum and one year goal for general engineering design. The quality of data from individual satellites tends to deteriorate with time, and short-term instrumental malfunctions and/or power restrictions sometimes occur also.

Problems of comparison between data from two or more satellites are at least equally important and difficult to resolve. Winston (1969) observed that such problems have been especially acute with data from infra-red radiation sensors, so much so that reliable *absolute* values of the radiation budget are basically unobtainable from the widely utilized Tiros data. He went on to affirm, however, that notwithstanding these problems much useful climatological information has become available from satellite sources.

Presentation of data

Since satellite data have their own unique characteristics the final major problem to be solved before they can be applied to the solution of broad climatological problems centres on the choice of advantageous modes of presentation of those data and their derivatives. Isopleth maps of mean cloud cover, earth/atmosphere albedo, surface temperatures, etc., are all obvious, but useful, climatological products. However, real progress is often made in science when the usefulness of the not-so-obvious becomes apparent. Hence climatology from satellites seems to be an exciting field thick with the promise of new discoveries of no small potential significance. Ingenuity and inventiveness are required to open it up.

Summary

We may conclude that the advantages and potential advantages of remote sensing of the earth's atmosphere from satellites in the context of climatology are most considerable, notwithstanding incidental problems that inevitably arise. The launch of the first Tiros satellite in 1960 marked the beginning of a decade in which rapid progress was made towards obtaining greatly increased meteorological observations on a global scale. At the beginning of the second decade of meteorological satellites, the applications of the new data and data types are already expanding rapidly. Leese *et al.* (1970) explain this increasing use by the operation of two factors. First, the long operational effectiveness of recent meteorological satellite systems which have produced lengthy and consistent data records; and, second, the development of appropriate processing methods to facilitate comparisons between satellite data and entities more familiar to meteorological analysts.

The biggest problem of all, as hinted earlier, remains that enunciated by Lehr in 1962. He observed that 'the large quantities of data being acquired and already acquired by . . . satellites have given rise to both enthusiasm and apprehension within the meteorological community. . . . Prior to the launch of Tiros I, and during the operation of each succeeding meteorological satellite, the archiving and retrieval procedures were and are a prime consideration.' Leese and his co-authors (1970) concluded their own review of archiving methods and climatological applications of satellite data by suggesting that the most efficient means of obtaining operationally useful information from satellite data might be through a so-called 'total systems approach'. In this, the full range of data from various sensors would be conserved, and the total information to be obtained from the data would be extracted in a unified processing sequence. For example, simultaneous measurements in the visible and infra-red wavebands would be analysed for information about several parameters from each set of measurements in turn before interrelations among them were sought.

Later chapters in this book indicate something of the possible scope of total systems approaches with respect to parameters specially amenable to satellite analysis and geographic coverage. They illustrate also the usefulness of satellite data in extending, completing or correcting earlier work in many diverse branches of climatology, especially in its regional, synoptic, physical and dynamic wings, and principally at the macroscale. The later chapters underline the need for more complete and better coordinated climatological analyses of all available data beyond the dominantly piecemeal and fragmentary studies to which the text refers. Much has been achieved in the field of climatology from satellites, but the best is yet to be.

PART TWO
The satellites

2 American meteorological satellite systems

Introduction: original objectives

The earliest experiments to obtain meteorological data from satellites were included in the American Vanguard and Explorer satellite systems of 1959, but the first satellite designed specifically for meteorological observations was Tiros I, launched on 1 April 1960. This began a series of ten Tiros satellites, of which the last was launched on 2 July 1965 (see table 2.1). Tiros, the Television and Infra-Red Observation Satellite, was superseded early in 1966 by the Tiros Operational Satellite system referred to as Tiros-Tos. Following a reorganization within the U.S. scientific administration a little earlier, the operational satellites were named Essa (Environmental Survey Satellites) by the Environmental Science Services Administration, although the satellites themselves were very similar to a previous Tiros type. In the meantime, another family of specialized satellites had been begun with the launching of Nimbus I in August 1964. The essential purpose of this continuing series is research and development. The chief Nimbus objective is the testing and qualification of advanced sensors for possible use later in operational series. More recently, from December 1966, the first family of truly 'second generation' meteorological satellites has been orbited under the designation A.T.S., the Applications Technology Satellite system. These satellites are multipurpose, whereas most of the members of the other families have been designed solely for obtaining and transmitting meteorological data directly to ground based Command and Data Acquisition (C.D.A.) stations or Automatic Picture Transmission (A.P.T.) reception facilities.

The next significant advance in operational satellite meteorology was heralded by the pre-launch testing of Tiros-M in 1969, this being the National Aeronautics and Space Administration (N.A.S.A.) prototype of the Improved

Table 2.1
Successful American weather satellites, 1960–70

Satellite	*Launch date*	*Angle of Orbit to Equator* (°)	*Av. height* (km)	*Cameras*	*Radiation experiments*
Tiros I	1 Apr. 1960	48	742	2 VCS	
Tiros II	23 Nov. 1960	48	676	2 VCS	√
Tiros III	12 July 1961	48	764	2 VCS	√
Tiros IV	8 Feb. 1962	48	777	2 VCS	√
Tiros V	19 June 1962	58	782	2 VCS	
Tiros VI	18 Sept. 1962	58	698	2 VCS	
Tiros VII	19 June 1963	58	649	2 VCS	√
Tiros VIII	21 Dec. 1963	58	753	1 VCS 1 APT	
Nimbus I	28 Aug. 1964	99	675	3 AVCS 1 APT	√
Tiros IX	21 Jan. 1965	96	1640	2 VCS	
Tiros X	1 July 1965	99	797	2 VCS	
Essa 1	3 Feb. 1966	98	769	2 VCS	
Essa 2	28 Feb. 1966	101	1384	2 APT	
Nimbus II	15 May 1966	100	1125	3 AVCS 1 APT	√
Essa 3	2 Oct. 1966	101	1436	2 AVCS	√
A.T.S. I	7 Dec. 1966	Geosync. orbit (151 °W)	35900	SSCC	
Essa 4	26 Jan. 1967	102	1381	2 APT	
Essa 5	20 Apr. 1967	102	1387	2 AVCS	√
A.T.S. III	5 Nov. 1967	Geosync. orbit (44–95 °W)	35900	IDCS MSSCC	
Essa 6	10 Nov. 1967	102	1445	2 APT	
Essa 7	16 Aug. 1968	102	1448	2 AVCS	√
Essa 8	15 Dec. 1968	102	1436	2 APT	
Essa 9	26 Feb. 1969	102	1465	2 AVCS	√
Nimbus III	14 Apr. 1969	100	1100	IDCS	√
Itos 1	17 Jan. 1970	102	1141	2 AVCS 2 APT	√
Nimbus IV	8 Apr. 1970	100	1090	IDCS	√
Noaa 1	11 Dec. 1970	102	1140	2 AVCS 2 APT	√

Tiros Operational Satellite (Itos), planned as the Essa replacement type for the early and mid 1970s. The first such satellite was launched on 17 January 1970 and followed by a second, designated Noaa-1 on 11 December 1970, for the National Oceanic and Atmospheric Administration. The observational capability of Itos is considerably in advance of the Essa satellites, and the new system notably incorporates a number of the subsystems tested earlier through Nimbus based research.

In an official N.A.S.A. report, *Significant Achievements in Satellite Meteorology, 1958–1964* (N.A.S.A. 1966), it was indicated that the earliest of all the American meteorological satellite objectives – namely the extension of observations over and above those yielded through conventional techniques at ground based locations – had by the end of that period become broader and more explicit. By 1964 the immediate objectives had expanded to include:

1 The development of satellite system equipment and techniques and satellite launchings in search of both an improved understanding of the atmosphere and a continued improvement of an operational meteorological satellite system.
2 Cooperation with the U.S.W.B. in the establishment and support of an operational meteorological satellite system.

Following the marked success achieved by N.A.S.A. towards achieving these two objectives, success most obviously manifest through the establishment of a truly operational satellite system under the auspices of the appropriate U.S. government body, E.S.S.A., still more recent objectives have been even broader in their scope and implications. At the W.M.O. headquarters in Geneva during March 1966 a meeting was held to examine the potential role of meteorological satellites in the global observing system of the World Weather Watch. It was suggested then that the main lines of W.M.O. activity in this direction should be:

1 The fostering of continued studies of the meteorological observing possibilities of space platforms to establish the lines along which effort should be directed in order to obtain the best observational return. Such studies must reflect the requirements of the world's meteorological services both as to the parameters concerned and the desired accuracy, resolution, frequency and timing of observations (see table 2.2).

Table 2.2
Immediate satellite objectives of the W.W.W. (after W.M.O. 1967)

Parameter	*Accuracy/Resolution*	*Frequency*
Cloud Ice and snow	2 km at subpoint (day) 8–15 km at subpoint (night)	Quasi-continuous 24-hourly
Surface (including cloud top) temperature	Approx. ±1 °C	12-hourly
Radiation and heat budget data	*c.*100 km (±2–5%)	12-hourly
Vertical temperature profile	±4 °C or better	12-hourly
Vertical humidity profile	±10%	12-hourly
Precipitation intensity	Light, moderate, heavy	12-hourly

2 To examine methods for coordinating meteorological satellite programmes so as to reduce unnecessary duplication of effort on the part of participating members of the W.M.O.
3 To take steps to see that the meteorological information provided by operational satellites is promptly available in suitable format, and that the advantages of this means of observation are made freely available to all members.

Thus the American Tiros-Tos system became more than a purely national programme through its role as the major contribution of the United States to the World Weather Watch during 1968–71, the initial planning period through which a global scale integration of satellite and conventional data was to be developed. In other words, research subsequent to the Geneva meeting has been geared largely to international requirements. The third point above has been accommodated by the U.S.A. in the sense that, as a matter of declared policy, data from operational satellite systems of the United States are made available routinely and freely to other nations for immediate operational use, and at the cost of reproduction if archived materials are required for research purposes.

It is more difficult to assess the progress made under 2 above, largely because the U.S.S.R., the world's second most important designer and operator of specialized meteorological satellites, is generally less forthcoming with respect to its achievements, and less categorical concerning its commitment to cooperative effort. Table 2.3 summarizes the extent of the Russian effort in satellite meteorology during the 1960s, the basic satellite configuration employed being similar to the American Nimbus (see fig. 2.1 (a)–(e)). When Russia's plans for the development of her Meteor meteorological satellite system until the mid 1970s are compared with America's, it appears that a great deal of overlap and even duplication of effort is likely to occur notwithstanding the intentions of these two nations as expressed jointly through the World Weather Watch statements of intent. Since the great majority of climatological studies of relevance to the present account have been based on American data, and since the principles of data collection and processing practised by the U.S.A. and the U.S.S.R. are basically alike, an outline account of American weather satellite systems will suffice as the necessary technological background to the studies discussed later in this book. Useful summaries of Russian hardware and planning philosophy are to be found in the W.W.W. Planning Reports 18 and 30 (W.M.O. 1967, 1970), and in the book edited by G. E. Wukelic (1968). Various other nations are at earlier stages in outline planning, feasibility or development studies involving appropriate satellites and supporting meteorological observations. France, Japan, the United Kingdom and the Federal Republic of Germany are prominent among such nations.

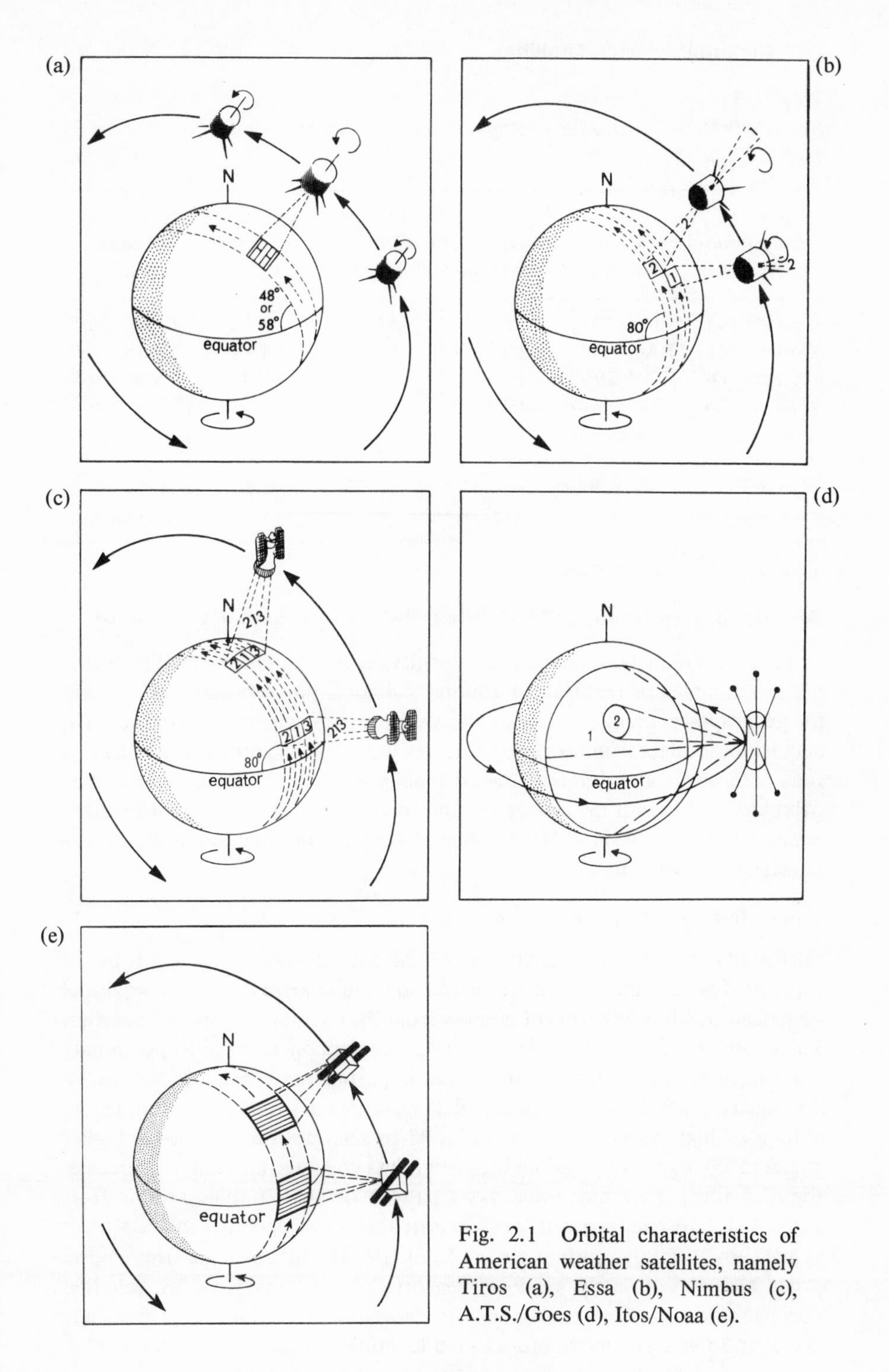

Fig. 2.1 Orbital characteristics of American weather satellites, namely Tiros (a), Essa (b), Nimbus (c), A.T.S./Goes (d), Itos/Noaa (e).

Table 2.3
Soviet Meteor system meteorological satellites in the 1960s (from W.M.O. 1970)

Designation of satellite	*Date of launching*	*Height of orbit (km)*	*Angle of inclination to Equator (°)*	*Period of revolution (min)*	*Date operation ceased*
Cosmos 144	28 Feb. 1967	625	81·2	96·92	29 Mar. 1968
Cosmos 156	27 Apr. 1967	630	81·2	97·0	22 Aug. 1967
Cosmos 184	25 Oct. 1967	635	81·2	97·14	23 May 1968
Cosmos 206	14 Mar. 1968	630	81·0	97·0	18 May 1968
Cosmos 226	12 June 1968	603 * 650 †	81·2	96·9	21 Feb. 1969
Meteor I	26 Mar. 1969	644 * 713 †	81·2	97·9	Still in operation

* perigee † apogee

American meteorological satellites: their orbits, sensors and data

From the point of view of the climatologist desiring to utilize satellite data in his programmes of research or routine statistical presentation and analysis, the geographical area of satellite data coverage is of primary importance. The acquisition of data from satellites requires not only the placing of a workable remote sensor in an orbit, but, at least equally importantly, its placement in an orbit that will permit the sensor to acquire data over carefully predetermined areas. There are several factors that determine the geographical area of coverage of a satellite.

The inclination of the satellite's orbit

This limits the latitudinal excursion of the satellite north and south of the Equator. Fig. 2.1 and table 2.1 illustrate and summarize the orbital angles of American satellites in terms of degrees from the equatorial plane. The earliest Tiros satellites (I–IV), intended to yield information relating to potentially destructive and harmful tropical vortices in particular, were orbited at 48° to the Equator, thereby giving a useful data coverage of low latitudes, but telling nothing of high latitudes. The next four Tiros satellites were orbited at higher angles (58°), thereby spreading the useful data coverage into middle latitudes. Fig. 2.2 illustrates an average day's pattern of photographic swaths from Tiros VII. Most subsequent weather satellites have operated in near polar orbits (96–100°), permitting a completely global data coverage from appropriate and appropriately arranged sensors. The successful A.T.S. satellites occupy so-called geosynchronous or geostationary orbits, approximately 'fixed' at 35,900 km above pre-selected locations on the earth's Equator.

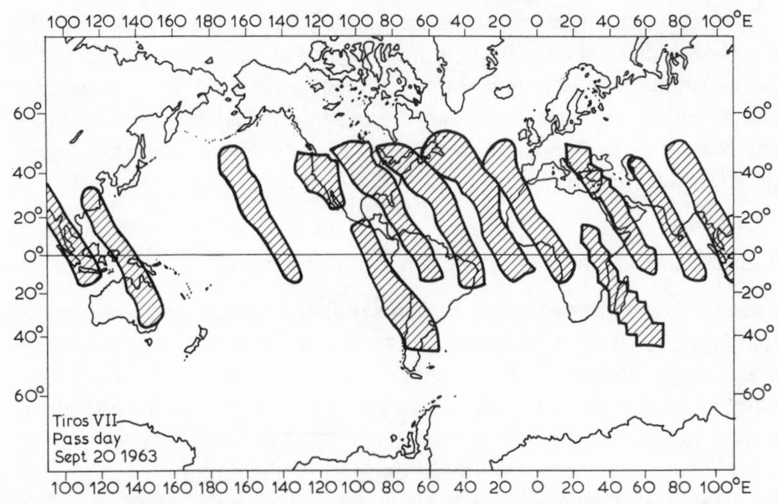

Fig. 2.2 A day's photographic coverage from Tiros VII orbiting at 58° to the Equator. (From Barrett 1967)

The properties of an ideal geosynchronous orbit may be summarized as follows:

(*a*) A prograde nature (i.e. the direction of motion is in the same direction as the earth's rotation).
(*b*) An equatorial plane (i.e. the inclination of the satellite orbit to the earth's Equator is zero).
(*c*) Essentially zero eccentricity (i.e. the orbit is circular).

An orbit possessing all three properties enables a satellite to 'hover' over a chosen point on the Equator. Small deviations from a perfect orbit cause the daily execution of a small 'figure eight' subpoint track crossing the Equator at the nominal subpoint. The latitudes of the northern and southern orbital extremities (e.g. 0·2° at launch with A.T.S. I and 0·4° with A.T.S. III) equals the angle of the orbital inclination (see A.T.S. Meteorological Data Catalogs). These angles tend to increase slowly as the satellites age.

Precession of the satellite's orbit

Widger and Wood (1961) give a detailed explanation of the limitations of coverage provided by Tiros satellites as a result of cyclic changes in their orbital precession measured in terms of the local solar times at which they crossed the Equator from day to day. The Essa and Nimbus satellites travel in much more nearly sun-synchronous orbits. That is to say, their orbits maintain a constant position with respect to the sun while the earth rotates beneath them. Hubert and Lehr (1967) provide an excellent general description of this matter. Orbital precession and the action of the earth's magnetic field on the satellite cause a daily change in its attitude to the earth. For example, in the early Tiros satellites (whose cameras were space-orientated, i.e. pointed in a constant direction with respect to absolute coordinates) the viewing areas changed from the northern to the southern hemisphere and back again through a ten week cycle.

The requirements of satellite sensors for solar illumination

Those meteorological sensors that measure radiation in the visible waveband obviously operate over sunlit areas, whereas sensors measuring outside the visible are more tolerant of the absence of current solar illumination. Low light intensifier cameras now under development for operation down to about one quarter moon illumination will not be subject to such limitations. The significant relationship to be optimized during the planning of a sun-seeking satellite orbit is that between the daylight terminator (separating day from night) and the satellite track. This determines the frequency of coverage of areas with direct solar illumination. This should be borne in mind when considering:

The attitudes of the satellite sensors

The attitudes are the orientations of the sensors with respect to their targets (usually the earth/atmosphere system). Since the early satellites Tiros I–VIII and Tiros X were space-orientated with sensors viewing the earth and its atmosphere through their base plates, they actually observed their targets for rather less than half of the daylight portion of each orbit. No observations were normally possible around the night time stretch of each orbit. Thus, the attitudes of Tiros satellite sensors, combined with other factors, limited the geographic coverage of the earth on average to about 20% per day. Enthusiasm for Tiros cloud cover pictures has, therefore, always been tempered by the fact that pictures taken from uniformly vertical or near vertical viewing positions are much easier to interpret and rectify than those from viewing positions that changed sequentially along every orbit, as suggested by fig. 2.1 (a). This realization led to the adoption of a different arrangement in the design of Nimbus. This provided a continuously earth-orientated platform so that cameras would maintain a fixed orientation with respect to the earth's surface beneath. A further bonus was added in that Nimbus was designed as a 'stable' spacecraft, not rotating about a spin axis as in the case of space-orientated Tiros, whose stable rotation rate was between nine and twelve revolutions per minute. Thus Nimbus data are not only simpler to assemble and interpret but also more complete than those from Tiros satellites.

An alternative method of maintaining constant picture orientation was demonstrated first by Tiros IX. A similar method was employed later through the entire Essa operational satellite system. These satellites were turned on their sides during the first few orbits, so that their central spin axes became aligned perpendicular to their orbital planes and tangent to the earth's surface beneath. Thus, in their final orbital attitudes, these satellites rolled along in the direction of their motion in so-called 'cartwheel' orbital configurations. The sensors, arranged to observe through the satellites' side walls, each pointed down towards the earth's surface once during each rotation of the satellite. By detecting the passing of the earth horizon with an infra-red sensor, each camera could be triggered to photograph only when pointing downward (see fig. 2.1 (b)). Because of the low, 650–800 km altitude of Essa 1, a single camera viewing vertically could not have viewed a field wide enough to overlap the pictures taken on preceding and succeeding orbital passes within the Tropics, so, to eliminate the gaps, two cameras were employed, mounted 26° 30′ to the right and left of the orbital plane. As the satellite spun, pictures obtained from the two cameras overlapped directly beneath them, but had sufficient width lateral to the orbital track for swaths from adjacent tracks to overlap also.

Thus, with a near polar, sun-synchronous orbit, it was possible to obtain pictures once every day over the entire sunlit portion of the earth, a basic requirement of a fully operational weather satellite system (Tepper and Johnson 1965). Later Essa satellites, viewing the earth from higher altitudes (generally about 1450 km up), have required only one radially mounted camera each to provide a breadth of coverage comparable to two at half that altitude, but they have been fitted with two apiece to provide full system redundancy. Early failure of the first to be used in each satellite is thereby covered. Thus improvements have been made possible in the lengths of their operational lifetimes. It is from Essa satellites that the climatologically most useful photographic data have been received. These form the basis of a whole range of computer rectified presentations discussed in some detail later in this chapter.

Itos satellites, designed to replace the Essa system, bear rather more resemblance to the Nimbus orbital configuration than to the cartwheel Tiros. These cube-like satellites are earth-stabilized, and their sensors, which view through the base plates, are continuously earth-orientated. As they are still subject to the usual minor orbital fluctuations and irregularities due to pitch, roll and yaw (see fig. 2.3), corrective adjustments are made by proven means to maintain the axes of the earth sensors within about one degree of the local vertical when the actual orbit is as planned, namely sun-synchronous and circular at an altitude of about 1250 km.

Lastly under this heading we turn to the question of the rather dissimilar attitude of the A.T.S. family. A.T.S. satellites were designed to spin on their long axes in space, parallel to the earth's axis of rotation. Hence, each A.T.S.

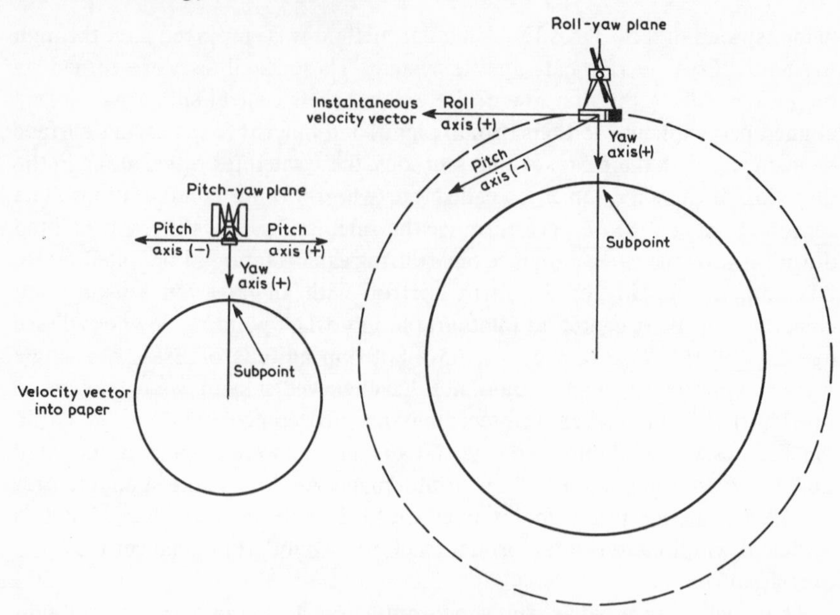

Fig. 2.3 Pitch, roll and yaw illustrated by Nimbus I: the *yaw* axis is the local vertical through the satellite, the *roll* axis is in the orbital plane, and the pitch axis is perpendicular to both yaw and roll axes. (Courtesy ARACON)

satellite acts as a free gyro, spin-stabilized at a nominal 100 r.p.m., although its stabilization is possible at any spin rate between approximately 50 and 150 r.p.m. Therefore, during any twenty-four hour period the spin axis of the satellite can be considered essentially fixed in inertial space at some angle relative to the axis of the earth, although, viewed from the subpoint on earth, that angle rotates through 360°. The earth-orientated sensors, viewing through the satellite side walls, scan across the target area as the satellite spins in space. So the satellite spin is accommodated as part of the means whereby observational data can be derived from the entire visible disc of the earth through the telescopic photometers which are used instead of conventional cameras. How individual A.T.S. 'photographs' are built up is examined in more detail later, within a discussion of various successful types of satellite remote sensors.

The satellite data storage capabilities

Data storage capability must exceed a single orbit if some orbital passes do not come within range of a read-out station. If on-board data storage cannot handle the observations for all the passes not read out immediately, a selection process must decide which passes will be stored. Within the operational

Essa system, two primary C.D.A. stations were established, at Fairbanks, Alaska, and on Wallops Island, Virginia, on opposite sides of the North American continent. These locations were chosen to facilitate acquisition of data from all the orbital passes of the Essa television satellites with minimum delay. The Fairbanks site was selected as the optimum location within the U.S.A. for acquiring the most passes per day from satellites in polar or near polar orbits while remaining within practical reach of supply and communication links. Meanwhile the site at Wallops Island is very favourably located for acquiring passes missed by Fairbanks. Together, these two stations miss only one orbital contact with Essa satellites per day on twenty-eight out of thirty days. It was planned that the information recorded during that missing orbit should be telemetered to the ground station on the next pass together with current pass information.

Summary and example

Summarizing briefly the sequence of operational events within the Essa photographic system, reference may be made to a specific satellite, Essa 7, by way of example. Pictures taken by its one vertically viewing television camera every 260 seconds were usually obtained in sequences of twelve frames to each pass, each picture covering approximately 10 million km^2, large enough to ensure a complete coverage even within the Tropics each day from twelve to thirteen orbits (each orbit lasting nearly 115 minutes.) When Essa 7 entered radio range of one of the two principal C.D.A. stations, satellite address and command sequences were provided in turn by the correctly directed station antennae. After receiving the proper address and commands the satellite would begin to transmit previously acquired, recorded, television data to the ground. There the information was displayed on a kinescope for monitoring and permanent recording on magnetic tape. Each frame could be electronically gridded by computer with lines of latitude and longitude. Then geographic outlines were merged with the picture.

An identification legend underneath each picture provided information relating to the year, month, day, hour, minute and second of picture taking time; the satellite track and (geographic) zone numbers; the relevant station initial; the satellite number; the mode of observation (in this case, television photography); the camera number; the longitude of a labelled grid intersection near the centre of the picture; the latitude spacing of grid in degrees; the longitude of a labelled grid intersection near the centre of the picture, followed by the longitude spacing of a labelled grid in degrees; the satellite pass number, and frame number; and the position (if any) of sun glint (see plate 1).

So the final prerequisites of global coverage by satellite are an adequate on-board storage facility for data gathered from remote areas of the world, and an appropriate distribution of data read-out stations.

Within the operational Essa system it was planned that only one of two satellites operational at any time should be of the television type, feeding globally comprehensive data to a central processing facility at the National Weather Satellite Center, Suitland, Maryland, prior to processing for incorporation into subjective and numerical analyses at the N.M.C. and digitizing for computer processing and archiving. The second satellite should employ a different photographic system, pioneered by Nimbus I and Tiros VIII, namely Automatic Picture Taking (A.P.T.), designed to give a direct indication of local meteorological patterns to local stations. This system functions without a comparable on-board storage facility, and it is therefore of great interest to local forecasters, but of little help to atmospheric scientists interested in the broader, hemispheric or global, contexts of local weather. Basically, the A.P.T. is a method of transmitting actual, detailed pictures immediately to any station in line of sight of a satellite, where appropriate (relatively cheap and simple) receiving equipment has been installed. Usually, images can be acquired from a single Essa-A.P.T. or Itos satellite each day during two or three consecutive passes near a station (fig. 2.4). From these, composites can be prepared to cover an area with a radius of approximately 2800–3800 km around a station in temperate latitudes. Since A.P.T. data are not gathered at a central facility, however, archiving is impossible, and climatologically their use must also be restricted to the area immediately around a local receiving station. Although the current Noaa satellites carry scanning radiometers instead of cameras, storage and read-out practices are essentially unchanged.

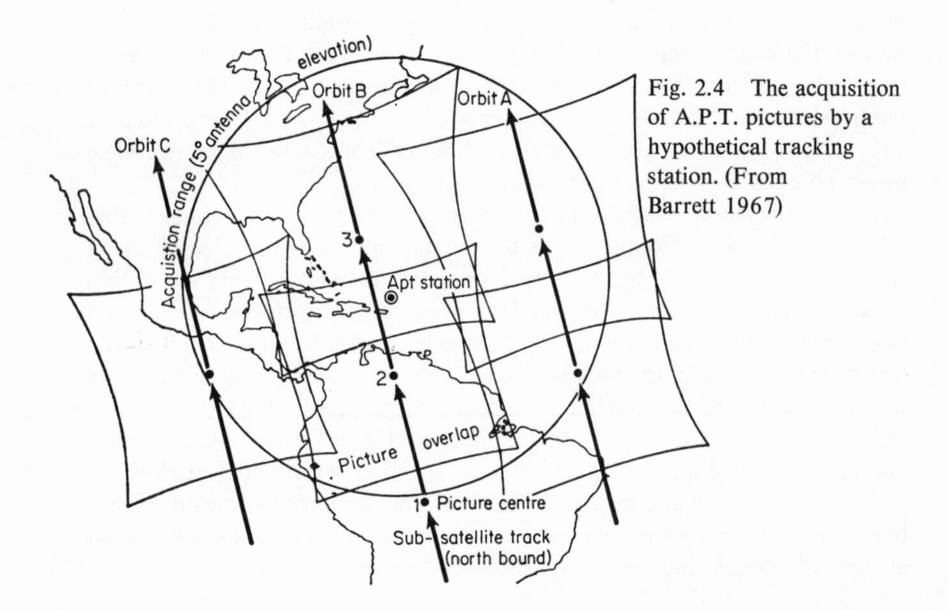

Fig. 2.4 The acquisition of A.P.T. pictures by a hypothetical tracking station. (From Barrett 1967)

Finally it should be noted that February 1966 (when the fully operational Essa system was inaugurated) marked the effective end of the experimental period of piecemeal meteorological observation from space, and the beginning of the period through which the global coverage has been largely complete. Reference to Essa Satellite Data Catalogs prepared by the Environmental Data Service of the U.S. Department of Commerce (1966 ff.) indicates that only a few days since the end of February 1966 have been marked by a total absence of photographic data for the world as a whole, or by incomplete data coverage in any given region. Occasionally, some data deficiencies have resulted from the temporary operation of various adverse factors, but, viewed realistically, they have been remarkably few when the complexity of the total operation is borne in mind. Areas of polar night are still 'ultimate blanks' on Noaa hemispheric mosaics compiled from visible waveband data, but such areas are very small compared with those unaffected by them. In any case, the problems of distinguishing ice and snow from clouds are extremely difficult to resolve. Thus the loss of photo definition in certain seasons in very high latitudes is not of great importance at the present.

Data acquisition: the basis for remote sensing from satellites

Much has been written about the technical details of observing instruments flown on meteorological satellites, but there has been relatively little concerning the basic principles of remote sensing generally. Hence the present account, concerned with data interpretation rather than satellite technology, emphasizes the types of meteorological observations made from platforms in space.

Fundamentally, information on various properties of the atmosphere may be obtained indirectly by measuring the intensity of the electromagnetic radiation reaching the satellite-borne sensor. Electromagnetic radiation may be defined as a propagation of energy in which, according to classical 'electromagnetic theory', a transverse wave motion exists in the form of periodic fluctuations of the strengths of electric and magnetic fields occurring at right angles to the direction of propagation of the energy. Such energy may be propagated either through some medium, or a vacuum – e.g. through the earth's atmosphere or through space respectively. The speed of propagation is generally about 3×10^{10} cm/sec.

One of the most valuable concepts in basic radiation theory is that of the 'black body', a body that absorbs all the radiation falling on it in all wavelengths, and which emits, at any temperature, the maximum amount of radiant energy. Such bodies, while they do not exist in nature, provide useful bases for radiation and absorption laws. Before outlining these, however, it should be pointed out that the term 'black' in this context may have little to do with the actual colour of such a body; for example, snow appears white because it scatters the visible light falling on it, but is, nevertheless, effectively a black body for certain longer wavelengths of radiant energy. Four laws are especially significant in the radiation context:

(1) Stefan's (or Stefan–Boltzmann's) law. This states that the intensity of the radiation emitted from a body is proportional to the fourth power of its temperature. Thus

$$E = \sigma T^4 \qquad (2.1)$$

where T is expressed in °K, and σ is a constant ($8{\cdot}31 \times 10^{-11}$ cal/cm^2/deg^4/min) relating to unit time from unit area of a black body.

(2) Wien's law. This says that the maximum output is in a wavelength that is inversely proportional to the temperature of the radiating body, i.e.:

$$\lambda_m = \frac{2900}{T} \qquad (2.2)$$

Note that the wavelength for maximum energy decreases with increasing temperature. Thus radiation from the sun, whose surface temperature is about 6000 °K, has an output peak at about 0·475 μm, whereas, by comparison, the earth's surface, with a temperature of about 285 °K, radiates most at about 10 μm. Thus, comparatively speaking, radiation from the sun is short-wave (or high temperature), while that from the earth is long-wave (or low temperature) radiation. This distinction is of considerable importance to climatology from satellites.

(3) Kirchoff's law. This states essentially that, if a body at a given temperature strongly absorbs radiation of a certain wavelength, it also radiates this wavelength strongly, provided it is present in the radiation spectrum for the temperature. This proviso is clearly crucial in the case of the earlier example of fresh snow, which absorbs very little direct, high temperature, short-wave solar radiation, but acts very nearly as a black body to lower temperature, long-wave radiation from the earth's atmosphere.

(4) Planck's law. Historically, the laws of Wien and Stefan preceded a far more general law derived from quantum theory, known as Planck's law. The following equation developed in 1900 by W. Planck united the other two laws and other information about radiation. It represents the distribution of energy with temperature (T) and wavelength (λ) for a perfect radiator as:

$$E_\lambda = c_1 \lambda^{-5} / (e^{c_2/\lambda T} - 1) \qquad (2.3)$$

where E_λ is the energy emitted in unit time from a unit area within unit range of wavelength centred on λ, and c_1 and c_2 are constants. The corresponding energy curve (fig. 2.5) holds good for all four radiation temperatures indicated at the base. It indicates, among others, the fact that the spectrum of solar radiation barely overlaps that of earth surface radiation. The area bounded by the curve, the wavelength axis and any selected pair of

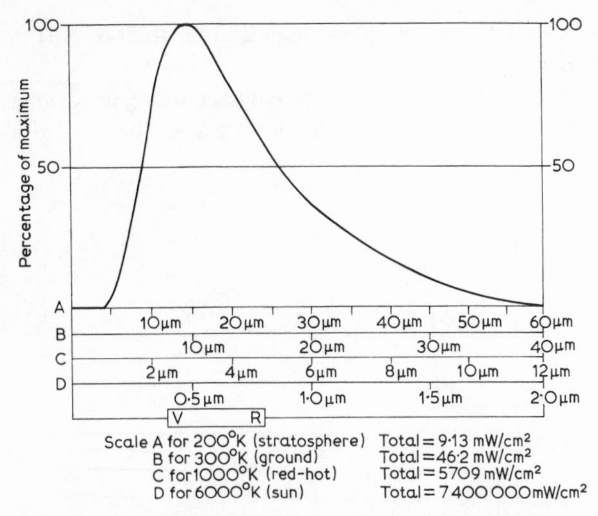

Fig. 2.5 The distribution of energy in black body spectrum. (Courtesy H.M.S.O.)

wavelengths indicates the relative amount of energy contained in the corresponding portion of the spectrum. Planck's law also deals, however, with the variation in total energy emitted over different wavelengths, and energy distributions for 200, 250 and 300 °K derived from his equation are given by the curves in fig. 2.6. The areas beneath the curves represent the quantities of radiant energy involved. Remembering Stefan's law, it is not surprising that the energy for 300 °K is much more than that for 200 °K (higher by a factor of 5), while Wien's law is illustrated by the fact that the wavelengths of maximum energy become progressively longer as lower temperatures are

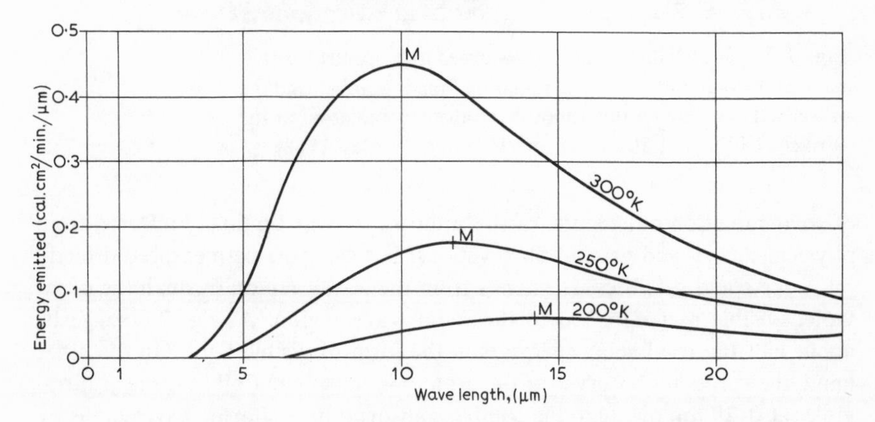

Fig. 2.6 Energy emitted by a black body radiating at (*a*) 300 °K, (*b*) 250 °K and (*c*) 200 °K. M represents the energy maximum in each case. (After Longley 1970)

involved, from 300 °K (shortest wavelength) through 250 °K to 200 °K (the longest of the three wavelengths portrayed).

Electromagnetic radiation is classified conventionally according to various wavebands, differing in wavelengths, as illustrated by fig. 2.9. The spectrum

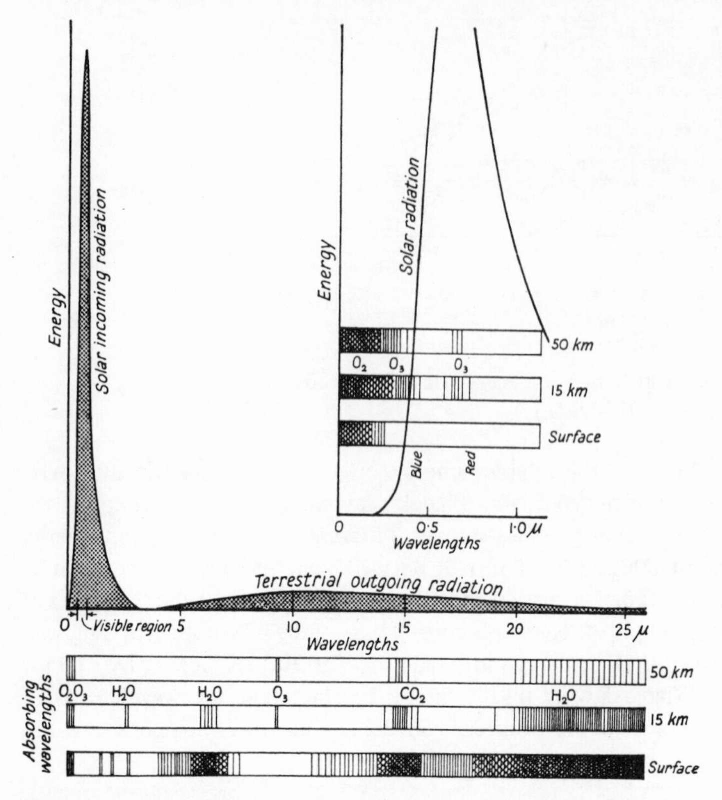

Fig. 2.7 The striking contrast between the spectral distributions of solar and terrestrial radiation (upper graphs), and the wavelengths in which the atmosphere absorbs radiation at the surface, 15 km and 50 km. (From Barry & Chorley 1968)

of solar radiation outside the earth's atmosphere was originally inferred from physical theory and ground observations, but can now be measured directly by rockets and satellites. It extends from the X-ray region through the ultra-violet, visible and infra-red to the radio wave region. As fig. 2.7 suggests, about half the total solar energy is in the form of visible light. On the other hand, the spectrum observed at the ground is sharply cut off in the near ultra-violet at 0·29 μm owing to the complete absorption of shorter wavelengths by gases (especially ozone) in the high atmosphere.

Absorption, essentially the opposite to the emission of radiation, involves removal of radiation from an incident beam, with conversion to another form of energy. Absorption of radiation by gases in the atmosphere is highly selective in terms of wavelength, and may depend also on pressure and temperature. Fig. 2.7 illustrates the wavelengths in which the atmosphere absorbs radiation at different altitudes. By comparison with the curves of solar and terrestrial radiation it becomes clear that, while some incoming solar radiation is absorbed at all three levels, the bulk of the atmospheric absorption occurs in longer wavebands, associated with radiation from the earth's surface, i.e. almost entirely with reradiated solar energy. About 15% of the incoming energy is absorbed directly by ozone and water vapour, ozone absorbing all ultra-violet radiation below 0·29 μm and water vapour absorbing to a lesser extent in several narrow bands between about 0·9 and 2·1 μm. Nearly 49% is immediately reflected back towards space by the atmosphere, by clouds and by the earth's surface, depending largely on the reflecting power (albedo) of each surface concerned. Thus albedo is of fundamental importance in energy and heat balance studies of the earth and its atmosphere, and, as later studies show, has been one of the parameters

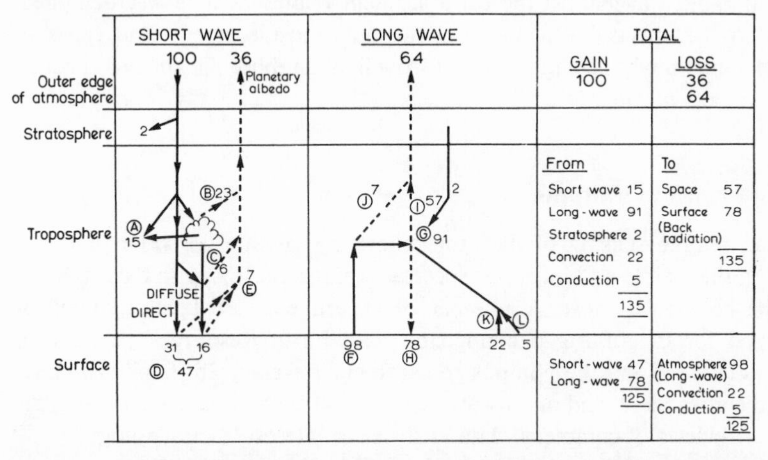

Fig. 2.8 An earth/atmosphere energy budget. Solid lines indicate energy gains by the atmosphere and surface in the left-hand diagram and by the troposphere in the right-hand diagram. The exchanges are referred to 100 units of incoming solar radiation at the top of the atmosphere. (From Barry & Chorley 1968)

investigated most usefully from satellite viewpoints outside the earth/atmosphere system. Most non-reflected insolation eventually heats the atmosphere, but indirectly, via the earth's surface, as fig. 2.8 suggests. The earth itself absorbs directly some 27% of the incoming short waves, together with an indirect 20% of energy reflected down or conducted from the atmosphere. Subsequent terrestrial radiation, of wavelengths mostly in excess of 3 μm, is partially absorbed by certain constituents of the atmosphere. Among these, the prominent absorbers include water vapour, carbon dioxide and ozone. The remainder of the terrestrial radiation escapes between their absorption bands, i.e. through 'atmospheric (or radiation) windows', back towards outer space.

Taking all these various processes into account, some 47% of an assumed 100% of energy arriving at the top of the earth's atmosphere is absorbed by the surface of the earth itself. Fig. 2.8 summarizes all these considerations in a so-called energy budget, generalized for the whole globe, and averaged through both day and night. Many satellite radiation studies have helped to add empirical reality to 'classical' climatological models of this type.

Information on various aspects of the atmosphere and atmospheric behaviour may be obtained indirectly, therefore, by measuring the intensities of selected wavelengths of electromagnetic radiation reaching satellite-borne sensors, whether in wavebands of visible light energy, infra-red or longer wave radiations. Fig. 2.9 depicts the relationship between the electromagnetic spectrum and meteorological phenomena as schematized by Tepper (1967). The two innermost segments show the spectrum itself ranging from ultra-violet to radio wavebands; the third segment relates various weather phenomena to the spectral regions; and the outermost segment lists types of sensors which might be employed to throw light on those phenomena through measurements of radiation.

Remote sensing systems

To date, remote sensing of the atmosphere by satellite has been predominantly in the visible and infra-red spectral regions. So it is with these that we shall be concerned most. A number of experiments are being planned or developed for the future in connection with Earth Resources Technology satellites (ERTS) and the Nimbus research and development family to measure in the ultra-violet and microwave regions, but such matters fall outside the scope of reviews of completed data analyses of interest to climatology.

A comprehensive account of the relationship between specific atmospheric parameters and the electromagnetic spectrum has been presented in the form of a COSPAR status report (1967). This specified two preconditions before measured radiation quantities might be related uniquely to desired meteorological parameters. Put differently, there are two basic prerequisites to be met before satellite measurements can be translated accurately into meteorologically useful data:

1 Suitable spacecraft sensory systems, capable of providing acceptable accuracy, resolution, calibration and geographic coverage in sets of measurements.
2 Adequate theoretical or empirical models to permit rigorous analytical

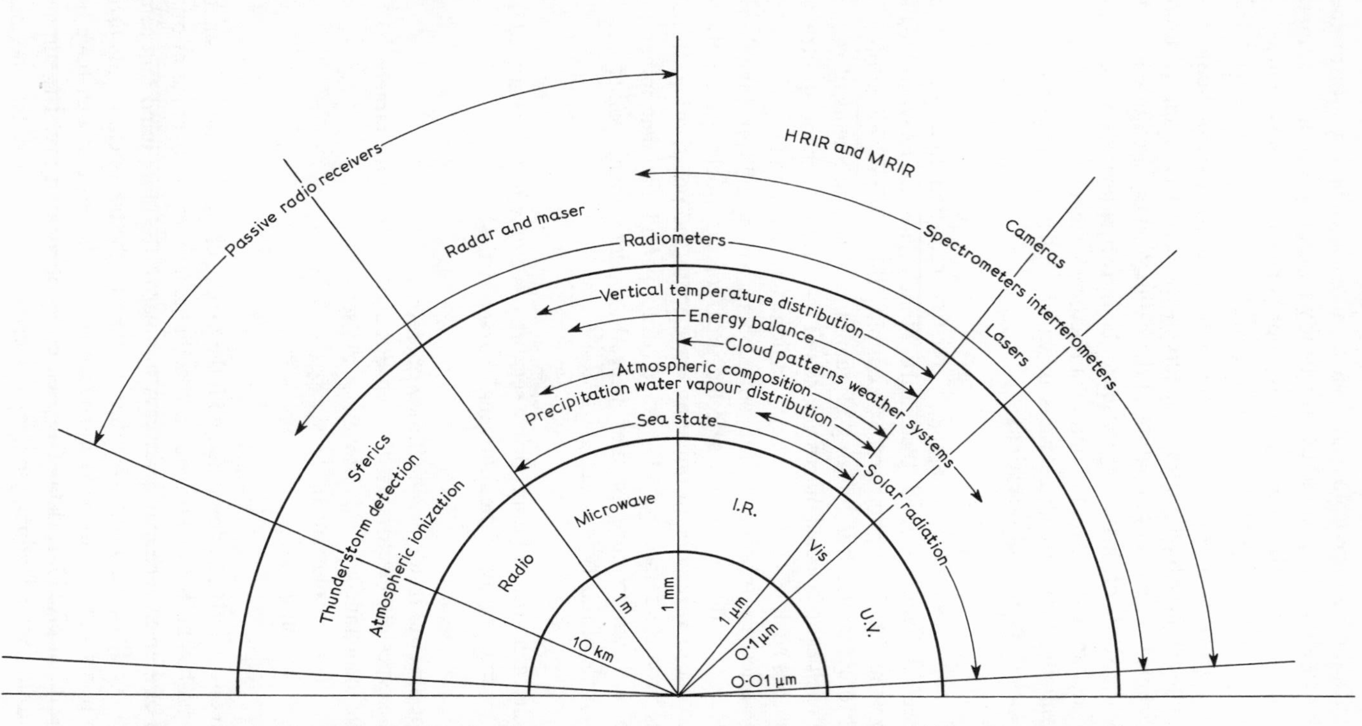

Fig. 2.9 Characteristics of the electromagnetic spectrum for remote sensing. (After Tepper 1967)

derivation of meteorological data from those radiation measurements. The validities of such models frequently depend on various assumptions which can be tested only by analysis or correlative observations.

Point 2 receives closer attention in chapters 3 to 5; however, some further comments are necessary at this juncture with respect to point 1, although finer technical details (e.g. relating to the calibration of satellite sensors) are outside the scope of this book; these can be drawn, if necessary from appropriate satellite 'User's Guides' and manufacturer's manuals.

Essentially, there are three ways in which radiation sensors can obtain useful meteorological observations:

1 Through continuous imaging of relative radiation intensities in wide spectral bands around the entire globe. Satellite television camera systems for cloud photography exemplify this mode of operation.
2 Through continuous imaging of quantitative measurements of radiation intensities in selected, narrow bands of the electromagnetic spectrum, by sensors such as certain scanning radiometers (see p. 41).
3 By selective probing of the atmosphere for significant variations in received radiation across selected areas of the world, especially with respect to variations in radiation wavelength and polarization. Infra-red spectrometers (see p. 43), designed to permit the determination of vertical temperature profiles beneath satellite flight paths, are one example of this type of sensor.

Within each group of appropriate sensors some may be differentiated from others by some or all of the following characteristics:

1 The spectral bands in which they operate.
2 The types of detectors used to sense radiations from the targets.
3 The instantaneous fields of view of the sensors.
4 The linear resolutions of the sensors.
5 Their scanning rates.

Such characteristics, combined with the orbital characteristics of a satellite at a given height, help to provide a specific geographical coverage of observations of a given type and quality. A number of the more important sensor systems are discussed below. Neither the list nor the treatment is exhaustive, but the notes should suffice to introduce briefly the systems that have been most useful in providing data of proven climatological utility. The summary is organized generally along the lines suggested by the three radiation sensing methods listed above.

Satellite camera systems

Most of the photographic satellites during the 1960s employed television-type ('vidicon') cameras as their principal meteorological sensors. These were modified in certain cases to provide extended image storage capabilities linked with the slower picture readout required by A.P.T. In principle, vidicon cameras are similar to television cameras in that they include a photo-sensitive face on which an image is exposed initially, and an internal scanning device which converts the picture of relative radiation intensities in the visible portion of the spectrum to an analogue signal. By varying the field of view (which is dependent on the camera lens angle from a given altitude), several different areas of coverage were achieved with Tiros-borne cameras, as indicated by table 2.4. The related resolutions ranged from about 2·5–3 km (wide angle cameras) to about 0·3–0·8 km (narrow angle cameras).

Table 2.4
Tiros Vidicon Camera characteristics (from Leese *et al.* 1970)

	Wide angle	*Medium angle*	*Narrow angle*
Field of view	104°	80°	12·7°
Area coverage (vertical field of view from average height of satellite)	1200 km square	725 km square	120 km square
Lens speed	*f*1·5	*f*1·8	*f*1·8
Shutter speed	1·5 milliseconds	1·5 milliseconds	1·5 milliseconds
Lines per frame	500	500	500
Resolution (per raster line pair, zero nadir angle)	2·5–3 km	2 km	0·3–0·8 km

Nimbus I and II, and odd-numbered Essa satellites from 3 to 9 inclusive, flew advanced vidicon camera systems (A.V.C.S.), incorporating larger vidicons (25·4 mm in diameter compared with 12·7 mm in the V.C.S. systems) and improved scanning arrangements (giving 800 scan lines to each picture compared with the earlier 500). From similar altitudes A.V.C.S. pictures enjoy significantly higher resolutions. The primary contrasts between the A.V.C.S. systems of Nimbus I and II on the one hand, and those of later Essa satellites on the other, lay in the numbers of cameras deployed, and their angles of view. In Nimbus I and II fan-like arrays of cameras were deployed, consisting of one vertically viewing camera in each case, and two lateral cameras angled at 35° to the vertical. In the Essa system, a single A.V.C.S. camera points vertically down at each instant for photography. A.P.T.

cameras in both types of satellites were 25·4 mm vidicons, modified for A.P.T. operations as outlined earlier.

Rather different photographic systems were installed in Nimbus III and IV and the A.T.S. family of satellites. The Nimbus system is known as the Image Dissector Camera System (I.D.C.S.) which feeds both real-time transmissions (i.e. the A.P.T. facilities), and videotape storages (for global scale archiving via a principal C.D.A. station). The image dissector is a shutterless electronic scan mounted behind a wide angle lens in such a way that each earth scene contained in a single frame is not the result of an instantaneous exposure made from one point in space, but rather a composite of scan lines built up perpendicularly to the direction of satellite motion as the satellite progresses along its orbital path. A 'step tube' mechanism ensures that each scan line exposes a new strip on the photo-sensitive tube. Hence each picture is built up by the continuous operation of scanning and stepping functions as a satellite orbits the earth. Current Noaa satellites employ a similar system for daytime and night-time imagery and no longer have any conventional cameras.

In A.T.S. satellites, spin-scan cloud cameras (S.S.C.C.) have likewise involved motions of their satellites in space in conjunction with telescopic photometers and stepping principles. More precisely, the multicolour S.S.C.C., responsible for the first colour photographs received from weather satellites, contained the following components:

1 A high resolution telescope.
2 Three photomultiplier light detectors.
3 A precision latitude stepping mechanism.

The latitude step motion, combined with the spinning motion of the satellite, permitted scanning of a complete earth disc (about one third of its surface). In A.T.S. III, for example, such a system gave a single west-to-east line scan with each, or every third, revolution of the satellite. Each successive scan line abutted the preceding line through a picture building sequence lasting about twenty-four minutes. This covered the visible disc of the earth from north to south. Upon reaching its southern limit, the telescope retraced directly to its northern limit in 2·4 seconds to commence the next picture building sequence. Thus the complete cycle, involving 2400 line steps, required 26·4 minutes when the camera stepped at one step per revolution if the satellite rate of spin was 100 r.p.m.

The three photomultipliers operated in conjunction with colour signals of different wavelengths including blue (centred on 0·43 μm), green (0·52 μm) and red (peak at 0·58 μm). On the ground, appropriate satellite signals were combined and Ektacolor film exposed to give colour pictures of the earth and its cloud cover.

However, in climatological and meteorological research, the chief contributions of A.T.S. photographs have been due more to their short-term repetitive coverages of selected regions than to their multicoloured appearances. Indeed, by the third month of A.T.S. III operations the red colour channel had failed, so the green channel has been used to provide black and white photographs through the exposure of a polaroid-type film instead of Ektacolor.

Satellite infra-red sensor systems

Climatologically, the most useful radiometers measuring radiation intensities in selected, narrow bands of the electromagnetic spectrum have been infra-red radiometers of the medium resolution infra-red (M.R.I.R.) and high resolution infra-red (H.R.I.R.) types. The functions of these two radiometer families may be differentiated as follows:

1 The M.R.I.R. systems were designed chiefly to provide broad resolution data for general radiation balance studies. The best linear resolution for Tiros M.R.I.R. radiometers was approximately 64 km. Characteristically, whether flown on Tiros, Nimbus or Cosmos satellites the M.R.I.R. have been multispectral sensors, capable of investigating simultaneously in several wavebands. Lesser (low resolution) radiometers for even more general radiation balance studies have been flown in various weather satellites, but their data have figured little in published work in climatology.

2 The H.R.I.R. systems have been designed to perform two major functions: (*a*) to provide data from which the earth's cloud cover may be mapped at night, and (*b*) to provide detailed measurements of cloud-top and earth-surface temperatures. Generally operating in the narrow, but relatively clearly defined 'atmospheric window' waveband from *c*. 3·5 to 4·1 μm, such radiometers with instantaneous fields of view of about 0·5 degrees of arc have yielded data with subsatellite ground resolutions of about 8 km from altitudes of 1100 km.

Fig. 2.10 summarizes the chief wavebands investigated by various medium and high resolution infra-red radiometers. It indicates also the main purposes for which various measurements are made. Despite the numerous problems associated with interpreting the observations correctly in meteorological terms, such observations in wavebands of energy on the long wave side of the visible portion of the spectrum have been enormously useful as base data for much research in climatology and meteorology. Some of the most significant results are referred to in later chapters of this book. However, a number of cautionary points must be listed to underline the even greater care that their analysis demands compared with the analysis and interpretation of satellite

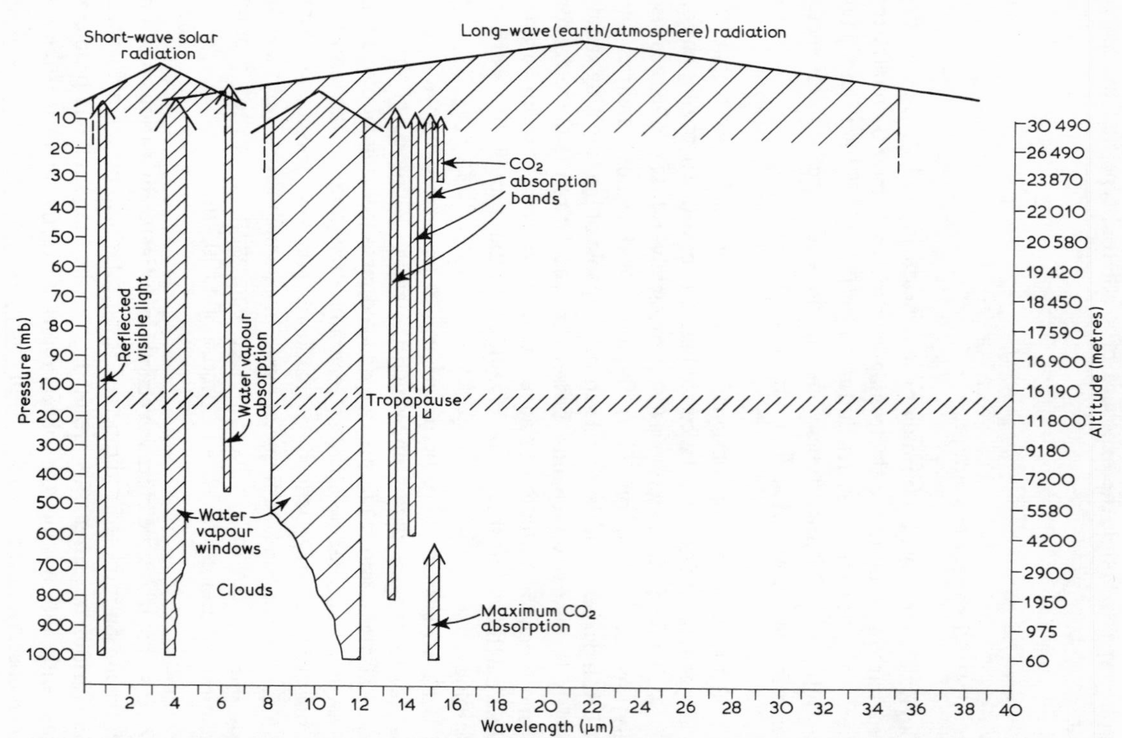

Fig. 2.10 The more important infra-red wavebands investigated by American weather satellites. (Medium and high-resolution radiometers)

photographs. The chief limitations on the uses of infra-red radiation data have included:

1 The uncertainty of the absolute values of such measurements due to (*a*) radiometer calibration, (*b*) data conversion procedures and (*c*) post-launch degradations of instrument responses.
2 Theoretical difficulties in interpreting the measurements.

For example, earth surfaces and cloud layers do not emit exactly in accorddance with black body principles touched on earlier. Furthermore, the absorption and re-emission of energy by atmospheric gases and aerosols can change markedly the characteristics of outgoing radiation from such surfaces.

Consequently the best uses of such data are found in relative studies covering wide areas through short periods of time, though careful processing of infra-red data can filter off many of the undesirable irregularities. Observations from Nimbus H.R.I.R. sensors have been made available not only in digital form, but also photographically, thereby permitting them to be interpreted both qualitatively and quantitatively. In these cases, electrical signals of radiometer outputs (proportional to observed radiation levels) have been converted, line by line, by facsimile recorders on the ground into continuous strip pictures using 70 mm film. In these pictures the blackness of each picture element is directly proportional to radiation sensed by the radiometer: the *darker* areas represent areas of high black body temperatures, and the *paler* areas regions characterized by relatively low black body temperatures. These data displays are useful in the interpretation and qualitative analysis of meteorological information relating, for example, to the heights of radiating cloud tops. The Frontispiece illustrates some routine Itos infra-red products.

Current Noaa satellites have very high resolution radiometers (V.H.R.R.) giving a ground resolution of about 1 km.

Sensors designed to determine vertical profiles through the atmosphere

Perhaps the most significant innovations in Nimbus III were its instruments designed to provide vertical soundings of temperature through the atmosphere. Unlike the radiometers introduced above, the Satellite Infra-red Spectrometer (SIRS) and the Infra-red Interferometer Spectrometer (IRIS), flown first on Nimbus III, have the capacity to provide data for vertical temperature profiles. More recently in Nimbus IV a third such instrument, namely the Selective Chopper Radiometer (S.C.R.) has been flight-tested in a satellite for the first time. Let us examine each of these three briefly in turn. Although they represent very recent developments from the point of view of the climatologist especially, it will be seen later that they are already providing data of considerable climatological significance (see chapters 3 and 6).

The satellite infra-red spectrometer

Essentially, SIRS is a medium resolution multichannel radiometer designed to examine its target area in a number of chosen spectral intervals simultaneously. SIRS-A, flown in Nimbus III, measured radiances in eight such intervals, and SIRS-B in fourteen. Here we will concern ourselves mostly with the former. SIRS-A measured seven radiances in the 13–15 μm CO_2 absorption band, and one in the 11 μm atmospheric water vapour window waveband. Since carbon dioxide in the atmosphere is uniformly mixed to within 1% or 2%, its emission spectrum when viewed from above is dependent only on the vertical distribution of temperature. In effect, therefore, the temperature structure of the atmosphere can be graphed from a simultaneous set of measured radiances on the assumption that each measured radiance emanates from a particular level. Some understanding of the source regions of the radiances observed by SIRS-A may be gained from fig. 2.11, which represents the so-called 'weighting functions' of the eight channels. These indicate the height distribution of the sources of radiation received by each one. The peak of each weighting function occurs at a height determined by the spectral response of the radiometer. The carbon dioxide channels thus record

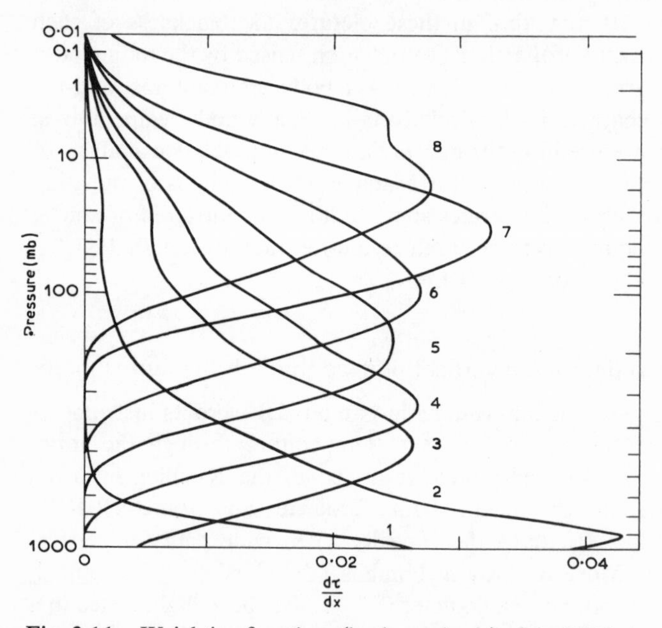

Fig. 2.11 Weighting functions (horizontal axis) of the SIRS-A channels versus pressure, using temperature and humidity measurements from the 1800 G.M.T. 16 April 1969 radiosonde at Nashville, Tennessee. (From Wark 1970)

radiation corresponding to temperatures at various heights in the atmosphere averaged over the vertical extent of the weighting functions.

In practice, some difficulty has been experienced in translating measured radiances into appropriate atmospheric temperatures, and various methods have been tried. These have been summarized by Wark (1970). The most 'direct' method involves the solution of the equation of radiative transfer (Wark and Fleming 1966), but this requires a more exact knowledge of the transmissions of the SIRS channels. Operationally, therefore, a partly empirical approach has proved more satisfactory. Soon after the launch of Nimbus III regular comparisons were made between satellite-derived profiles of temperature and geopotential height and those from radiosondes, where geopotential height (Z) has the following relationship with geometric height (x):

$$Z = \frac{gz}{9 \cdot 80} \tag{2.4}$$

Therefore, where the gravitational acceleration (g) has its near average value of $9 \cdot 80$ m/sec^2, heights expressed in these two terms are the same. For $g < 9 \cdot 80$ m/sec^2 the geopotential height is less, and for $g > 9 \cdot 80$ m/sec^2 it is greater. Most upper-level synoptic charts (e.g. the 500 mb maps on the British Daily Aerological Record) are drawn in terms of geopotential units in preference to geometric units for physical reasons that need not be explored here. Suffice it to say that the pressure-height distribution expressed in these terms is the basic parameter for weather prediction.

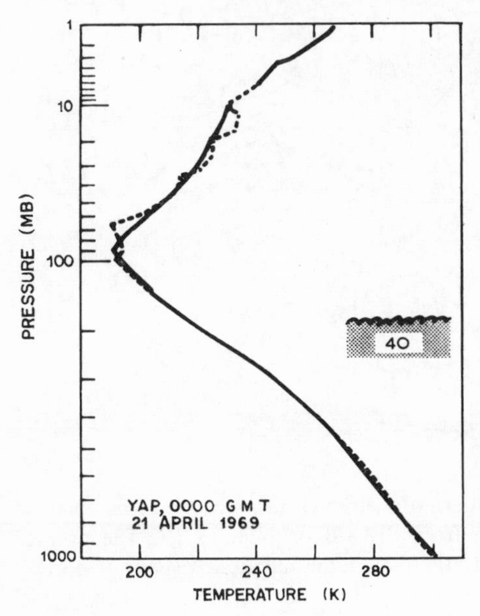

Fig. 2.12 The first temperature sounding obtained from satellite measurements. The SIRS sounding (solid line) is compared with the Kingston radiosonde sounding (dotted line). (From Wark 1970)

Statistical analyses have suggested that the geopotential height distribution can be specified very accurately from SIRS radiation data even without prior knowledge of surface pressure (Smith 1969). Smith's approach was to relate thirteen levels of the atmosphere as observed by radiosondes with the eight SIRS radiances. The multiple correlations obtained from a statistical sample are then used to relate other SIRS data to an unknown profile. In areas lacking surface data, this technique is used to obtain geopotential heights independently; where surface data are available, the geopotential heights are obtained from the temperatures. Fig. 2.12 portrays the first retrieval of a temperature profile from a satellite, obtained on the very day of launch of Nimbus III. Viewing vertically towards the satellite subpoint, SIRS-A had a

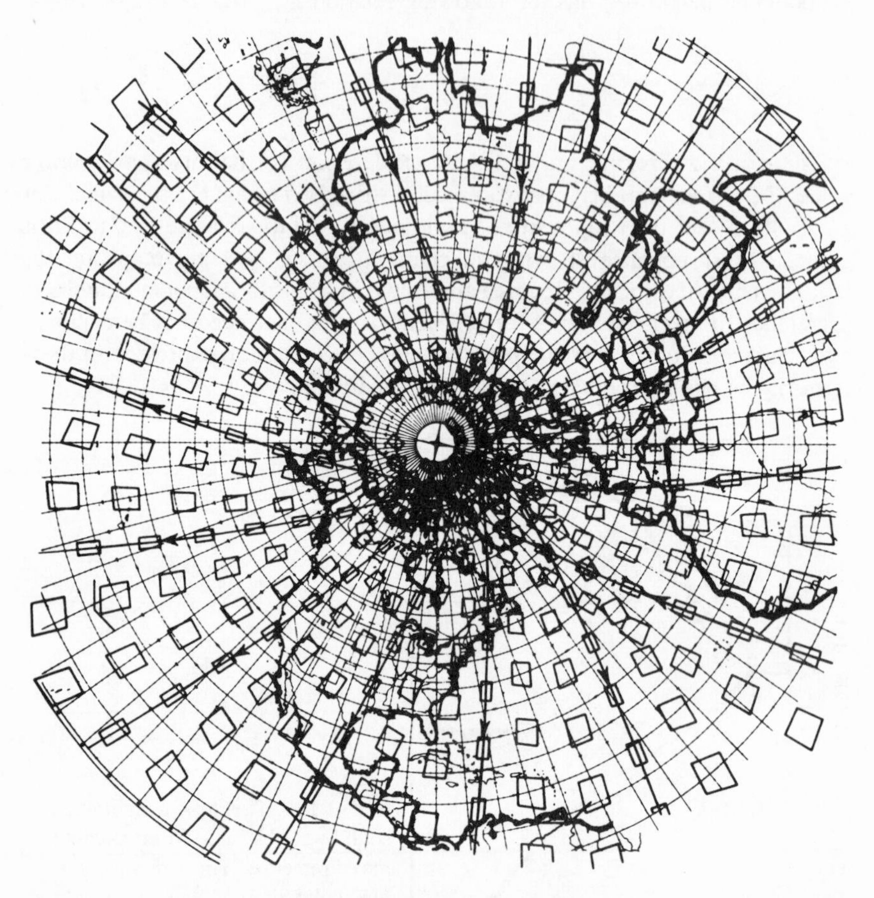

Fig. 2.13 A northern hemisphere pattern of viewed areas which would be obtained from twelve hours of complete data from SIRS-B on Nimbus IV. The average separation of the viewed areas in about 500 n.mi. in the sub-arctic regions. (Courtesy ARACON)

field of view of 11° 30′ on each side, giving a linear resolution of about 350 km on a side square from about 1100 km up. Lastly, fig. 2.13 represents a northern hemisphere pattern of viewed areas which may be obtained from twelve hours of complete data from SIRS-B on Nimbus IV. Compared with SIRS-A this has the added capability of spatial scan for greater area coverage.

Infra-red interferometer spectrometer

The data obtained from this instrument, described in detail by Hanel and Conrath (1969) have not yet been employed very fully in a lower atmospheric climatological context. Consequently, some brief notes will suffice to outline its operation and intended use. IRIS measures intensities of radiation across a broad waveband from 5 to 25 μm, registered upon a thermistor bolometer in the form of an interferogram. Through Fourier transformations spectra are obtained of the earth's infra-red radiance.

IRIS was designed and intended primarily to investigate the vertical distribution of atmospheric humidity through water vapour absorption patterns, but its spectral range also covered CO_2 and O_3 bands and other spectral features associated with minor atmospheric constituents such as methane and nitrous oxide. Hence, the original list of objectives for IRIS included the following in particular:

1 The derivation of atmospheric temperature and humidity structure on a global scale to feed numerical studies of the general circulation.
2 The global observation of temperature, water vapour and ozone fields for synoptic meteorological studies.
3 The collection of spectra for research studies in meteorology, including radiative transfer studies, as well as for other, non-meteorological, uses.

Selective chopper radiometer

This six-channel radiometer also monitors emission from CO_2 in the 15 μm band by means described by Abel *et al.* (1970). The translation of measured radiances into temperature values for different levels in the atmosphere is achieved by similar means to those outlined earlier for SIRS. The unique structure of the instrument, however, has the advantage of narrowing the appropriate weighting functions. It is thought that this may lead to the development later of a radiometer capable of extending indirect soundings into the mesosphere. Some early stratospheric temperatures patterns from the S.C.R. on Nimbus IV are illustrated in the following chapter.

Still further approaches to the vertical sounding problem are planned for the future. Table 2.5 summarizes many of the additional observation experiments planned for testing during the scope of the 1972–5 plan of the W.W.W. Doubtless many of these additional experiments are destined to make important contributions to the rapidly expanding field of climatology from satellites.

Table 2.5
World Weather Watch requirements for an operational geostationary weather satellite system: specifications for Primary Sensor Instruments planned for 1972–1975 period (Goes satellite) (after W.M.O. 1970)

	Scanning Radiometer	*Very High Resolution Radiometer (VHRR)*	*Vertical Temperature Profile Radiometer (VTPR)*	*Goes Radiometer Telescope*
Spectral Intervals (micro-metres)	1) 10·5–12·5 2) 0·52–0·73	1) 10·5–12·5 2) 0·60–0·75	1) 14·96 2) 14·8 3) 14·4 4) 14·1 5) 13·8 6) 13·3 7) 18·8 8) 12·0	1)10·5–12·5 2) 0·55–0·70
Sensitivity	1) 1 °C at 300 °K 2) 4 °C at 185 °K	1) 1 °C at 300 °K 2) 3 °C at 185 °K		1) 1 °C at 300 °K 2) 4 °C at 200 °K
Horizontal Resolution (km)	1) 6 2) 3	1) 1 2) 1	45	1) 6 2) 1 to 6 adjustable
Coverage	Global	Global	Global	55° great circle arc from sub-satellite point
Availability	Direct and stored	Direct	Stored	Central readout, rebroadcast direct

The processing of satellite data

Lastly in this chapter consideration must be given to those data forms that have proved most useful to date in studies of a climatological nature. They fall into three main categories: images, numerical presentations and manually compiled nephanalyses.

Images

Perhaps the key steps here involve the computer processing of original data displays to yield either gridded or mapped representations. In gridding, the image projection of the earth as seen by a satellite sensor is kept constant, and latitude and longitude grids are computed to fit this projection (see plate 1). The reverse is true in the mapping process, where image data are transformed to fit a common type of map projection such as a polar stereographic or a Mercator (see plate 2). The chief advantages of gridding over mapping stem from the amounts of digital data that need to be handled. The grids themselves are relatively simple to compute from the satellite orbital and altitude information; the actual image data need not be handled at all. Unfortunately, the comparison of gridded but unprocessed image data for different times across a single area is often a difficult and tedious business. On the other hand, although mapping of image data is very time-consuming and requires the facility of an advanced data processing system, it has even greater ultimate advantages than data gridding, particularly since further computer processing is much simpler. Furthermore, manual analysis and interpretation is easier, especially where comparisons of image data for different times are involved. Since the climatologist is usually concerned with time-averaging procedures at some stage in his inquiries, processes that facilitate a range of automatic analyses of large volumes of satellite data must be particularly attractive to him.

Since the inauguration of the Essa operational satellite system, vidicon image data from the appropriate satellites have been gridded and mapped operationally at the Data Processing and Analysis Facility of N.E.S.C. (later N.E.S.S.) in Suitland, Maryland. The major steps involved in the mapping of vidicon data are shown diagrammatically in fig. 2.14. The digitized cloud maps that result from the routine application of that procedure to raw image data through a high speed digital computer are listed below. In each case,

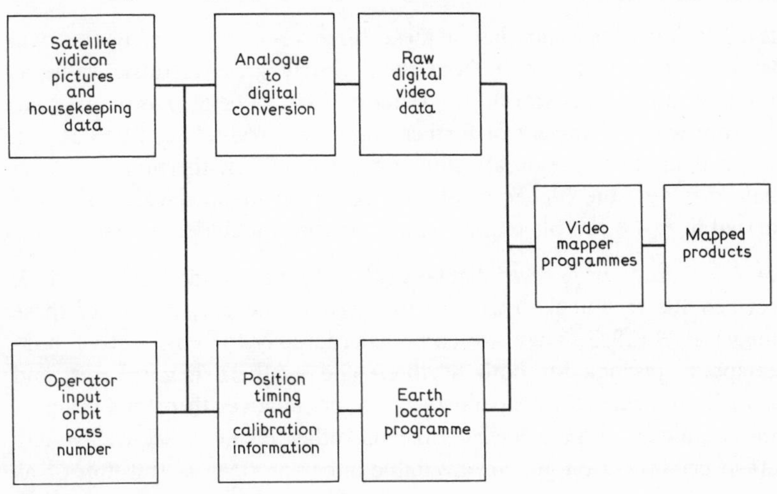

Fig. 2.14 Major steps in computer mapping of satellite data. (From Leese *et al.* 1970)

signals comprising a picture taken by a satellite are assigned numerical values to indicate the relative brightness of each picture element. These data are subjected next to a 'brightness normalizing' process, designed to remove or reduce picture variations due not to earth–cloud variations but to variations caused by technical and instrumental factors, including the efficiency with which earlier picture responses are erased from the vidicon tube, and the operational behaviour patterns of the optical system (filter, aperture and lens). Illumination variations due to solar angle (including extreme local sun glint problems) may be significant too. Once the data have been brightness normalized, as far as this is possible, they are then earth-located and repositioned on standard map projections. Magnetic tapes are produced for input to a cathode ray tube film display device. Let us take the case of Essa 7 again as an example. Digitized cloud maps from its raw vidicon data were so constructed that the overlap of day, twenty-four hours apart, was positioned between 60° and 80° E. A more detailed description of the procedures used to produce digitized cloud maps has been published by Bristor *et al.* (1966). A condensed and edited version of the same article appears in the Essa 3 data catolog (U.S. Dept of Commerce 1967). The basic daily computer products include:

(*a*) *Two hemispheric polar stereographic displays*, one for each of the northern and southern hemispheres.
(*b*) *Eight polar quadrants*, including four for each hemisphere. These are larger-scale subsections of the polar stereographic hemispheric displays.
(*c*) *Six or seven Mercator segments* together encircling the globe between about 30° N and 30° S. These are especially useful for studies of tropical weather and climate.

Plates 3–5 illustrate examples of these three very important operational satellite products, which, apart from their clearly great intrinsic value as bases for climatological research programmes, have value also as data organizations from which a variety of further computer products can be derived, including a number of mesoscale summary products. In the preparation of mesoscale displays, the higher resolution computer-mapped video data are summarized in 8 × 8 sample clusters. Such products include:

(*a*) *Five-day average brightness* displays (plate 6). Booth and Taylor (1969) described the technical procedure followed in the preparation of these displays at N.E.S.C. They are compiled routinely for consecutive, non-overlapping periods for both northern and southern hemispheres, and essentially indicate relative persistences of cloud cover through a range of photographic brightness levels. Like the other displays below, five-day average brightness photos are available either on 35 mm roll film or as 1 : 50 m photoprints. They are constructed on polar stereographic projections.

(*b*) *Thirty-day average brightness* products (plate 7) for both northern and southern hemispheres. These indicate the relative persistence or mobility of cloud cover over periods long enough to interest both long-range weather forecasters and climatological research workers organizing their studies on monthly bases.

(*c*) *Ninety-day average brightness* products (plate 8) for both northern and southern hemispheres. These throw light on seasonal characteristics of major components of the general circulation of the earth's atmosphere.

(*d*) *Five-day minimum brightness* composites (plate 9) for both hemispheres. Since these save and display the dimmest (i.e. the least bright) responses from each sample area in the original data, areas of ice and snow can be identified as those of relatively high brightness levels. Similarly, particularly persistent cloud patches can be seen clearly away from the regions of frozen surfaces, e.g. preferred tracks of mobile storms.

(*e*) *Five-day maximum brightness* composites (plate 10) for both northern and southern hemispheres. These save and display the brightest responses from each sample area in the original data. Features such as the centres of persistent anticyclones can be inferred from the resulting brightness distributions.

Various other products are produced daily by the N.E.S.S., as summarized by Leese *et al.* (1970, appendix C), but most are of more direct application to meteorology than climatology and need not be discussed here. Computer processed photo displays (either mapped or gridded photographs) have become available from certain satellites other than those in the Essa and Noaa series, but generally on rather *ad hoc* bases. For example, some pictures from A.T.S. satellites were mapped on both polar stereographic and Mercator bases, and some Nimbus radiation data have been available on similar projections in a limited way. Earlier Tiros photographs were obtainable as sequences of individual frames, while most Nimbus A.V.C.S., I.D.C.S., H.R.I.R. and M.R.I.R. swaths or strips are available in gridded forms. Generally speaking, gridded image data have been of more direct use to the meteorological than the climatological community, whereas mapped data have been regarded with particular favour by climatologists. It should be pointed out, however, that the experimental nature of the Nimbus programme has much to do with the sporadic time coverage of the Nimbus archives, as well as with the varied types of data they contain. Many of the frequent Nimbus innovations may be expected to be incorporated into future operational programmes after the manner of past practice. For example, the operational processing of satellite cloud pictures by computer outlined above evolved essentially along lines projected first for Nimbus (Bristor 1966).

Numerical data presentations

As well as the visual computer products, digital data relating to many of the above mentioned formats are available on magnetic tapes for specialized computer treatment or for re-formating to fit other visual display devices. Taylor and Winston (1968) used digital-averaging and display techniques in constructing maps of monthly and seasonal global brightness averages, presenting their results in Mercator digital map forms (fig. 2.15). In certain circumstances these may yield useful climatological information, though in some ways they are more difficult to interpret unambiguously than the corresponding thirty-day or ninety-day average brightness photographs. Their chief disadvantage in cloud climatology stems from the problem of 'background contamination', i.e. the different and variable contributions to local brightness values afforded by earth surface reflectivities. The chief utility of digital brightness maps is probably in albedo climatology (see p. 155).

Perhaps the most appropriate satellite data for numerical display are radiation measurements from narrow band radiometers or vertically profiling spectrometers, whose measurements can be interpreted fairly directly in terms of important atmospheric parameters such as temperature and humidity. Medium resolution infra-red data from Tiros and Essa satellites have been computer processed, together with orbital position, altitude information and calibration data, in the preparation of 'final meteorological radiation' (F.M.R.) tapes. These are the basic repositories of the radiation measurements in terms of equivalent black body temperatures or effective radiation emittances. Grid print maps can be produced for single orbits, or composites for several orbits, composed of computer printed and contoured data maps referenced to square mesh grids on the usual polar stereographic or Mercator map bases. Radiation catalogues were produced for Tiros II and III, containing a selection of grid print maps for different regions and at different scales, presented in some cases alongside conventional synoptic charts and Tiros photographs for purposes of comparison.

Similarly, radiation tapes and grid print maps have been archived in the Nimbus programme, with the useful addition of data population maps indicating the numbers of individual measurements contributing to each grid point average; sample latitude–longitude overlays for locating the data geographically are provided generally with each grid map.

Nephanalyses

In a sense, a discussion of nephanalyses, the simplified cloud charts compiled routinely from A.P.T. and computer-rectified A.V.C.S. satellite photographs, is out of place in this chapter since subjective human interpretation of data

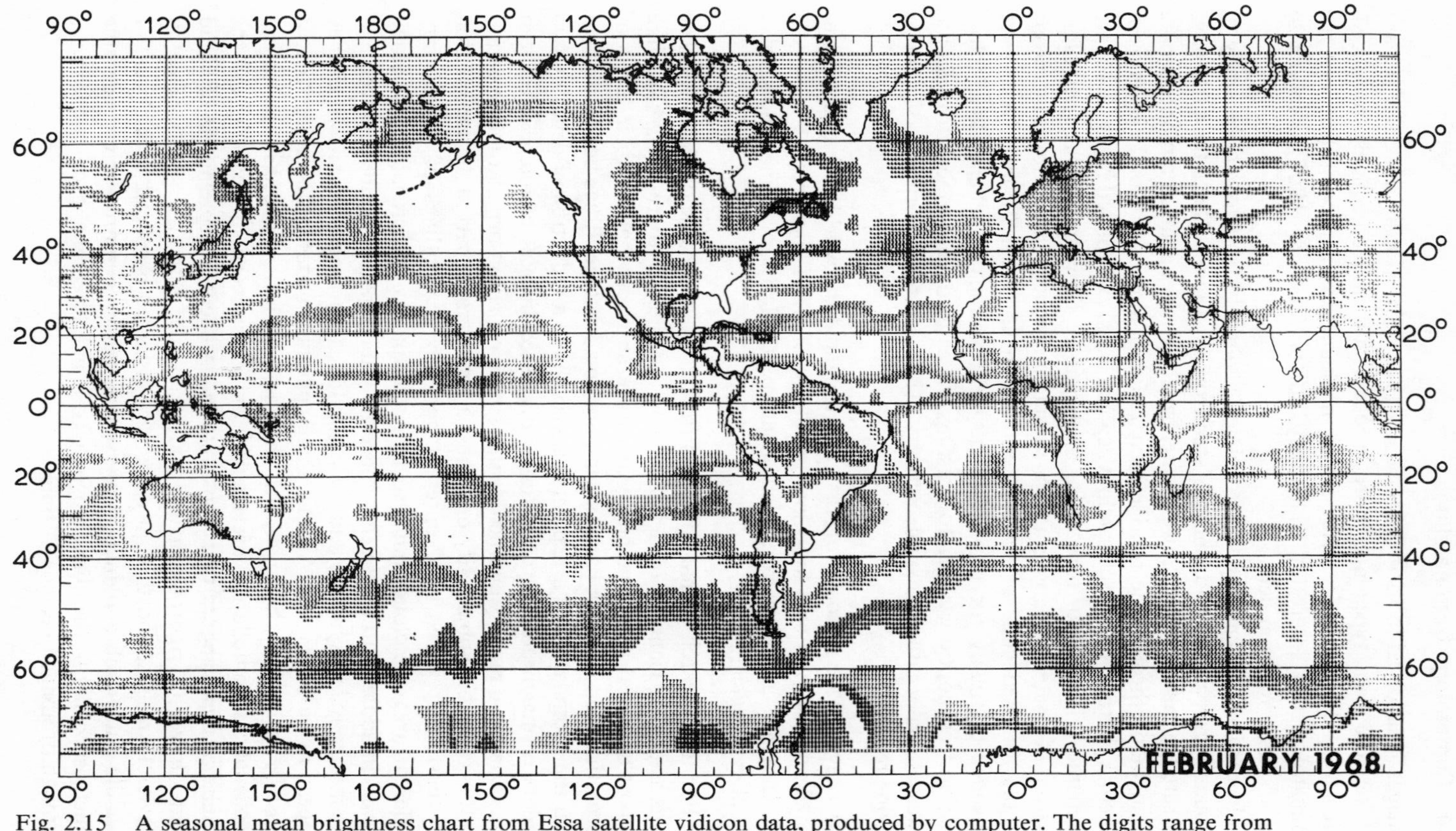

Fig. 2.15 A seasonal mean brightness chart from Essa satellite vidicon data, produced by computer. The digits range from 0 (darkest) to 10 (brightest). (From Leese *et al.* 1970; from Taylor & Winston 1968)

images is involved in their compilation; the two previously discussed groups of satellite data types are more mechanically and objectively produced. However, the regularity with which nephanalyses are compiled, and the climatologically interesting data they portray (Barrett 1967) combine to justify some mention here.

In earlier, conventional, meteorology 'nephanalysis' involved the study of cloud types, patterns and amounts seen from the ground. Since the early days of satellite meteorology, however, nephanalysis has acquired a more specific connotation through the routine preparation of simplified cloud charts from satellite photographs. Such satellite nephanalyses, exemplified by fig. 2.16, are prepared primarily for forecasting applications. They present the rich information of a satellite photograph or mosaic in a much simpler form by means of the accepted code of symbols set out in fig. 2.17. Nephanalyses were intended originally for near real-time transmissions of salient cloud information to synoptic weather forecasting offices throughout the world via the international weather facsimile network. They are still important today in two major respects:

(*a*) They provide national forecasters throughout the world with independent checks on their own local interpretations of A.P.T. pictures.
(*b*) They are useful research documents particularly for some studies of a climatological nature.

For the research climatologist, as for the operational short-term weather forecaster, the richness of detail in most satellite photographs (which, like almost all aerial photographic systems are unselective above their levels of resolution) can be an embarrassment, not an advantage. It has been estimated that the compilation of a nephanalysis results in a reduction of detail by two orders of magnitude. Several studies discussed in later chapters have been based on these much simplified displays rather than on original image data.

It is clear that subjectivity must play a considerable part in the compilation of charts that purport to represent not only the outlines of significantly different areas of cloud, but also by-eye estimates of percentages of cloud cover, in addition to statements concerning the types of clouds involved and the more obvious synoptic and subsynoptic scale arrangements in which they appear to be organized. Against this apparently undesirable subjectivity, it must be recognized, however, that the photo analysts engaged in the task of nephanalysis compilation are trained and experienced in this job, and the results of their labours thus constitute independent interpretations of cloud fields drawn to rigorous operational standards. Although a loss of detail inevitably results, and this may not be considered beneficial by all users of satellite data, such a reduction of content can be a source of considerable relief, especially to the climatologist who is interested in appropriate features through prolonged periods of time.

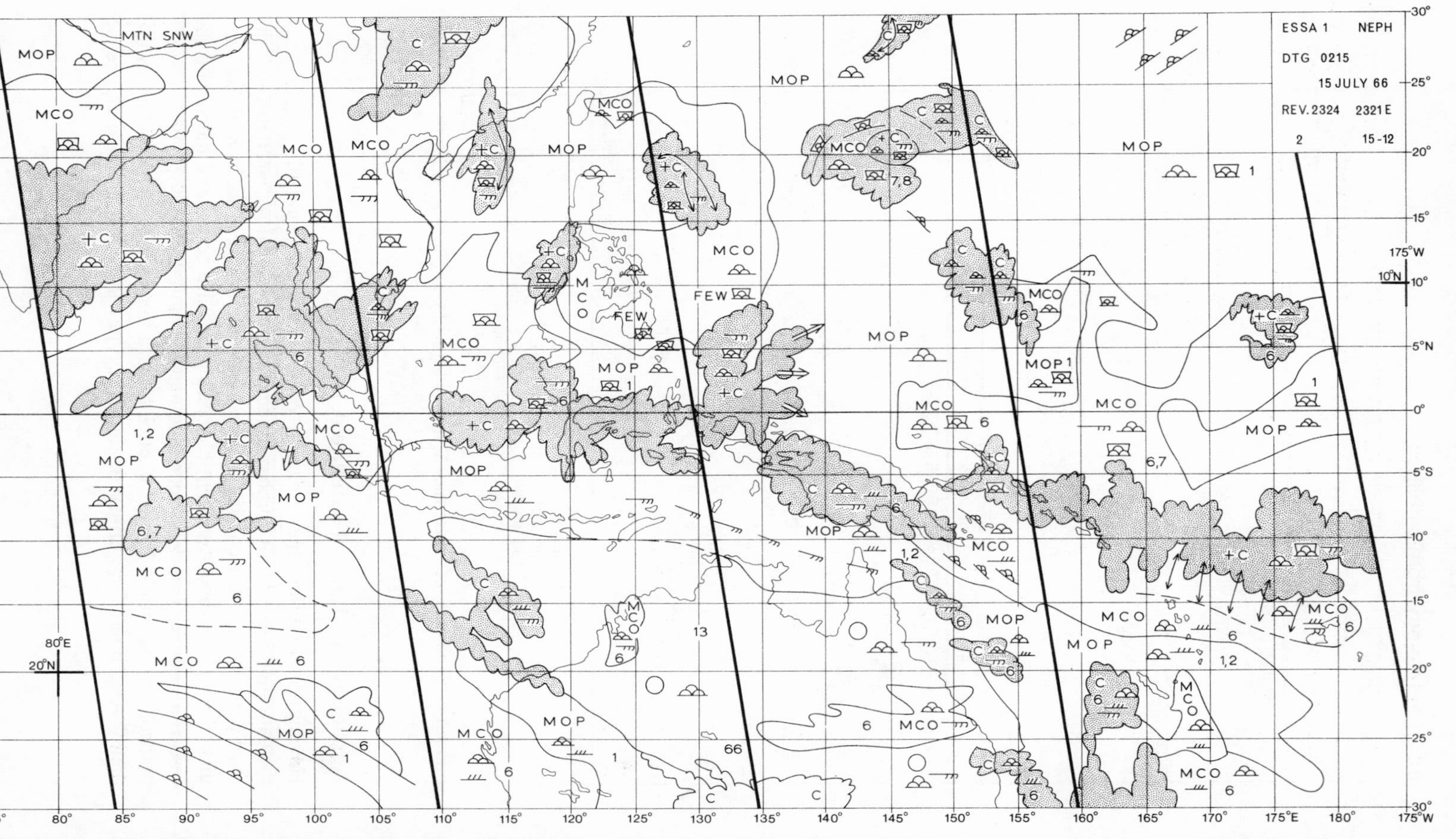

Fig. 2.16 A sample nephanalysis, depicting the chief features of cloudiness across the tropical Far East, 15 July 1966. (From Barrett 1970)

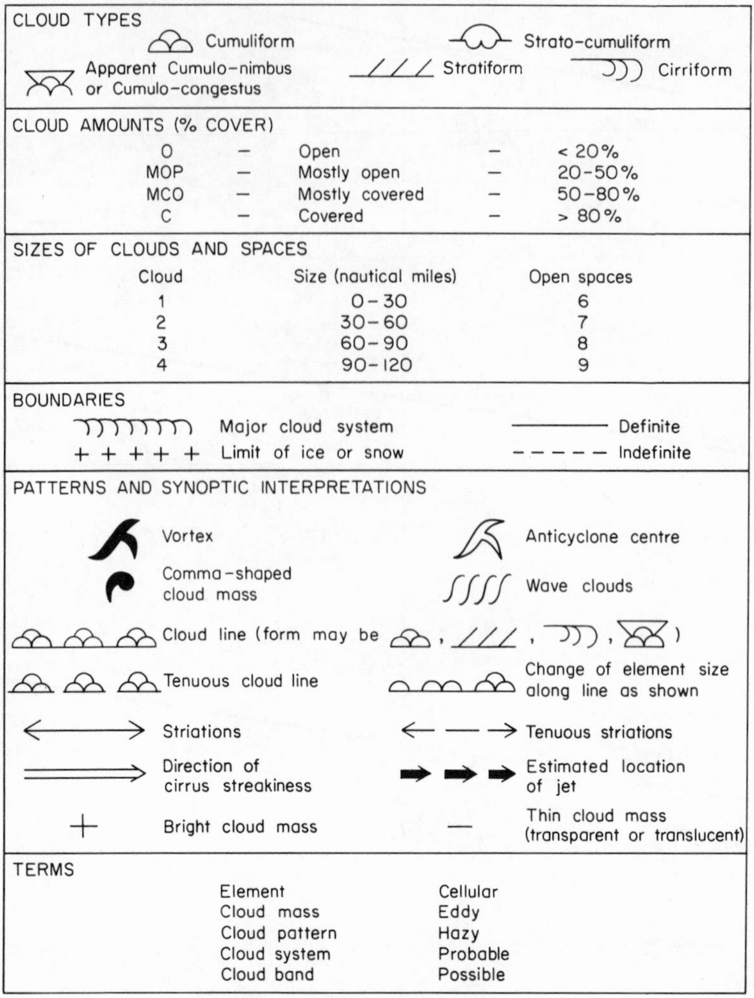

CLOUD TYPES

Cumuliform — Strato-cumuliform

Apparent Cumulo-nimbus or Cumulo-congestus — Stratiform — Cirriform

CLOUD AMOUNTS (% COVER)

O	–	Open	–	< 20%
MOP	–	Mostly open	–	20–50%
MCO	–	Mostly covered	–	50–80%
C	–	Covered	–	> 80%

SIZES OF CLOUDS AND SPACES

Cloud	Size (nautical miles)	Open spaces
1	0–30	6
2	30–60	7
3	60–90	8
4	90–120	9

BOUNDARIES

Major cloud system — Definite

\+ + + + + Limit of ice or snow — - - - - - Indefinite

PATTERNS AND SYNOPTIC INTERPRETATIONS

Vortex — Anticyclone centre

Comma-shaped cloud mass — Wave clouds

Cloud line (form may be , , ,)

Tenuous cloud line — Change of element size along line as shown

Striations — Tenuous striations

Direction of cirrus streakiness — Estimated location of jet

\+ Bright cloud mass — — Thin cloud mass (transparent or translucent)

TERMS

Element	Cellular
Cloud mass	Eddy
Cloud pattern	Hazy
Cloud system	Probable
Cloud band	Possible

Fig. 2.17 The code of conventional nephanalysis symbols. (From Barrett 1967)

Application and use of satellite data

Having reviewed the general natures of various weather satellite systems and the data they provide for forecasting and climatological research through the chief processing and archiving procedures, some bridge must be built between this chapter and the subsequent chapters which are concerned more with the research results. Two key questions may be posed, therefore, which every potential climatological user of satellite data may ask initially: When to use them? And how? The utility of the data types introduced above can range from little or none in the seeking of solutions to some problems, to very great in trying to solve others. The main criteria by which the utility of satellite data may be assessed involve their availability, quality and suitability for a particular application.

Data availability involves much more than simply whether or not observations are available for a specific place and time. Of equal importance to the climatologist is the format or formats in which the data have been stored. Furthermore their compatibility with conventional data may be critical, since satellite and conventional data are often interpreted best in conjunction with each other; the advent of globally complete weather satellite observing systems must not be expected to render redundant the pre-existing conventional station networks. Rather, an expanded, improved surface network is necessary to permit the greatest possible benefits to be drawn from satellite observations. By the same token, of course, conventional climatology itself is given a considerable fillip by the growth of climatology from satellites.

Of further importance still is the compatibility of satellite data with a processing procedure a research worker wants to use. So many satellite data have been accumulated already that the task of transforming the archived data into quantitative climatological presentations is already formidable; hence the significance of the question 'How can satellite data be accommodated most readily and beneficially within existing climatological frameworks?'

Next, the quality of the archived satellite data sometimes limits their use. Schwalb and Gross (1969) have reviewed the whole problem of obtaining stable quantitative data from vidicon sensors, based on experience with the Essa satellites. Many influences have a potential bearing upon the internal consistency of such data pools.

Finally, the question of the suitability of satellite data for particular applications is a broad question indeed. It can be answered properly only within the limits of each individual study. Perhaps sometimes in the past satellite observations have been used inappropriately. Be that as it may, there is no doubt that only the merest surface of the total fund of satellite data has been scratched as yet in the furtherance of climatology. This, then, is but the beginning of an era.

Let us proceed forthwith to review the underlying relationships between satellite data and the fundamentals of the earth's atmosphere, summarized under the headings 'atmospheric energy', 'atmospheric moisture' and 'atmospheric circulation', before considering some early results of actual climatological studies from satellites.

PART THREE
Principles of weather satellite data analyses

3 Patterns of atmospheric energy

Introduction

In pre-satellite days two general approaches were made in search of an adequate climatological understanding of the global energy budget. In both, satellite observations are already proving their worth. The two approaches are summarized below (after Newell *et al.* 1970).

Studies of energy fluxes through the earth/atmosphere system

Such studies depended basically upon estimates of the net radiation made available to that system by the sun. Commonly, the mean flux (i.e. the mean energy crossing a unit area in unit time) of insolation energy perpendicular to the earth's surface was taken to be about 0·5 cal/cm^2/min. This value is subject to five major variable factors: atmospheric interference, the output of radiation by the sun, the altitude of the sun in the sky, distance from the sun and the length of day. Thus the energy balance diagram, fig. 2.8, represents a mean climatological statement for the earth and its atmosphere from which an almost infinite variety of departures may be expected locally in the short term. The classical studies in pre-satellite days of necessity invoked *estimates* of several important energy balance components – for example, the solar radiation flux at a surface normal to the sun's beam outside the earth's atmosphere at the earth's mean distance from the sun (the 'solar constant'), and fluxes of outgoing radiation towards space (see, for example, London 1957; Davis 1963; Budyko and Kondratiev 1964).

Interest in questions relating to the net radiation available to the system as a whole, the radiative fluxes at the earth's surface and other components of the net radiation and energy budgets has quickened with the advent of satellites. These are able to measure quite accurately several components of these budgets. Thus unique opportunities are arising for comparisons between theories and observations. At last it is becoming possible to prepare a secure empirical platform for such energy flux studies.

Inquiries into the uses of energy traversing the earth/atmosphere system

In general, these were concerned with evaluations of the amounts of energy involved in transformations from one form to another. In particular, several of the more important studies investigated transformations between potential energy and kinetic energy (see e.g. Margules 1903; Lorenz 1967).

Satellite evidence is providing more accurate pictures of the relevant distributions of atmospheric energy, moisture and circulation patterns, and is helping to establish the chief sources and sinks of various energy forms. Hopefully, the end result will be a much better understanding of the organization and operation of the atmosphere as a whole.

Satellite radiation measurements

General comments

We have examined already (in chapter 2) some of the complex geometrical factors that may influence radiation measurements made by satellite-borne sensors. Before proceding to review some of the more useful types of measurements that have been made by the time of writing, it is worth noting that these are often affected also by a range of atmospheric factors. These include air temperature, the size and composition of included particles, and the nature and temperature of the surface at the base of the atmosphere. Any scheme designed to derive climatologically – or meteorologically – significant data from the original observations must make appropriate allowances for such factors.

Unfortunately, as Nordberg (1967) pointed out, their effects upon sensed radiation levels are extremely complex and cannot be stated explicitly for all cases. For example, in the case of infra-red emission by a surface on earth, the intensity (I) of the radiation measured by a satellite in a wavelength interval (λ_1) to (λ_2) is generally related to the temperature (T) and emissivity (e) of that surface by Planck's law. This states that

$$I \sim \int_{\lambda_2}^{\lambda_2} e(\lambda) B(\lambda, T) d\lambda \qquad (3.1)$$

where λ is the wavelength of the measured radiation and $B(\lambda, T)$ the Planck function. However, some of this radiation may be absorbed and re-emitted by the atmosphere, depending on a variety of factors, including radiation wavelength, atmospheric temperature, pressure and composition, and the absorption coefficients of the various gases in the column through which the radiation has been transmitted. As a consequence, many simplifying assumptions must be made before radiation measurements from earth-orbiting satellites can be translated into estimates of meteorologically significant parameters. Further inquiry into this area is outside the scope of the present account. For present purposes, it is more pertinent to ask what the translated measurements may reveal climatologically than how the initial data were obtained or how they were rendered in their translated forms. Since water vapour and carbon dioxide absorption and re-radiation patterns are more appropriate to the contents of chapters 4 and 5 respectively, the following remarks pertain to the basic structure of the radiation budget only. The interrelations between all the significant components of earth/atmosphere energy and radiation budgets are discussed in chapter 6.

Incoming solar radiation

Pre-satellite studies were plagued by the fact that, despite its name, the solar constant is thought to be a variable. Not only is the total flux subject to short-period fluctuations, but also the composition of that flux in terms of certain wavebands, in particular the ultra-violet in association with sunspot cycles. The establishment of a dependable value for the solar constant is an essential first step towards the solution of many radiation and energy budget studies in climatology. Since some incoming solar radiation is absorbed by the earth's atmosphere, while more is reflected or back-scattered towards space, the solar flux at the top of the atmosphere can be measured reliably only from outside the atmosphere by rockets or satellites. Such observations have been made intermittently by a variety of American rockets since 1947, and more continuously in recent years by satellites like the American OSO (Orbiting Solar Observatory). Measurements extending over many months are necessary to quantify the mean magnitude of the solar radiation flux and also the range of variations characterizing the visible and near ultra-violet wavebands that affect the lower atmosphere most directly.

A significant improvement in the satellite capability for measuring ultra-violet variations was achieved on the launch of Nimbus IV in 1970. This satellite was the first to be equipped with a Monitor of Ultra-violet Solar Energy (MUSE: see Nimbus IV User's Guide, ARACON 1970). The purpose of this new type of sensor is to search for time changes in the ultra-violet solar flux within five narrow wavebands, and to measure the attenuation (i.e. depletion) in these wavebands as the satellite views the setting sun. Instruments for more detailed examinations in many visible and ultra-violet spectral ranges are being tested for future satellite flights (Drummond 1970).

Currently, on satellite evidence, a value of 1·95 cal/cm^2/min is often invoked for the solar constant (see chapter 6).

Albedoes and reflected radiation

The albedo or reflectivity of an object may be defined as the percentage of its daily flux of incident solar radiation which is returned to space without altering the temperature of the surface of that object, and without alteration to its own wavelength. Consequently, albedo (A) can be expressed:

$$A(\lambda, \phi, d) = \frac{R(\lambda, \phi, d)}{S(\lambda, \phi, d)} \times \frac{100}{1} \tag{3.2}$$

where S is the flux of incoming solar radiation and R the flux of reflected radiation crossing a horizontal element of an area outside the atmosphere. The geographical coordinates of that area are (λ, ϕ). The letter d designates the day for which the computations are made, and involves a certain declination of the sun and a certain distance from the earth to the sun. Most albedo data from satellites have been obtained directly by viewing the sun and the earth with the same instrument (Vonder Haar and Suomi 1971).

It has long been known that, in general light-coloured surfaces such as cloud tops, ice and snow are good reflectors of incident solar radiation whereas dark surfaces reflect less efficiently. Typical values for different surfaces range from 80% for fresh snow, through some 50 to 65% for clouds depending on their type and thickness, 55% for old snow, 25% for grass, 20–30% for sand, 10–15% for rock, 10% for wet earth and 5–10% for forest surfaces. The albedo of a water surface may range from about 5% at high solar elevation to 70% when the sun is low in the sky, though surface roughness is an important factor too. However, attempts to combine such varied figures in appropriate proportions in order to estimate mean earth/atmosphere albedoes on a global or regional basis must be fraught with enormous difficulties. Small wonder that the satellite estimates of global albedo in table 6.3 (p. 155) differ significantly as a group from those compiled before the first satellites were launched. In the meantime, of course, a satellite albedo map like fig. 3.1 surpasses any pre-satellite map of estimates both in its reliability and its breadth of coverage.

Unfortunately, from the heat budget point of view, a satellite automatically reports the total albedo of each area whether this is due primarily to one factor such as a sheet of cloud or due to several factors in combination. Hence the desert areas of north Africa, Arabia and Australia possess albedoes in excess of 30% in fig. 3.1, suggesting contributions simultaneously from rock, sand and also cloud. Conversely, in middle latitudes, the albedoes of 40–50% seem rather low for cloudy regions dominated by temperate depressions, but may be accounted for by probable breaks in the cloud fields. On a more general level, it is interesting to note that the geographical and climatological dissimilarities of northern and southern hemispheres help to prompt quite different albedo patterns north and south of the Equator. Whereas the

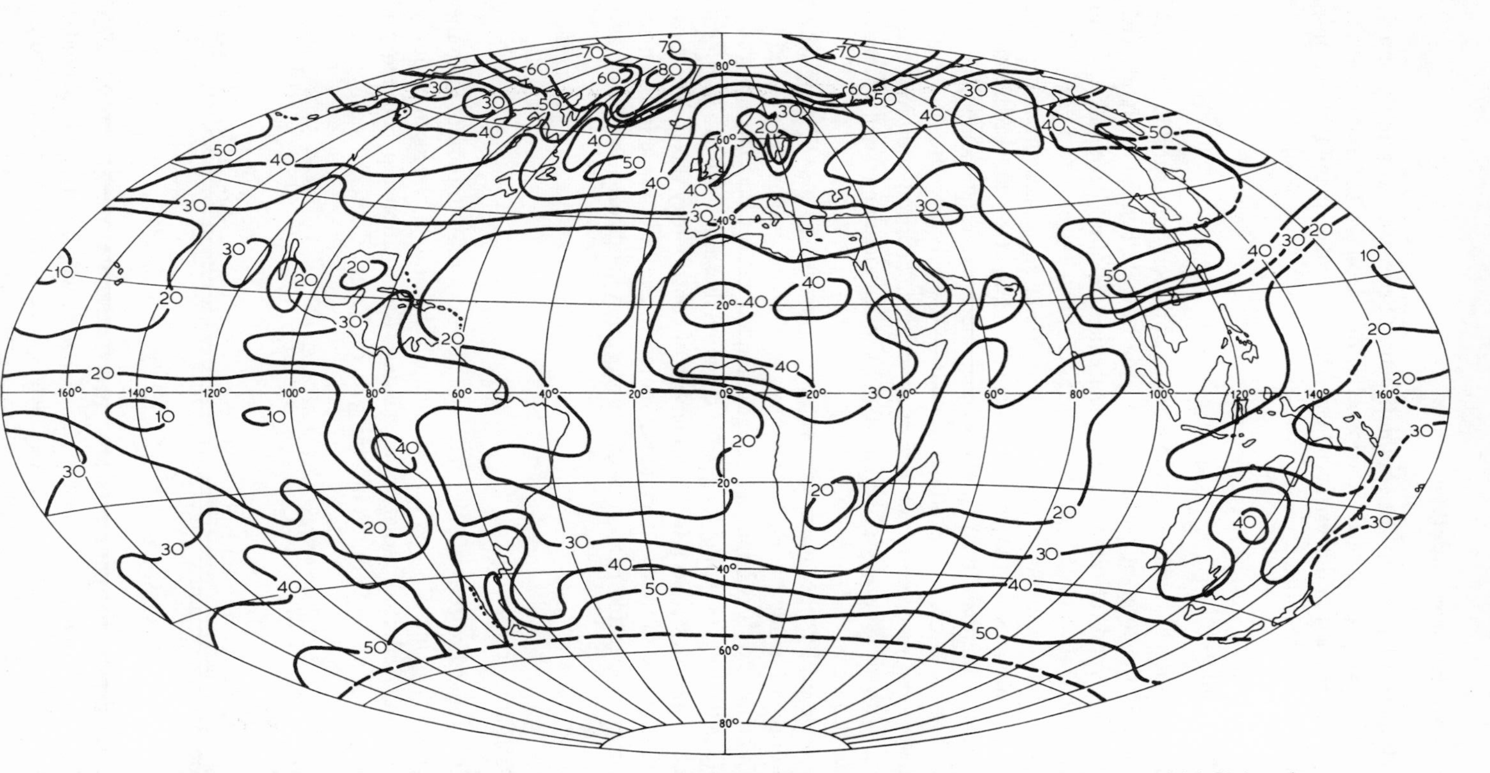

Fig. 3.1 The albedo of the earth–atmosphere system from Nimbus II evidence, during the period 1–15 June 1966. Values in per cent. (From Raschke & Pasternak 1967)

preponderance of water in the southern hemisphere leads to a fairly simple pattern of zonally orientated albedo belts, the stronger contrasts between areas of land and sea in the northern hemisphere stimulate many irregularities in its albedo distribution.

Terrestrial radiation

One of the principal radiation parameters most directly measurable by satellite means is the total outward flux of long-wave terrestrial radiation (E). A significant fund of long-wave measurements, falling generally within the waveband from 4 to 100 μm, has been accumulated over recent years. Fortunately in this case the problem of translating such measurements into acceptable estimates of the radiation actually emitted by the earth and its atmosphere is comparatively slight so that the intermediate analytical and empirical models are quite straightforward (Wark *et al.* 1962). Existing radiometers can provide measurements of emitted earth radiation in a given direction with an accuracy of better than 1% at an angular resolution better than one degree.

Climatologically the most important and useful forms of long-wave radiation data include the maps and mean meridional graphs exemplified by those in chapter 6. It is important to radiation and energy budget studies that not only is the total outgoing long-wave radiation flux amenable to assessment by satellite means, but so also are a number of contributory fluxes. The most significant of the supplementary measurements have been made in the principal water vapour and carbon dioxide absorption bands (see chapters 4 and 5).

Absorbed solar radiation

The geographical pattern of solar radiation absorption, so important from the climatological point of view, can be established through considering jointly incoming and reflected radiation intensities. The relationship between the absorbed portion (ABS) of the incoming flux (S) can be determined by:

$$ABS(\lambda, \phi, d) = \left(1 - \frac{A(\lambda, \phi, d)}{100}\right) S(\lambda, \phi, d) \qquad (3.3)$$

where A is the earth/atmosphere albedo. Substituting for A by invoking equation 3.2, we obtain:

$$ABS(\lambda, \phi, d) = S(\lambda, \phi, d) - R(\lambda, \phi, d) \qquad (3.4)$$

Fig. 3.2 depicts the pattern of absorbed radiation contemporaneous with that of reflected radiation shown by fig. 3.1. It is worth noting in particular that, were the albedo of the northern hemisphere to be uniform, the North

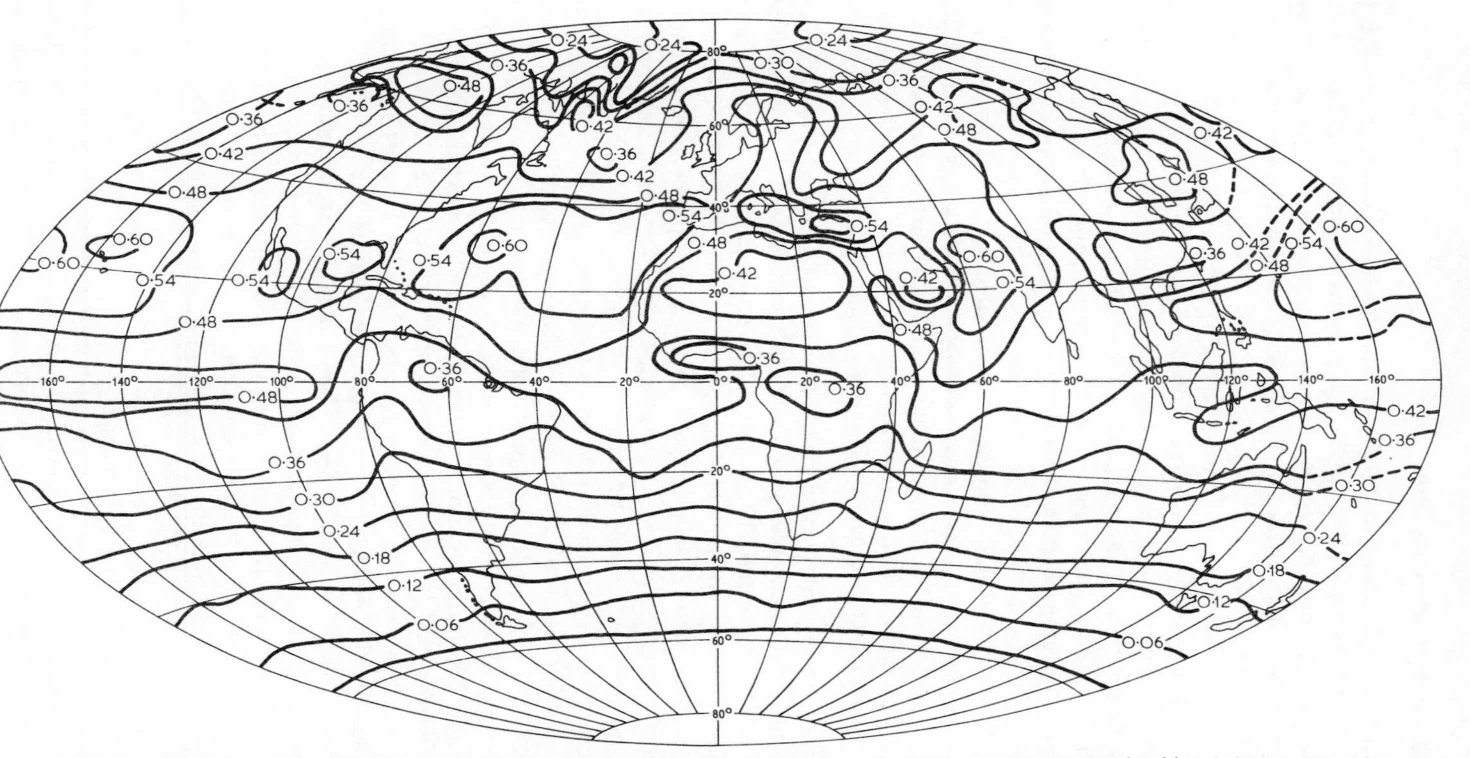

Fig. 3.2 The global pattern of absorbed radiation evidenced by Nimbus II from 1–15 June 1966. (In cal/cm²/min) (From Raschke & Pasternak 1967)

Pole would absorb more energy daily in summer than the subtropics. This is principally due to the greater length of day in that season in high latitudes. It is the albedo pattern that prompts the zone of maximum absorption of solar radiation to fall within about 10° to 35° N. Note, however, that this zone is interrupted over North Africa and Arabia, and over south-east Asia, where deserts and clouds respectively have high albedoes. Away from 10° N and 35° N levels of absorbed solar radiation decrease towards both poles. The northward decrease is due primarily to the increasing albedoes. The southward decline owes most to the declining flux of incoming solar radiation. It is from the zone of maximum absorption of solar radiation that much energy is exported by the atmosphere and oceans to the regions of absorption minima. Meridional profiles compiled from maps like fig. 3.2 show that more energy is absorbed by oceans than by continents on the same latitudes. This underlines the fact that the oceans act as great stabilizing reservoirs of energy which stabilize significantly the general circulation of the atmosphere, especially its seasonal fluctuations. Atmospheric absorption involves three gases in particular, whose climatological significances are quite distinct:

1 Absorption by water vapour. It has been remarked that 'the water vapour content of the atmosphere represents a vast storehouse of heat, in latent form' (McIntosh and Thom 1969). That energy can, of course, be traced back ultimately to the sun. Latent heat may be added to, or released from, that storehouse principally through changes of state of atmospheric water as explained in chapter 4. Energy in latent form may be transported horizontally or vertically by the movement of moist air.
2 Absorption by carbon dioxide. Since the carbon dioxide content of the air is relatively small (about 0·03% of the volume of dry air) this gas is climatologically important since it absorbs much terrestrial radiation which would be lost otherwise to space. In satellite climatology, carbon dioxide becomes significant in attempts to analyse more thoroughly the circulation patterns in the stratosphere, in whose lower layers carbon dioxide is the only significant energy-absorbing gas. Its other significance in satellite climatology is in vertical profile analysis.
3 Absorption by ozone. So far as lower atmospheric climatologies are concerned ozone is important only indirectly, since its maximum concentration occurs 15–35 km above the earth/atmosphere interface. To man its chief significance is as a filter of most harmful ultra-violet radiation waves before they reach the levels of human habitation.

Surface temperature characteristics

The principles employed in mapping surfaces of the earth and/or its cloud cover from infra-red measurements are basically quite simple. Essentially, all objects omit electromagnetic radiation with spectral distributions and intensities that are unique functions of their temperatures and surface configurations. For black bodies, however, the intensities of their radiation must be functions of temperature alone, since surface variations are absent from such ideal cases. Assuming at first that terrestrial and cloud surfaces act like black bodies the next task is to choose an appropriate waveband in which to investigate the intensity of emitted radiation (E).

For useful measures to be made of surface temperatures, Planck's law (equation 3.1) can be invoked so long as the atmosphere is transparent, or nearly so, between the limiting wavelengths λ_1 and λ_2. Hence surface radiation measurements have been made most commonly in the well-known 'atmospheric window' wavebands from 3·4 to 4·2 μm and 8 to 12 μm. Within these limits little absorption of energy occurs in the earth's atmosphere. Thus, in the absence of clouds, radiation emitted by the earth's land or water surfaces reaches a satellite with only minor interference by a cloudless atmosphere. Any interference that does occur can be minimized through methods described by Kunde (1965) and Wark *et al.* (1962) respectively for the two wavebands specified above. For example, if hot black body surfaces (say at 300 °K) are seen through a warm, moist (tropical) atmosphere the T_{BB} values derived from the radiation measurements must be corrected by as much as 2–4 °K. For a dry atmosphere the required correction is less, while for cold surfaces (<280 °K) no correction is necessary.

Where earth surfaces are concerned it should be appreciated that the black body approximation holds good more nearly for some types of surfaces (e.g. water and heavily vegetated areas) than others (e.g. certain types of soil). Where the approximation is inappropriate weighting functions are essential for the translation of radiant emittances into acceptable temperature estimates. Unfortunately, the emissivities of many soils are not known precisely. Therefore satellite-derived temperature maps of continental surfaces must be interpreted and utilized with care (see e.g. Pouquet 1969; Barrett 1967, ch. 9). Potentially such displays are of interest climatologically for meso- and small-scale heat budget studies.

So far as cloud surfaces are concerned, some of the principal problems associated with the temperature evaluation of cloud tops are discussed in the following chapter.

The net radiation balance at the top of the earth's atmosphere

Especially at the hemispheric and global scales the parameters outlined above can be combined in evaluations of a most important quantity, the global net radiation balance (Q). This represents the difference between total incoming

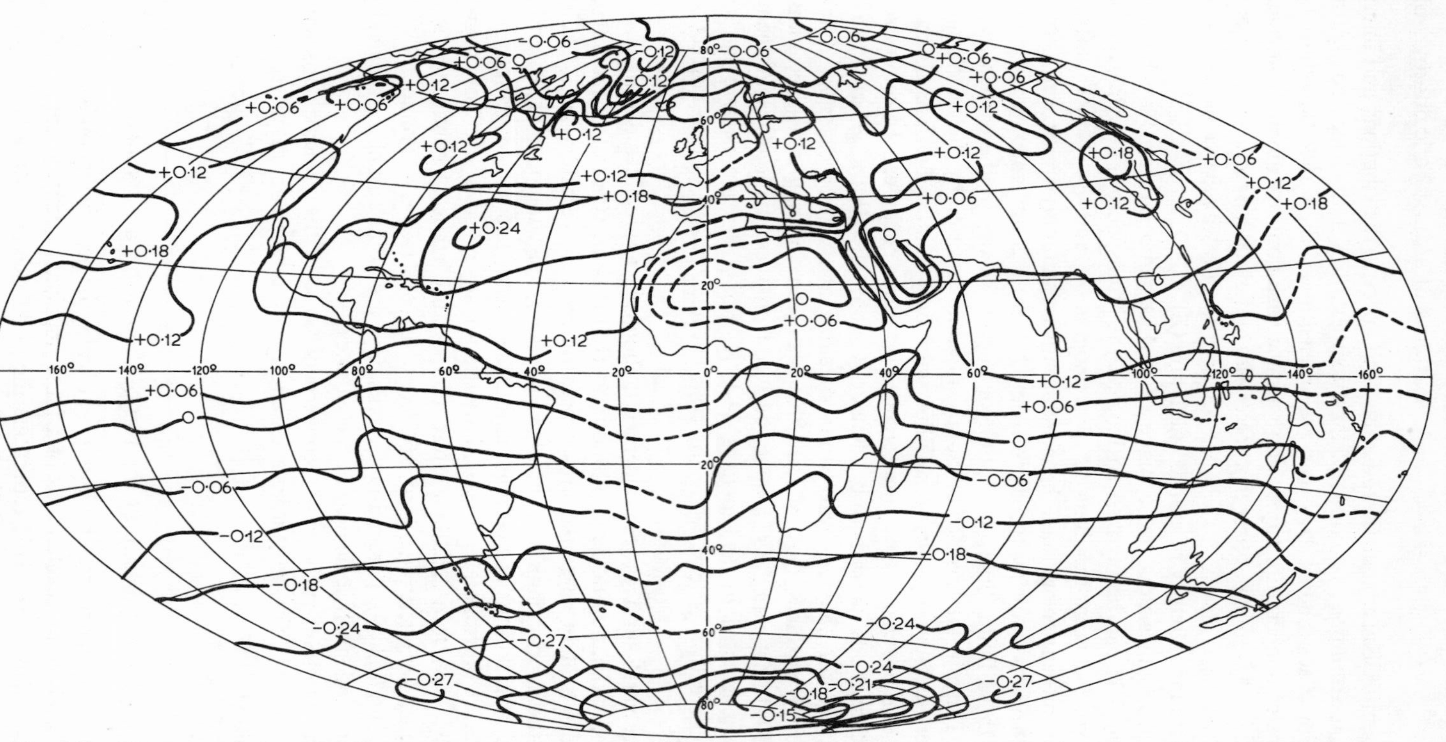

Fig. 3.3 The net radiation balance, Q, at the top of the atmosphere, from Nimbus II evidence, 1–15 June 1966. (In cal/cm^2/min) (From Raschke & Pasternack 1967)

and total outgoing radiation (fig. 3.3). It figures prominently in the discussion in chapter 6. The net radiation balance can be expressed by:

$$Q(\lambda, \phi, d) = S(\lambda, \phi, d) - R(\lambda, \phi, d) - E(\lambda, \phi, d) \qquad (3.5)$$

Earlier estimates of Q (e.g. by Houghton 1954, and Budyko 1963) were based – through necessity – on incomplete climatological data. Satellite observations of the types explained above support much more depêndable estimates. The general conclusions that have emerged from the first of these may be summarized as follows:

1 There is an excess of incoming radiation in low latitudes and a deficit in high latitudes.
2 The amplitudes and locations of the maxima and minima of incoming radiation are seasonally variable.
3 There is, in consequence of (1) and (2), an energy transport by the general circulation of the atmosphere, and by ocean currents.
4 The yearly global average of Q tends to zero, indicating that the earth/atmosphere system is probably in energy equilibrium over periods of several years.

Closer examinations of the sources and sinks of atmospheric energy, their seasonal variations, and the associated patterns of horizontal energy transfer, all established on satellite evidence, form the bases of chapter 6. We turn lastly in the present context to a review of radiation work seeking to sound the atmosphere vertically, not horizontally as in the previous cases.

Vertical temperature profile analyses

Climatologically, two byproducts of particular importance may be expected to emerge from the routine collection, processing and analysis of data from vertical sounding instruments such as SIRS, IRIS and S.C.R. (see chapter 2).

(1) Once-daily soundings through a considerable depth of the atmosphere from orbiting satellites in near polar orbits will support mean profiles for different localities through selected periods of time. Figs. 3.4 and 3.5 exemplify the forms in which vertical temperature data are now being made available to the meteorological and climatological communities. Over extended periods these will aid the recognition, description and final explanation of anomalous conditions at all levels, and assist our understanding of the transmission of the influences of such anomalies either upward or downward.

(2) Daily maps in the horizontal plane will provide the basic data necessary for the compilation of maps of mean distributions of chosen parameters in the horizontal, e.g. temperature and humidity. Already such maps represent a considerable advance over conventional maps, especially in regions of poor radiosonde coverage. Numerical predictions for the northern hemisphere

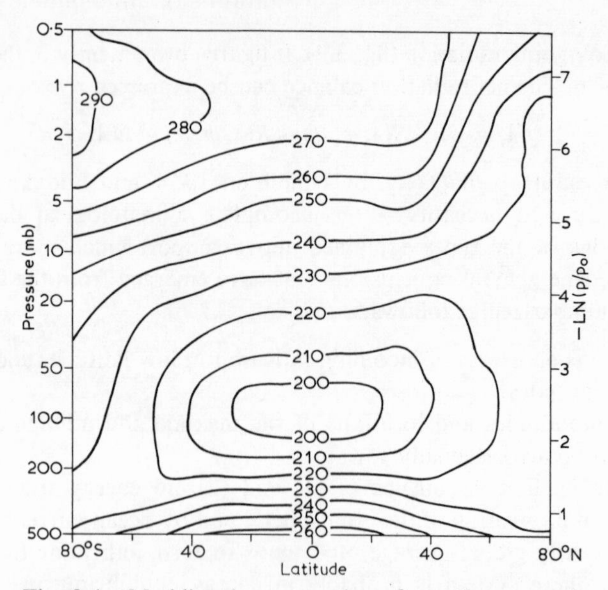

Fig. 3.4 Meridional cross-section of zonally averaged temperatures. (°K), 16 July 1970, from Nimbus IV Selective Chopper Radiometer. (From Barnett *et al.* 1972)

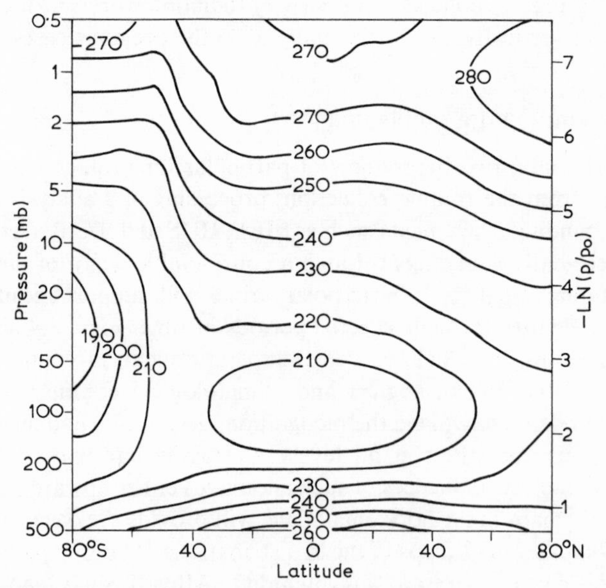

Fig. 3.5 As fig. 3.4, for 21 January 1971. (From Barnett *et al.* 1972)

have already been improved significantly through SIRS charts of the 500 mb surface, while for the southern hemisphere, which is notoriously devoid of adequate meteorological data, Smith *et al.* (1970) have produced what may be considered the first complete upper air charts (fig. 3.6). These 300 mb patterns seemed very reasonable on theoretical grounds. For the northern hemisphere, it has emerged that the 300 mb pressure surface height contours derived jointly from conventional and SIRS data show good agreement except where conventional data were sparse. Although this largely suggests that the SIRS data improve the analyses in these areas, the correlation techniques currently used to translate satellite radiances into temperatures necessitate the following steps to improve those analyses still further. These

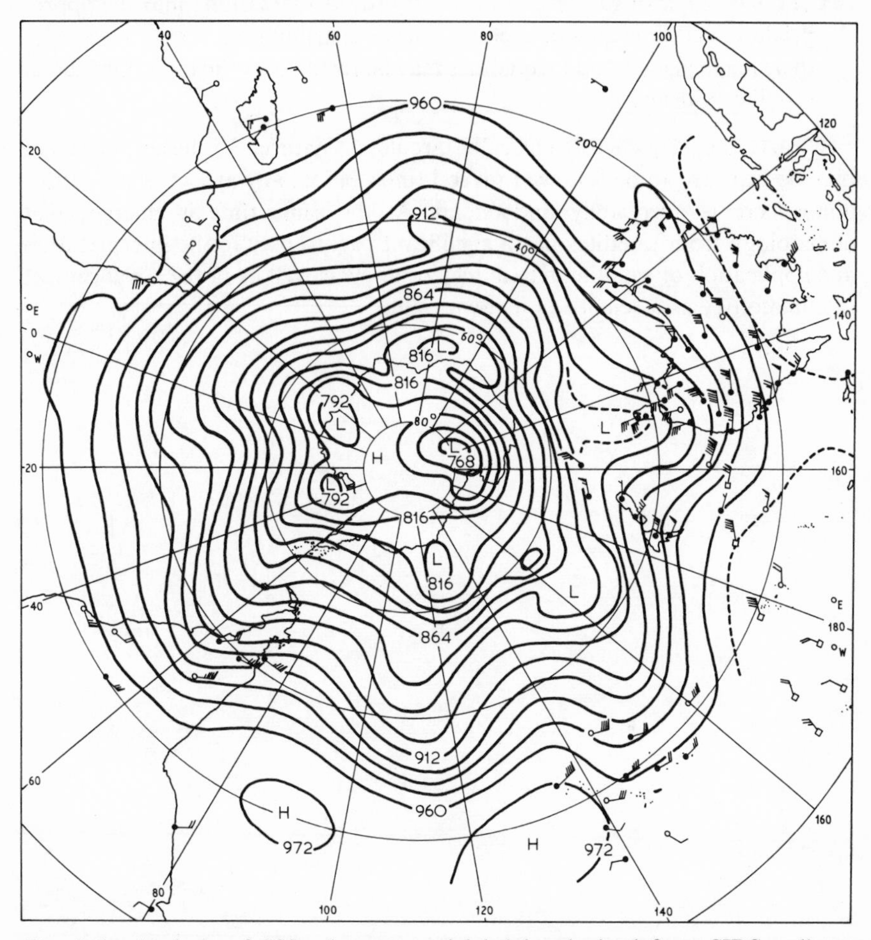

Fig. 3.6 Analysis of 300 mb geopotential height obtained from SIRS radiance observations. The analysis is based on 24 hr of satellite data centred around 00 G.M.T., and approximates to a 00 G.M.T. synoptic chart. (From Smith *et al.* 1970)

are commonly followed in satellite meteorology and climatology where improved interpretations (of satellite data) are required.

(*a*) The establishment of climatological patterns with various degrees of homogeneity from the available instrumental records.
(*b*) Regressions of satellite data against related climatological data as soon as theoretical relationships between the two have been set out in experimental form.
(*c*) Repeated applications of the suggested regression relationships to assess the weaknesses of the correlation equations, which can then be modified and improved.
(*d*) The feed-back of the results of the improved regressions into the appropriate climatological statements of mean distributions. Some, at least, of their inadequacies and inequalities may be rectified by the incorporation of satellite evidence.

Clearly this sequence is basically circular in nature. So these four stages may be expected to be followed several times before any aspect of the earth's atmosphere is adequately understood. At the same time it emerges that climatology from satellites has a significant part to play in the interpretation and application of satellite data in meteorology as well as in the advancement of climatology as a science in its own right.

4 Distributions of atmospheric moisture

Introduction

Water, on account of its peculiar characteristics, its changes of state within the normal range of atmospheric temperature and its distribution through the troposphere, is of considerable significance to the climatologist. Indeed, it can be claimed that many changes in weather are attributable largely to changes in the state of atmospheric water. Similarly, many contrasts in climate from one geographical location to another are both obvious and economically important by reason of differences in the density and state of water in the atmosphere. Since this water may exist in any of three states – solid, liquid and gaseous – within the normal range of tropospheric temperature, and changes of state may take place with ease under common tropospheric conditions, it plays a large part also in the energy balance of the earth/atmosphere system.

It is convenient to discuss the contributions made by satellites to our climatological knowledge of atmospheric water under three headings: satellite cloud observations, satellite measurements of atmospheric water vapour, and precipitation estimates from satellite data. It will emerge that such contributions relate more to assessments of the amounts and distributions of atmospheric moisture than to formulations of revolutionary theories. However, since moisture is so largely involved in the general workings of the atmosphere, more accurate evaluations of its quantities and distribution patterns cannot be too highly rated.

We begin with the interpretation of satellite-viewed cloudiness, since upon analyses of the cloud fields depicted by photographs and infra-red displays were based some of the earliest routine operational usages of these new data in atmospheric science. As a consequence, much climatology and meteorology from satellites still depends upon cloud characteristics.

Satellite cloud observations

Moisture in liquid form lends itself particularly well to investigation from satellite altitudes. Aggregates of liquid water droplets organized into different forms of clouds comprise the only regularly visible natural phenomena within the atmosphere affording clear indications of such patterns and organizations as atmospheric pressure, temperature, and air circulation, as well as the distribution of moisture itself. In the hydrological cycle of the earth the atmosphere acquires moisture by evaporation from oceans, lakes, rivers and damp soils, and by moisture 'breathed', or transpired, from plants. Condensation, the reverse process, may take place either on the surface of the earth itself in the form of dew, fog or frost, or within the atmosphere in the presence of sufficient condensation nuclei. Contact cooling over cold surfaces is the chief factor involved in the former cases, while mixing of different layers or masses of air is an important process leading to condensation within the atmosphere. The effects of a cold earth/atmosphere interface may be influential too, as may additions of more moisture to an already near-saturated atmospheric layer.

The most effective cause of all, however, is the dynamic process of adiabatic cooling, due not to a change in the energy in the system, but to an increase in the volume of a mass of air due to a decrease in pressure. This usually occurs when air is lifted to a lower pressure level causing its temperature to fall towards dew point. Consequently, we may say that the chief cause of moisture condensation in the atmosphere is uplift, whether free (related to heating from below and consequent expansion permitting a less dense air bubble to rise) or forced by physical features (e.g. mountain ranges lying across the paths of moist air currents).

From conventional viewpoints, clouds are classified best on the joint bases of their altitudes and general appearances. Together, these criteria help to indicate the chief stimuli to the development of different types of clouds. Altitude separates the lower, often thicker, water droplet clouds from higher, generally more diffuse and shallow clouds of ice crystals. On the basis of general appearance, we may differentiate sheet-like clouds from tower-like clouds, with a transitional category between the two. Sheet clouds are usually associated with slow, widespread uplifts like those characteristically found in frontal zones between contrasting air masses. Tower clouds, meanwhile, form in association with strong vertical currents in the atmosphere, especially those accompanying a markedly unstable weather situation. The intermediate group of clouds develops in response to moderate uplift and/or turbulence. Fig. 4.1 depicts the principal cloud types in relation to such an approach.

This scheme, though very simple, is basic to much that follows in this chapter. It is capable of considerable elaboration. A useful classification which has been developed in much greater detail is laid out in the *International Cloud Atlas* (W.M.O. 1956). Table 4.1 explains some of the more important basic terms figuring in contemporary cloud classification schemes.

Table 4.1
An introduction to cloud nomenclature: some basic cloud terms

Root	*Meaning*	*Adjective*	*Meaning*	*Cloud types to which qualifying adjectives mostly apply*
Cirrus, cirro- (Ci)	Feathery, fibrous (ice crystal composition)	*Uncinus*	Hook-shaped, comma-like	Cirrus
Stratus, strato- (St)	Stratified, in layers	*Castellanus*	Turreted	Altocumulus, cirrocumulus
Cumulus, cumulo- (Cu)	Piled, heaped up	*Lenticularis*	Lens-shaped	Stratocumulus, altocumulus; occurs where air currents undulate abruptly in the vertical, e.g. on lee sides of mountains
Alto- (Al)	Middle altitude	*Fractus*	Broken, ragged	Stratus, cumulus
Nimbus, nimbo- (Ni)	Rain-yielding	*Humilis*	Lowly, humble, poorly developed in vertical	Cumulus
		Congestus	Heaped, cauliflower-like	Cumulus

Many meteorological characteristics of a cloudy air mass can be inferred from the nature and form of its included cloud. For example, it is often possible to assess at least the lateral extent, the type and the degree of vertical motion from the three categories in fig. 4.1. The depths of unstable layers may be suggested by the heights of the cloud tops; inversions may be assumed where marked flattenings of cloud tops indicate the subsidence of warmer, drier air from aloft; the vertical configurations of clouds and their different rates of movement at different levels may permit evaluations to be made of changing wind speeds and directions of flow with height; freezing levels may be estimated from changes in cloud outlines as sharply defined water droplet layers give way to more diffuse layers comprised of ice crystals.

A decade of analysis of satellite cloud photographs has proved that the

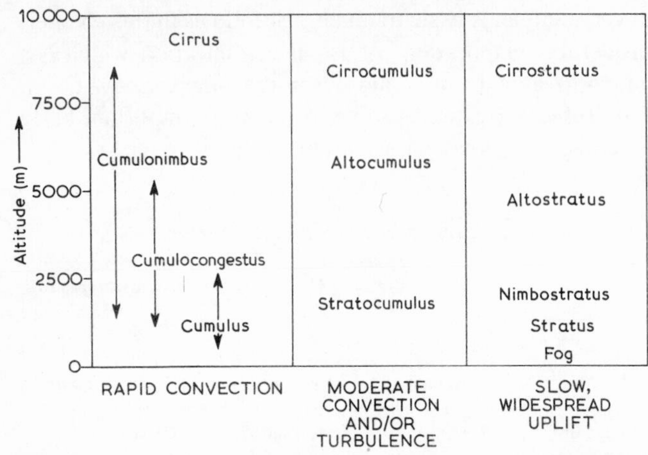

Fig. 4.1 A simple classification of clouds, based on altitude and mode of development.

forms, patterns and compositions of large cloud masses and organized cloud fields can be just as meaningful at the larger, synoptic and global, scales. Before proceeding to discuss such matters in more detail, it is instructive to review the climatological significance of cloudiness in general. Four points are significant:

1 Cloudiness shields the underlying air layers and surfaces of the earth from certain wavebands of solar radiation and increases the proportion of insolation that is reflected back to space. Thus temperatures below cloud layers are depressed during daytime.
2 Clouds are effective barriers to the transmission of longer-wave energy radiations towards space. Their lower layers re-radiate energy back towards the energy sources on earth and in the air beneath the cloud. So surface temperatures tend to be somewhat higher at night where cloud is present than where it is not.
3 Clouds act as temporary reservoirs of latent heat, and absorb a proportion of the radiation incident upon them. As mobility is a characteristic of most clouds, they play a very significant part in transporting energy across the surface of the earth.
4 There are clearly visible links between the stability characteristics of the atmosphere and the geographical distribution of precipitation. Unstable air is conducive to cloud development, and the nature of the prevalent cloud type or types in any region plays a large part in the differentiation of precipitation regimes.

Satellite cloud observations have greatly increased scientific interest in cloud organizations at the meso- and macroscales, precisely those at which pre-satellite studies were scant and incomplete. Most of this account is concerned with photographic rather than infra-red cloud evidence, though future developments in satellite instrumentation are likely to favour studies in the non-visible portions of the electromagnetic spectrum. To date, analytical and interpretational procedures have been developed to a much higher degree in connection with satellite photographs. On the whole the infra-red displays pose more difficult problems of interpretation. With respect to the current usefulness of infra-red cloud evidence, therefore, a few general principles will suffice before we turn to discuss the photographic evidence at much greater length. Infra-red data have been invoked in two ways in particular.

The first of these is in mapping cloud distributions through atmospheric window waveband measurements of surface temperatures. When a satellite-borne radiometer views a cloud covered region and a uniform cloud or terrestrial surface fills its instantaneous field of view, its average black body temperature can be derived from equation 2.3. Fig. 4.2 exemplifies the computed grid print maps of temperature prepared from Nimbus II H.R.I.R. observations. It indicates that it is possible to process the raw data automatically for presentations of absolute values of temperature in their approximate geographical locations. Fig. 4.3 shows that acceptable maps of synoptic scale can be constructed from consecutive satellite orbits. This is an isotherm map drawn from the evidence of fig. 4.2. At least some of the Temperature Humidity Infra-red Radiometer (T.H.I.R.) data from Nimbus IV are being processed likewise for approved users in the following map options: (*a*) polar stereographic, scale 1 : 30,000,000; (*b*) polar stereographic, scale 1 : 10,000,000; (*c*) multi-resolution Mercators, down to a scale of 1 : 1,000,000. These are proving useful not only in meteorology, but also in the geography of the earth's surface (see e.g. Sabatini *et al.* 1971). When sufficiently long runs of data have become available for specific regions, various more truly climatological investigations will become possible.

Secondly contingent upon mapping temperatures from infra-red data is the possibility that acceptable estimates of cloud top altitudes may be made from cloud top temperature maps. The earliest attempts to achieve this end were made by Fritz and Winston (1962) using Tiros II data, and by Bandeen *et al.* (1964) using Tiros III data to map the cloud patterns associated with hurricane Anna. Unfortunately, operational verifications of Tiros cloud top estimates from M.R.I.R. data indicated frequent gross errors, sometimes as great as 3000 m. These were ascribed to the rather poor resolution performance of that radiometer for such an exercise. Cloud fields are commonly intricately structured in terms of the altitudes of their components. Worse still, from the point of view of a medium resolution radiometer, clouds are frequently broken or discontinuous.

The Nimbus H.R.I.R. systems, designed to view areas only a few kilometres across, have supported much more fruitful research in this direction

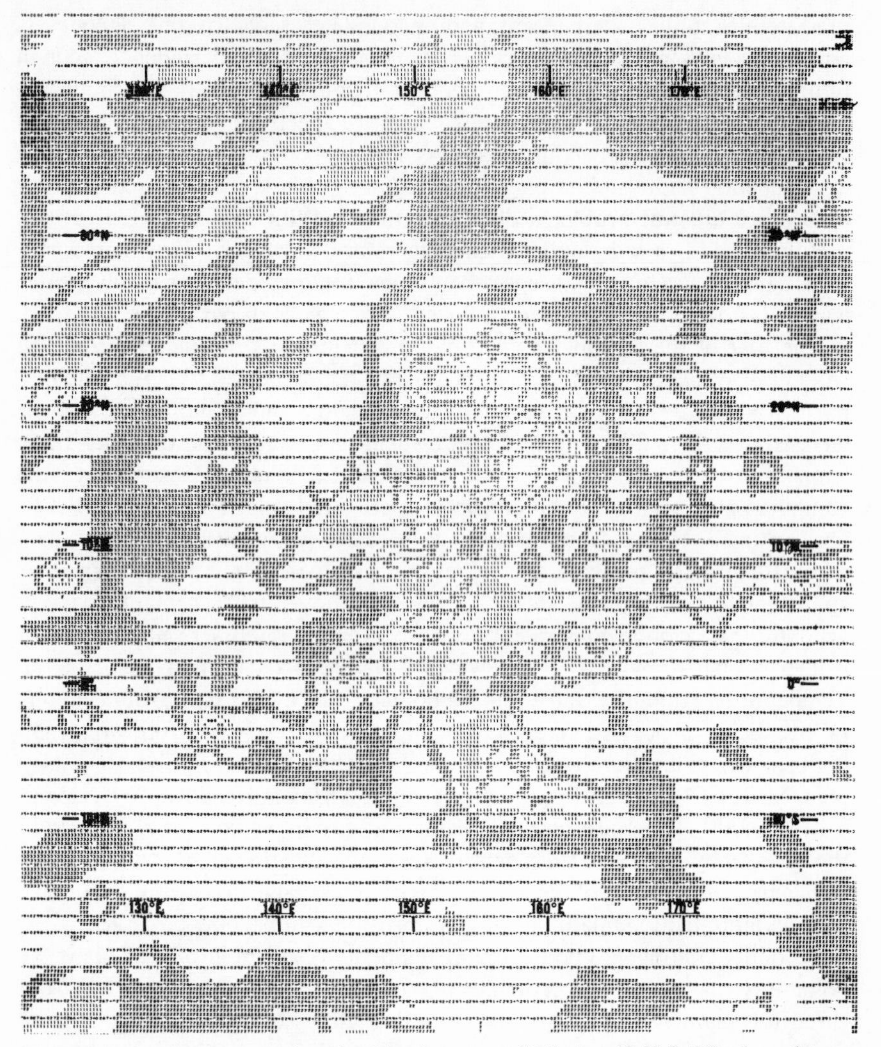

Fig. 4.2 A computer-produced grid print map of Nimbus II H.R.I.R. data. (From Barrett 1972: N.A.S.A. map)

(Widger *et al.* 1965). On the bases of climatological temperature profiles through the local atmosphere, or balloon sonde temperature measurements, unique relationships can be established between radiation temperature estimates and height above the ground.

It has been claimed that such an approach to cloud climatology has wide potential applications (Nordberg 1967). However, it must be admitted that few research reports have contained satellite-derived maps of cloud heights, and none of a climatological, as distinct from a meteorological, nature. This

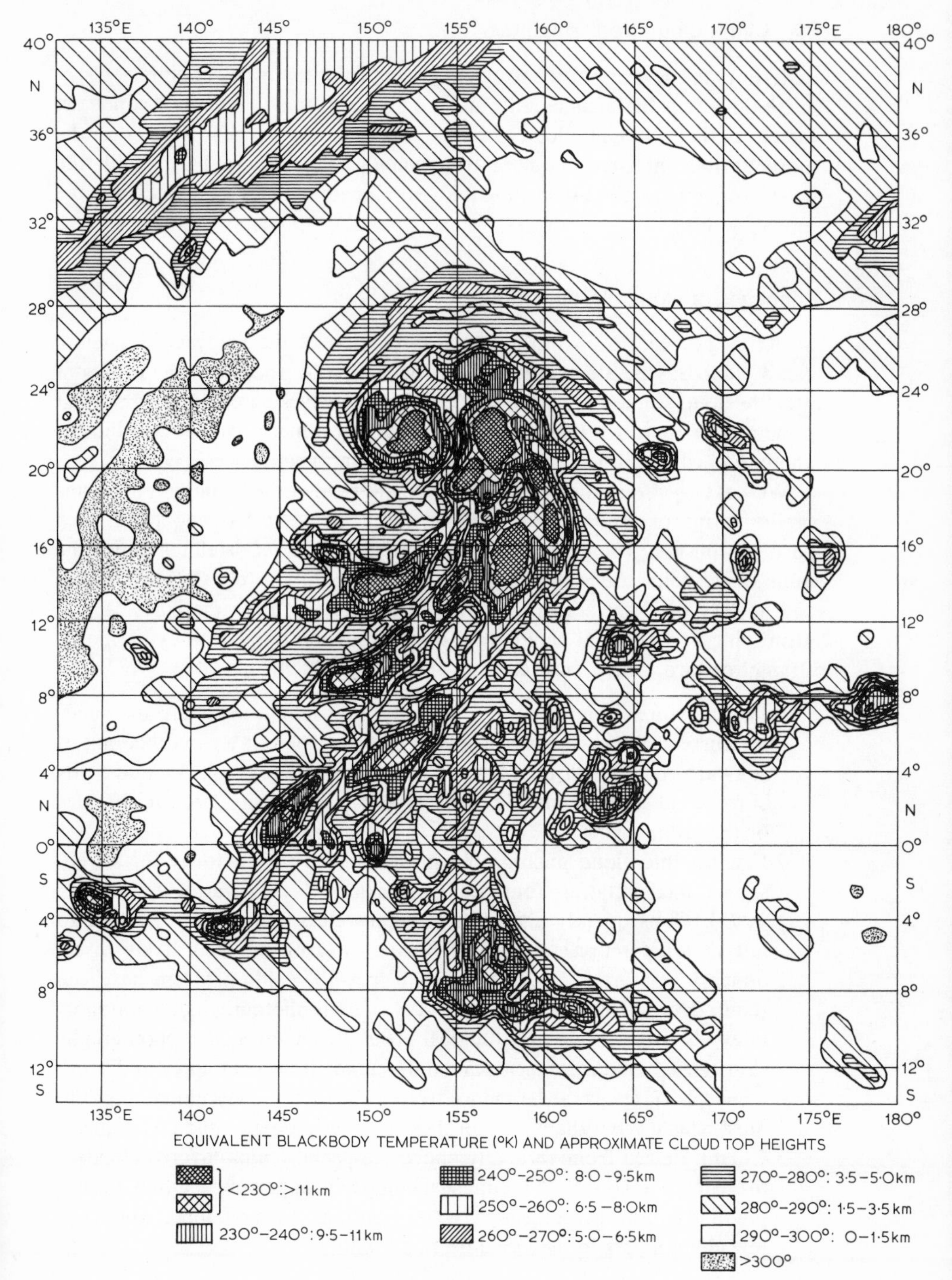

Fig. 4.3 An isotherm map of a hurricane over the western North Pacific, drawn from the data in fig. 4.2. (From Barrett 1972: N.A.S.A. map)

underlines the severity of the problem arising from the rarity of simple, clear-cut boundaries between cloudy and non-cloudy areas. We may conclude, therefore, that although cloud analyses may eventually depend mostly on the more flexible infra-red measurement techniques, photographs are still the most profitable satellite data to employ in climatological cloud studies at the time of writing.

Principles of cloud classifications

So let us proceed to an introductory discussion of the analysis of satellite cloud photographs. Since much of this book is concerned with the results of satellite cloud studies, we need only concern ourselves here with the broadest principles of cloud recognition and classification. Finer points of detail in the photographed cloud fields will be dealt with later where necessary.

We have seen that different cloud types were identified routinely in terms of the forms of individual clouds or cloud elements as seen from fixed positions on the ground, prior to the advent of weather satellites. Since the same cloud fields are viewed by the satellites from entirely different angles and at very different scales, it is not surprising that even the basic task of identifying conventional cloud types on satellite photographs is by no means straightforward. Three facts contribute to this problem:

1 The individual components of many types of cloud fields are too small for current photographic systems to resolve. Therefore careful comparisons are required to establish which satellite-photographed cloud fields correspond to each specific cloud type classified from aircraft altitudes or the ground.
2 Few satellite cloud photographs have resolved satisfactorily for stereoscopic interpretation (but see Ondrejka and Conover 1966; Shenk 1971). Consequently most photo interpreters have been required to interpret flat representations of clouds in terms of their likely three-dimensional forms. In operational meteorology this problem has been reduced recently by the advent of A.T.S. satellite photographs taken at short intervals over certain geographical areas. From successive photographs short-term cloud movements and differential air flows at upper and lower tropospheric levels can be deduced (see p. 136). Ultimately, climatology too must benefit from the results of these new technologies and techniques.
3 Clouds viewed from vertically above – especially multilayered clouds – present the analyst with an unusual composite appearance. Often it is not possible to assess confidently the contributions made by upper, middle or low cloud layers to the cloud field as a whole.

So, it is not surprising that some of the earliest analytical work based on satellite photos sought to establish an identification system involving a pre-existing cloud terminology, categorizing the appearances of satellite-viewed cloud fields in terms of conventional cloud classifications. Fortunately, some of the earliest of such attempts proved highly successful, notably those of Conover (1962, 1963), whose modified scheme has received wide acceptance since its introduction in 1963. Being essentially a *generic classification* it rested on six descriptive criteria (see plates 11*a–f*).

Cloud brightness

The brightness of cloud areas portrayed by meteorological satellite photographs has been shown to be affected by several meteorological factors in addition to various technical aspects of the satellite/ground system hardware complexes. The latter have been amply outlined by Jones and Mace (1963) and need not be detailed here. As far as meteorological factors are concerned, cloud brightness has been shown to be affected both by the depth of cloud and the nature of cloud constituents. The albedo of clouds increases rapidly with depth and reaches its maximum when clouds are some 200 m deep, provided that the constituent cloud droplets are large. For clouds of given depth, the smaller the cloud drops, the brighter the cloud for the same water content. Further comparisons have shown that liquid (i.e. water droplet) clouds generally appear brighter than ice crystal clouds of similar thickness under similar conditions of illumination. Cirrus varieties and cirrostratus, for example, are frequently translucent when each occurs alone, and may be visible only when seen against a dark background surface. Thin stratus sheets are predominantly pale grey and contrast markedly with the very bright appearances of cumulonimbus clusters, especially in the Tropics. Again, the angle of solar illumination is a factor of significance; it has been demonstrated that, under cloudless conditions and with a low water vapour content in the atmosphere above a horizontal surface near sea level, illumination increases fourfold as the angle of incidence of the sun's rays rises from 20° to 90° (see the W.M.O. Technical Note No. 75 1966).

Cloud texture

Cloud surfaces, when viewed from above, often vary in degrees of smoothness. This variable characteristic is referred to as the texture of the clouds. Conover distinguished several textural appearances, including smooth or fibrous (cirriform clouds), smoothly opaque (low stratus and fog), mottled or irregular (cumuliform or stratocumuliform clouds), smooth but ragged with interdigitating brightness bands (stratiform clouds) and amorphous patches of different brightnesses (various cloud types together), and various hybrid forms. Plate 11 includes examples of several different cloud textures. The differences are due mainly to mesoscale contrasts in the horizontal structures of the cloud fields, especially at cloud top levels.

Vertical structure

This can be deduced from some oblique photographs of the earth-orientated Tiros variety, portraying, for example, the edges of mid-latitude fronts where layered clouds often overlap one another in the vertical. On more modern vertical photographs, cloud shadows and highlights are often quite helpful in interpreting the structure and/or the heights of different clouds viewed. Under oblique illumination, areas of deep convection may be identified frequently through the shadows cast by tall cumuliform and cumulonimbus towers; jet stream cirrus clouds may cast shadows on lower undercasts of cloud or on the surface of the earth (see Whitney 1966); and high cirrus may be translucent, permitting outlines of lower cloud shapes to be seen through it, as is often the case, for example, where cirrus shields extend beyond the circumferences of hurricane cloud vortices (see p. 125).

The form of cloud elements

Cloud elements, the smallest distinguishable units on satellite photographs, are often helpful in cloud type recognition in areas of cumuliform, cumulonimbus and stratocumuliform cloudiness, but rarely so across more continuous cirriform or stratiform cloud fields. Hubert (1963) explained how the forms of cloud elements are often closely related to the directions and strengths of the thermal winds (see p. 133) through the depths of the atmosphere in which the clouds are developed. The nature and degree of instability or turbulence may be important factors too.

The patterns of cloud elements

The smallest cloud elements visible in a satellite picture are either quite randomly distributed or are organized into mesoscale cloud formations with some regularity of pattern (Anderson 1969). Such patterns (e.g. those in plates 11*a* and 11*e*) are quite important both for identifying certain cloud types and for explaining the physical processes that produce them. The variety of cloud patterns at the subsynoptic scale is, however, so rich that their classification is still incomplete. Such patterns develop in response to a wide variety of atmospheric and/or topographic factors, including, for example, the strength of horizontal air motion, the intensity of convection and its organization, vertical and horizontal variations in wind speeds, and the roughness or smoothness of the underlying surface of the earth.

The size of the elements and/or patterns

Such a criterion is necessitated by the fact that organizations within the earth's atmosphere are not infinite in form despite their rich variety in detail, but are composed instead of a small group of basic shapes which may recur across a wide range of scales.

Examples of cloud classifications

Fig. 4.4 illustrates the generic cloud classification developed by Conover (1963). This is based on cloud appearance with little or no recourse to the way in which different shapes and forms are produced. It demonstrates the important fact that many types of clouds identifiable from below can be differentiated also on photographs from space. More recently, Lee and Taggart (1969) of the Meteorological Service of Canada have prepared a most detailed procedure for satellite cloud photo interpretation. Part I (see table 4.2) sets out the types of evidence that they suggest may be brought to bear in initial interpretation stages. Part II (see table 4.3) summarizes characteristic appearances of clouds from satellite altitudes in terms of criteria closely similar to the six discussed above.

Once a would-be satellite climatologist has become familiar with the appearances and identities of clouds in satellite displays, his next problem may be to build up an expertise in the genetic interpretation of the same cloud fields. That is to say, his new interest goes beyond the appearance of a mass of cloud, extending to the atmospheric organization to which it is related. Here it is axiomatic that clouds are related to the various atmospheric processes that stimulate their development (Barrett 1968). Hence, cloud organizations must be expected both to indicate and integrate patterns of organization of other atmospheric elements. Perhaps the first serious attempt to develop 'an ultimate standard (genetic) classification that will be widely accepted by the meteorological community as a whole' was formulated by Hopkins (1967). This was designed not only as a contribution to academic knowledge but also as a framework for a computer-operated satellite photograph retrieval system to be operated at the American National Weather Records Center (N.W.R.C.). It was employed for a while to assist researchers interested in studying specific organizations of weather photographed by Nimbus II (see the Nimbus II User's Guide, ARACON 1967). Unfortunately, Hopkins's scheme (fig. 4.5), which subdivided cloud images into 'vortical cloud features', 'major cloud bands' and 'general cloud features', did not do justice to some features, in particular to anticyclonic cloud organizations (see figs. 4.6 (a), (b), (c)). This was a pity since these are still the least examined and least well-documented and understood.

The basis of an alternative scheme was, therefore, proposed later by Barrett (1970b). Since genetic cloud classifications must involve both cloud arrangements and the atmospheric organizations that give rise to them, any detailed, fully comprehensive scheme must be complex indeed. Clearly such a classification would presuppose a good understanding of meteorological processes related to subsynoptic and synoptic scale cloud patterns, and, at present, such understanding is incomplete. Hence the classification portrayed in figs. 4.7 and 4.8 is only a logical beginning of an answer to the problem,

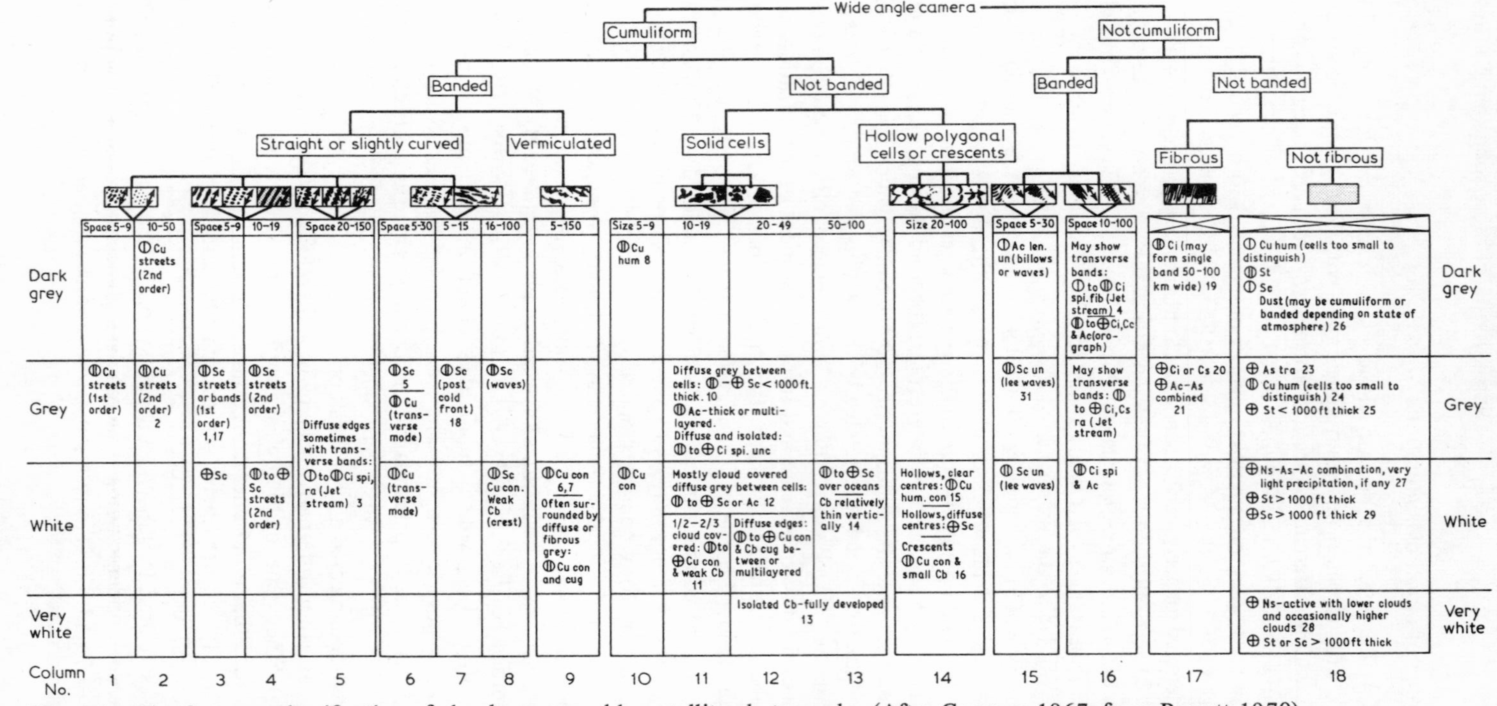

Fig. 4.4 The Conover classification of clouds portrayed by satellite photographs. (After Conover 1967; from Barrett 1970)

Table 4.2

A procedure for satellite cloud photo interpretation
(from R. Lee and C. I. Taggart 1969)

Cloud type	*Tentative cloud identification by photo interpretation techniques*	*Direct evidence to confirm tentative interpretation*	*Indirect supporting evidence*
Cirrus Cirrostratus	Cirrus can almost always be positively identified by photo interpretation techniques. Cirrostratus may be distinguished from stratus and fog by its characteristic translucence	1 Surface observations 2 Aircraft observations	1 Orientation of cloud streaks generally parallel to 300 mb winds 2 Occurrence of jet stream core parallel and to the left of cloud edge, relative to downwind direction 3 Positive vorticity advection at 300 mb where cloud occurs, although this is not necessarily conclusive evidence since only 85 per cent of cirriform clouds occurs in areas of positive vorticity advection at 300 mb
Anvil Cirrus (detached from Cumulonimbus)	Can be positively identified by photo interpretation techniques	1 Surface observations 2 Aircraft observations	1 Orientation of cloud streaks generally parallel to 300 mb winds 2 Showalter stability indices less than +2 within or upstream from cloud area
Altostratus Altocumulus	Can usually be identified by its size, shape, very bright tone and uniform texture although overcast layers of stratocumulus in areas of high reflectivity may have a similar appearance, other evidence will help to distinguish the two; when it occurs over an extensive ice shield, there are generally pronounced shadows	1 Surface observations within or near outer boundary of middle cloud layer 2 Aircraft observations 3 Radar observations of precipitation over extensive area	1 Occurrence of extensive precipitation within area of cloud 2 Cloud coincides with main isotherm ribbons at 850 mb and 700 mb 3 Cloud coincident with areas of large scale ascent on vertical motion charts 4 Cloud pattern occurs between trough line and downstream ridge line at 700 mb or 500 mb 5 Coincidence of cloud and positive vorticity advection at 500 mb 6 Dew point depressions less than 2 °C in layer 850 mb to 600 mb 7 Cloud area occurs between 850 mb and 600 mb frontal contours on frontal contour chart 8 Coincidence of cloud with ascending motion inferred from wind shear hodographs

Table 4.2 (*cont.*)

Cloud type	*Tentative cloud identification by photo interpretation techniques*	*Direct evidence to confirm tentative interpretation*	*Indirect supporting evidence*
Wave clouds Cirrus Altocumulus Stratocumulus	Wave clouds as a class are conclusively identified by appearance and organization. However, cloud generally can only be distinguished by appeal to direct observations or indirect supporting evidence	1 Surface observations 2 Aircraft observations	1 Occurrence of wave clouds over or downwind from mountain range or other high terrain features 2 Cirrus wave clouds will be found to be transverse to the 300 mb wind direction 3 Altocumulus wave clouds will be transverse to the 700 mb wind direction 4 Stratocumulus wave clouds are distinguished by cloud bands transverse to the boundary layer wind 5 Lifting condensation level of surface air likely occurring within layer of turbulent mixing will confirm stratocumulus type
Cumulus Towering Cumulus	These convective cloud forms can normally be identified directly by size, shape, shadow, tone, texture and pattern	1 Surface observations 2 Aircraft observations	1 Occurrence of instability in lower troposphere as deduced from tephigrams 2 Inferred instability from interpolated surface, 850 mb and 700 mb temperatures in cloud area 3 Likely occurrence of strong surface radiational heating or heating over relatively warm band or water surface, notably over the Gulf Stream
Cumulonimbus	Positive identification of cumulonimbus clouds is normally possible because of their characteristic size, shape, pronounced shadow, very bright tone and texture. However, other cloud areas of similar dimensions and pattern or even small lakes in sun glint may be misinterpreted as cumulonimbus because of enhanced brightness	1 Surface observations 2 Radar observations 3 Aircraft observations	1 Clouds appear in unstable areas on stability index charts 2 Unstable air mass characteristics, as inferred by tephigram analysis in cloud area 3 Alignment of clouds with surface or upper level fronts 4 Cyclonic curvature of mean sea level isobars in cloud area 5 Occurrence of clouds in forward part of upper wave trough 6 Evidence of strong surface heating 7 Evidence of differential advectional cooling, or differential moisture advection patterns favourable to cumulonimbus developments

Stratocumulus	Can usually be identified by tone organization and texture. However, extensive layers of overcast stratocumulus clouds over oceans may appear similar to middle cloud, hence their differentiation may only be possible by referring to other evidence	1 Surface observations 2 Aircraft observations	1 Alignment of cloud streets in general direction of the boundary layer winds 2 Surface wind speeds greater than 10 to 15 miles per hour together with low lifting condensation level in cloud area 3 Occurrence of a low level turbulence inversion in cloud area 4 Clouds occur in cold air mass to rear of surface cold front with cloud streets bearing an appreciable angle to the alignment of frontal cloud
Stratus	Cannot be normally distinguished from fog by appearance alone, although pictures having good definition and tonal range show boundaries of stratus layers to be more diffuse than fog boundaries; cloud shadow also a distinguishing characteristic, may conform to terrain features	1 Surface observations 2 Aircraft observations	1 Surface winds in 12–15 mph range with low lifting condensation level 2 Occurrence of a low level turbulence inversion in cloud area 3 Evidence of surface cooling by radiation, motion over colder surface or upslope motion 4 Absence of middle cloud, i.e. low moisture content in middle levels or descending motion
Fog	Cannot normally be distinguished from stratus by appearance alone; although sharp boundaries and absence of shadow are useful characteristics to look for, usually conforms to shape of terrain	1 Surface observations 2 Aircraft observations	1 Surface temperature and dew point equal in cloud area 2 Surface winds less than 5 mph in cloud area and evidence of radiational cooling in case of radiational fog 3 Surface air motion conductive to advectional cooling over colder land, water or snow surface 4 Pronounced upslope motion with light or even moderate winds in case of upslope fog 5 In case of Arctic sea smoke, evidence of surface air temperatures appreciably lower than water temps. with offshore flow 6 In case of low temperature fog, occurrence of fog near inhabited area with surface temperatures below $-30\ ^{\circ}F$

Table 4.3

Appearance of clouds from satellite altitudes
(from R. Lee and C. I. Taggart, 1969)

Cloud type	*Size*	*Shape (organization)*	*Shadow*	*Tone (brightness)*	*Texture*
Cirrus Cirrostratus	Typical lengths of organized bands hundreds of miles, widths of single bands may be 25–50 miles; extensive layers of cirrostratus may also cover large areas	Long bands, parallel to upper tropospheric winds, often having sharply defined left boundary, relative to an observer facing downward; right boundary sometimes well-defined, but is more frequently indistinguished when it appears over a middle cloud layer	Normally present as a dark line along one edge; most noticeable when shadow is cast on a lower cloud layer or a smooth surface with high reflectivity	Typically light grey, but tone is dependent on sun angle; translucent; lower clouds and geographical features are usually only partly obscured by cirriform cloud	Cirrostratus normally has a uniform texture, while cirrus tends to be more fibrous. Cloud bands perpendicular to the wind indicates wave structure
Anvil Cirrus (detached from Cumulonimbus)	May be quite extensive, covering areas around five hundred miles or more in length and width	Chaotic appearance with alignment of cloud streaks parallel to upper tropospheric winds; diffuse or poorly-defined edges	Only detectable when cloud layers are sufficiently thick and shadow falls on an illuminated lower cloud layers or brightly reflective land or water surface	Light grey or white, depending on cloud thickness and sun angle	Fibrous with numerous streaks, or more uniform texture when dense cirrus layers are concentrated within a small area
Altostratus Altocumulus	Extensive sheets or bands covering areas tens to hundreds of thousands of square miles; bands may be two or three hundred miles across	Organized into vortices, bands, lines or large comma-shaped areas associated with cyclones and fronts; characterized by persistence in form over periods of 12–24 hours or more, since cloud is associated with synoptic scale motion systems, usually well-defined boundaries	Often present along one edge, shadow enhanced if it appears on a layer of lower cloud	Very white, one of the two brightest cloud forms, the other being cumulonimbus—due to great vertical depth of cloud; whitest cloud layers are often associated with nimbostratus and precipitation at the ground	Stratiform cloud with uniform top surface has uniform texture, if convective clouds are present or the middle cloud is not solid. Variations in texture will appear as a result of shadows, breaks, or thickness variations

Wave clouds Cirrus Altocumulus Stratocumulus	Narrow parallel bands of the order of ten to a hundred miles in length; uniform spacing of cloud bands is characteristic of these cloud forms	Uniformly spaced, parallel bands, more or less perpendicular to the wind direction at cloud level, most often found to be of hills and mountains, notable examples appear over the Rockies, Appalachians, Labrador, and other ranges	Not usually discernible	Grey, occasionally white, depending on sun angle and vertical thickness of cloud	Continuous or broken parallel bands, may be vermiculated
Cumulus Towering cumulus	Individual cumulus cloud cells are normally too small to be discernible at 800 miles; rather what appears similar to individual cumuli as seen from the ground are groups of clouds having a regular organization or pattern not normally detectable from surface observations; dimensions of cloud groups 3 to 10 miles	With light winds, cloud groups present a uniform cellular pattern or may be organized in single or parallel streets, straight or gently curved, generally parallel to the winds. Occasionally hollow polygonal cells, crescents or solid cells will appear in the overall pattern; usually lumpy appearance	Usually present with towering cumulus, destructive shadows or down sun side; shadows not so evident with smaller clouds or cloud groups	Broken dark grey, grey or white depending on dimensions and thickness of cloud groups as seen from satellite altitudes; smaller cloud groups are darker in tone, while cumulus cells smaller than the threshold resolution of the camera (2 miles) will not be visible if separation is greater than 2 miles; areas of small cumulus will appear in broken grey tone	Non-uniform, alternating pattern of white, grey and dark grey, often having great regularity; due to contained shadows; hollow centers may be present in a ring of cells
Cumulonimbus	Individual isolated cumulonimbus clouds are of the order of tens of miles in diameter, combined clusters of such clouds may present a pattern as large as a hundred miles across due to merging of cirrus anvils	Isolated cells have sharply defined edges on one side with cirrus anvil spreading out on the opposite side in the presence of pronounced wind shear; otherwise they appear as isolated, white, nearly circular cells	Shadows usually present and well-defined with cumulonimbus	Very white, particularly tops have characteristic brightness	Uniform texture, sharply defined edges, although cirrus plumes are often quite diffuse beyond main cells

Table 4.3 (*cont.*)

Cloud type	*Size*	*Shape (organization)*	*Shadow*	*Tone (brightness)*	*Texture*
Stratocumulus	Apparent size of cells 2–10 miles although layers will have no distinctive size	Streets or bands aligned with the boundary layer winds, or extensive areas with well-defined boundaries	Shadows may show striations along the wind	Small cloud groups are mostly grey over land, thick overcast strato-cumulus layers over oceans often appear white due to contrast in reflectivity	Overcast stratocumulus cloud layers often shows hollows with diffuse centers
Stratus	Variable	Variable, except when stratus cloud is lower than surrounding terrain, in which case it assumes the shape of a valley, mountain or coast line, etc.; boundary well-defined but may have a ragged edge	Normally not discernible, but presence is dependent on height of stratus layer above ground	White or grey, depending on vertical cloud thickness and sun angle	Uniform
Fog	Variable	Variable, irregular, but in the case of fog over bodies of water, shape conforms to that of surrounding land; boundaries sharply-defined and may be the only distinguishing characteristic from stratus	Normally not discernible	White or grey, depending on thickness of fog layer and sun angle; normally if depth of fog layer exceeds 1000 ft. it appears white	Very uniform

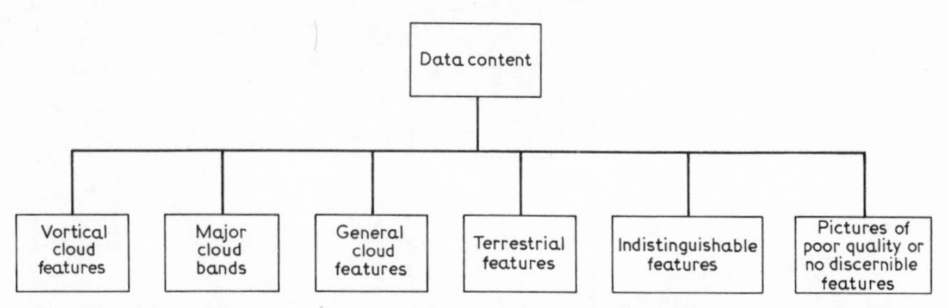

Fig. 4.5 The basis of a genetic classification scheme to be used with a computerized Nimbus data storage and retrieval system. (From ARACON 1966)

not its final solution. Basically, it separates those cloud arrangements that seem to owe most to meteorological processes in the free atmosphere (fig. 4.7) from those that are governed more strongly by processes at work in the friction or boundary layer of the atmosphere. Boundary layer clouds possess features of detail that reflect earth surface patterns, among them the nature of the surface (whether land or water), surface temperature patterns and the topography of land areas (fig. 4.8).

In fig. 4.7 the first-order subdivisions depend on the fundamental contrasts between clockwise and anticlockwise circulations, i.e. cyclonic and anticyclonic arrangements. Cloud arrangements of basically different types characterize positive and negative vortices respectively (see chapter 5). These usually involve spiraliform arrangements of clouds coiling into or away from their respective centres of circulation. The intermediate group of cloud arrangements includes frontal and instability bands, i.e. cloud organizations that are linear (either markedly curvilinear or roughly rectilinear) in plan, not obviously coiled.

Genetic classifications could be developed much more fully than in figs. 4.7 and 4.8, but for present purposes these general outlines suffice to illustrate at least something of the wide range of different atmospheric weather systems that may be recognized and identified in satellite image data presentations. That many atmospheric systems can be identified through cloud evidence is one of the foundations upon which much climatology from satellites depends. Whereas the regional climatologist may be more interested in map-form statistical presentations of radiation, cloud cover, etc., compiled from satellite sources, dynamic and synoptic climatologists are more interested in the distributions of significant organizations of the weather, especially in regions that have long been relatively data-remote. More details concerning suggested classifications of families of weather systems in equatorial, tropical and midlatitudinal zones are introduced where relevant in the regional discussions later.

Ultimately analyses of photographs for clouds classified either generically or genetically may be conducted automatically by appropriate computers

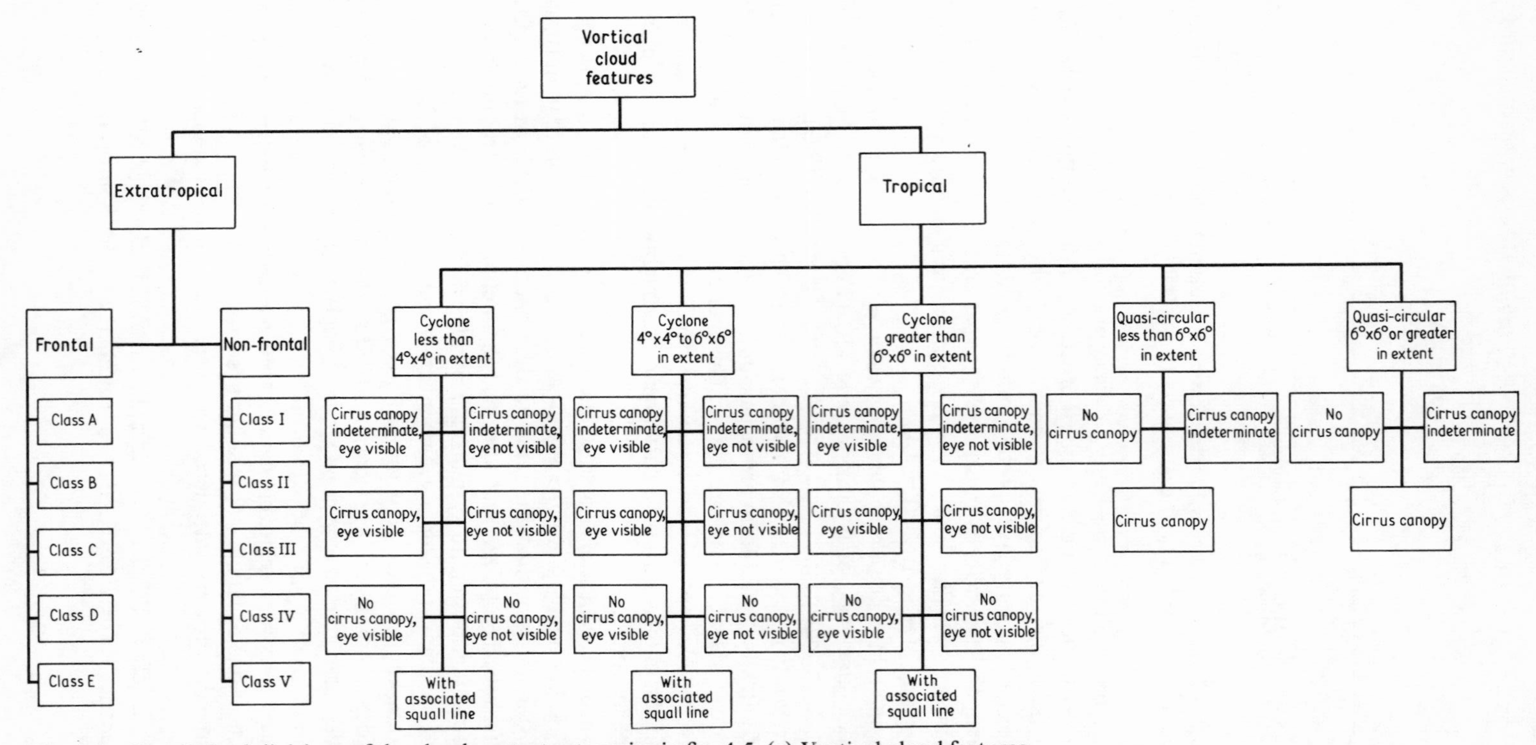

Fig. 4.6 Detailed subdivisions of the cloud content categories in fig. 4.5. (a) Vortical cloud features

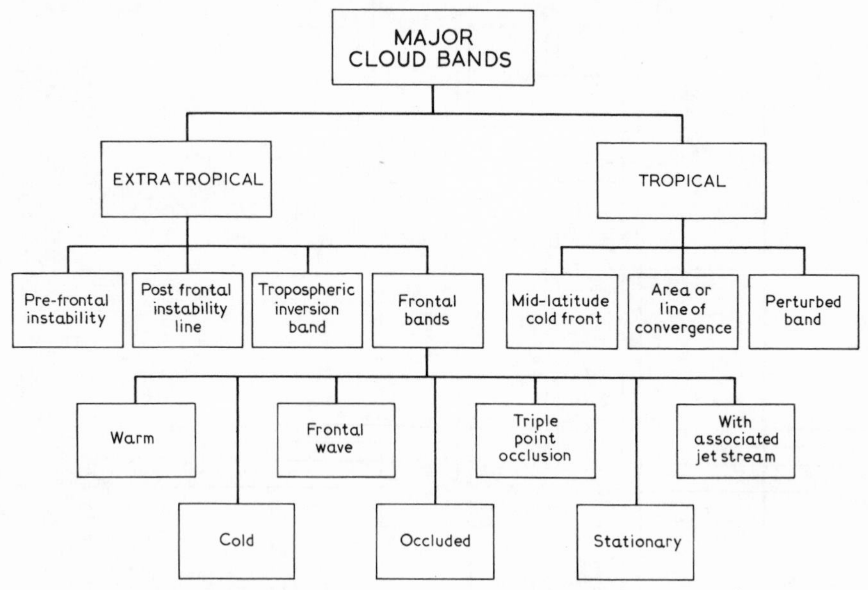

Fig. 4.6 (b) Major cloud bands

and/or pattern matching processes (see e.g. Rosen 1967). However, certain difficult problems must be solved before such approaches could become routine. Apart from the fundamental problem of establishing appropriate cloud models for automatic means to match, other problems arise since specific cloud systems may occur (*a*) at a wide variety of scales, and (*b*) at various orientations through all 360° of the compass.

Thus it would be necessary for cloud photographs to be scanned parallel to the *x* and *y* axes, as well as along diagonals, and perhaps even spirally from their centres. Basically suitable procedures have been developed already for general purpose pattern recognition (Rosen, 1967) and for weather forecasting (Hu 1963), but the consensus of present opinion is that the likelihood of developing operational hardware especially for such complex tasks is further distant than anticipated earlier. For the immediate future, therefore, human recognition and identification of clouds and cloud organizations remains the main key to many uses of satellite photographs in meteorology and climatology.

Cloud cover studies

Lastly in this review of basic methods of analysis and interpretation of climatologically significant cloud parameters, let us summarize the variety of approaches for assessing global strengths of cloud cover. Such studies all involve evaluations of satellite data in terms of proportions of overcast sky as

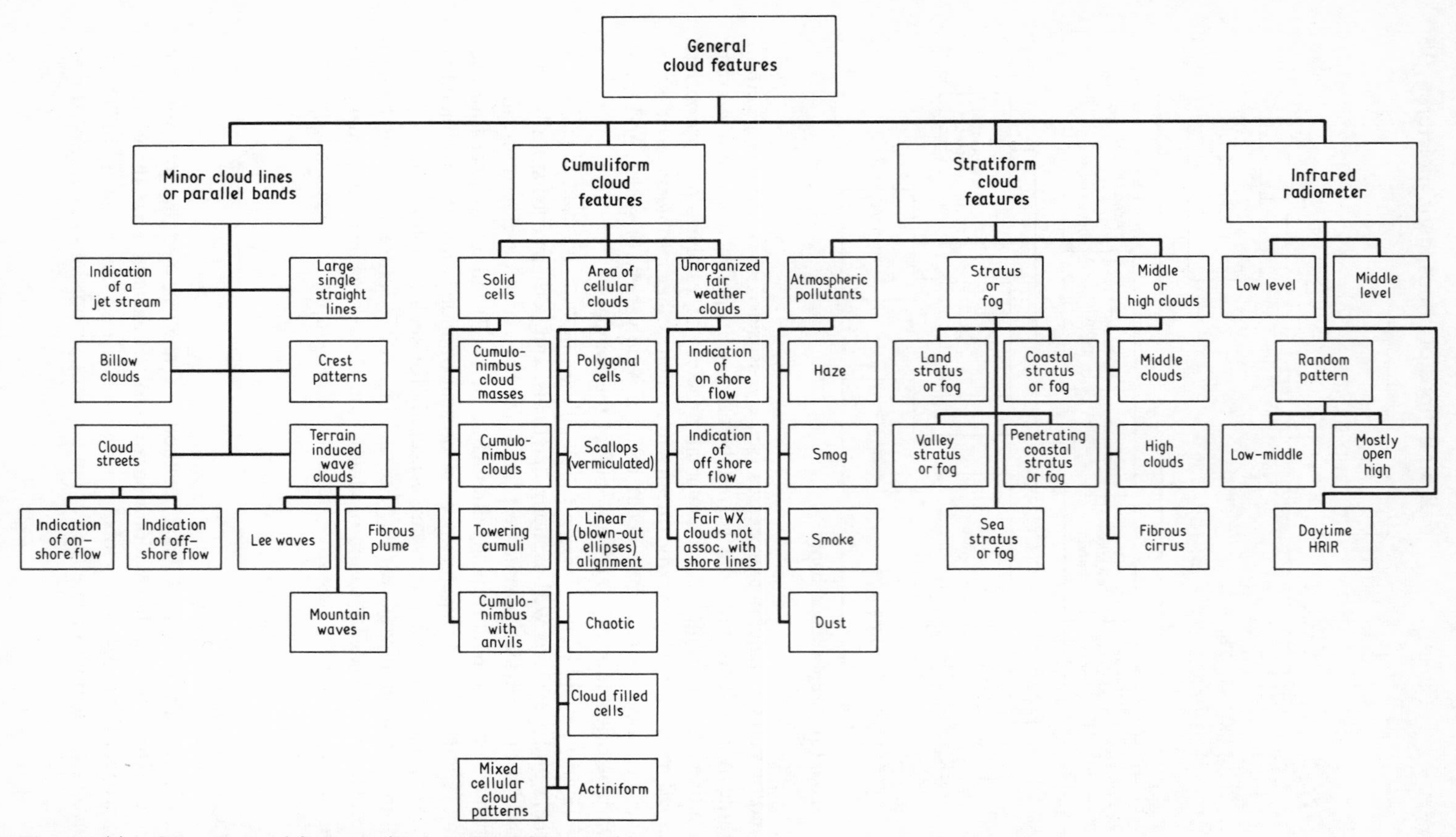

Fig. 4.6 (c) General cloud features. (From ARACON 1966)

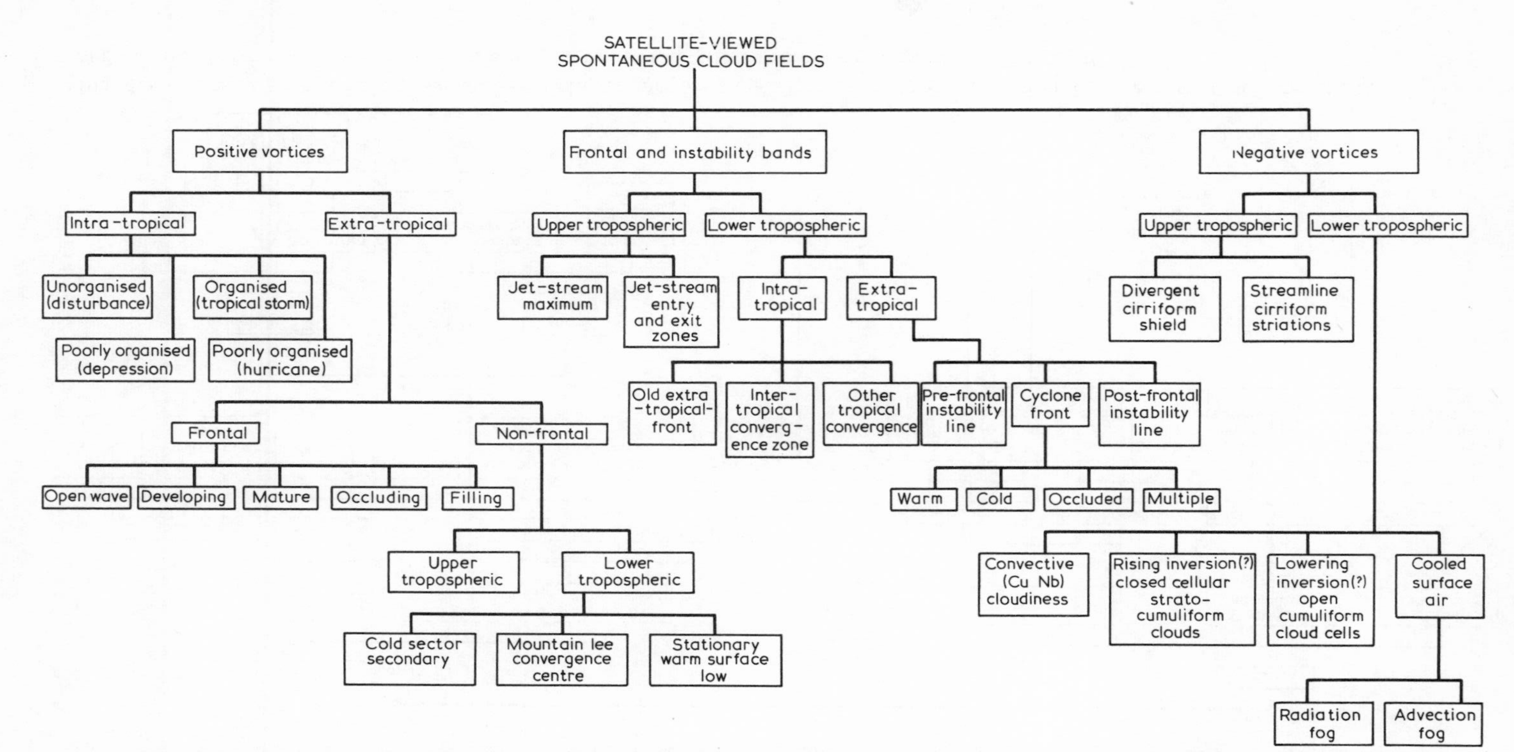

Fig. 4.7 A basis for a genetic classification of clouds portrayed by satellite photographs. Section 1 deals with cloudiness organized dominantly by processes in an atmosphere affected little by earth topography. (From Barrett 1970b)

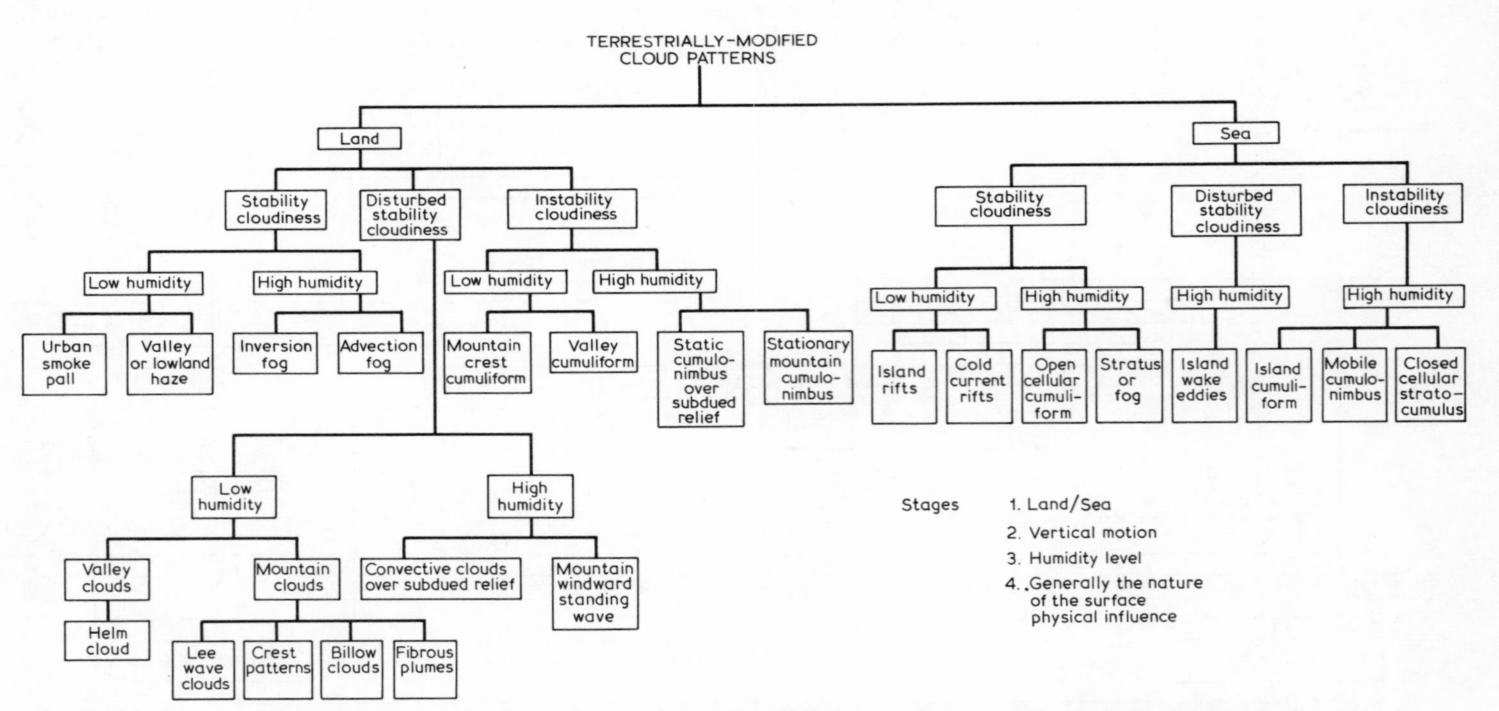

Fig. 4.8 A basis for a genetic classification of clouds portrayed by satellite photographs. Section 2 suggests a subdivision of clouds and cloud rifts stimulated largely by variations in terrestrial surface characteristics. (From Barrett 1970b)

indicated by various photographic brightnesses, infra-red temperature estimates or nephanalysis cloud distributions. The more promising suggestions are given below.

Photographic analyses

Cloud cover determinations have been based directly on photographed cloud fields following comparisons between corresponding satellite and surface observations. In this way the proportions of cloud-covered sky indicated by satellite cloud fields can be established (Barnes and Chang 1968).

Digitized brightness data from Tiros satellites were employed in mean cloud mapping by Arking (1964), who assigned numerical values to the brightnesses of each of 250,000 spots within a given television picture (500 on each of the 500 raster lines). Threshold brightness values were determined subjectively to separate clouds from non-cloudy surface backwards. Classified information was then computer processed to yield average percentages of cloud cover in each 10° latitude interval between 60° N and 60° S.

Taylor and Winston (1968) computer processed Essa photographic data to give mean brightness values for each 5° grid square. They then prepared monthly and seasonal mean brightness maps showing brightness patterns by eleven categories from 0 to 10 inclusive (see p. 52). Their results are subject to similar extraneous variables to those encountered by Arking; the visible radiation recorded photographically is not dependent on physical variations in clouds alone, but also on instrumental factors and contaminating influences of surface backgrounds of various and varying albedoes.

More recently, Booth and Taylor (1969) have produced digital multi-day brightness averages and composites on polar projections. In their method, photo data are recorded on magnetic tape and displayed by means of a cathode ray film device, giving photographic products (see p. 50).

The mesoscale archive of satellite brightness data currently growing through the routine application of Booth and Taylor's method to Essa digital products yields daily estimates of cloud amount (N_S) using an automated technique described by Miller (1971a). Mesoscale brightness data arrays (see p. 50) are produced by reducing full resolution mapped arrays so that each mesoscale spot represents an 8 × 8 block of full resolution data. In Miller's method, the original relative brightness range of fifteen values is divided into five equal classes and a frequency count is made of each class from the original sixty-four spot population. The resulting histograms comprise the daily mesoscale archive (fig. 4.9). Next, a weighting scheme (see table 4.4), derived empirically through comparisons between satellite and ground observations, is used to translate the original brightnesses into estimates of total cloud amount using the equation:

$$N_s = \sum_{i=1}^{5} w_i f_i / 64 \qquad (4.1)$$

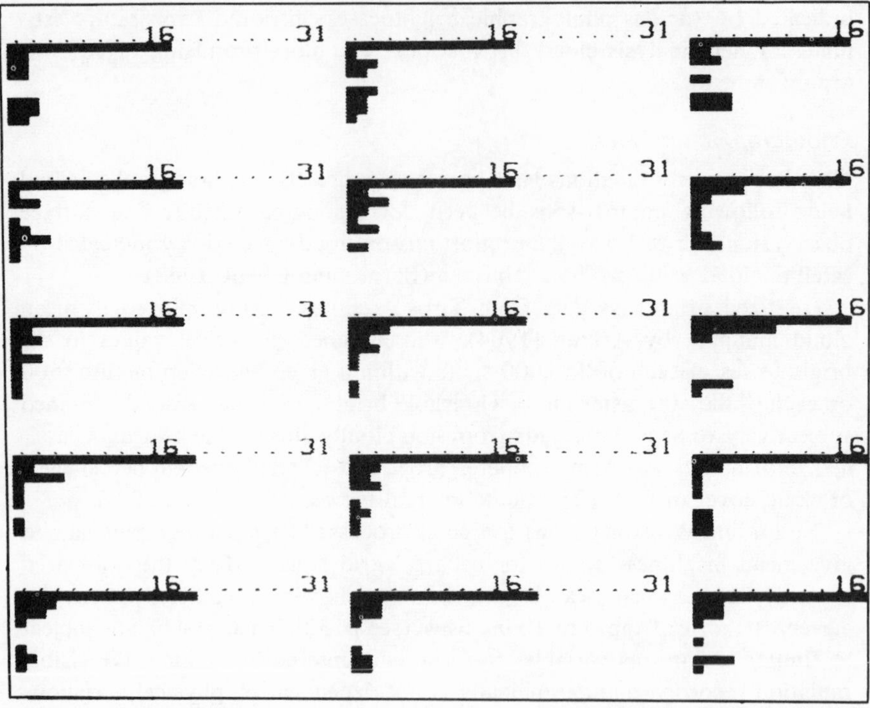

Fig. 4.9 Histograms of total-cloud amount for Washington, D.C. area, May 1969. The centre histogram is Washington itself, the others are 30 n.mi. apart with north at the top of the figure. The frequencies (abscissa) are in days and the cloud amounts increase from 0 to 8 oktas vertically down the page. (From Miller 1971a)

Table 4.4

A 'probabalistic' or weighting scheme to weight the five-class histograms comprising the daily mesoscale archive of digitized video data, according to each one's most likely contribution to the estimate of total cloud amount. The weighting scheme was derived empirically by comparing N_S values with visual estimates from satellite photographs, and by comparing N_S values with surface observations (N_t) (after Miller 1971a)

Original brightness range	*Class*	*Contribution to total cloud amount*	*Weights (Essa 7 and 9)*
0, 1, 2	1	0%	0
3, 4, 5	2	25%	2
6, 7, 8	3	88%	7
9, 10, 11	4	100%	8
12, 13, 14	5	100%	8

where N_s are the total cloud amounts in oktas, f_i are the frequencies of the five brightness value classes and w_i are the weights in table 4.4. Next, steps are taken to reduce contamination due to background brightness (ice, snow, sand, sun glint, etc.). This is effected by retaining minimum daily brightness values through multiday periods to give a composite minimum (C_m) field representing background brightness. By subtracting C_m from daily fields of $N_s(N_s - C_m = N_s^*)$ values remain that may be considered as being due to clouds alone (N_s^*).

This method seemingly underestimates low cloud amounts since the vidicon camera system cannot sense small amounts of small cumulus or thin cirrus cloud. However, Miller claims that the automated N_s values are at least as good as those that can be derived by eye from satellite pictures, and the method is being used now for processing the available Essa record (from January 1967) for a climatology of total cloud amount (see Miller 1971b). More specialized products can be prepared also, for example maps of cloud cover exceeding a selected okta level.

Other photographic representations of large-scale mean cloud distributions have been achieved by Kornfield *et al.* (1967) through photographic compositing techniques. Such products have been called 'multiple exposure averages' (see plate 13 and pp. 210–12). Unfortunately this method results in a qualitative, not quantitative, display.

Infra-red assessments of global cloud cover

It has been shown that a high correlation exists between vertical motion in the lower troposphere and a satellite-derived quantity, *TD*, the difference between the actual surface temperature and the radiation temperature measured by a satellite (Shenk 1963). In general, *TD* is large where thick clouds exist at high and middle levels, and is small where skies are relatively cloud free. The advantage of this parameter over the satellite-measured temperature itself is that small values of window temperatures can occur over cold surfaces (including snow and ice) under clear skies as well as over thick and high clouds, thus making it difficult to differentiate between cold surfaces and cloud tops at middle and high latitudes. The computation of *TD* obviates this problem. Since surface temperature changes are small from day to day climatological information can be used where surface temperatures cannot be obtained from satellite data. Lethbridge and Panofsky (1969) have shown that *TD* can be written as:

$$TD = \frac{\partial T}{\partial Z}\Delta Z \tag{4.2}$$

where $-\partial T/\partial Z$ is the lapse rate (relatively invariant), and ΔZ the difference in height between the radiating surfaces and the ground. Thus *TD* can be related to the mid-tropospheric vertical motion, and can be used to locate the regions

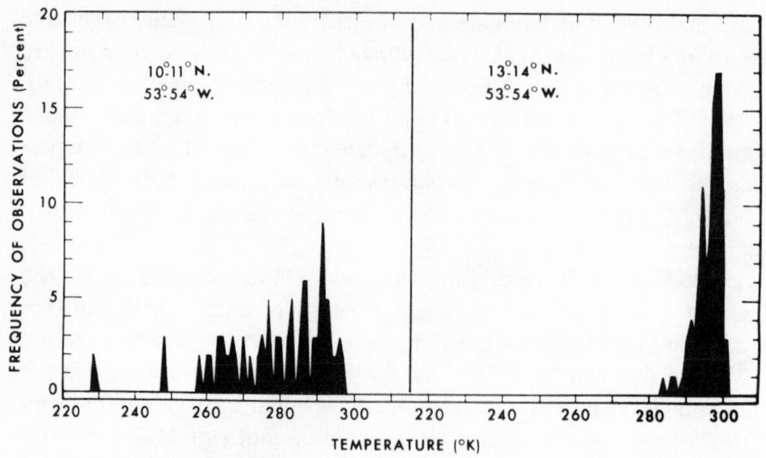

Fig. 4.10 A typical histogram for Nimbus III night-time H.R.I.R. temperatures for two selected areas, one relatively clear (*right*), the other relatively cloudy (*left*). Low temperatures on the left side indicate cloudiness over the area under consideration. (From Rao 1970)

of cloudiness and the associated ascent. Lo and Johnson (1969) have estimated cloud amounts from Nimbus II M.R.I.R. data through values of *TD*.

Rao (1970) has outlined a method for estimating cloud amount and height from H.R.I.R. data, exemplifying its application with respect to Nimbus III observations. Rao showed first that the shapes of H.R.I.R. temperature histograms for selected areas indicate the types and amounts of their cloud covers (see fig. 4.10). For example, thick and high overcast clouds produce a sharp peak on the low temperature side. As the height of the overcast decreases, the peak shifts towards the high temperature side. A flat distribution generally indicates multilayered cloudiness.

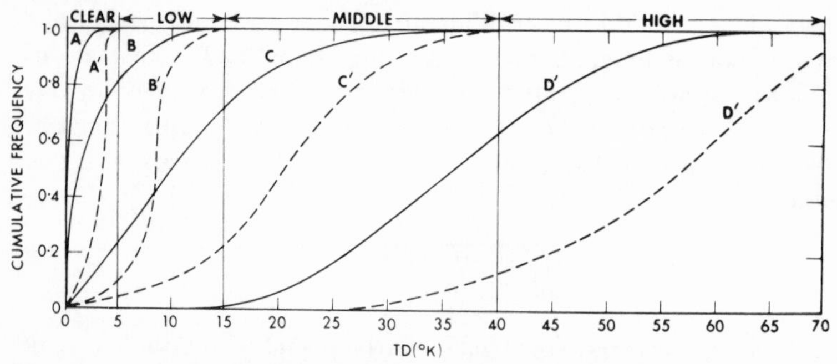

Fig. 4.11 A hypothetical cumulative frequency distribution of *TD* in °K, for various cloud amounts and heights. (From Rao 1970)

His next step obtains the temperature difference *TD* (see above) and generates a cumulative frequency distribution of *TD* values for each grid square (see fig. 4.11). By comparing curves based on operational data with the hypothetical set of ideal shapes, first the actual cloud heights can be differentiated into low, middle and high categories and their proportions quantified; second, the cloud amounts in each category can be established: the percentage of values between each *TD* interval gives the amount of cloud in that category.

This method is designed for a completely automated production of cloud analyses for climatological studies on a global scale.

Nephanalysis-based methods

Nephanalysis patterns, based on 'eyeball' methods of estimating the strengths of cloud fields portrayed by satellite photographs (see fig. 2.17), have been used as data sources for cloud cover information by many workers using simple grid intersection sampling methods. Comparisons of results obtained by different workers for similar areas should be undertaken with care, however, since (i) the nephanalysis code has been changed from time to time, and (ii) different weightings have been given by various investigators to the cloud cover categories within those codes (see, for example, Clapp 1964; Allison *et al.* 1969; Barrett 1971). The simplest translation relates Open conditions (O) to 10% cloud cover, Mostly Open (MOP) to 35%, Mostly Covered (MCO) to 65% and Covered (C) to 90%, these values being the medians of the nominal ranges covered by each category.

Using the same type of data, Sadler (1967) assessed mean cloudiness in each $2\frac{1}{2}°$ grid square across his areas of interest through the use of an octadic

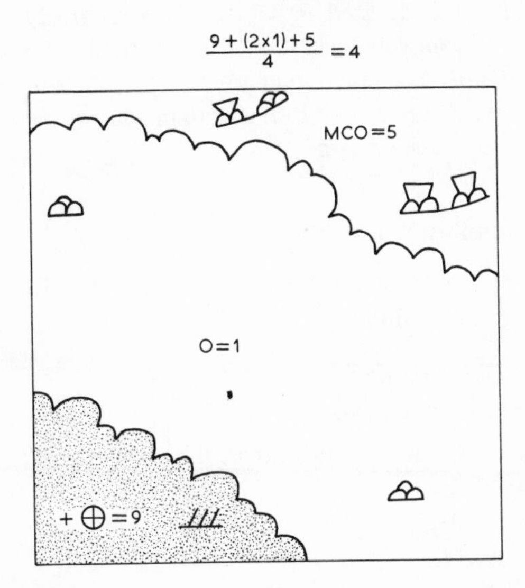

Fig. 4.12 An example of Sadler's averaging process for a grid square containing three nephanalysis categories of cloudiness. (After Sadler 1967)

scale (see table 4.5), and mental averaging of the cloud cover depicted. By this means he established a representative cloud cover value for each grid square each day (fig. 4.12). Mean cloud cover estimates were made by summing and averaging over longer periods. Choropleth-type maps were drawn to depict the results. While this approach is more comprehensive areally it is more lengthy to pursue, yet does not yield necessarily more acceptable results.

Table 4.5
Relationships between nephanalysis categories and cloudiness as suggested by Sadler (1969)

Nephanalysis category (symbol)	*Range of cloudiness (%)*	*Assigned value*	*Approximate cloudiness (oktas)*
Open (O)	<20	1 2	0–1 2
Mostly Open (MOP)	20–50	3 4	3 4
Mostly Covered (MCO)	50–80	5 6	5 6
Covered (C)	>80	7 8	7 8
Heavily Covered (+C)	>80	9	8

Salomonson (1969) underlined the importance of such techniques by observing that, whereas ground-based weather observations still offer the best source of data from the length of record standpoint, their geographical coverage and representativeness is quite limited, especially over sparsely populated oceans. So the much more complete spatial coverage provided by polar orbiting Essa and Nimbus satellites has much to commend it for broad-scale cloud cover studies, to some degree because of rather than in spite of the comparatively low average resolutions of their data.

Satellite measurements of atmospheric water vapour

In addition to statements of cloud cover and precipitation, the moisture content of the atmosphere may be expressed meteorologically in a variety of ways.

The mass mixing ratio (or humidity mixing ratio)

This relates the admixture of water vapour with dry air in the atmosphere. Hence

$$r = \frac{M_v}{M_d} \tag{4.3}$$

where r is the mixing ratio expressed in terms of the mass of water vapour (M_v) to the mass of dry air (M_d) within which the water vapour is contained. The units are usually grammes of water vapour per kilogramme of dry air. The bulk of the moisture in the atmosphere is in the 1000–500 mb layer.

Vapour pressure measurements

These quantify that part of the total atmospheric pressure which is exerted by water vapour. The vapour pressure (e') of water vapour in moist air at total pressure (p) and mixing ratio (r) is defined by:

$$e' = \frac{r}{0{\cdot}62197 + r}\,p \qquad (4.4)$$

Vapour pressure is established indirectly from dry- and wet-bulb temperature readings with the aid of an humidity slide rule or tables. At mean sea level atmospheric pressure, water vapour generally contributes about 10 mb.

The relative humidity of a sample of air

This expresses the actual moisture content of an air sample as a percentage of that contained in the same volume of saturated air at the same temperature. The relative humidity of moist air (U) is defined by:

$$U = 100\,\frac{e'}{e'_w} \qquad (4.5)$$

where e'_w is the saturation vapour pressure of the air at the same pressure and temperature. Relative humidities are measured conventionally by hair hygrometers (the direct method) or from wet- and dry-bulb temperature readings interpreted by appropriate humidity tables (an indirect method). Relative humidity assessments, from satellite or conventional sources, must be used with care; relative humidity has a definite diurnal variation opposite in sense to that of temperature. Hence, relative humidity is often at a maximum just before dawn, and a minimum in the afternoon. Thus the local time of satellite overflight is significant to the results from such a source. Through the year as a whole there is a similar inverted response to temperature. This is more subdued than the daily effect, but is important in climatology.

Dew-point temperature

This (T_d) is the temperature at which saturation occurs if air is cooled at constant pressure and constant mass mixing ratio without addition or removal of water. Thus

$$e' = e'_w \text{ at } T_d \qquad (4.6)$$

Therefore the dew point temperature is that temperature at which moist air becomes just saturated with respect to water. Once again a hygrometer, or wet- and dry-bulb temperatures, may furnish the necessary basic information.

Precipitable water

This rather less widely used assessment of the moisture content of the atmosphere concerns the depth of water (or total mass of water), M_w, that would be obtained were all the water vapour in a column of air of unit cross-sectional area condensed out. An approximate measure is given by the equation:

$$M_w = \frac{1}{g}\int_{p_1}^{p_2} r \,.\, dp \qquad (4.7)$$

where p_1 and p_2 are the pressures in mb at the bottom and top of a column respectively, r is the mixing ratio in g/kg as before, and g is the gravitational acceleration. Precipitable water may be expressed either in cm/cm^2, or in g, these being numerically equal. The average total moisture content above Britain, for example, is about 1 cm in January, rising to 2·4 cm in July.

Analyses of satellite data yielding acceptable measures of the amounts and distributions of water vapour in the atmosphere may be grouped conveniently under four headings, relating to the types of satellite data that are involved: (1) nephanalyses, (2) satellite cloud photographs, (3) infra-red measurements in selected wavebands and (4) spectrometer soundings of the atmosphere in depth. The more promising methods are outlined below to exemplify the kinds of principles involved.

Nephanalyses

A nephanalysis approach was followed by McClain (1966) with special reference to distributions of mean relative humidities through the 1000–500 mb layer across the U.S.A. McClain suggested that relationships between cloud cover in any given nephanalysis and some measure of the vertically integrated relative humidity could be established best through consideration of the so called 'saturation deficit'. This quantity, which may be taken as a measure of the mean relative humidity in the layer 1000–500 mb, is a parameter currently used in operational quantitative precipitation forecasting, as explained by Younkin *et al.* (1965). The saturation deficit (h_d) employed by McClain is defined by:

$$h_d \equiv h_5 - h_s \qquad (4.8)$$

where h_5 is the 1000–500 mb thickness, and h_s is a quantity termed the 'saturation thickness', i.e. the value of h_5 that would obtain in a layer having a uniform relative humidity of 70%, a moist adiabatic lapse rate and a mass of water vapour equal to the observed precipitable water.

When the saturation deficits and their corresponding mean relative humidities were related to both satellite cloud cover and cloud conditions (see table 4.6), it emerged that a strong positive correlation exists between improved nephanalysis – depicted cloudiness and layer mean relative humidity below 500 mb. As with most similar programmes of work, the intention is that eventually the entire process of relating thickness patterns to satellite-viewed cloud patterns may be automated, from the satellite raw data input to a moisture data output in a suitable form for further use in numerical weather analysis and forecasting, and then for further climatological processing.

Table 4.6
The relation of 'covered' cloud conditions on satellite pictures to saturation deficit, as established from 916 cases by McClain (1966) (modal classes underlined)

Relative humidity (U.S. standard atmosphere (%))	Saturation deficit (10 m units)	Cloud cover from satellite nephanalyses			
		Covered (80%)	Mostly Covered (50–80%)	Mostly Open (20–50%)	Open (20%)
84–70	−6–0	19·0	7·8	2·5	0·3
69–58	1–6	24·5	14·7	9·0	6·3
57–48	7–12	21·6	19·2	22·2	16·4
47–40	13–18	17·4	20·0	18·4	18·2
39–33	19–24	12·7	26·9	19·8	26·2
32–21	25–36	4·8	11·4	28·1	32·1

Satellite cloud photographs

Photographic approaches have been developed most notably by Smigielski and Mace (1970), dependent partly upon an earlier study by Thompson and West (1967). As with nephanalysis approaches, photographic studies have been basically subjective, involving mean relative humidity estimates from the earth's surface to 500 mb through joint considerations of various cloud types and the meteorological processes that produce them. Smigielski and Mace assert, however, that 'a satellite meteorologist, by applying his knowledge of meteorology and satellite picture interpretation, can arrive at such estimates with a considerable degree of accuracy', a statement for which operational support was forthcoming. Their main criteria for estimating mean relative humidities from satellite pictures were the amounts, types and vertical developments of clouds. These characteristics were established through the applications of photo interpretation principles outlined earlier in this chapter.

It may be concluded that a reasonably accurate analysis of the moisture fields can be accomplished over oceanic areas especially by combining knowledge of synoptic meteorology, the climatology of the area in question and satellite photo interpretation. Since the standard numerical forecasting model enters a nominal 45% humidity value for data-deficient areas, estimates of relative humidity based on any real observational evidence must be preferred by meteorologists and climatologists alike.

Infra-red analyses

Probably the most useful approaches involving infra-red observations are those based on 'evaluation diagram' techniques relating mean relative humidities of the troposphere simultaneously to infra-red measurements in more than one waveband. The foundation for more recent work was lain effectively by Möller (1961), who showed that the outgoing radiance from the earth/atmosphere system in the interval 5·7–6·9 μm is an acceptable measure of the mean relative humidity of the troposphere provided that the temperature distribution is also known. Since some radiation from boundary surfaces, in the form of cloud tops, also penetrates to space within this interval however, the temperatures of these surfaces must be known as well. Hence, Möller (1962) and, later, Raschke (1965) combined simultaneous radiation measurements from the 5·7 to 6·9 μm and 8 to 13 μm channels to account for the temperatures of the cloud surfaces. Subsequently, Raschke and Bandeen (1967) employed a similar technique to produce quasi-global maps of tropospheric water vapour for a five month period from February to June 1962 portraying the mean relative humidity distributions of the upper troposphere, and, in conjunction with radiosonde temperature data, the mass of water vapour above 500 mb. The more abundant presence of cloud in the lower troposphere denies the effective application of such methods to atmospheric layers nearer the earth's surface.

In the evaluation of the Tiros data for moisture parameters, three other variables had to be accommodated:

(*a*) The degradation of the response of radiation sensors in orbit.
(*b*) The angles from which radiations were measured. Three groups of measurements were considered, covering viewing angles between 0° (a vertical viewpoint) and 45°. All other measurements were disregarded.
(*c*) Different 'model atmospheres' for different climatological regions. These models presupposed standard mean temperature profiles, and a cloudless, dust-free upper troposphere bounded below by a black emitting surface.

Thus a range of evaluation diagrams (see fig. 4.13) was constructed on the basis of water vapour absorption and atmospheric window waveband radiations to yield relative humidities and radiating surface temperatures for chosen grid elements. Fig. 4.13 is the evaluation diagram for radiation emerging vertically from the tropical atmosphere. It appears that, with increasing surface temperatures, the curves for the 6·3 μm (water vapour absorption channel) bend markedly, while those for the window channel change direction only slightly. Hence, on the left hand side of the diagram the indications are that the upper troposphere is nearly saturated through being filled with cloud. Unfortunately, this assumption does not always seem to hold good, since it is not possible to differentiate between measurements of upward radiation between a high, cold, thin cirrus layer that is not opaque, and a warmer, thicker and more opaque cloud at a lower level. Hence the

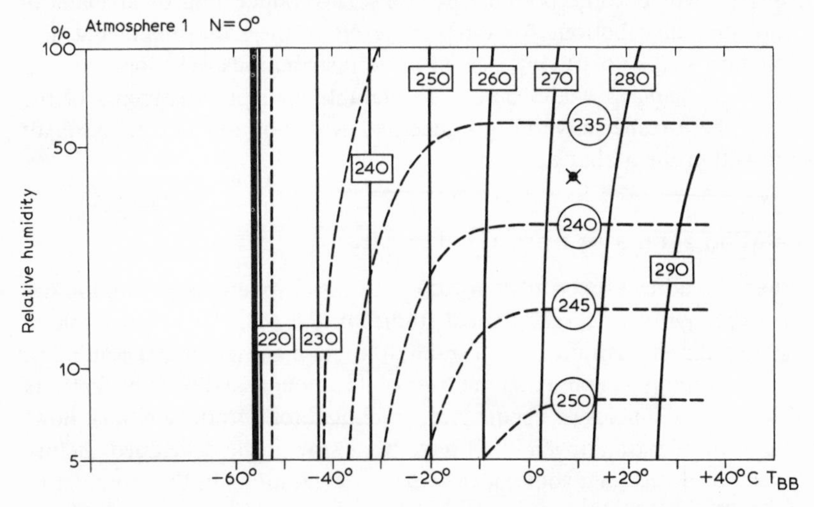

Fig. 4.13 Tiros III evaluation diagram for the tropical atmosphere. This permits the determination of tropospheric mean relative humidities, and surface temperatures of the ground or of clouds, through the simultaneous evidence of M.R.I.R Channel 1 and Channel 2 measurements. For example, the intersection of Channel 2 (260 °K) and Channel 1 (235 °K) indicates a mean relative humidity of 60%. (From Bandeen *et al.* 1965)

diagram should be used with more caution there than on the right-hand side where the higher angled intersections of the two sets of radiation curves facilitate reasonable relative humidity assessments.

Climatological results of the use of this method of moisture estimation are presented in chapter 7. It should be noted that they comprise values of a weighted average of the relative humidity above the clouds, excepting where the clouds are very high, giving temperatures <228 °K through either channel. The types of errors that arise as a result of deviations by actual atmospheric columns from those proposed by model atmospheres have been discussed extensively by Raschke (1965). They include the variable presences of high, cold, but semitransparent clouds (see Fritz and Rao 1967), dust layers and aerosols, and variations in the emissivities of cloud surfaces and the ground.

Spectrometer soundings

Spectrometer evaluations of vertical moisture profiles may be mentioned for the sake of completeness, though, at the time of writing, these are still under investigation, and have not yet reached an operational stage. Suffice it to say that the determination of humidity profiles is a basic objective of the IRIS programme as outlined already in chapters 2 and 3 (see pp. 47 and 71). Hopefully, it will become possible not only to produce improved maps of daily moisture distributions for different levels in the troposphere but also climatological maps portraying related mean distributions over long periods of time. Such displays would be of considerable interest to students of the free atmosphere, especially those concerned with horizontal transports of moisture within the global circulation.

Precipitation estimates from satellite data

More than the other aspects of moisture in the atmosphere, precipitation has figured prominently in climatological studies in the past. It is impossible to gainsay the direct significance of rainfall to human life and economics, whether that rainfall is expressed in terms of its amounts and distributions, its intensities or frequencies of occurrence. In climatology from satellites, however, methods of estimating rainfall have been slow to be developed. Before we proceed to investigate some of the earliest suggestions in this field, let us review briefly the main types of precipitation, classified in terms of the mechanisms which prompt the condensation of moisture in the atmosphere (see Barry and Chorley 1968):

1. 'Cyclonic type' precipitation (often steady, and moderate in amount). This is related primarily to the gradual uplift of air over a wide area in association with atmospheric low pressure and convergence.
2. 'Convective type' precipitation (characteristically this is short lived but heavy or very heavy in intensity). It is linked with thermal convection, in turn dependent on insolation heating. Buoyancy forces operate to effect vertical mixing through the agency of convection cells.
3. 'Forced convective' precipitation (this is usually tied to particular relief features, and may be steady and prolonged). Vertical air motion is prompted by mechanical forces, especially the passage of air over rough and/or high ground. Frequently, enhanced precipitation of this type results in a 'rain shadow' effect on the lee of the upstanding relief feature. This effect results partly from the depletion of atmospheric moisture by precipitation on the windward side, and partly through the effects of subsident motion.
4. 'Orographic' precipitation, of which type 3 may be a part. This rain is

caused mostly by the forced uplift of air over high ground. Apart from the influence of topography in increasing forced convective rainfall locally, cyclonic-type rainfall may also be increased as moist air streams are retarded as they approach the crest line of the ridge. The funnelling effects of valleys in upland regions may be important too.

The climatologist has two main reasons for wishing to be able to map gross rainfall distributions as accurately as possible from satellite data.

First, precipitation is an outstanding example of a climatological parameter that has not been assessed adequately around the globe as a whole due to the complexity and variability of its distribution and the inadequacy of the conventional stations recording it.

Second, an accurate knowledge of rainfall, and the associated latent heat releases as atmospheric water vapour condenses out, would be extremely useful to students of energy budgets related to the working of the global circulation.

Additionally, of course, more detailed pictures of precipitation amounts and distributions would be of great benefit to meteorologists and hydrologists in poorly instrumented regions for purposes of daily forecasting, catchment research, flood water control, etc.

In a Cospar (1967) status report on the application of space technology to the W.W.W. three possible types of methods were suggested for inferring precipitation from radiation measurements:

(1) 'Passive' methods, by which precipitating clouds could be distinguished owing to differences in emitted radiation from the rain drop clouds and background surfaces. Since most precipitation is accompanied by cloudiness, wavelengths must be selected carefully to permit the penetration of clouds, or, at least, a differentiation to be made between precipitating cloud masses and non-precipitating cloud. This is not possible in visible wavebands, or in the commonly used atmospheric window wavebands. However, as Thaddeus (1966) has indicated, at wave lengths longer than 1 cm thin, non-precipitating clouds are fairly transparent, permitting the desired distinctions to be drawn from comparisons of patterns made in both regions of the electromagnetic spectrum simultaneously.

(2) 'Active' methods, involving the distinguishing of precipitating clouds against underlying water or land surfaces through differences in reflection characteristics between the raindrop clouds and their background surfaces. Dennis (1963) investigated the possibilities of finding rain with satellite-based radar, and, although difficult technical and interpretational problems would have to be overcome before such an approach were successful, this may become feasible eventually.

(3) 'Delayed response' methods, in which infra-red investigations are aimed at differentiating surfaces recently exposed to precipitation. Rain

falling on a cold ocean leaves a warm, non-salty layer on the surface, a relatively stable layer that seems to persist for some time. Over land, rain-soaked terrain may be recognized by its decreased emissivity in the window wave-bands or by decreased reflectivity in the visible portion of the spectrum. This last relationship was used to detect areas of prior precipitation from Gemini space photographs (Hope 1966).

In practice, the earliest methods of estimating rainfall with moderate success have been of a fourth type, statistical in nature. Such methods have used comparisons between selected weather characteristics as seen by satellites, and contemporaneous surface observations. It is noteworthy that the longer-term (monthly) estimates have been more accurate than the shorter-term (daily) estimates. This is scarcely surprising, however, since the climatological approaches have aimed to evaluate the mean performances of precipitation clouds rather than to specify their probable performances on individual days. Of the four rainfall estimation methods outlined below only the first two are based on radiation data; the other two employ cloud photographs and nephanalyses as their data bases.

The first infra-red method was suggested by Lethbridge (1967) who attempted to estimate the probability of precipitation events from infra-red data. She found that there was a good relationship between atmospheric window radiation temperatures and the probability of precipitation. The colder cloud areas within an overcast, i.e. the deeper cloud areas, were usually areas of greater precipitation. A good relationship also appeared between cloud brightness on satellite photographs and precipitation probabilities. The best relationship, however, involved combinations of window radiation data and cloud brightness on the one hand, and the probability of precipitation on the other.

Later, in a second radiation based method, Clapp and Posey (1970) related American precipitation to brightness levels and surface albedoes through contingency table techniques described by Panofsky and Brier (1958). Table 4.7 summarizes their results. It relates seven classes of brightness level to five of surface albedo, and indicates selected observed precipitation characteristics in each box.

The top number in each box represents mean monthly precipitation in tenths of inches. Note that precipitation tends to increase with brightness level within each class of surface albedo. However, in each case the highest category of brightness shows a decrease in average precipitation, a fact for which no simple explanation is evident. The most likely explanation may involve changes in cloud type; the brightest cloud masses, when averaged over a whole month, may be comprised of relatively long-lasting thick stratiform types from which rain falls may be intermittent and/or relatively gentle.

Another interesting finding reported by Clapp and Posey is that a sharp

Table 4.7
Mean monthly precipitation for certain classes of cloud brightness level and surface albedo observed across the U.S.A. by Essa satellites. The top number in each box is precipitation in tenths of inches; middle number is range within which central one third of cases fell; and bottom number is the number of cases (after Clapp and Posey 1970)

Range in surface albedo (per cent)	Range in brightness level (BL x 10) 0–20	21–30	31–40	41–50	51–60	61–70	71–90	Average or total
8–12	15 4	27 20–28 35	36 25–40 46	41 32–41 18	39 3			33 106
13–17	5 2–6 11	11 7–11 22	10 9–12 13	34 25–36 19	33 5			18 70
18–22	2 0–2 6	8 4–10 18	9 7–9 20	14 8–18 13	22 5	16 2		10 64
23–32		19 1	5 1	15 7–10 6	10 8–12 7	13 10–20 7	10 4	12 26
33–52					8 1	12 4	8 4	10 9
Average or total	6 21	17 76	25 80	30 56	22 21	14 13	9 8	21 275

decrease in average precipitation accompanies increases in surface albedoes. This may be due to the association between higher surface albedoes and dry grasslands or semideserts, or with cold, snow-covered areas where total precipitation tends to be small. What does emerge is that other environmental parameters enter the picture beside cloud brightness alone.

The middle number in each box indicates the range within which the central one-third of mean monthly precipitations fall. The evidence of table 4.7 is that, despite the large variability indicated in most cases, and considerable overlapping of the ranges for adjacent boxes, the separation of precipitation between classes is probably real and significant. No statistical test for significance was applied, however, because of its questionable validity for data having strong autocorrelations in space and time.

The bottom figure in each box indicates the number of cases considered, each from 5° latitude/longitude grid squares across the U.S.A.

A somewhat similar, though simpler, approach was followed by Johnson *et al.* (1969), who examined relationships between an index of cloudiness (based upon satellite observations for a single 5° grid square covering East Pakistan) and the average daily rainfall at stations within the area. Their cloud indices were derived from Essa nephanalyses specially drawn to show more detail than those available for operational uses. The indices represent the percentage coverage by cloud forms normally associated with appreciable precipitation; in terms of the international conventions this is the percentage coverage of cloud masses designated C+, C or MCO with cumulonimbus cloud, along with other C areas that the analysts considered significant. Fig. 4.14, which displays the results for the 1967 season of the south-west monsoon suggests fluctuations in the intensity of monsoon weather activity over periods from ten to twenty days. The correspondence between the maxima and minima of the two curves on that time scale is reasonably good. Sawyer (1970) has

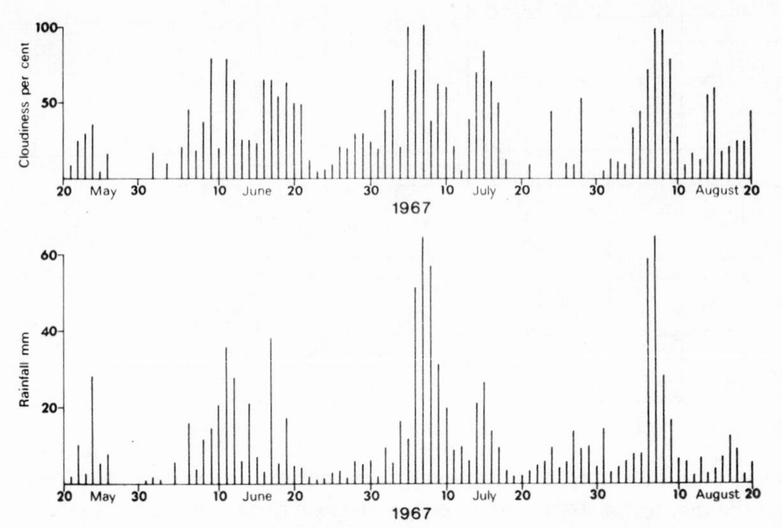

Fig. 4.14 Relation between an index of cloudiness based upon daily satellite observations for a 5° square over East Pakistan and the average daily rainfall at stations within the area. 24-hour rainfalls were available from between 15 and 22 stations. (From Johnson 1970)

published a similar diagram relating satellite-observed cloudiness to rainfall at Gan Island in the south Indian Ocean for January and February 1967; variations in cloudiness and rainfall on a similar scale were observed. It seems likely that in these cases discrepancies between cloud cover and regional rainfall measurements may have been due mainly to the following:

1 Meteorological factors, including differences of rainfall intensities and durations from different types and organizations of clouds.
2 Topographic factors, which influence precipitation distributions especially at the mesoscale in a region like East Pakistan.
3 Instrumental factors, especially exposure, at the surface recording stations from which the conventional data were drawn.

To reduce the effects of such factors it is desirable to improve the theoretical bases upon which such rainfall estimates depend. A beginning was made in this direction by Barrett (1970a), who initially applied his method to the tropical Far East between 90° and 180° E, and 30° N and 30° S (Barrett 1971). In summary, it involved the following stages:

1 The formulation of a 'precipitation coefficient equation' established as follows through physical theorizing and trial and error to quantify monthly rainfall indications from nephanalyses:

$$K_r = \bar{C} \sum \frac{(Mp_i i_i c_i + Np_2 i_2 c_2 + \ldots Rp_6 i_6 c_6)}{100} \tag{4.9}$$

where $\bar{C}$ is the mean monthly cloud cover percentage (a measure of the mobility of rain-bearing systems); $\sum$ represents the summation of available data, weighted where necessary to the full length of the calendar month in question; $p_i \ldots p_6$ are assigned probabilities of rainfalls from each nephanalysis category (see tables 4.8 and 4.9); $i_i \ldots i_6$ are assigned intensities of rainfalls from each cloud category (as in table 4.9); $c_i \ldots c_6$ are the six nephanalysis cloud categories; and $M \ldots R$ are the numbers of occurrences of $c_i \ldots c_6$.

Table 4.8
Nephanalysis indications of strength of cloud cover (from Barrett 1971)

Nephanalysis cloud cover categories	*Key letters*	*% range of cloud cover*	*Mean cloudiness in each category* (%)
Closed (or covered)	C	80–100	90
Mostly closed (or mostly covered)	MCO	50–80	65
Mostly open	MOP	20–50	35
Open	O	0–20	10
Clear skies	Clear	—	—

2 The comparison through statistical means of recorded monthly rainfall totals at selected stations with corresponding evaluations of the precipitation coefficient equation. Relationships between recorded data and nephanalysis indications of rainfall can be established best through regression analysis. By choosing surface stations from all latitudes, altitudes and topographic situations within the region studied, it becomes reasonable to suppose that the best-fit regression curve relationship can be used throughout that region to translate precipitation coefficients

evaluated across a uniform grid into useful precipitation estimates. Thence the main structures of 'precipitation fields' may be established.

3 Testing the validities of those estimated precipitation patterns by contingency methods. Actual monthly precipitation totals at selected stations within the region of study are tabulated against estimated totals derived by extrapolation from the maps of estimated rainfall.

In stage (1), the absence of generally accepted ranges of values for p and i necessitated the construction of relative scales ranging from 0·00 to 1·00 in

Table 4.9
Rainfall probabilities and intensities related to dominant nephanalysis cloud type categories (from Barrett 1971)

(1) Cloud type category	*(2)* Relative probability of rainfall	*(3)* Relative intensity of rainfall	*(4)* Rainfall indicator compounded from columns (2) and (3)	*(5)* Rainfall indicator in area of 'synoptically significant' cloudiness
Cumulonimbus	0·90	0·80	0·72	2·88
Stratiform	0·50	0·50	0·25	1·00
Cumuliform	0·10	0·20	0·02	—
Stratocumuliform	0·10	0·01	0·001	—
Cirriform	0·10	0·01	0·001	—
Clear skies	—	—	—	—

either case, with values ascribed to each nephanalysis 'state of the sky' as seemed best, bearing in mind the relative rainfall characteristics of different families of clouds. Since especially heavy and prolonged rain may be expected from major nephsystems within the meteorological tropics, areas stippled on nephanalyses on account of their 'special synoptic significance' were accorded values four times those listed in table 4.9, column (4).

Fig. 4.15 illustrates the regression of evaluated coefficients againt recorded rainfalls for four separate months at twenty-nine stations in the Australia/New Guinea region during 1966. The computed cubic regression line, used again in later extended studies in the same region (Barrett 1971; and see pp. 188–94), has the specific form:

$$Y = 0{\cdot}002x^3 - 0{\cdot}051x^2 + 1{\cdot}301x - 0{\cdot}328 \quad (4.10)$$

This cubic polynominal had a multiple correlation of 95·27%, and coefficients significant at 5% and 1% levels, considerably better than for the best-fit linear regression.

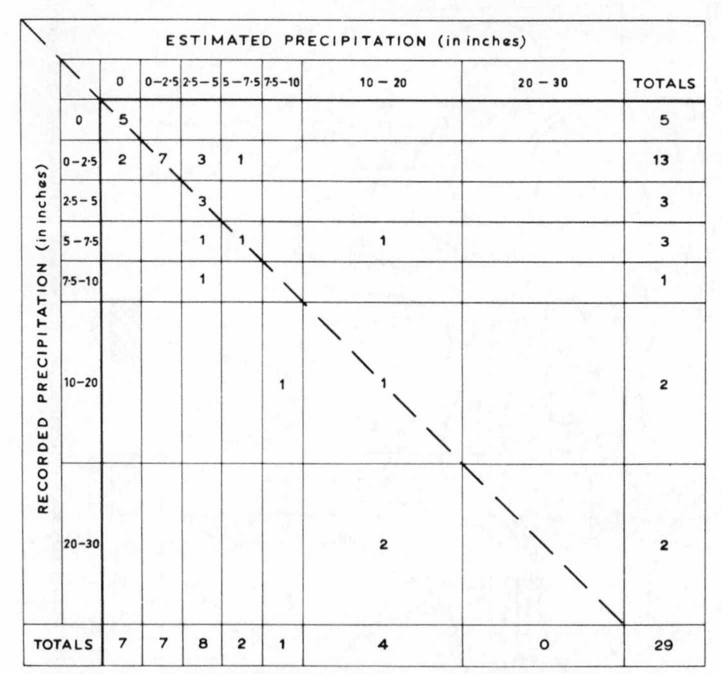

RECORDED PRECIPITATION (in inches) \ ESTIMATED PRECIPITATION (in inches)	0	0–2·5	2·5–5	5–7·5	7·5–10	10 – 20	20 – 30	TOTALS
0	5							5
0–2·5	2	7	3	1				13
2·5–5			3					3
5–7·5			1	1		1		3
7·5–10			1					1
10–20					1	1		2
20–30						2		2
TOTALS	7	7	8	2	1	4	0	29

Fig. 4.15 Evaluated rainfall coefficients for 29 rainfall stations in Australia and New Guinea, March to June 1966, plotted against rainfall measurements. (From Barrett 1970)

Fig. 4.16 portrays the initial results in the form of a climatological map of rainfall in the area chosen, from 90° to 180° E and 15° N to 30° S, for July 1966. Unfortunately, the map is open to some criticisms similar to those levelled against earlier estimates of climatological precipitation, especially since precipitation processes over land and sea areas are not identical. The methods employed in the compilation of fig. 4.16 assume that they are. Most earlier approaches to wide-scale precipitation mapping have been based on comparative frequencies of cloud types and hydrometeors observed by ships and land-based stations (see e.g. Jacobs 1951; Tucker 1961). The difficulties with which they were fraught were so great that comparable assessments of precipitation mapped by different workers have shown only small measures of agreement with each other.

Meanwhile, as far as the comparison between precipitation processes over land and sea is concerned, the relevant literature is not categorical. The general consensus of opinion is that maritime falls of rain tend to be more random at any time than those over land, but less complex in the pattern of their distribution. Over land, topography plays a major part in modifying the details of rainfall distributions. As cloud features depicted on nephanalyses

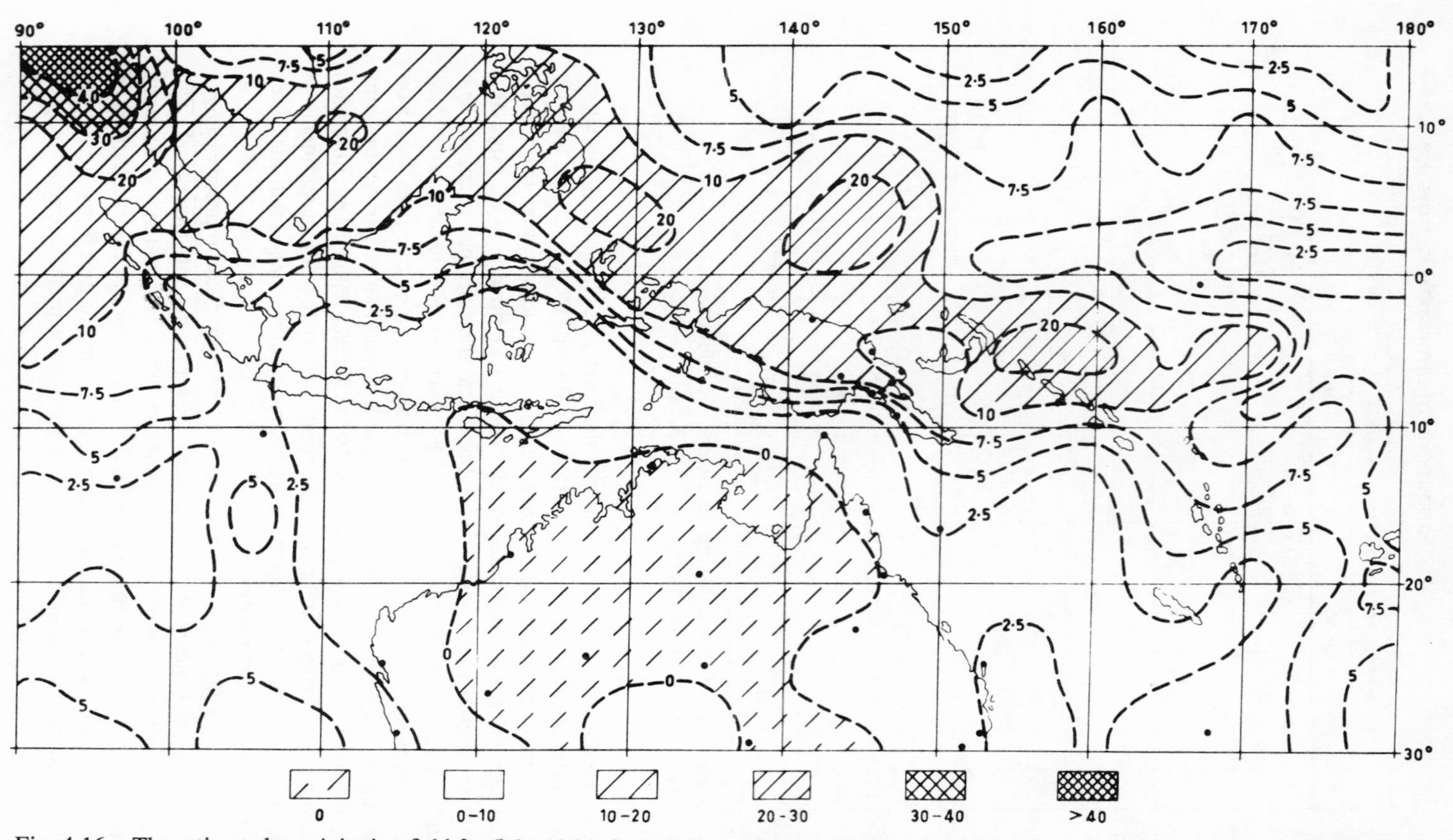

Fig. 4.16 The estimated precipitation field for July 1966, derived from daily nephanalyses sampled at 5° intervals of latitude and longitude, and the regression line in fig. 4.15 (in inches). (From Barrett 1970a)

are no more detailed than the subsynoptic scale, more local, topographically induced cloudiness does not appear at all on these cloud charts. Hence, nephanalysis-derived rainfall maps may be less detailed in better documented regions than others compiled from conventional sources, but their patterns are more related to stimuli operating at uniform scales across both land and sea. The most useful application of the above approach may prove to be for data inputs in computer studies of the earth's water budget. Modifications of the method have been made for the estimation of daily rainfall for shorter-term meteorological purposes (Follansbee 1973), but these require that much more careful consideration be paid to local atmospheric and topographic peculiarities. A purpose-built method for daily rainfall forecasting has been proposed recently by Barrett (1973).

Perhaps the standard approach of the mid-1970s may involve surface weather radar and geosynchronous satellite data along lines similar to those suggested by Woodley and Sancho (1971). Their study involved the superimposition of contoured precipitation echoes on synchronized A.T.S. III cloud photographs that had been brightness-contoured with a colour densitometer. In the more distant future it may become possible to measure rainfall from space by microwave radiometry (as data from prototype microwave sensors on Nimbus 5 suggest), while multichannel radiometers may provide data for calculating integrated totals of liquid water through the depths of the clouds. In general areas of high atmospheric water content should correspond with regions of heaviest precipitation.

5 Wind flows and air circulations

Background matters

O. G. Sutton, a former Director-General of the British Meteorological Office, once observed that 'circulation is a fundamental feature of our atmosphere'. It is indeed a fact of everyday observation that the air around us is seldom still: it is in a state of almost perpetual motion. Within the earth's atmosphere it is usual to differentiate between circulations at the global or primary, the secondary, and the tertiary, scales. Unfortunately for the student of climatology from satellites atmospheric circulations are not evidenced directly by weather satellite data. Fortunately for him, however, natural 'tracers' of wind flows and organized air circulations do exist in the forms of clouds. With the notable exceptions of those cloud phenomena that develop more or less tied to topographic features, most cloud fields move under the influence of winds and/or organized circulation patterns. Since air movements arise in response to still more basic physical stimuli, these must be reviewed briefly before we turn to discuss the values of various satellite contributions to studies of winds and wind systems.

According to Newton's first law, for a body of unit mass to change its state of motion it must be acted upon by an unbalanced force. Two dissimilar groups of forces operate upon the atmosphere: (1) those that exist regardless of the state of motion of the air, and (2) those that arise only as motion begins.

In a word, the first group may be termed driving forces, the second steering forces. The chief driving forces in the vertical include gravity (acting downwards) and the vertical pressure gradient force (acting upwards). Since these act in opposite directions vertical motions are consequently gradual. In the horizontal, however, the horizontal pressure gradient force (H.P.G.F.) which arises in conjunction with regional differences in atmospheric pressure is not offset by gravity and stronger accelerations of air from high to low pressure may occur. The H.P.G.F. may be written:

$$\text{H.P.G.F.} = -\frac{1}{\rho}\frac{\partial p}{\partial n} \qquad (5.1)$$

where ρ is air density and $\partial p/\partial n$ the horizontal gradient of pressure.

The chief steering force (diverting winds from straight line flows at right angles to the isobars) is caused by the rotation of the earth about its polar axis. In 1844 the effect of the earth's rotation on all moving objects on its surface was formalized by G. G. Coriolis, a French mathematician. This deflecting effect can be expressed by:

$$\text{Co.F.} = -2\omega V \sin\phi \qquad (5.2)$$

where ω is the angular velocity of spin of the earth (15°/hour), ϕ the latitude in question, and V the velocity of the moving object. Clearly the magnitude of the so-called Coriolis force (Co.F.) is proportional to the speed of motion, as well as latitude: it is zero at the Equator and maximum at each pole. Its sense is to the right of the H.P.G.F. in the northern hemisphere, and to the left in the southern hemisphere.

Near the surface of the earth a further force is significant too, namely friction. The strength of this effect is related to the degree of roughness of the surface over which air flows. Although the frictional drag retarding a solid object pushed across a solid surface is more marked than that retarding a gas flowing across the same surface, lower tropospheric winds are slowed markedly by the earth's surface, particularly where the topography is hilly or mountainous. The most immediate effect is a reduction of wind speed. This, in turn, leads to a reduction in the Coriolis force, which, as we have seen already, is dependent partly upon the velocity of the moving mass of air. Under such circumstances, frictionally affected wind flow is generally across the isobars at low angles from high to low pressure, since the H.P.G.F. is not affected by the presence of friction and now outweighs the weakened Coriolis force. Such cross-isobaric flow is normal in the surface layer of air, at the bottom of the atmosphere.

One last steering force must be mentioned here. It is a fact of casual observation that daily synoptic weather charts (e.g. those included in the British Daily Weather Report) contain more curved isobars than straight. Where curved flow is found, a further deflecting force, the centripetal acceleration, is necessary to 'explain' the pattern. Such a force is, strictly speaking, a mathematical requirement rather than a physical reality. The centripetal acceleration may be conceived as the force acting upon a body rotating in a circle, acting inwards in the direction of the centre of rotation (sometimes in meteorology a corresponding effect acting outwards from the centre of rotation, namely the centrifugal acceleration, is invoked instead; this is equal and opposite to the centripetal). The centripetal acceleration is expressed mathematically in either of two terms – namely ($\omega^2 R$), where ω is the angular velocity and R the radius of curvature of the path, or $\pm(V^2/R)$, where V represents an instantaneous linear velocity.

Until the very recent development of vertical sounding techniques, all satellite observations of the earth's cloud fields were essentially horizontal in nature. Thus it is more necessary for present purposes to be well acquainted with the combined effects of such forces in the horizontal than the vertical plane. Horizontal wind flows can be subdivided into two fundamental families, equilibrium and inequilibrium winds:

Table 5.1
Summary of important equilibrium winds

Wind	*Balance*	*Equation*
Geostrophic wind	Coriolis force = horizontal pressure gradient force	$V_G = \frac{1}{2\omega \sin \phi \rho} \frac{\partial p}{\partial n}$
Gradient wind	Centripetal force = horizontal pressure gradient force + Coriolis force	$V_{gr} = V_G \pm \frac{Vgr^2}{2\omega \sin \phi R}$
Cyclostrophic wind	Centripetal force = horizontal pressure gradient force	$V = \frac{R}{\rho} \frac{\partial p}{\partial n}$
Inertial wind	Centripetal force = Coriolis force	$V = 2\omega \sin \phi R$

Balanced or equilibrium winds

These would prevail if the atmosphere were frictionless and behaved ideally. The most important balanced wind is the geostrophic wind, since, in the 'free atmosphere' above the shallow 'friction layer' overlying the surface of the earth, geostrophic approximations are often close to reality. This fact has long been basic to upper level climatology and meteorology. It is invoked frequently in upper tropospheric wind analyses from satellite altitudes. Strictly speaking the geostrophic wind concept applies only to straight line flow. Where curved flow patterns are involved the gradient wind concept is more appropriate. A third balanced wind, namely cyclostrophic flow, occurs in limited meteorological contexts where large horizontal pressure gradient forces prompt sufficient centripetal acceleration for balanced flow to result, parallel to the isobars. This may arise:

(*a*) In low latitudes, in intense cyclones (see chapter 10). These are marked by strong horizontal pressure gradients in regions where Coriolis effects are slight.

(*b*) Locally in mid-latitudes in intense circulations of the tornado type. Since strong pressure gradients are found here in circulation systems only a few hundred metres in diameter the centripetal forces can be exceptionally strong.

Finally, mention may be made in passing of the inertial wind. Theoretically a 'circle of inertia' is followed by air in horizontal motion which is subject to no driving force but gravity. Under such circumstances the centripetal force is balanced perfectly by the Coriolis force. Strictly speaking, circular flow cannot be followed in reality since the Coriolis effect varies with latitude. However, inertial flow has been invoked to explain certain cloud field peculiarities on satellite photographs, e.g. the curious lee eddy patterns sometimes observed to the lees of islands in the trade wind zones (see Chopra and Hubert 1965).

Inequilibrium (cross-isobaric) winds

The winds outlined above all presuppose steady conditions of balanced air flow. Clearly these are not to be identified too frequently in an atmosphere characterized by kaleidoscopic growths, decays and movements of its included weather systems. However, the concepts and equations of balanced flow constitute a most useful frame of reference in which actual winds may be adjudged. One of the most useful yardsticks for comparing equilibrium and non-equilibrium winds is the ageostrophic wind, illustrated by fig. 5.1. The ageostrophic wind (or geostrophic departure) is fundamentally important on account of its association with net accumulations (convergences) or depletions (divergences) of air and their related motions in the vertical plane.

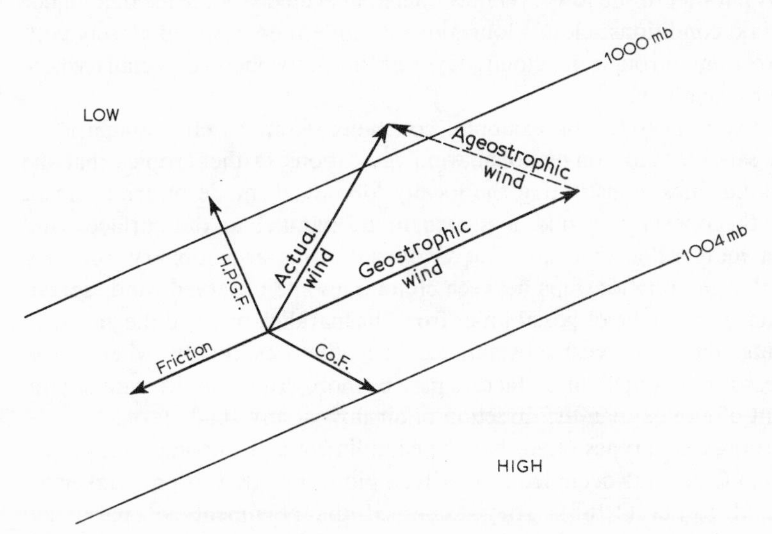

Fig. 5.1 The vector relationship between geostrophic, ageostrophic and actual winds.

Both involve the rate, as well as the pattern, of wind flow. Therefore two types of wind field analyses are necessary to distinguish the one from the other:

(*a*) Isotach maps. These portray areas of equal wind speed.
(*b*) Streamline maps, which indicate directions of horizontal wind flows. The streamlines themselves are constructed everywhere tangent to the instantaneous wind flow.

Together such maps comprise a complete analysis of any wind field. Since their application to climatological studies is increasing as interest in dynamic and synoptic climatology grows, it is fortunate that under certain circumstances quite acceptable isotach and streamline maps can be constructed from satellite cloud photographs. Let us proceed to examine in more detail the wind flow and circulation patterns that can be derived from such evidence.

Wind flow analyses from satellite data

The direction of lower level winds

From once-daily photographs, most interpretations of cloud features in terms of lower level winds have entailed the propensity for clouds to become elongated down the direction of air flow. Even in pre-satellite days it was recognized by Kuettner (1959) that, in the organization of clouds in rows down to the mesoscale, from the Tropics as far north as the Arctic, most cloud bands line up in the direction of flow. Similarly, Malkus and Riehl (1964) affirmed that, at least in the Tropics, 'the common mode of (cumulus) organization is parallel to the low level flow'. Satellite evidence confirms that, under appropriate conditions, cloud elongations do indeed correspond closely with wind directions through the cloudy layer of the atmosphere, especially where that layer is shallow.

Thus Gaby (1967), for example, concluded from careful comparisons between satellite and conventional wind indications in the Tropics that the larger cloud lines, constituting the locally 'dominant' mode of arrangement *vis-à-vis* the observed winds, appeared to lie parallel to the surface wind direction more often than not. But Gaby did strike a cautionary note. He showed that the relationships between cloud lines and observed winds spread across the entire range of possibilities from the 'parallel' mode to the 'normal', supporting the now well-substantiated fact that, especially where deep cloudiness is concerned, other factors may be more critical in determining the alignment of clouds than the direction of air flow at any single level.

Regarding cloud types other than the cumuliform and cumulonimbus types with which Gaby was occupied primarily, a pioneer study was undertaken by Lyons and Fujita (1968). They examined the alignments of mesoscale striations in oceanic stratus sheets across parts of the north Pacific. On account of their scale, similar striations have only been identified infrequently from conventional observations while often appearing as patterns in otherwise rather amorphous stratiform cloud fields photographed by satellites. Experimental work in the laboratory by Faller (1965), enabled him to simulate parallel bandings in continuous or nearly continuous 'cloud fields' of smoke in a rotating tank. Faller suggested that the cloud bands they represented in the atmosphere should be aligned to the left of the geostrophic wind in the northern hemisphere at angles ranging from 10° to 17°, with an average of 14°. Putting his conclusions in a different way, it seems that such bands may also be expected to lie preferentially along streamlines near the surface of the earth, provided once again that the clouds are low and relatively shallow. The empirical work by Lyons and Fujita supports this suggestion; the stratiform rolls they examined on satellite photographs generally crossed the isobars at about 16°. This angle approximates to the mean angle of lower level wind flow across isobars from high to low pressure in mid-latitudes.

Aloft, similar relationships again seem to hold good. Numerous studies of cirrus streamers and belted or strap-like arrangements of jet stream cirrus indicate in general that streamline analyses can be inferred under good conditions. In the Tropics, for example, cirrus clouds often top tall cumulonimbus towers at heights generally above 200 mb. Strong winds aloft (especially $\geqslant 20$ knots) often blow out the cirrus plumes into long tails that reflect quite faithfully the upper level directions of wind flow (see, for example, Jager *et al.* 1968). Similarly the directions of flow along high-level wind maxima may be indicated by long, narrow bands of jet stream cirrus, with or without cross-striations, and/or by the edges of strong cirrus shields (Whitney 1966).

Lastly, qualitative evidence from trade wind anticyclonic areas suggests that cloud bandings in stratocumuliform cloud fields also become organized along the dominant directions of wind flow especially beneath well-marked trade wind inversion levels (Barrett 1970).

Together such observations all suggest, therefore, that acceptable streamline analyses may be drawn from satellite cloud photographs of good resolution providing that the cloud elements and patterns upon which such analyses are based are chosen with considerable care. In particular, deep clouds should be disregarded, and cloud elongations at different levels must not be confused.

Since a whole phalanx of terms exists to describe different kinds of cloud elongations (e.g. cumulus 'rows' or 'streets', cirrus 'streamers', stratiform 'striations', and stratocumuliform 'stripes' or 'bands'), it has been suggested that all satellite cloud elongations be referred to corporately as 'nephlines'. Thus a tracing of appropriate cloud lines from a photograph becomes a 'nephline analysis', and, from this a 'smoothed nephline' map may be prepared, whose components should resemble the streamlines at the selected level. Figs. 5.2 and 5.3 illustrate these types of maps (from Barrett 1970).

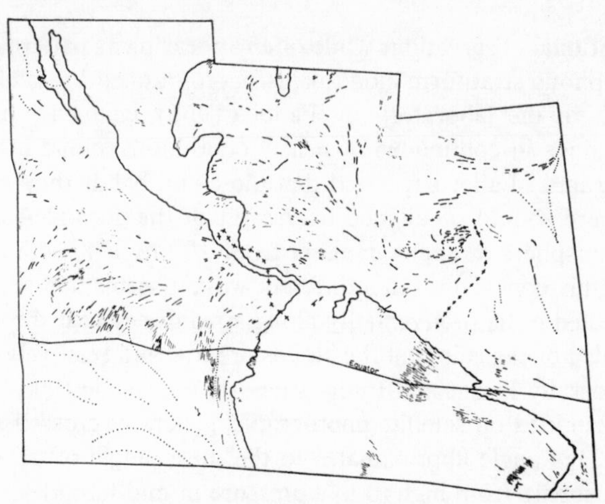

Fig. 5.2 Lower tropospheric nephline analysis of Central American region from Nimbus II photo-mosaic, 24 July 1966. Instability axes shown by lines of crosses, cumuliform nephlines by continuous lines and stratocumuliform nephlines by dotted lines. (From Barrett 1970b)

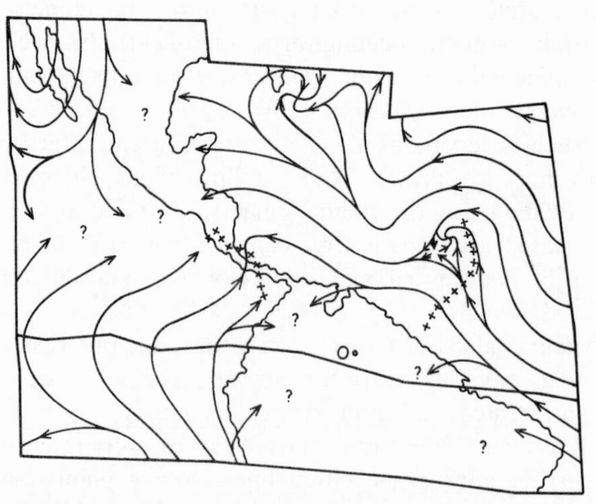

Fig. 5.3 Smoothed nephlines (approximate streamlines) drawn from the evidence of fig. 5.2. (From Barrett 1970b)

Similar maps were prepared from all available Nimbus II A.V.C.S. photographs of Central America from the launch date of that satellite on 15 May 1966 until the malfunction of its recording equipment after 31 August. They were then classified in terms of their circulation patterns. The day-to-day sequences of five main types that were identified threw interesting light upon the summer circulation climatology across the isthmus and its adjacent seas (see chapter 9).

The speed of wind

Although a parameter as variable as the speed of the wind is difficult to assess meaningfully even from conventional observations, some climatologically significant studies have involved the assessment of wind speeds from satellite evidence. Four may be included here, to illustrate the range and diversity of potentially useful approaches.

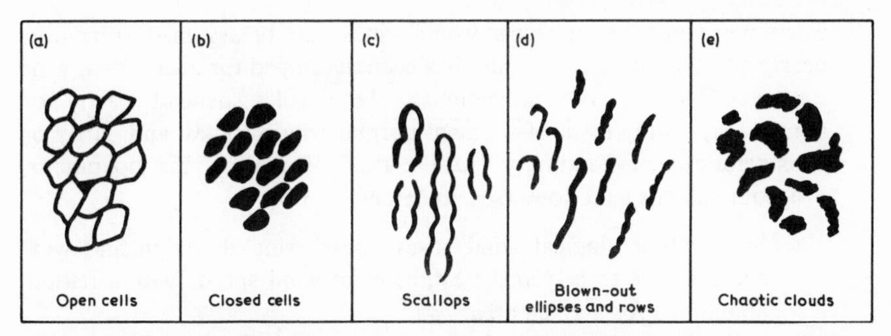

Fig. 5.4 Sketches of polygonal cellular clouds ((a) and (b)); vermiculate clouds (c); linear cloud arrangements (d); and chaotic (non-organized) clouds (e). All in cumuliform and/or cumulonimbus cloud fields. (After Merritt & Rogers 1965)

(*a*) It was suggested by Merritt and Rogers (1965) that relationships might exist between certain types of cloud fields and rates of atmospheric motion. Their analyses of satellite photographs, for selected areas in low latitudes, revealed the following facts (fig. 5.4):

(i) Polygonal cellular arrangements of cumuliform and/or cumulonimbus clouds seemed to be associated mainly with lower tropospheric winds of less than 20 knots.

(ii) Vermiculate ('worm-shaped') cells (i.e. polygonal cells with some cross-wind links missing) indicated wind speeds of 20–25 knots.

(iii) Linear arrangements of cloud elements, and blown-out ellipses, occurred mostly where higher wind speeds were prevalent, namely in excess of 13 knots.

(iv) Chaotic (non-organized) clouds occurred most frequently where wind speeds were up to 30 knots.

Unfortunately, of course, the wind speed categories associated with different cloud field patterns were not mutually exclusive. For this reason approaches of this kind have not proved very useful in operational contexts. In middle latitudes little comparable work has been attempted, apart from some preliminary analyses of lee wave situations (see, for example, Fritz 1965, and chapter 12). It has been shown that the spacings of the subparallel cloud bands that form under optimum conditions on the leeward sides of mountain ranges depend largely upon the mean wind speed at the cloud altitude, as well as the vertical profiles of winds and temperatures. In summary, the longest wavelengths seem to be associated with strong winds and a rapid decline of temperature with height (Conover 1964). Other studies involving topographically modified clouds, and the atmospheric parameters associated with them, have been reviewed by Barrett (1967, ch. 7).

(*b*) For upper tropospheric levels, where winds may be assumed to be more nearly geostrophic, a set of rules has been developed for operational wind estimates from cirrus indications. These rules depend partly on climatology in the form of seasonal normal wind profiles, and partly on considerations of continuity. At the time of writing, the operational procedure stands as follows (see Anderson 1969):

(i) Study climatological wind roses, mean wind direction and wind speed charts to help make estimates of wind speeds within reasonable limits for a given area.
(ii) In dealing with wind maxima, make corrections where necessary, for the change of speed with height. The use of a climatological atlas to determine a normalized wind profile gives the best results.
(iii) Westerly flow originating in equatorial areas generally becomes progressively stronger with increasing latitude.
(iv) Check continuity and nearby wind reports from recent upper air charts to assure against making unreasonable estimates.
(v) If a cirriform cloud formation suggests the proximity of a jet axis, the winds are likely to be strongest along the axis of that formation.
(vi) The presence of transverse bands in cirrus indicates strong winds.
(vii) Very generally, the longer the plumes or cloud trails from cumulonimbus clouds, the stronger the winds.
(viii) Over the Tropics the presence of diverging filaments of cirriform cloud from areas of solid convective cloudiness is generally indicative of weak high level outflow and light winds.
(ix) Streaky cirrus extending in one direction for hundreds of kilometres is associated with large-scale currents of moderate speed. This cloud

formation is made up of rather long plumes or cloud tails orientated in the same or nearly the same direction.

(x) Areas exhibiting a change in orientation of cloud tails in short distances indicate the presence of weak winds.

(xi) Empirical rules for various local areas are helpful. For example, some Indian work has indicated that strong high level easterlies over the Bay of Bengal accompany greater than usual convective activity, and weaker winds diminished convection.

(xii) If in doubt about wind direction, do not attempt an estimate of wind speed.

These satellite 'rules of thumb' will probably be modified and improved with time, but they have proved useful in the preparation of wind field analyses within the Tropics since March 1967. Values of the speed and direction of the wind at both 300 mb and 200 mb levels have been made daily for use in automated tropical analyses (Bedient *et al.* 1967). Although they are dependent in part upon existing climatological knowledge, insufficiencies in that knowledge are being made good especially in sparse-data regions.

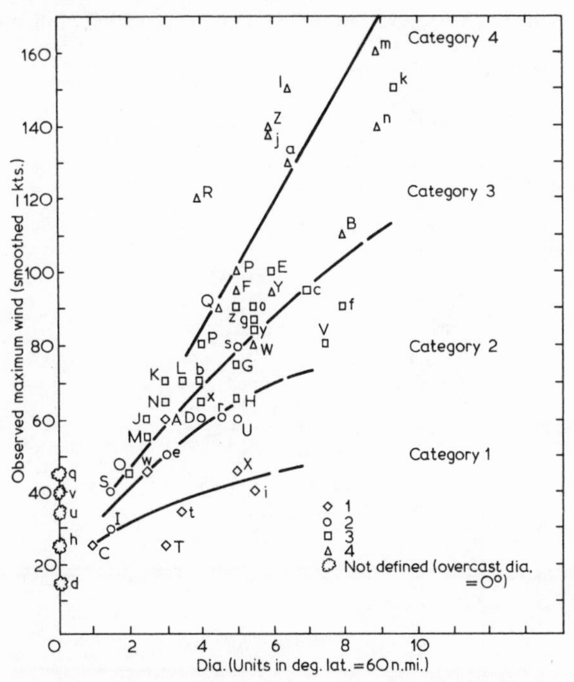

Fig. 5.5 Observed hurricane wind speeds related to the sizes of certain classes of tropical vortices. (After Fritz *et al* 1966; from Barrett 1967)

(*c*) For several years operationally valuable estimates of wind speeds have been made from hurricane synoptic scale cloud organizations. The economic impact of hurricanes is so great, especially in the American region, that early work led to the development of detailed and comprehensive classifications of such storms from satellite altitudes (Oliver 1969a; see also chapter 10 of this volume). Subsequently schemes have been formulated for the derivation of maximum wind speed estimates from the appearances of hurricane cloudiness in satellite photographs. In brief, the storms are classified in terms of:

(i) The diameters of their overcast areas.
(ii) The degree of circularity of their spiral cloud bands.
(iii) Whether the centre of the spiral band structure is located inside or outside the major cloud mass of the overcast.

The earliest hurricane wind speed estimates were made from Tiros evidence in 1962. They involved the plotting of aircraft observations of hurricane wind speeds against satellite-viewed cloud vortices classified on the bases of their sizes and organizations (Timchalk *et al.* 1965; Barrett 1967, ch. 5). Contingency tables summarizing the estimates of hurricane wind speeds derived from fig. 5.5 plotted against contemporaneous observations by aircraft during the next hurricane season revealed that most

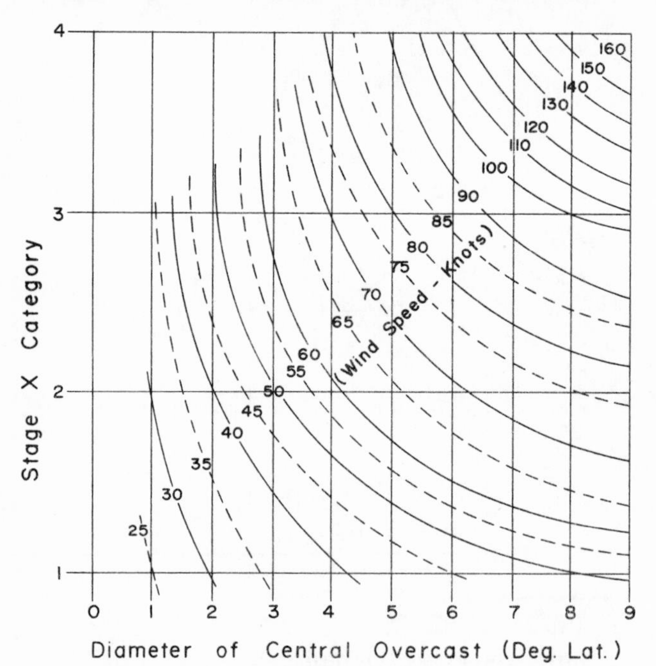

Fig. 5.6 A nomogram for obtaining maximum winds of Stage X tropical disturbances. (From Anderson 1969)

of the estimates were within 15 knots of the actual maximum wind speeds. This result was considered sufficiently good for the method to be employed routinely in storm forecasting through the next few years. Recently Hubert and Timchalk (1969) have redrawn the appropriate regression lines using computer techniques instead of free-hand approximations, and benefiting from years of accumulated data and experience. Fig. 5.6 illustrates a recent nomogram for obtaining maximum wind speed estimates from tropical disturbances of 'Stage X' (see fig. 10.2, p. 266), i.e. tropical storms or hurricanes in mature stages of development.

(*d*) Lastly, a newer method designed to estimate maximum wind speeds in

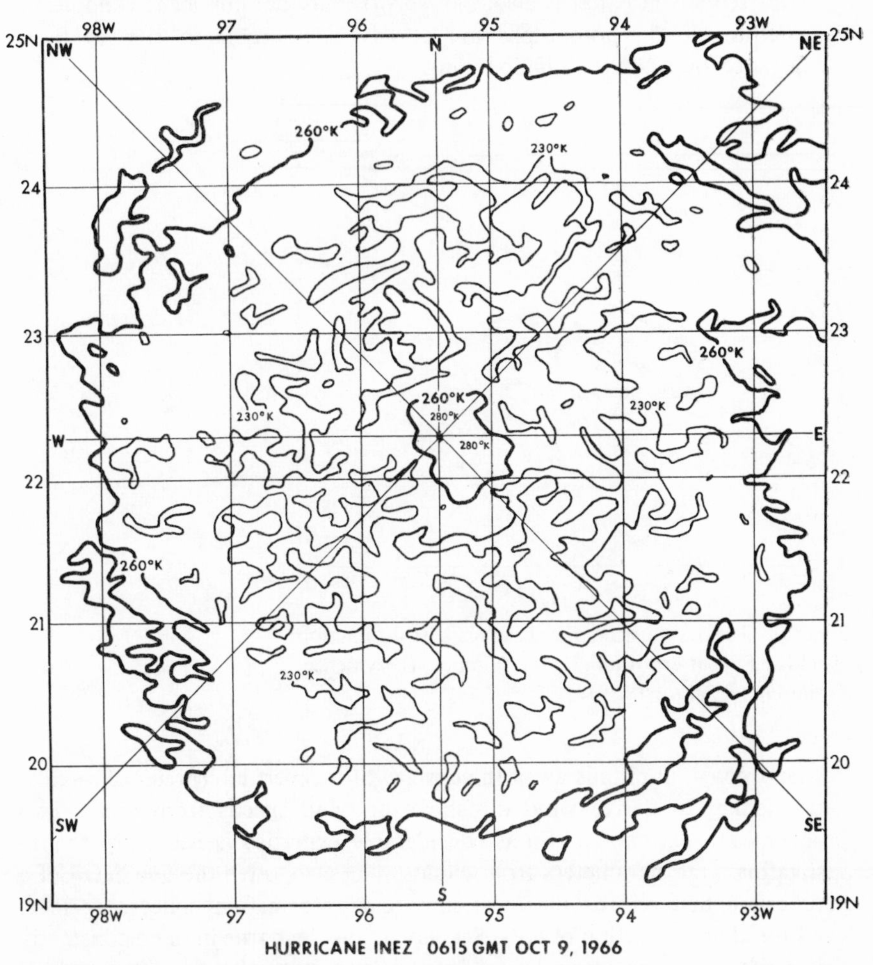

Fig. 5.7 Analysis of a digital map of H.R.I.R. data for hurricane Inez, 9 October 1966. The inner 260 °K isotherm encloses the eye; the outer one the overall canopy. (From Hubert *et al.* 1969)

hurricanes has been based on radiation patterns measured by Nimbus H.R.I.R. equipment. This involves two features of the radiation pattern of a storm vortex – namely the overall canopy size (r_o) and an inner radius associated with the storm's eye (r_i) (see fig. 5.7).

Preliminary investigations suggest that the ratio (r_o/r_i) was the function pattern seemingly related best to maximum wind speeds in an initial sample of storms (fig. 5.8). If further tests with independent data substantiate this relationship, a usable method for estimating maximum wind speeds from infra-red data will have been established. Alternatively it seems possible that the previously outlined photographic method might apply to infra-red storm patterns also. Not surprisingly the hurricane canopies in satellite cloud photographs are closely comparable in size to those depicted by satellite radiation maps.

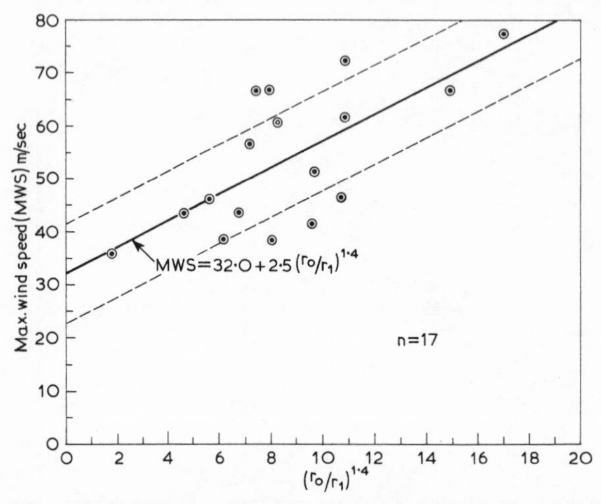

Fig. 5.8 Maximum wind speed versus best-fit function of H.R.I.R. pattern radii, from Nimbus II evidence. (From Hubert *et al.* 1969)

Many potential pitfalls await the unwary or inexpert interpreter of satellite data wishing to derive wind directions or wind speeds from them. Two problems are paramount. First, considerable expertise is necessary for investigating wind parameters from clouds since it is often the fine detail of a photograph that affords the most important information concerning local wind speed and direction of flow. Second, it must be borne in mind constantly that clouds are three-dimensional entities, and especially those that extend through a considerable depth of the atmosphere may owe some of their appearance on a flat photograph to wind speed and direction at one level of the atmosphere and some to another. So it is necessary to have some knowledge of thermal winds and wind shears before proceding further.

The thermal wind direction

If two pressure surfaces, say 1000 mb and 500 mb, are contoured and their patterns superimposed, a chart depicting the spatial variations of the 1000–500 mb thickness can be compiled. Such a chart separates centres of relatively warm and cold air by isopleths of equal layer depth, i.e. by thickness lines. Obviously thickness must be proportional to the mean temperature of an air column. If the lower-level geostrophic wind is represented by V_L, and its upper-level counterpart by V_U, the wind variation through the

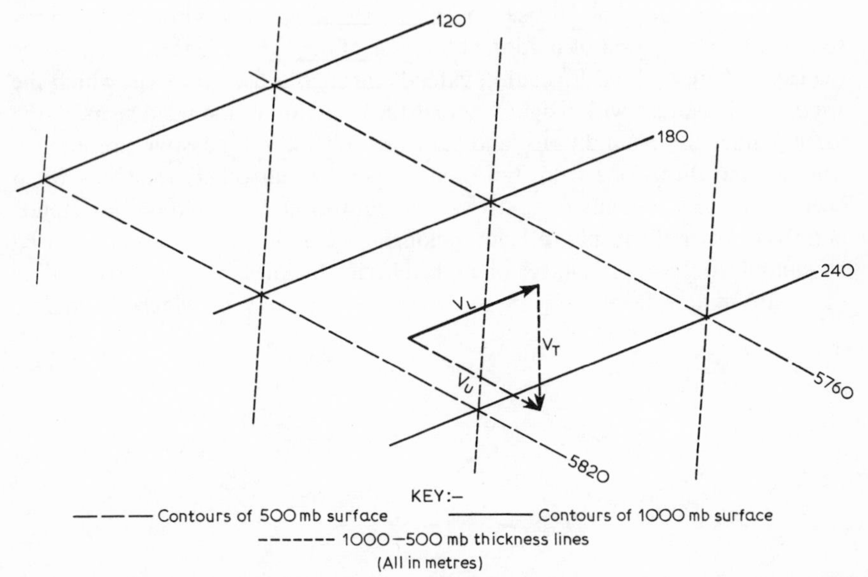

Fig. 5.9 The thermal wind, V_T, in relation to a low level geostrophic wind, V_L, and an upper level geostrophic wind, V_U. (In plan)

layer bounded by those two levels is V_T, as shown by the vector triangle in fig. 5.9. This represents the so-called thermal wind. This stands in the same relation to the thickness lines as the geostrophic wind does to the contours of selected pressure surfaces. The thermal wind is defined by the relationship:

$$V_T = V_U - V_L \tag{5.3}$$

It must blow along the thickness lines with cold air to the left in the northern hemisphere and to the right south of the Equator. Finally, to fix ideas, gradients of wind velocity in the vertical between V_L and V_U are known as the vertical wind shear, and in the horizontal as horizontal wind shear. Such changes are measured in specified directions normal to the reference directions of wind flow.

Returning to the question of satellite analyses, we note that the shapes and forms of many cloud elements and organizations in satellite photographs are determined importantly by vertical wind shear through the cloudy layer of the atmosphere (Oliver 1969b). Thus many cloud lines and bands are not aligned along the wind direction at any single level, but rather along the thermal wind through the layer from cloud base to top. Where very tall clouds are concerned, such as stray cumulonimbus, vertical wind shear may be strongly marked (Erickson 1964). No inference of the direction of the wind at any single level can be drawn from the simple two-dimensional appearance of such a cloud viewed from vertically above.

Just occasionally, of course, the direction of vertical wind shear corresponds to the direction of motion at the base of the cloudy layer. This must be the case where a cloud formation extends through a deep layer in which the wind speed changes with height whereas the direction of the wind remains the same. Under these conditions, and very few others, it is possible to establish simple wind directions from tall cloud types. On all other occasions when deep cloudiness prevails (e.g. strong cumuliform and cumulonimbus clouds and frontal stratiform cloud layers) elongated cloud features should be used to compile or correct charts of atmospheric thickness and not streamline maps for unique levels. Fig. 5.10 exemplifies the way in which, in middle

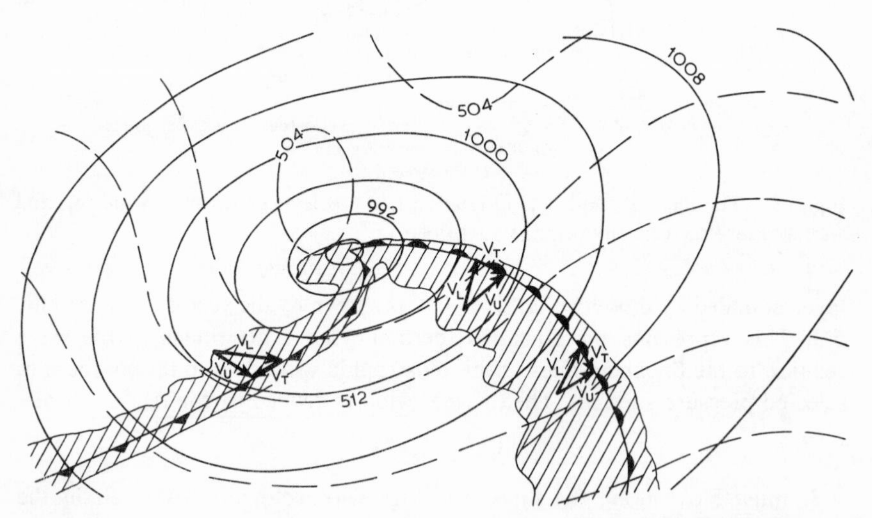

Fig. 5.10 The usual relationships between deep frontal cloud bands, upper (V_U), lower (V_L) and thermal (V_T) winds in a northern hemisphere extratropical cyclone.

latitudes, belts of extratropical cyclone frontal clouds commonly lie along the direction of the thermal wind.

Air circulations

General theory

Daily synoptic weather maps are dominated not by simple straight line wind flows but rather by organized air circulations, especially vortices or rotations in the aptly named 'sea of air'! Before we proceed to review some most important recent developments in satellite photo analysis, we may profitably summarize some background meteorological theory in order to explain why climatology from satellites is concerned more with air circulations than the pressure patterns with which they are associated.

Essentially, the rotational spin of air in the horizontal plane may be described in terms of relative and/or absolute vorticity.

Thus the vertical component of relative vorticity (ζ) – that is to say, spin additional to vorticity imparted by the rotation of the earth and its atmosphere about the polar axis – may be expressed by:

$$\zeta = \frac{V}{R} + \frac{\partial V}{\partial n} \qquad (5.4)$$

where ζ is the vertical component of vorticity in terms of velocity (V), radius of curvature of the streamlines (R), and differentiation along a direction normal to the streamlines ($\partial V/\partial n$). Thus the vertical component of vorticity may be regarded as the sum of components due to curvature of horizontal flow (V/R) and to horizontal wind shear ($\partial V/\partial n$).

Absolute vorticity is the sum of the relative vorticity and the vorticity of the earth appropriate to the latitude in question. The earth's vorticity (f) is twice its angular velocity in latitude ϕ:

$$f = 2\omega \sin \phi \qquad (5.5)$$

i.e. the Coriolis parameter. This spin is cyclonic in sense, the earth rotating from west to east viewed from above the North Pole. Thus, in latitude ϕ, the absolute vorticity (ζ_a) of air may be expressed as:

$$\zeta_a = (\zeta + f) \qquad (5.6)$$

ζ, the relative vorticity, has the same sign as f when it is cyclonic, and may be referred to as 'positive relative vorticity'. Conversely, the sign is different when it is anticyclonic, in which case negative relative vorticity is indicated. Corresponding values of f (functions of latitude only) can be added algebraically to each value of ζ if fields of absolute vorticity are required (see, for example, p. 140). Because cyclonic ζ has the same sign as f, and also has numerically higher values than anticyclonic ζ, positive values of $(\zeta + f)$ are almost universal. However, small negative values of $(\zeta + f)$ do occur in regions of strong anticyclonic shear of wind aloft, and both ζ and f tend towards zero at the Equator.

Now in a fluid medium such as the atmosphere, vorticity is often accompanied by deformation in the horizontal plane, especially if the rotating system has been set in motion only recently. Changes of shape of volumes of air are also noted when atmospheric systems are investigated in three dimensions. These become obvious when the relationships between divergence and vorticity are investigated a little further (note that it has become customary to regard convergence as negative divergence when mathematical treatments of these meteorological concepts are required). Thus the general relationship between divergence and vorticity is expressed by:

$$\frac{1}{\zeta + f} \frac{d}{dt} (\zeta + f) \simeq -\mathrm{div}_H V \qquad (5.7)$$

This states that horizontal convergence of the wind velocity approximately equals the fractional increase of its absolute vorticity with time. Put differently, accumulations of air in tubes of lessening diameters are accompanied by increased rates of spin. This principle is employed by the ice skater who achieves a more rapid spin by drawing in his arms to his body than by extending them. Meteorologically, it helps to explain the extremely rapid circulations observed in the smaller, more intense atmospheric systems such as hurricanes and tornadoes. Climatologically, it can be argued that vorticity distributions are in some ways more meaningful than pressure patterns (see chapter 14), largely because of the relationships between vorticity and divergence, and the sense of divergence and atmospheric stability.

Positive vortices encourage vertical cloud growth and associated weather; negative vortices encourage subsidence and low sheet clouds and/or inversions near the surface of the earth. Against this, maps of atmospheric pressure are difficult to subdivide into areas of high and low pressure because of the subjective nature of the process of choosing a suitable boundary value. That often chosen – 1000 mb – may, for example, be within cyclonic circulations in some regions, and anticyclonic circulations in others.

This line of reasoning, coupled with the greater ease with which wind field analyses can be made from satellite evidence compared with analyses of pressure distributions, dictates that, for a while, climatology from satellites must be more concerned with patterns of atmospheric circulation and vorticity and with patterns of atmospheric pressure.

Satellite studies of air circulations

Suggestions have been made already as to ways in which not only wind flow characteristics but also features of organized air circulations can be derived from satellite photographic data. A basic requirement for such procedures, however, is photography of a given region much more frequently than one a day. Fortunately geosynchronous satellites are meeting this need. The ground methodological work has been undertaken principally by Suomi and his co-workers in the University of Wisconsin utilizing movie film loops of A.T.S. spin-scan camera pictures. The film loops not only illustrate cloud motion generally, but also permit the separate identification of upper- and lower-level air flow patterns through the differential movements of their visible tracers, the included clouds.

The validity of wind estimates from geosynchronous satellite cloud pictures has been investigated carefully by Hubert and Whitney (1971). From comparisons between A.T.S. wind estimates at cloud levels and rawin-derived wind vectors it emerged that the estimated velocities of the lower clouds corresponded best with actual winds at 3000 feet (*c.* 900 m) while upper cloud velocities correspond best with actual winds at 30,000 feet (*c.* 9000 m). Hubert and Whitney concluded that directional deviations between estimated and actual winds were modest, and median vector deviations of estimated from observed wind speeds were only 9 and 17 knots at the lower and upper levels respectively. Uncertainties as to the heights of the clouds in the satellite photographs seemed to be the principal sources of error. So, despite the inevitable problems, wind data from A.T.S. pictures have been found to be useful operationally as reasonable estimates of actual conditions in both the lower and upper tropospheres at around 900 mb and 300 mb in particular.

We may conclude by noting that a whole range of atmospheric motion parameters can be evaluated from short-interval sequences of geosynchronous satellite photographs for a selected region (Fujita *et al.* 1969). These include not only air flow parameters ((*a*)–(*d*)) but also wind circulation characteristics ((*e*)–(*g*)). All depend on the identification of particular elements on successive photographs, since these are the tracers by which air motion is assessed. Where necessary, allowances have to be made for changes in cloud shape, brightness and development.

(*a*) Cloud motion vectors are constructed to represent the observed changes in the geographical locations of cloud elements identified on successive photographs. The directions and amounts of cloud displacements are depicted in plan by arrows of appropriate orientations and lengths.

(*b*) From (*a*) streamline maps can be constructed to show the direction of air flow.

(*c*) Also from (*a*) isotach maps can be compiled to show the rates of air flow corrected to m.p.h., knots, km/h or m/s according to taste.

(*d*) From (*b*) and (*c*) divergence patterns can be established if changes in the wind speeds and directions are known. Here, this equation is appropriate:

$$\zeta = \frac{\partial v}{\partial x} - \frac{\partial u}{\partial y} \qquad (5.8)$$

This expresses horizontal divergence in terms of local changes of velocity components (u and v) in east–west and north–south directions (x and y) respectively. Within the tropics it might be evaluated for each intersection in a 2° 30′ grid of latitude and longitude for synoptic scale purposes.

(*e*) Standard meteorological equations then permit the computation of relative vorticity patterns. So the satellite data input can be processed to yield spatial distributions of ζ, the local spin of the air.

(*f*) Absolute vorticity patterns may be elucidated next by adding the Coriolis parameter (f), which varies with latitude alone, to establish values of ζ_a.

(*g*) The centres of organized circulations at the level in question become obvious from this set of photographic byproducts. The natures of those circulations are easy to identify.

Thus comprehensive, detailed, and usually quite accurate assessments can be made of the meteorological condition of the lower troposphere in particular with respect to its motion. Five maps from an early set illustrating the above sequence are included here as figs. 5.11–5.15 (see also plate 14). The

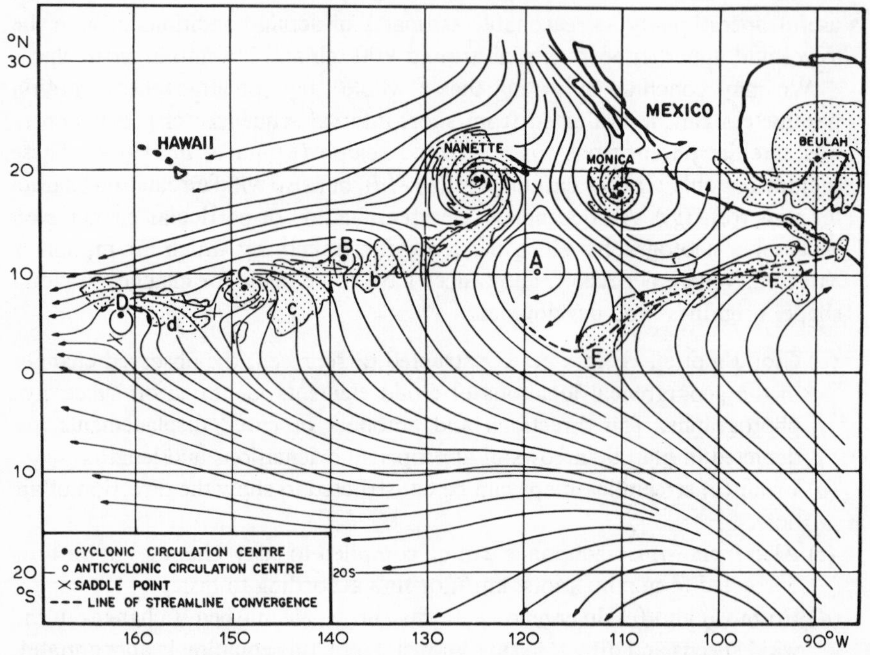

Fig. 5.11 Streamlines of low-cloud velocities drawn from velocity vectors in plate 14, an A.T.S.-I picture, with velocities obtained from six such pictures at about 25-minute intervals on 17 September 1967. (From Fujita *et al.* 1969)

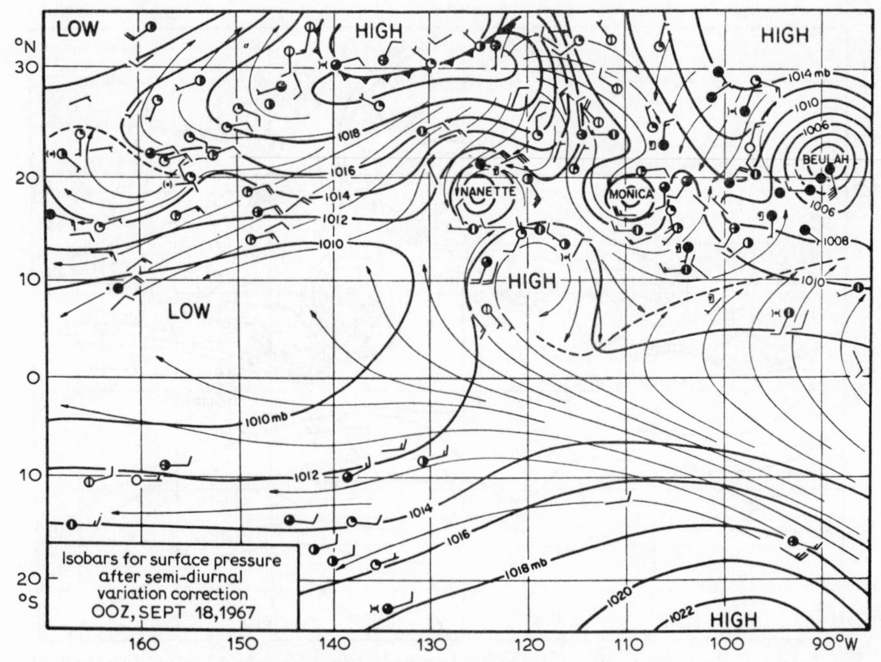

Fig. 5.12 Surface chart for 0000 G.M.T., 18 September 1967, about 3 hours after the time of the velocity computation in fig. 5.11. An anticyclone, indicated by streamlines in fig. 5.11, appears in this surface map. (From Fujita *et al.* 1969)

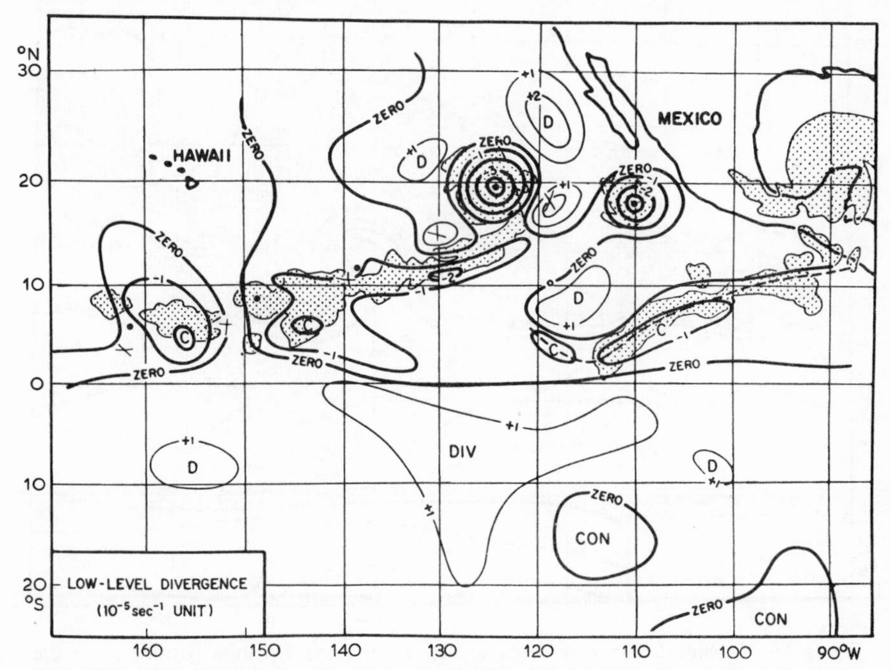

Fig. 5.13 The divergence of low-cloud velocities in plate 14. As expected, the inter-tropical nephsystems are characterized by fields of convergence. (From Fujita *et al.* 1969)

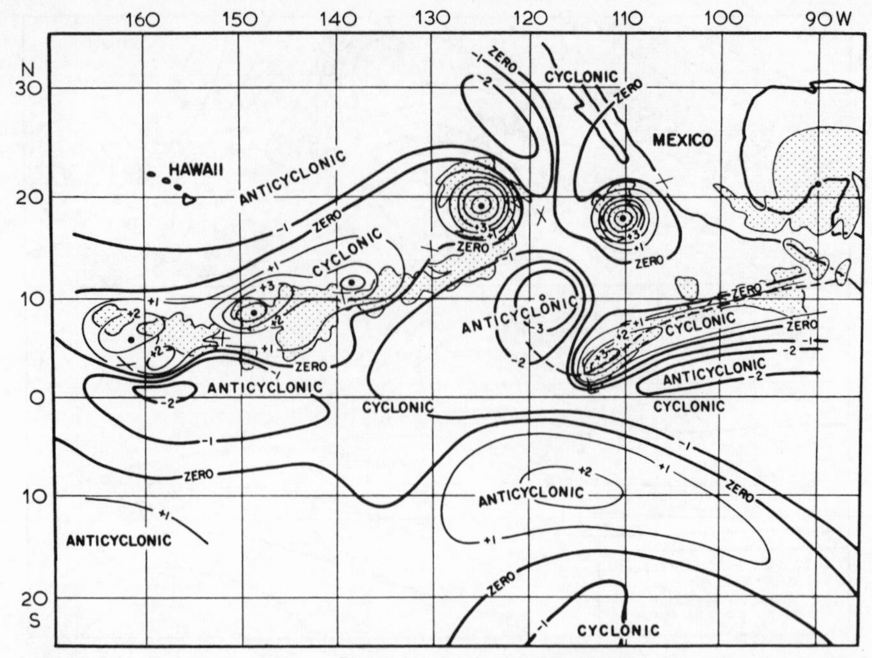

Fig. 5.14 Relative vorticity of low-cloud velocities in plate 14. Note that the areas of nephsystems correspond to those of large relative vorticity. (From Fujita *et al.* 1969)

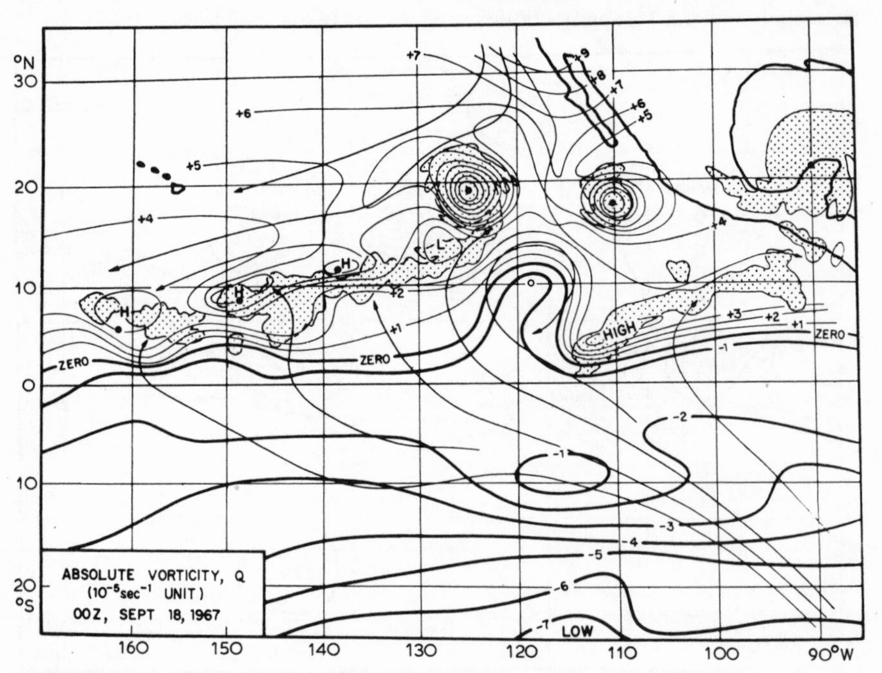

Fig. 5.15 Absolute vorticity, obtained by adding the Coriolis parameter to the relative vorticity of fig. 5.14. (From Fujita *et al.* 1969)

analysis of such maps has already revealed new circulation features in the Tropics, of significance in climatology (see p. 257). When sufficient runs of such maps have been compiled, new climatic maps of considerable intrinsic interest may result.

While the work carried out by Fujita and his co-workers depended upon a manual tracing of clouds in order to establish the initial products (*a*) and (*b*) above, present efforts are being directed towards devising suitable means for automating such procedures so that the whole methodology may be executed ultimately by computer means. Some technical aids have been developed to assist the manual interpreter in the analysing of A.T.S. photographs for cloud motion, for example loop projectors (Fujita *et al.* 1968) and television display complexes (Evans and Serebreny 1969; see plate 15). The former permits rapid replaying of appropriate geosynchronous film loops, by which means the general mobility of a cloud field can be established. The latter, more sophisticated, set of hardware resembles the loop projector in its requirement for a human operator, but it is more advanced in that it gives him a controlled access to picture sequences, plus a 'zoom capability' for the examination of features at different scales. Its television display consoles facilitate an accurate geographical registration of pictures by matching landmarks. Finally, it enjoys a facility for direct output of position data to a computer. If, however, computer methods could be devised not just to aid, but to simulate and eventually replace entirely the human operator in making cloud motion measurements, then near real-time usage of enormous volumes of very rich geosynchronous satellite data would be facilitated even further.

One such computer method has been tested recently by Endlich *et al.* (1971). First, their approach involves the representation of a cloud pattern by a relatively small number of objectively determined centres of brightness. Next, the brightness centres computed within one geographical region for two successive times are paired appropriately, and vector displacements through the time interval separating the two pictures are calculated. Once the direction and speed of cloud motion has been established, a variety of other computer products are possible from the list outlined above.

A second computer method full of promise for the future is based upon the establishment of cross-correlation coefficients (see Panofsky and Brier 1958) for pairs of A.T.S. photographs taken twenty-four minutes apart, and mapped on Mercator projections (Leese and Novak 1971). The speed necessary for a real-time application of this procedure is accomplished by using the 'fast Fourier transform' (Cooley and Tukey 1965). This is an algorithm permitting cross-correlation coefficients to be computed much more rapidly by three Fourier transforms than by the direct cross-correlation method. A.T.S. pictures are available in digital forms as arrays of data digitized to sixty-four grey-scale values related to the observed brightness of the original camera target. Cloud displacements can, therefore, be computed from comparisons of

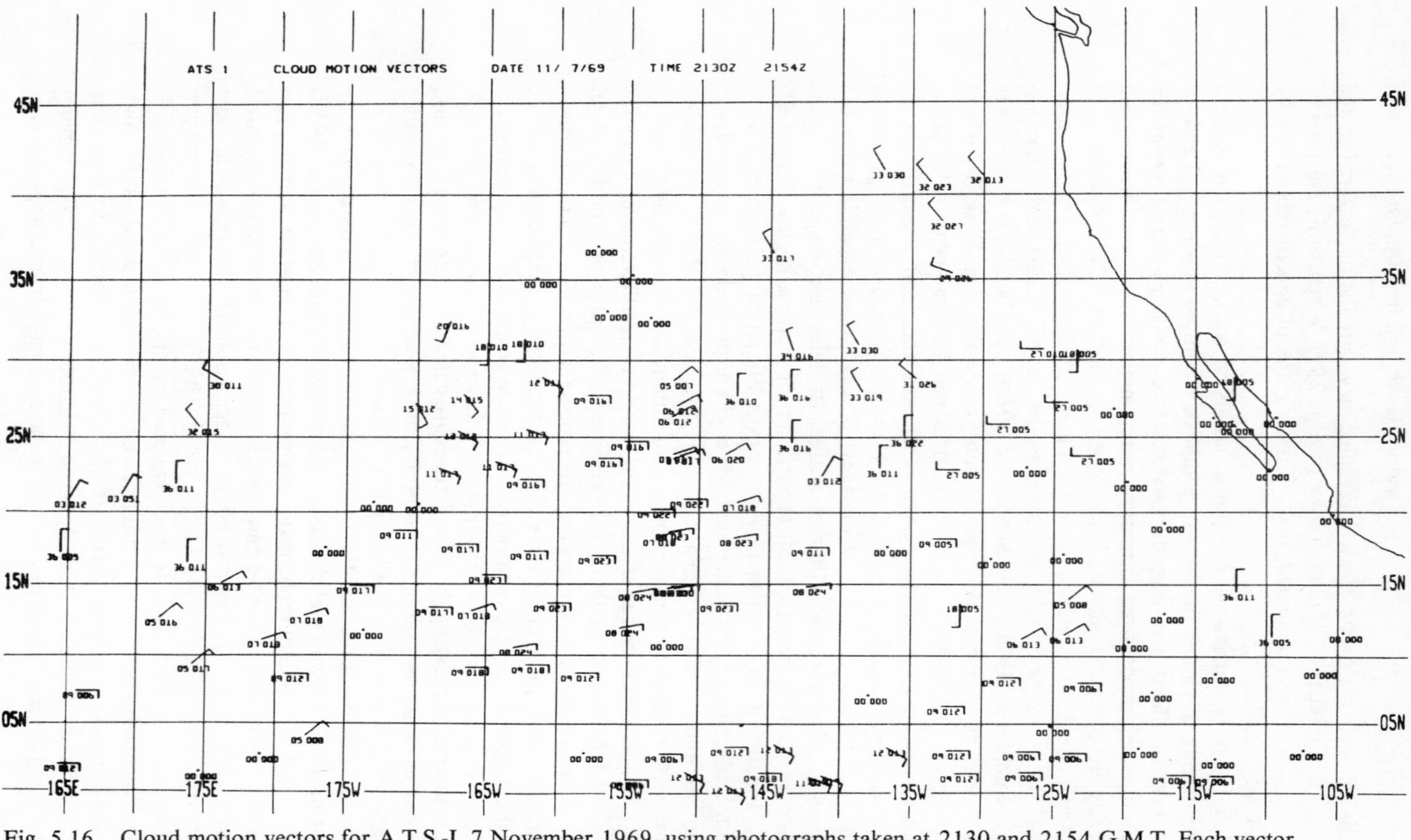

Fig. 5.16 Cloud motion vectors for A.T.S.-I, 7 November 1969, using photographs taken at 2130 and 2154 G.M.T. Each vector is represented by a five-digit number. The first two digits give the direction from which the cloud moved in tens of degrees, and the last three the speed in knots. (From Leese & Novak 1971)

brightness levels on sequential pairs of digitized satellite pictures. In a photograph like plate 16, some 1800 matrices must be investigated to yield data commensurate with the grid spacing of 2° 30′ latitude currently used in numerical weather prediction. Clearly that total alone substantiates the need for a rapid automated technique for real-time operational use.

In fig. 5.16, a map of cloud motion vectors, each vector is represented by a five-digit number. The first two digits give the direction from which the cloud moved in tens of degrees, and the last three the speed in knots. The barbs at each location show the direction only, in the manner customary to meteorologists. The cloud motion vectors depict a consistent wind field except where no motion was detected (i.e. five zero digits on the map). These usually occurred where the cross-correlation technique integrated the motion of two superimposed layers to yield a resultant vector near zero. Although further automatic processing could screen some such complications, early test results indicate that a combination of manual and automated techniques provides the best immediate solution to the separation of upper- and lower-level flows, and, therefore, to the whole problem of obtaining cloud motion vectors from geosynchronous satellite data.

Without doubt, the future is bright with possibility so far as A.T.S. photo interpretations of wind fields for both meteorological and climatological purposes are concerned. Furthermore, it seems inevitable that interest in clouds can only grow now that satellite data are becoming well accepted as routine inputs to atmospheric data pools. Not only are clouds important indicators of atmospheric moisture, but also of atmospheric circulation, a very fundamental feature of the atmosphere of the planet earth.

PART FOUR

Satellite data analyses in global climatology

6 Global patterns of atmospheric energy

Introduction: theoretical considerations

The sun may be described as an enormous atomic furnace constantly emitting radiation at the rate of about $5{\cdot}2 \times 10^{24}$ kilocalories, or $6{\cdot}15$ kilowatts, per cm^2 per minute of its surface area. Only a minute fraction of this radiation reaches the earth, 150 million km away, but solar energy comprises none the less almost the entire energy supply available for raising earth surface and atmospheric temperatures to their climatic means, and for driving circulations within an atmosphere whose mean flow is embroidered with rich spatial and temporal variation. Bearing in mind that energy considerations must, therefore, be vital to many atmospheric studies, it is fortunate indeed that meteorological satellites have done so much to bring the greater realism of observation into fields dominated previously by estimation.

Put in the simplest terms, the earth/atmosphere energy budget can be expressed as follows:

$$\underset{(1)}{\underset{\text{(or sinks)}}{\text{Energy sources}}} - \underset{(2)}{\underset{\text{(or release)}}{\text{Energy storage}}} - \underset{(3)}{\underset{\text{(or import)}}{\text{Energy export}}} = \underset{(4)}{0} \qquad (6.1)$$

Clearly since, on average, the terms balance perfectly, any short-term tendencies to imbalance prompted by variations in term (1) must be compensated by related changes in terms (2) and/or (3). Now it is important to note that energy transfers by radiation, rather than by other processes of heat transfer – e.g. convection, advection and conduction – comprise a subordinate semi-closed vertical interchange of radiant energy involving:

1 Insolation gains from the sun.

2 Radiation losses from the earth and its atmosphere.

Thus a net radiation budget may be investigated for its own sake, representing the prime forcing function that drives the atmospheric and oceanic circulations. An important principle upon which equation 6.1 in part depends is the principle of energy conservation. This requires that the total energy budget at every location should balance over all time periods. The radiation budget, which is a component of the energy budget, entails no such requirement. Indeed, we can go further than this and say that radiative imbalance is the rule rather than the exception in radiation budgets. This fact has far-reaching meteorological and climatological consequences.

Let us examine in more detail these two budgets.

Energy budget considerations

The total energy budget of an earth/atmosphere column can be expressed as the sum of two separate budgets, one for the interface between the earth and its atmosphere, and one for the atmosphere alone. Different authorities have employed several different notations to represent these energy systems. The present chapter will use generally that employed by Sellers (1965) and Vonder Haar (1968), excepting in carefully specified instances. Thus, energy balance equations can be constructed as follows:

(*a*) *For the earth's surface.* Here, four quantities must be involved if RN_E is to be evaluated, i.e. the net radiation balance at the surface of the earth. These are:

(i) Net energy gained or lost by the surface through changes in the state of water. This is expressed in terms of the product of E (the amount of water that leaves the surface and is added to the atmosphere) and L (the latent heat of phase transformation of water, i.e. 600 cal/g).

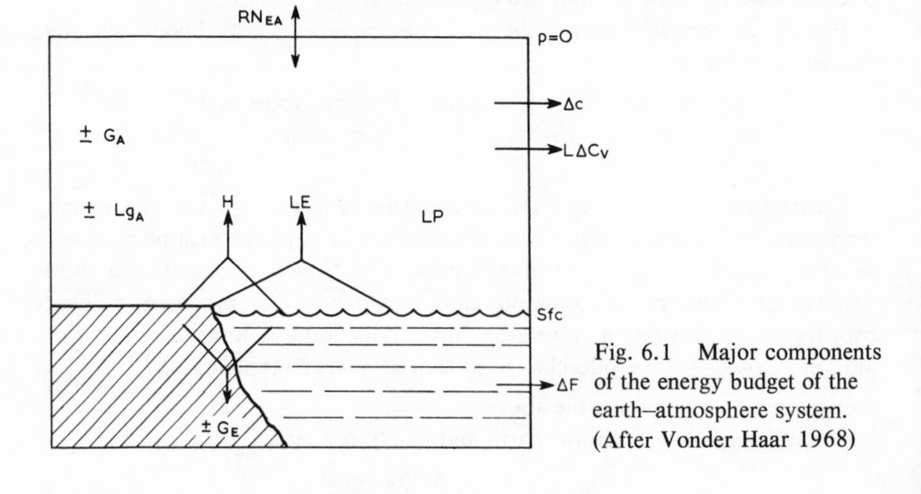

Fig. 6.1 Major components of the energy budget of the earth–atmosphere system. (After Vonder Haar 1968)

(ii) H, the net energy gained or lost by exchange processes at the atmosphere/surface interface.
(iii) G_E, the subsurface storage or release of energy.
(iv) ΔF, the energy transported horizontally beneath the earth's surface, e.g. by conduction through soils, or by ocean current transport. This may be a positive value for water bodies, but is negligible or zero for land surfaces due to the slowness of the processes involved.

So, within the energy budget for the earth's surface, a net radiation balance (RN_E) may be either positively or negatively compensated by various energy circulations. The surface energy balance equation may be written:

$$RN_E - LE - H = G_E + \Delta F \tag{6.2}$$

(*b*) *For the atmosphere.* This involves four similar types of components:

(i) A counterpart of *a*(*i*), namely LP, the net heat added to or removed from the atmosphere by phase transformation of water.
(ii) H, as above.
(iii) G_A, energy stored in, or released from, the atmosphere.
(iv) ΔC, representing horizontal energy transport within the atmosphere, predominantly by moving currents of air.

Thus RN_A, the net radiation balance of the atmosphere, is supplemented by other terms as follows in the energy balance equation:

$$RN_A + LP + H = G_A + \Delta C \tag{6.3}$$

(*c*) *For the combined earth/atmosphere system.* Equations 6.2 and 6.3 combine to represent the energy balance within the earth/atmosphere system. RN_{EA} is the net radiation balance within that system (fig. 6.1).

$$RN_{EA} + L(P - E) - G_A - G_E = \Delta C + \Delta F \tag{6.4}$$

So far, however, we have considered the part played by atmospheric moisture only in terms of transformations of energy related to phase changes of water. The transport of latent energy by atmospheric water is very significant also, and can be accommodated within a modified form of equation 6.4. Climatologically it is significant in many ways that water evaporated in one geographic region may be transported by moist air streams into another area before some or all of that moisture is redeposited as precipitation. The global distribution of precipitation/evaporation ($P:E$) ratios is extremely significant from many socioeconomic points of view; new climatic classifications are taking these into account increasingly, as indicated in chapter 14.

In equation 6.4, $L(P - E)$ may be replaced by two further terms, Lg_A, representing storage or release of water vapour in the atmosphere, and $L\Delta C_V$, the transport horizontally of water vapour, since

$$L(P-E)=L\Delta C_V-Lg_A \tag{6.5}$$

This equation states that any difference between water vapour evaporated into and precipitated out of an atmospheric column must be balanced over a period by an appropriate storage or release and/or removal horizontally of water vapour.

By substituting for $L(P-E)$ in equation 6.4, an alternative expression for the energy budget is obtained, being one in which latent heat transport and storage are included:

$$RN_{EA}-G_A-G_E-Lg_A=\Delta C+\Delta F+L\Delta C_V \tag{6.6}$$

This statement says that the total divergence of energy from a vertical system must be balanced by its net radiation balance and the release of sensible and latent heat within it. This is an expanded, slightly rearranged form of equation 6.1.

Radiation budget principles

Let us take out the net radiation balance term (RN_{EA}) from equation 6.6 for closer study (fig. 6.2). The net energy exchange across the top of the atmosphere can be expressed as follows:

$$RN_{EA}=I_0-I_0A-H_L \tag{6.7}$$

This states that the radiation budget of any atmospheric column depends on the direct incoming solar radiation (I_0), less reflected and scattered solar

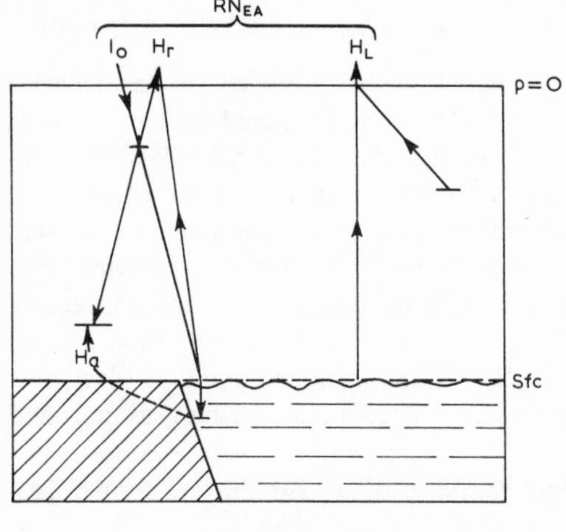

Fig. 6.2 Major components of the radiation budget of the earth–atmosphere system. (After Vonder Haar 1968)

radiation (I_0A, where A is the planetary albedo or reflectivity percentage) and outgoing long-wave radiation (H_I) lost to space from the earth's surface and its atmosphere. For many years scientists have tried to estimate the magnitude and variation of this energy exchange, represented diagramatically by fig. 6.2, on account of the meteorological and climatological significances of its variations with latitude from equatorial to polar regions. Lettau (1954) has reviewed most of the earlier work within this field. Those earlier studies were faced with two major problems (Vonder Haar and Suomi 1969): (*a*) the difficulty of computing the transfer of radiation in a cloudy atmosphere and (*b*) the deficiences in global observations of temperature, moisture and cloudiness.

More recently, complete sets of satellite measurements have been used in studies of the earth's radiation budget, for example by Vonder Haar (1968), who evaluated more than thirty months of observations for periods during 1962–5 from low resolution sensors flown on Tiros-type satellites. Winston (1967, 1969) described similar results derived from medium resolution data, and pointed to the usefulness of such studies as applied to more local areas. Nimbus medium resolution infra-red results were pioneered with evidence from Nimbus II, processed and interpreted by Raschke and Pasternak (1967). Most of these and other similar studies demonstrate the ability of satellite observations to yield inferences regarding atmospheric conditions over areas of the world where conventional meteorological stations are sparsely spread. Prominent amongst such inferences are those of atmospheric moisture, as explained in chapter 5. So new light is being thrown upon key questions relating to global distributions of weather and climate by the broadening of the empirical bases on which their answers may be formulated.

One cautionary note should be sounded before discussing radiation and energy budget investigations in more detail geographically: although satellite measurements represent a big advance on theoretical estimates of energy components, they are still far from ideal data. Godson (1958) observed that measurements of the earth/atmosphere radiation budget should be made optimally by continuously recording, accurate solar and thermal radiation sensors placed at all locations over the top of the atmosphere. In practice, the satellite data involved in the studies outlined later in this chapter were obtained from various sensors in several satellites whose orbits and viewing geometries were far from standardized.

Radiation budget climatology: mean global and hemispheric radiation patterns

Budget diagrams (like fig. 2.8, p. 35) displayed in basic textbooks of climatology can be misleading unless it is clearly understood that they represent long-term mean situations around the globe as a whole. Actual energy and radiation budgets differ and vary almost infinitely from place to place and from time to time. First, in a series of discussions dealing with radiation budgets for a range of geographical scales, let us examine global and hemispheric scale differences and variations. The first published results were based on Nimbus data, the Nimbus satellites being the earliest successful satellites to achieve a fully global data coverage through medium resolution infra-red channels (see e.g. Raschke and Bandeen 1970). Those results may be used to exemplify the satellite approach. More recent studies, notably by Vonder Haar and Suomi (1971), have been able to analyse considerably longer runs of data. Their general findings are summarized in the conclusion to this chapter.

One of the most interesting studies dealing with the radiation balance of the planet earth from radiation measurements of Nimbus II covered the period from 15 May to 28 July 1966. (See figs. 3.1–3.3.) Using an alternative notation to that employed above (but see the equations in chapter 3) Raschke and Bandeen expressed the radiation balance (Q) at each location specified by λ (longitude) and ϕ (latitude) as:

$$Q(\lambda, \phi) = S(\lambda, \phi) - R(\lambda, \phi) - E(\lambda, \phi) \qquad (6.8)$$

where S, R and E are, as before, the incident solar radiation, reflected radiation and long-wave (emitted thermal radiation lost to space) respectively. Q, S, R and E were considered to be time averages over periods of twenty-four hours or more.

Probably the most critical simplifying assumption that had to be made concerned the choice of a value for the solar constant (see p. 153). Estimates of incident solar radiation depend largely upon this term. The value adopted was 2·00 cal/cm^2/min, which yielded global averages of solar radiation across the top of the atmosphere ranging from – 0·488 to + 0·484 cal/cm^2/min during the northern hemisphere summer fortnight from 1 to 15 July 1966. The critical significance of the value chosen for the solar constant is underlined by table 6.1, which lists solar constants employed by various authorities during the last twenty years, and the variations in the partially dependent computations of other components of the radiation balance. Generally speaking, the higher the assumed solar constant, the greater the values of computed incoming radiation (S) and the net radiation balance (Q), but the lower the values for the planetary albedo (A), expressed in percentage terms. Clearly, accurate measurements of the solar constant are essential if radiation data are to be put to their best advantage in studies of the earth/atmosphere radiation budget.

From their own series of computations, Raschke and Bandeen (1970) arrived at new assessments of the various terms within the radiation balance equation. Table 6.2 summarizes their findings. It is interesting to compare

Table 6.1
The radiation balance of the planet earth, and its dependence on the selected level of the solar constant (after Raschke and Bandeen 1970)

Authorities	*Chosen solar constants*	*S*	*A*	*Q*
Stair and Johnston (1956)	2·05	0·496	29·4	+0·005
Johnson (1954)	2·00 ± 0·04	0·484	30·1	−0·007
Drummond *et al.* (1968)	1·95 ± 0·02	0·472	30·9	−0·019
Thekaekara *et al.* (1968)	1·936 ± 0·041	0·469	31·1	−0·023
Kondratiev and Nikolsky (1968)	1·918	0·464	31·4	−0·027

The solar constant *S*, and *Q*, are expressed in cal/cm²/min. *A* is a % value.

Table 6.2
The radiation balance of the planet earth from Nimbus II measurements. S represents the incident solar radiation, R reflected solar radiation, R/S the albedo, E long-wave radiation, S–R absorbed solar radiation and Q the radiation balance.

(1)	(2)	(3)	(4)	(5)	(6)	(7)	(8)	(9)
Period of Nimbus II measurements (1966)	*Area*	*S* (*cal/cm²/min*)	*R*	*R/S* ÷100 (A) (%)	*E* (*Equiv. B.B. temp.*) (°*K*)	*E*	*S–R* (*cal/cm²/min*)	*Q*
May 16–31	N.H.	0·660	0·215	33	255	0·345	0·445	*+0·100*
	S.H.	0·315	0·080	25	253	0·335	0·240	−0·095
	Globe	0·490	0·145	30	254	0·340	0·340	*+0·000*
June 1–15	N.H.	0·675	0·220	33	256	0·350	0·455	*+0·105*
	S.H.	0·295	0·075	26	253	0·335	0·220	−0·115
	Globe	0·485	0·150	31	255	0·340	0·335	−0·005
June 16–30	N.H.	0·675	0·215	32	257	0·355	0·460	*+0·105*
	S.H.	0·290	0·075	25	254	0·335	0·215	−0·120
	Globe	0·485	0·145	30	255	0·345	0·340	−0·005
July 1–15	N.H.	0·670	0·210	31	257	0·355	0·460	*+0·105*
	S.H.	0·300	0·070	24	254	0·335	0·225	−0·110
	Globe	0·485	0·140	29	255	0·345	0·345	−0·005
July 16–28	N.H.	0·655	0·205	31	257	0·355	0·450	*+0·095*
	S.H.	0·315	0·085	27	253	0·335	0·235	−0·100
	Globe	0·485	0·145	30	255	0·345	0·340	−0·005

(Values to the nearest 0·005 cal/cm²/min, 1 °K and 1%: these are levels of accuracy believed to be commensurate with the instrumentation and the computational methods used.)

them with those listed in table 6.1. A number of important points emerge from table 6.2, which may be summarized as follows:

Column 1 specifies the five successive periods of uniform length for which computations were possible.

Column 2 indicates that, for every period, separate calculations were made for either hemisphere. Thus comparisons are possible between a summer hemisphere (the northern), and a winter hemisphere (the southern).

Column 3, listing computed means of incident solar radiation, indicates the strengths of the contrasts between the summer and winter hemispheres. The southern hemisphere received rather less than half the radiation received by the northern hemisphere in every one of the five half-monthly periods.

Columns 4 and 5 list values of reflected radiation in cal/cm^2/min (column 4), and as percentages of incident radiation (column 5). The albedo values in column 5 show that the losses of radiation by reflection from the northern hemisphere were higher than those from the southern hemisphere on average by approximately one quarter. The relatively low albedoes of the southern hemisphere may be explained mainly in terms of the low winter losses by reflection across the southern oceans and Antarctica, where there were low intensities of insolation energy during the period of south polar winter night.

Column 6. The mean black body temperature of the planet earth appeared to range from 254 to 255 °K, with rather little difference separating values from the two hemispheres (2–4 °K). The global means of 254–255 °K are 3–4 °K higher than those proposed by London (1957) in pre-satellite days.

Column 7 lists the long-wave radiation measurements from which the temperatures in column 6 were derived.

Column 8 sets out values of absorbed radiation ($S - R$) in cal/cm^2/min. The hemispheric contrast appears to be roughly proportional to those inherent in column 3. This suggests that variations in received solar radiation have the biggest single effect upon this climatologically most important quantity ($S - R$). Absorbed radiation, as remarked earlier, has a big influence upon earth surface temperatures, while energy absorbed within the atmosphere drives circulations in both vertical and horizontal planes.

Column 9 represents the ultimate product of this Nimbus II radiation exercise, namely an assessment of the earth/atmosphere net radiation balance expressed in terms of separate hemispheres as well as the globe as a whole. In view of the various assumptions that were made en route to this quantitative destination, too much reliance cannot be placed on the absolute accuracies of the figures listed, nor even on the signs, i.e. whether the values are positive or negative. However, the relative levels indicated are certainly important. From these, the most significant fact to emerge is that the magnitude of the radiation gain in the northern hemisphere was usually less than the loss in the southern hemisphere. In general, therefore, it is concluded that the southern hemisphere acted as a heat sink during its winter season in 1966, and in doing so more than counterbalanced the radiation gain in the northern hemisphere heat source. Presumably if the same were true from year to year, a global net gain situation must prevail at the other high season in order to give an overall balance to the entire system in the course of a year as a whole. It is possible, however, that the calculated net balances are not completely accurate, and/or that variations of the earth's radiation field occur from year to year. Statistically it is improbable that June–July 1966 represents perfectly a normal net radiation picture.

Table 6.3
Estimates of planetary albedo in their historical sequence (after Vonder Haar 1968)

Date	*Albedo*	*Authority*
1917	50	Dines
1919	43	Aldrich
1928	43	Simpson
1934	41·5	Baur and Phillips
1948	34·7	Fritz
1950	35	Möller
1954	34	Houghton
1954	34	Lettau
1957	35	London
1962	33–38	Angström
1965	35	House
1965	32	Bandeen *et al.*
1967	30	Raschke and Pasternak
1968	29 ± 1	Vonder Haar

One other topic involved in the above analysis is worthy of further comment at this juncture, namely the observed global albedo. The values of 29–31% listed in table 6.2 are considerably lower than earlier estimates based on surface climatological data, aerological records or observations of earthshine reflected from the moon. Table 6.3, setting out planetary albedo estimates by various authorities at different dates across the last half century, indicates that such estimates have progressively decreased with time. The earliest estimates (*c.* 45–50%) gave way at first to lower values (*c.* 35%) primarily because of new information on the albedoes of clouds. The even lower values that have been calculated in satellite-based inquiries apparently result from better assessments of amounts of cloud cover. These were somewhat overestimated prior to the operation of satellite observing systems.

Fig. 6.3 depicts the observed mean seasonal variations of radiation budget components set out earlier in table 6.4 (Vonder Haar 1968). These

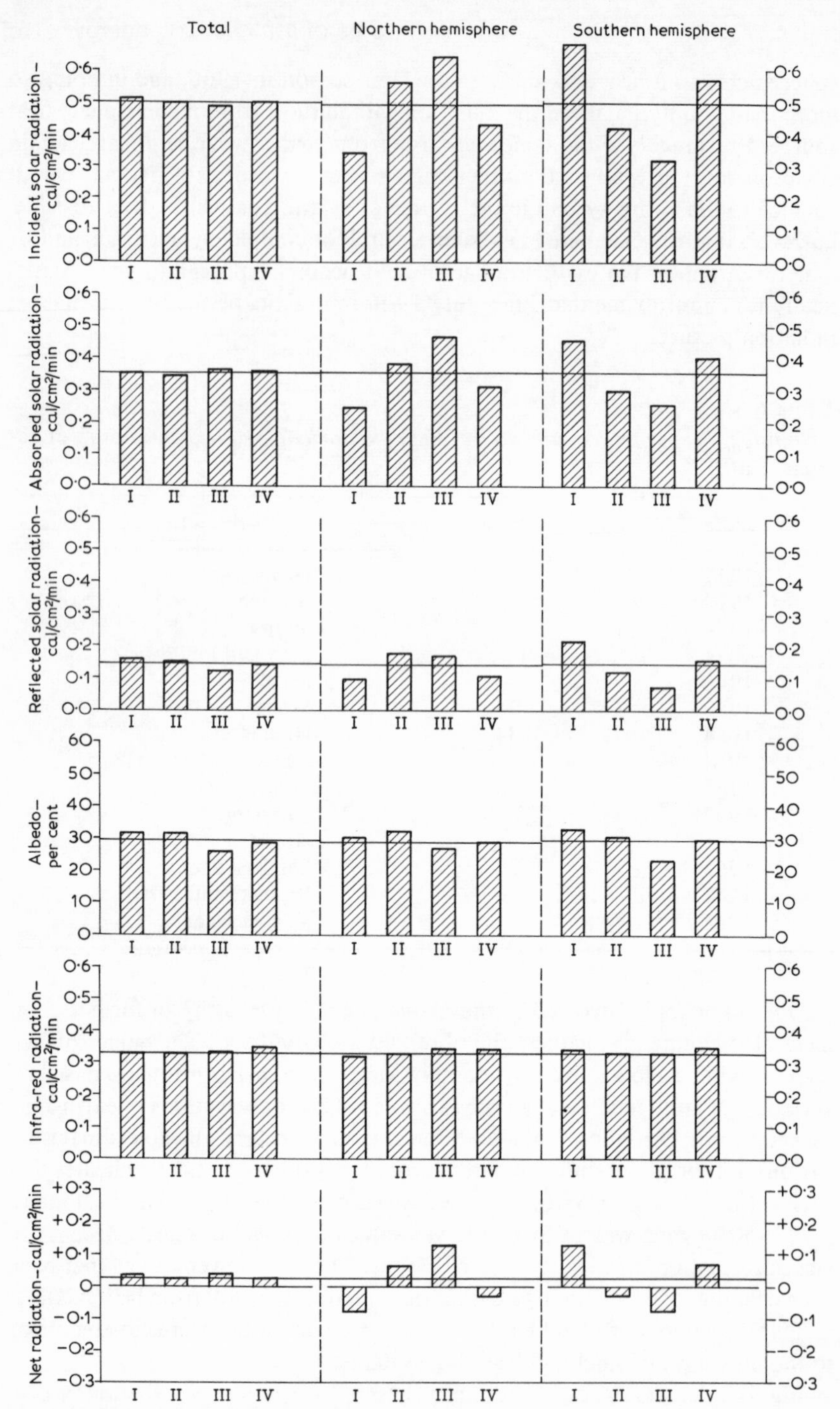

Fig. 6.3 The mean seasonal variation of major components of the radiation budgets of the entire earth, and either hemisphere, established from satellite evidence obtained in 1960–5. (From Vonder Haar 1968)

graphs, and table 6.4, were compiled from all the available data obtained from first generation weather satellites from 1960 to 1965. They illustrate the chief interrelations between the various components of the budget, and indicate two types of effect which, though operating at different scales, are certain to be elucidated further by satellite means:

(1) An overriding solar effect which operates at both global and hemispheric scales, stimulating in-phase variations in all the short-wave radiation budget components, i.e. in incident solar radiation (I_0), reflected solar radiation (H_R) and albedo (A). It should be noted that in this case

$$H_R = I_0 A \qquad (6.9)$$

In this notation (H_R) is equivalent to (R) in table 6.2, so that, through both, zonal averages of reflected short-wave radiation are expressed in cal/cm²/min. In sharp contradistinction, table 6.4 shows that H_L (or E), the outgoing long-wave radiation, varies differently. This is related additionally to the annual pattern of absorbed solar radiation, H_a, which tends to be more efficient in the second half of the year than in the first.

(2) Smaller, second-order effects result from mean seasonal changes in the physical state of the earth/atmosphere system, that is to say from changes in the types and/or amounts of cloudiness and the albedo of the earth's surface. For example, albedo values in fig. 6.3 are higher in both hemispheres in December–February than in June–August. Repercussions must be felt in other radiation budget components, but it is difficult to discern them in such a generalized display.

Table 6.4
Mean annual and seasonal radiation budget of the earth/atmosphere system observed from the first generation meteorological satellites (see fig. 6.3; from Vonder Haar 1968)

	Global Average					*Northern Hemisphere*					*Southern Hemisphere*				
	DJF	*MAM*	*JJA*	*SON*	*Annual*	*DJF*	*MAM*	*JJA*	*SON*	*Annual*	*DJF*	*MAM*	*JJA*	*SON*	*Annual*
I_0	0·51	0·50	0·49	0·50	0·50	0·34	0·56	0·65	0·42	0·50	0·69	0·43	0·32	0·58	0·50
H_a	0·34	0·35	0·37	0·36	0·35	0·24	0·39	0·48	0·31	0·36	0·46	0·30	0·25	0·41	0·35
H_r	0·16	0·15	0·12	0·14	0·15	0·10	0·18	0·17	0·12	0·14	0·22	0·13	0·07	0·17	0·15
A	0·31	0·31	0·25	0·28	0·29	0·29	0·31	0·26	0·27	0·28	0·32	0·30	0·22	0·29	0·29
H_L	0·32	0·33	0·33	0·34	0·33	0·32	0·33	0·34	0·34	0·33	0·33	0·32	0·32	0·34	0·33
*RN_{EA}	0·03	0·02	0·03	0·02	0·02	−0·07	0·06	0·13	−0·03	0·02	0·13	−0·02	−0·07	0·06	0·02

Where I_o = incident solar radiation (cal/cm²/min)
H_a = absorbed solar radiation (cal/cm²/min)
H_r = reflected solar radiation (cal/cm²/min)
A = planetary albedo (per cent)
H_L = emitted infra-red radiation (cal/cm²/min)
RN_{EA} = net radiation budget of the earth/atmosphere system (cal/cm²/min)
* Probable absolute error of ±0·01 cal/cm²/min.

The most immediate need for the future is for similar studies to be carried out for longer periods so that the interesting implications of these pioneer, globally complete, radiation surveys may be confirmed or corrected. Of one thing there can be no doubt, however: satellite observations of the earth/atmosphere radiation budget are removing the necessity for much of the scientific guesswork necessitated previously by this kind of climatological inquiry.

Radiation budgets of latitudinal zones

One of the commonest ways of depicting the earth's radiation budget has been traditionally the plotting of budget components as functions of latitude (see figs. 6.4 (a)–(f)). Latitudinally averaged budgets are useful for indicating generally where changing surface and atmospheric conditions modify the large-scale radiation pattern over and above modifications related to seasonal changes in insolation. Fig. 6.4 (a) depicts insolation changes season by season. This set of curves is the only one in fig. 6.4 to have been established by theoretical means. It is clear that the annual receipt of heat energy along the Equator is some 2·5 times higher than that at the poles. At the same time, fig. 6.4 (d) shows that outgoing long-wave radiation varies much less from the Equator to the poles. Without horizontal net energy flows from the radiation sources in low latitudes to the sinks in higher latitudes, the tropical belt would become progressively hotter and the polar regions steadily colder. Although most energy storage takes place in the world's oceans, it has been estimated that some four fifths of the total energy transferred polewards is carried by the atmosphere.

Fig. 6.4 (b) represents latitudinal variations of the reflecting capability of the earth/atmosphere system. The tight clustering of the seasonal curves demonstrates that this is a less variable parameter than most of those in the radiation budget. Although these curves, like the others in fig. 6.4 are tentative to some degree, their general shapes and forms are corroborated by others in published works by Winston (1969) and Bandeen *et al.* (1965). All these studies show that the earth's albedo is highest in middle to high latitudes in association with the strong cloudiness of the extratropical depression belts and high latitude ice and snow. The mean albedo over Antarctica (exceeding 60% in every season) seems substantially higher than that across the north polar region, probably in response to the different geographies of these ends of the earth. In the vicinity of the Equator albedoes rise to a minor, though definite, peak associated with stronger cloudiness in the equatorial trough. Meanwhile, the darkest, least reflective areas are astride the Tropics, those along the southern Tropic being a little lower. It is not immediately obvious, however, why it is that the albedoes of the Tropics and subtropics during June–August should be significantly lower than their counterparts in the rest of the year.

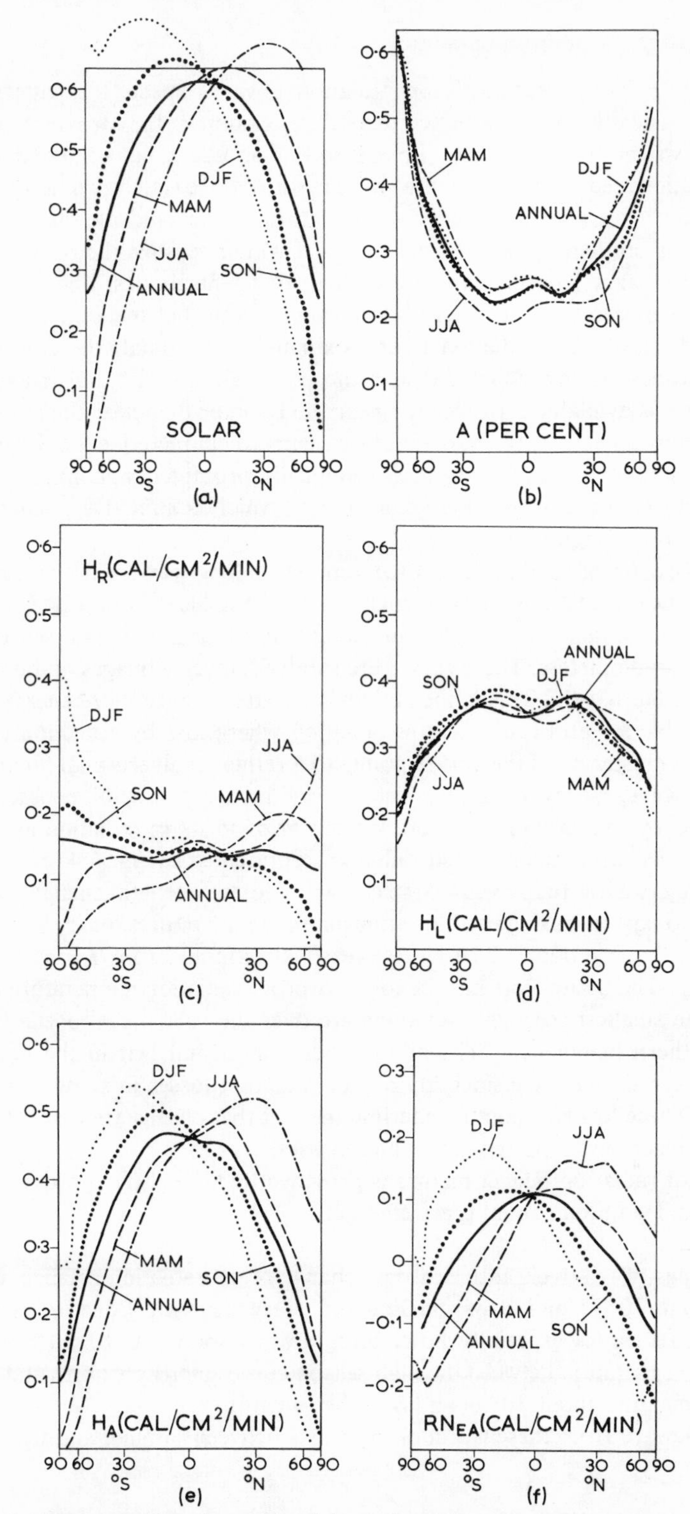

Fig. 6.4 Mean meridional profiles of the major components of the radiation budget of the earth–atmosphere system for each season, and the annual case (solid line), from satellite evidence. (After Vonder Haar 1968)

In fig. 6.4 (c), for short-wave radiation towards space, the patterns are forced basically by the sun (compare, for example, the two high season curves, which are near mirror images of one another), but some departures from anticipated ideal curves are evident. In each seasonal profile there are inflexions related to the cloudy, wet equatorial zones and the clearer, drier subtropics on their flanks. During the local winter seasons the polar regions lose little energy by reflection despite their high reflectivities, since very little insolation is received. It is also interesting to note that the annual curve of reflected short-wave radiation is approximately horizontal: the albedoes of low latitudes are low compared with those in high latitudes, but much more insolation is available for reflection near the Equator than near the poles. The largest annual short-wave radiation loss seems to characterize the South Pole (*c.* 0·16 cal/cm^2/min). Other, meteorologically prompted, maxima are found just north of the Equator, and about 35° N, which is an active polar frontal zone in some longitudes.

The next set of curves, fig. 6.4 (d), represents long-wave radiation loss, i.e. emitted heat radiation, drawn chiefly from evidence from L.R.I.R. disc measurements and the 7–35 μm 'total infra-red' waveband sensors in M.R.I.R. radiometers. These curves are subdued, inverse images of the albedo curves in fig. 6.4 (b). This is the anticipated pattern, since more energy must be available for absorption and re-radiation where loss by reflection is less. The tight clustering of the curves points to a rather small seasonal fluctuation in long-wave energy emission. The energy involved has, of course, been absorbed by the earth/atmosphere system prior to its re-radiation at longer wavelengths, while the reflected radiation represented by fig. 6.4 (e) consists of energy waves redirected to space with no wavelength change, having played no part in raising earth or atmospheric temperatures, or in the fuelling of atmospheric motions. The largest seasonal differences in H_L characterize the poles. The South Pole has the lower over-all radiation temperature of the two. The smallest seasonal variations are over the subtropical ocean belt in the southern hemisphere. The mid-latitude continental belt in the northern hemisphere displays considerably higher radiation losses in summer than in winter. These last two points underline the fact that oceans are more efficient storage reservoirs of energy than land masses.

A total radiation budget picture is provided by fig. 6.4 (f), through curves of RN_{EA}. Two broad zonal types emerge:

1 Zones of relatively little seasonal change of net radiation balance, within about 30° N and S of the Equator. Very near the Equator itself the budget varies especially little, being always about –0·10 cal/cm^2/min. The separation between the high season curves increases north and south to maxima about 30° N and S of the Equator.
2 Poleward from these latitudes, contrasts between summer and winter net

radiation balances are fairly uniformly large, of the order of 0·20 cal/cm^2/min.

It is interesting to note that the range in the Arctic is larger than that in the Antarctic. Wide areas of winter ice and snow in the north polar region thaw in summer and lead to an actual gain of energy through reduced albedoes. By contrast, most of the area within the Antarctic Circle remains frozen constantly throughout the year. Thus the dissimilar geographies of a nearly landlocked sea and a high, roughly symmetrical land mass are key factors in differentiating markedly the radiation budgets for north and south polar regions. Lastly, note that the net radiation deficits in the local winters are greater at about 60° N and S than across the poles themselves, an interesting fact discussed at greater length later in this chapter.

In concluding this section it should be stated that, when other energy budget effects are small, the net radiation profiles in fig. 6.4 (f) represent the dominant forcing functions driving circulations in the atmosphere and hydrosphere. Therefore the radiation sources and sinks depicted by fig. 6.4 (f) must be related physically to patterns of horizontal energy transport in the maintaining of the generally observed *status quo* of mean annual temperatures across the surface of the earth. Although the influence of incident solar energy is the most powerful overriding factor here, however, this does not preclude significant year-by-year departures from the depicted profiles. Studies summarized by Winston (1969) have indicated clearly that the same seasons may have rather different budgets in different years. The factors thought to cause such variations are primarily atmospheric, and can be examined through both conventional and satellite meteorological data. The most significant of such factors include the type, proportion and distribution of cloud cover through their fairly immediate influences upon zonal albedoes. However, earth surface 'feed-back' mechanisms also play their part, especially through variations in ocean currents and frozen surfaces. The detailed comparative analysis of net radiation curves for different years therefore comprises an intriguingly complicated piece of environmental detective work. Probable cause-and-effect relationships are singularly difficult to tease out with confidence. Fig. 6.5 illustrates and exemplifies some of the quite noticeable differences that may distinguish one year from another. From left to right, each time-latitude section depicts successive seasonal changes from mid 1963 to late 1965 along mean meridians. Two types of variations are apparent: (1) especially in fig. 6.5 (c), annual cyclic repetitions are quite obvious, but (2) less regular variations are evident also, especially in figs. 6.5 (a) and (b).

Interpreted together, these charts support inferences concerning the causes of those differences in annual radiation budgets that were obvious in fig. 6.4 (a). For example, consider the months of June, July and August. For 1963 and 1965 the net radiation profiles are very similar. Positive values of

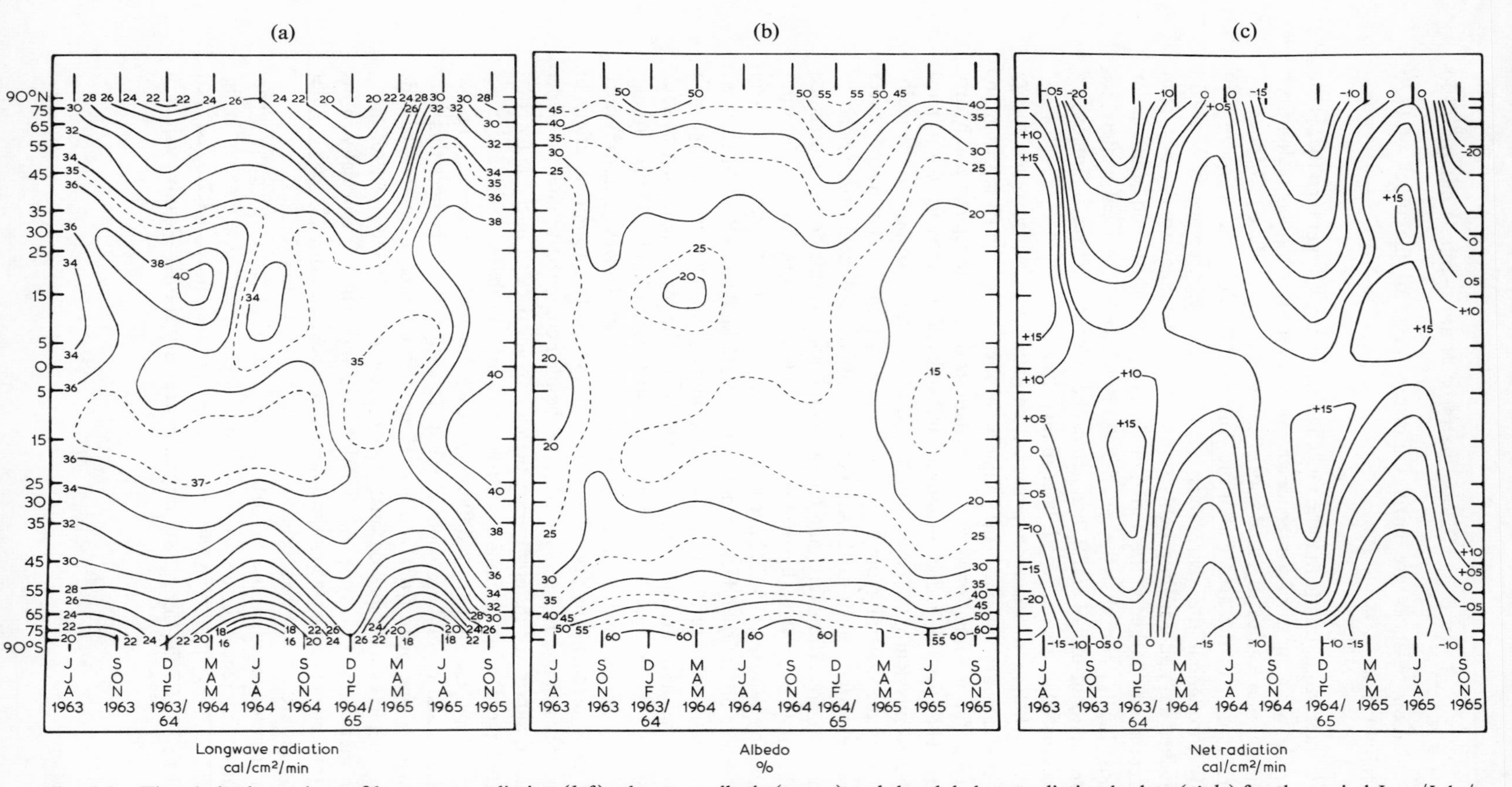

Fig. 6.5 Time latitude sections of long-wave radiation (*left*), planetary albedo (*centre*) and the global net radiation budget (*right*) for the period June/July/August 1963 to September/October/November 1965. (From Vonder Haar 1968)

net radiation extended from 15° S to the North Pole, and the maxima exceeded − 0·15 cal/cm²/min in both years. The albedo profiles for those periods include low values through the Tropics and subtropics on either side of the Equator, and the mean position of the cloudy zone at 15° N diminished long-wave re-radiation there. During June and August 1964, however, slightly higher albedoes about 15° N helped bring about a small reduction in long-wave radiation in middle latitudes in the northern hemisphere, and a marked reduction in the net radiation balance across the same latitudinal zone.

Thus it appears that circulations outside the Tropics might be expected to be influenced by variations in meridional energy gradients within the Tropics. Further satellite studies should attempt to locate the atmospheric causes of such anomalies, and seek to establish the chain reactions that they prompt. The immediate cause of the anomaly in June–August 1964 seems to have involved the equatorial cloud belt in the northern hemisphere summer, although global albedoes were generally higher than in 1963 or 1965.

Regional radiation patterns

Thus far we have discussed only global and hemispheric averages and the mean radiation budget within latitudinal zones. Such studies can, of course, be misleading because significant variations in the earth's radiation budget occur *within* latitudinal zones, especially the Tropics. Various workers have mapped mean radiation patterns from Tiros, Nimbus and Essa observations, including Rasool (1964), Winston and Taylor (1967) (Tiros M.R.I.R. data) and Macdonald (1971) (Essa observations). However, the most comprehensive study published by the time of writing has been that of Vonder Haar and Suomi (1971), who combined data from all those three satellite families in assessments of the earth's radiation budget through a period of five years, from 1962 to 1966 inclusive. Their results are summarized below. For more detailed pictures of both season-to-season contrasts, and relationships between radiation budget components and topography, reference should be made by the reader to the account by Winston (1969).

Planetary albedo (fig. 6.6)

Here an interesting contrast is evident between the mean reflectivities of middle and high latitudes in northern and southern hemispheres, polewards from 30° N and S. Albedo isopleths in the southern hemisphere follow a much simpler zonal pattern than those in the northern hemisphere, where the distribution of land and sea is more complex and seasonal variations in albedo much greater, especially with the Arctic Circle.

Within the Tropics, semipermanent features of the atmospheric circulation,

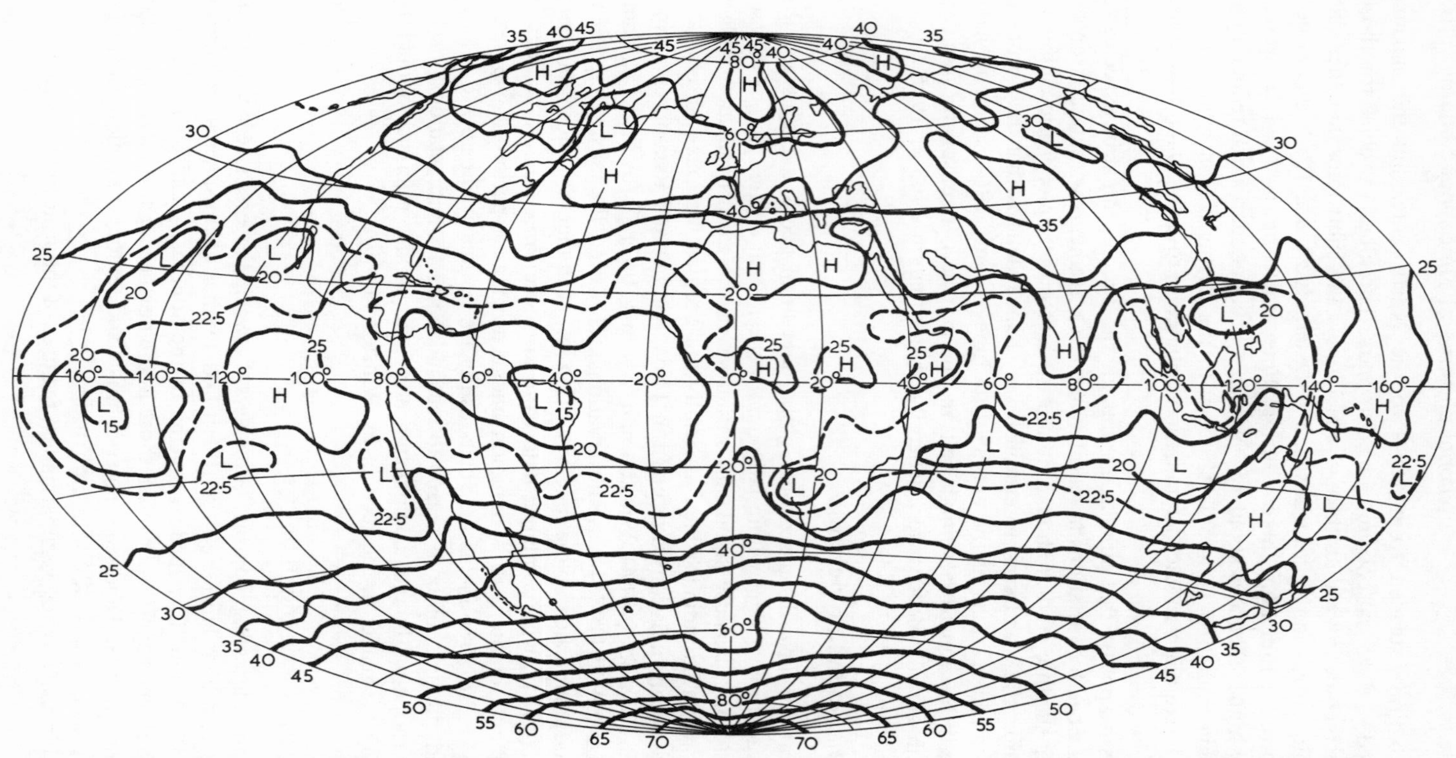

Fig. 6.6 The mean annual geographical pattern of planetary albedo (in per cent) measured from satellites during 1962–5. (After Vonder Haar & Suomi 1971)

terrain features, and/or the influence of special climatic conditions (e.g. the south Asian monsoon reversal) all cause significant departures from the annual mean along any chosen latitude. Thus the highest albedoes in the Tropics are associated with deserts, such as the Sahara, continental convective cloudiness, for example in central Africa, and ocean areas dominated meteorologically by convergence, as, for instance, in the case of the equatorial eastern Pacific.

Such a map necessarily smooths out most daily and even seasonal anomalies. These can be evidenced climatologically by maps of standard deviation of daily albedo averages (σ_A). Winston (1969) includes one such map in his radiation essay. The pattern derived from Nimbus II data in 1966 revealed clearly the geographical distribution of most frequent cyclonic activity and the preferred paths of other types of travelling disturbances through relatively high values of σ_A. Low values of σ_A indicated relatively non-variable albedoes associated with static geographical features such as deserts and ice caps, as well as the more permanently positioned anticyclones and areas of cloud. In this way glimpses may be afforded of the dynamic and synoptic weather variations that contribute to patterns such as those in fig. 6.6.

Outgoing long-wave radiation (fig. 6.7)

The long-wave radiation pattern must owe a considerable debt to the albedo pattern discussed already, but clearly statements about the earth/atmosphere system prompted by either one cannot be taken perfectly in lieu of statements of the other. While it is true that a high albedo locally must depress the local rate of outgoing long-wave radiation loss, it should be remembered that albedo is a percentage quantity whereas the local level of short-wave radiation received from the sun is measured in scalar terms. Especially in high latitudes, therefore, insolation effects override albedo influences in producing the observed pattern of long-wave radiation loss; so this pattern is much more similar in northern and southern hemispheres than was the case with their albedoes.

Meanwhile in low latitudes an inverse relationship between albedo and long-wave radiation is more strongly evident. Strong cloud covers shield the underlying surfaces to some extent and cause less radiation to be available for absorption and re-radiation. A notable exception to this inverse relationship principle is north Africa, where the high albedo was associated with brightly reflective desert surfaces from which high rates of long-wave radiation loss were permitted by the relatively dry and cloud-free atmosphere over the Sahara.

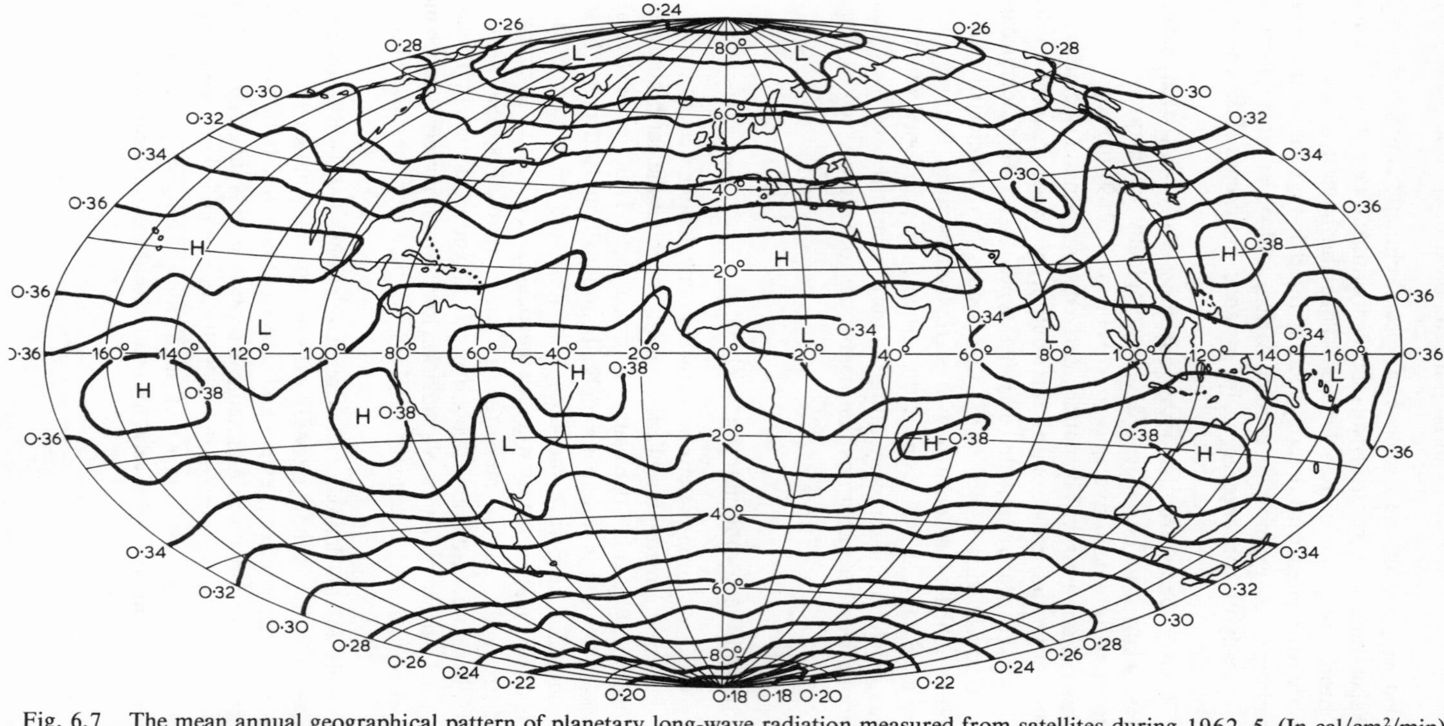

Fig. 6.7 The mean annual geographical pattern of planetary long-wave radiation measured from satellites during 1962–5. (In cal/cm²/min) (After Vonder Haar & Suomi 1971)

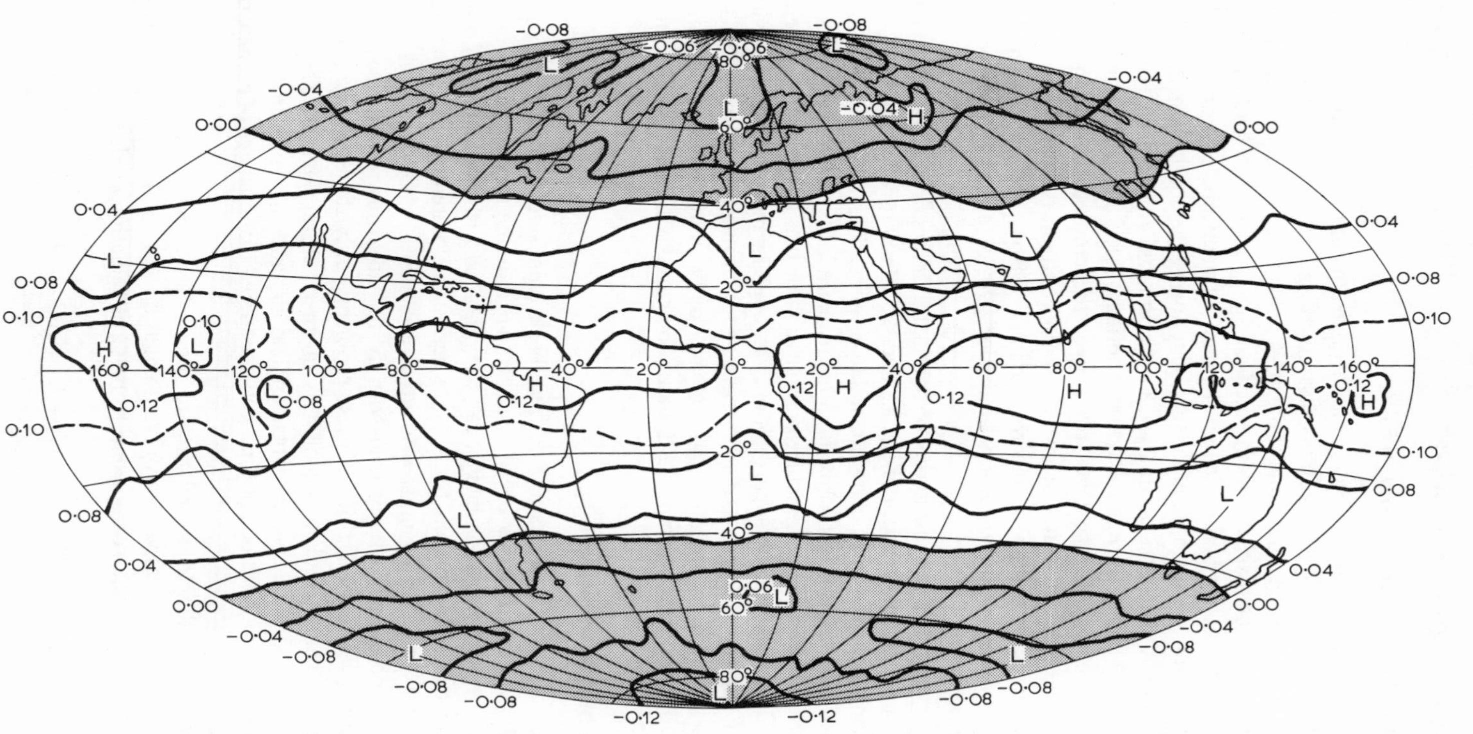

Fig. 6.8 The mean annual geographical pattern of planetary net radiation measured from satellites during 1962–5, assuming a solar constant of 2·00 cal/cm²/min. (After Vonder Haar & Suomi 1971)

The net radiation balance (fig. 6.8)

Net radiation gain is evident in fig. 6.8 through almost the entirety of the zone from 40° N to 40° S. This zone is flanked by radiation sinks which deepen generally towards the poles. One of the most unexpected features of the pattern in the southern hemisphere seems, at first sight, to be the broken ring of relatively large net radiation deficits around the hemisphere at about 60° S. The explanation seems to lie in the fact that, during the southern hemispheric winter, much higher rates of long-wave radiation loss characterize the coastal seas around Antarctica than the frozen continent itself (see p. 350, fig. 13.4). Very little energy storage takes place over the continent in summer, and insolation receipts in winter are very low. In contrast, the encircling oceans store much more energy during the summer season, and, in consequence, have more to lose as the zone of net radiation gain swings away northward in the local autumn.

Thus, in so far as the prompting of regional differences of radiation budget components are concerned, the shapes and distributions of the world's continents and oceans are of paramount importance. The relatively disjointed pattern of land and sea in the northern hemisphere leads to a higher degree of complexity in its radiation picture compared with that in the southern hemisphere. Geography plays its part not only through its mean effects on surface albedoes, energy storages and long-wave radiation emissions in that order, but also through the greater seasonal and daily complexity of organized weather distributions north of the Equator compared with those to the south.

Energy budget climatology

In conclusion, it is necessary to put back this satellite-based radiation climatology into a broader energy perspective. Satellite measurements of the earth/atmosphere radiation budget can be combined with evaluations of other energy terms in the search for a more acceptable statement of the energy budget as a whole. This must also involve energy sources and sinks, and profiles of polewards energy transport by the atmosphere and oceans. In view of the great complexities of these problems, only the mean annual case is discussed below, and no account is taken of storage terms. These two simplifications are reasonable when applied in tandem, since storages are negligible over long periods of time if it is assumed that changes of climate do not occur.

By removing the terms relating to atmospheric and subsurface energy storage (G_A and G_E respectively) from equation 6.4 (p. 149), Vonder Haar evaluated the following relationship from Tiros radiation evidence:

$$RN_{EA} + L(P - E) = \Delta C + \Delta F \qquad (6.10)$$

Here, the difference between precipitation and evaporation is considered the primary factor modifying the local radiation balance and prompting a net energy convergence ($+\Delta C$ and/or $+\Delta F$) or divergence ($-\Delta C$ and/or $-\Delta F$).

Mean annual patterns of RN_{EA} were constructed from all available Tiros radiation data to give the pattern illustrated by fig. 6.9. The precipitation (LP), evaporation (LE) and oceanic transport (ΔF) terms were estimated from conventional climatic data taken from Budyko (1963). This left one remaining term (ΔC) as the unknown. It represents the net energy transport in the atmosphere required to adjust the balance. Fig. 6.9 depicts Vonder Haar's results. These compare well with those established by Budyko from conventional data alone, though the absolute values of energy divergence indicated by the satellite-aided study are higher generally by 10–20 kcal/cm^2/yr, and are higher especially in low latitudes.

The primary centres of export and import of atmospheric energy (sensible heat plus potential energy) appear to be as follows:

Areas of export: The dominant areas are along the low pressure trough astride the Equator narrowing over the oceans and broadening across the continental areas of central Africa, South America and south-east Asia. Midlatitudinal areas only appear in part in fig. 6.9, but hint that extratropical depression belts embrace some lesser centres of energy export.

Areas of import: The regions of most marked energy convergence include the subtropical oceanic regions, even the relatively small Arabian Sea. Their anticyclonic natures and their large mean annual excesses of evaporation over precipitation require that they import energy, in view of their favourable net radiation budgets. The chief circulation link between the energy source regions along the Equator and the energy sinks astride the Tropics are upper tropospheric, although oblique penetration corridors of heat and moisture are found in the lower troposphere, notably along the eastern coasts of Asia and south-eastward from New Guinea.

In higher latitudes the contrasts between regions of energy convergence and divergence are less strong. The land masses of central Asia, the U.S.A., north Africa and Australia appear very close to a net energy balance in that their atmospheric energy transports in the horizontal are very small. In such regions small variations in either the radiation or water budgets could have important influences upon the required divergence of atmospheric energy.

In mean meridional form, the net energy transport by the atmosphere is depicted by fig. 6.10. Clearly the satellite estimates made by Vonder Haar compare well with earlier estimates by Rasool and Prabhakara (1965) based solely on conventional data, and by these same workers (1966) using some Tiros VII radiation data. These curves illustrate some of the best estimates available currently of the required energy transport established by the energy budget method of approach. They show that a southward transport of non-latent energy is required across the Equator, amounting to some 10^{19} kcal/yr.

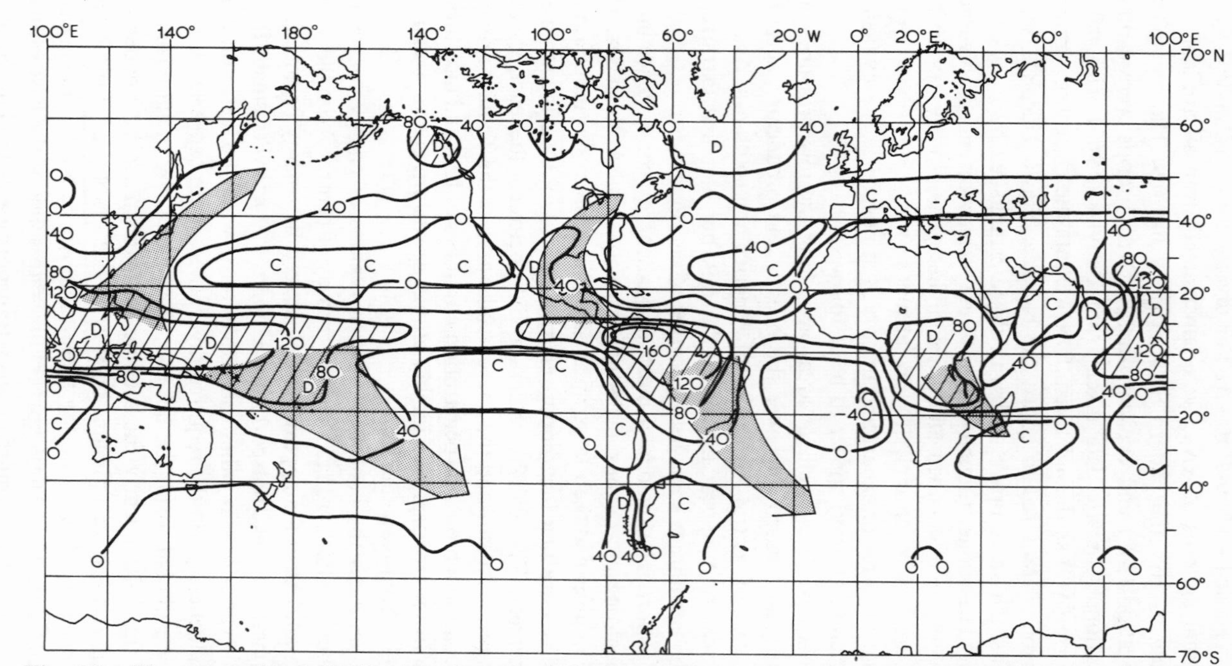

Fig. 6.9 The geographical distribution of required divergence of non-latent energy by the atmosphere for the mean annual case. The units are kcal/cm^2/yr, and the major energy exporting regions are shaded. (After Vonder Haar 1968)

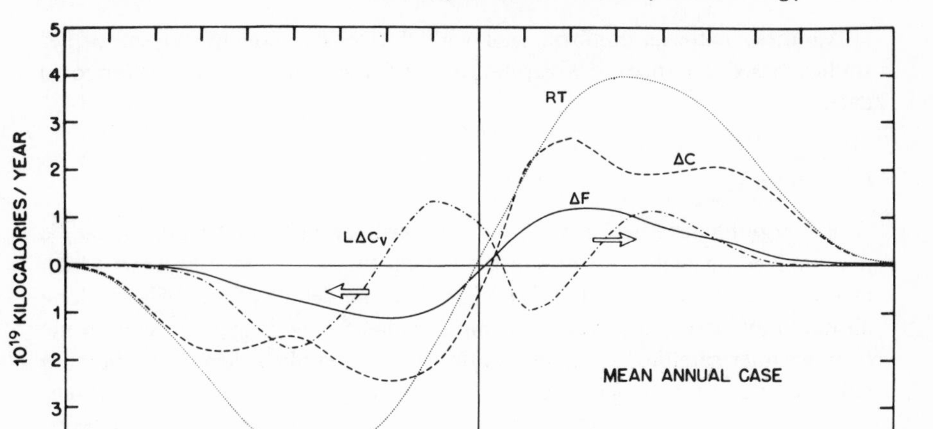

Fig. 6.10 The total poleward energy transport (RT) required by the radiation budget of the earth–atmosphere system. ΔF represents the contribution made by the oceans, $L\Delta C_V$ the atmosphere's transport of latent heat, and ΔC the atmosphere's sensible heat plus potential energy. (From Vonder Haar 1968).

Both the satellite curves show also that the net transport attains similar maxima in both hemispheres, being about $2{\cdot}2 \times 10^{19}$ kcal/yr. The zero net flux near 5° N marks the mean position of the meteorological equator.

Last of all, latent heat transport can be brought into the picture through an equation for required transport of energy (RT) derived from net radiation distributions:

$$RT = L\Delta Cv + \Delta C + \Delta F \qquad (6.11)$$

Thus the required transport is the sum of transports of both latent and non-latent energy in the atmosphere, as well as oceanic energy transport. The terms may or may not have the same signs as each other. Three particularly important points emerge from fig. 6.10:

(*a*) Energy transport within the oceans (ΔF) accounts for some 20–25% of the total transport in both hemispheres.

(*b*) Latent heat ($L\Delta Cv$) is transported both Equatorward and poleward from the evaporation maxima centred about 25° N and 25° S. Much moisture converges between 5° and 10° N, the mean location of the equatorial trough.

(*c*) Atmospheric energy transport (ΔC) shows double maxima in either hemisphere, mirror-imaging latent heat transport in middle latitudes.

All these patterns conform well with the general conclusions of earlier studies based on theoretical arguments and the evidences of conventional data.

Conclusions

'The energy balance between the sun, the earth, and the earth's atmosphere is the outstanding feature of one of the largest thermodynamic systems which man is called upon to study' (Sutton 1965). The details of this balance, long obscured by inadequate data, are being elucidated now through measurements from weather satellites. Although future studies may be based on much more uniform and complete satellite observations than those invoked above, their results should not be contradictory. Further studies will be of especial interest and importance in indicating the types and extents of time and space variations within radiation and energy budget patterns both globally and regionally.

In the meantime, the studies discussed at some length above have another kind of practical significance. The rapid development of numerical models of the atmosphere's circulation has played its own large part in the growth of the need for information about the earth's radiation budget (see e.g. Washington 1968). Numerical models do not require radiation observations as input parameters, but any model that properly describes variations of atmospheric conditions must contain a method to account for radiant energy transfer. Thus such models can yield computed values of the radiation budget at the top of the atmosphere over various intervals of time and for various locations. Since satellites provide observed values of this same budget, average values of satellite observations (particularly of the net radiation, RN_{EA}) are good controls for the developing numerical models. With this in mind, one of the special fields listed for research within the Global Atmospheric Research Program (GARP) involves the global distribution of radiation balance (net flux).

Finally the chief results from the first extended satellite studies of measurements of the earth's radiation budget may be summarized (Vonder Haar and Suomi 1971):

(1) Satellite observations consistently indicate that the earth is a warmer and darker planet than was previously believed; more solar energy is absorbed, primarily in the Tropics (see Vonder Haar and Hanson 1969).

(2) The Tropics (30° N–30° S) as a whole gain energy from space during all seasons, although significant longitudinal variations in the net input are noted, and some areas (e.g. the Sahara) may be areas of net radiation loss in some periods of the year.

(3) The net annual radiation budgets of the northern and southern hemispheres are both in balance (despite the great contrasts in geography), but

global symmetry with respect to radiative exchange with space is not exactly found. Geographically stimulated variations from a simple symmetrical pattern are noted, especially in subtropical and polar regions.

(4) The prime forcing function of the general circulation, the Equator-to-pole gradient of net radiation, has its greatest relative change between summer and autumn in both hemispheres. Significant differences of gradient seem to occur from time to time.

(5) Satellite measurements can be combined with independent surface observations to promote a better understanding of the global energy balance and the transport requirements of the atmospheres and oceans. Such knowledge will allow us to quantify better the effects of natural or inadvertent changes within our atmosphere upon the global radiation budget, and thus on worldwide weather and climate.

7 Global patterns of atmospheric moisture

Basic considerations

Atmospheric moisture is second only in importance to energy as a vital element in all the fundamental processes of the atmosphere (Rumney 1968). Indeed, air itself is often considered by meteorologists to be diluted water vapour. Although no special emphasis was placed on the role of atmospheric moisture in chapter 6, something of its significance was at least implicit in much that was said there. For example, the reflectivity of the earth varies more on account of changes in cloud cover than any other single variable, while water vapour, through its propensity to absorb and re-emit radiation, provides an important means for horizontal energy transport within the general circulation of the atmosphere.

Atmospheric moisture also has its significance in differentiating regions climatologically. For example, the nature and extent of cloudiness helps determine the degree of solar control of local climates; the humidity of the atmosphere has been an important factor affecting the location of certain of man's industries; amounts of precipitation are frequently critical for human decision-making of many kinds – for example, with respect to agriculture. Many aspects of the moisture patterns of the atmosphere affect the other environmental factors in the natural home of man. For example, the local type, intensity, duration, frequency and variability of rainfall play key roles in moulding the surface land forms of the earth; such factors are critical determinants of the type of vegetation and the class of soil.

Much more research is vital for the global picture of atmospheric moisture to become even as well known and as complete as that of either atmospheric energy or circulation. Unfortunately much of the satellite evidence for moisture distributions other than in the form of clouds is essentially indirect. Since so much attention is paid to cloudiness elsewhere in this book, the cloud notes that follow are very brief, while the treatments of humidity and precipitation are more complete. For convenience, the order followed below is substantially the same as that in chapter 4.

Evaluations of mean cloud cover from earth-orbiting satellites may seem at first sight to be statistics singularly easy to compile, but comparatively uninteresting except in an encyclopaedic sense. Experience reveals, however, that the first of these appearances is much more real than the second. Prior to the advent of meteorological satellites, the climatology of clouds had been but poorly developed. Hence, even the simplest satellite-derived cloud statistics are assuming considerable importance especially in three areas of atmospheric science:

1 In support of other programmes of space research.
2 In computer modelling the earth's atmosphere.
3 In tracing the circulation of the atmosphere.

Further comments are appropriate against these three points.

Global cloudiness and space research

Currently scientists and engineers are developing many instruments for use in earth-orbiting satellites to record a wide variety of earth resources (Salomonson 1969). In particular, remote sensors are being designed to map aspects of global mineralogy, hydrology, oceanography and agriculture from platforms such as the Earth Resources Technology Satellites (ERTS) and manned space stations like Skylab. Many such sensors investigate primarily through the visible, near infra-red, and far infra-red portions of the electromagnetic spectrum. As a result, satisfactory measurements can be made only when a large percentage of the target area is free from clouds. A detailed knowledge of the frequency and type of cloud occurrence is essential in appraising the probability of achieving satisfactory earth surface measurements in any area.

Detailed cloud information is also essential for planning the re-entry and recovery of manned spacecraft returning from a variety of missions.

Global cloudiness and computer modelling the atmosphere

Perhaps the most notable sequence of papers dealing with thermodynamic modelling of the lower atmosphere has been that constructed by Adem (1964a, 1964b, 1965, 1967). The primary object of Adem's work has been the establishment of computer-based methods founded on firm physical principles to replace eventually the present obviously unsatisfactory methods of predicting weather over periods ranging from a few days ahead to months, seasons or even longer. There are three stages in his approach:

(*a*) The development of a radiation model to specify and predict mean seasonal temperatures in the atmosphere and at the earth's surface (Adem 1964a, 1964b).
(*b*) The involvement of additional aspects of earth energy budgets to permit improved results. At this stage, such variables as storages of thermal energy, turbulent heat transfer and changes in the states of atmospheric moisture are also invoked (Adem 1965).
(*c*) The derivation of pressure, temperature and wind flow patterns from the predicted mid-tropospheric and surface temperatures. This is possible, for 'it is evident that the wind field is coupled with the temperature field, and with the heat sources and sinks, and that it must therefore be generated within the model' (Adem 1967).

Namias, another active worker in long-range atmospheric prediction, has pointed out that the energy feedback between the atmosphere and the earth surface is a most important factor influencing large-scale circulation patterns through its effects on the thermal field, especially in its surface temperature expression. Albedo anomalies are thus significant to all three stages of Adem's work. In its initial form, his model incorporated cloud cover as the only independently variable parameter. Later tests with the model confirmed that period mean temperatures are very sensitive to cloud amount: it appeared that a change by only one tenth in mean cloudiness led to changes in surface temperatures of several degrees. Clapp (1965) made the supplementary point that full knowledge of mean seasonal cloud cover will be useful in simulations of cloud amounts so that these too may be generated eventually within the model.

Before that becomes possible, however, much more information is required of mean cloudiness, cloud thicknesses, the distributions of cloud types, and their vertical arrangements.

Global cloudiness as a tracer of atmospheric circulation

This point is included here for the sake of completeness. It is discussed more fully in chapter 8, which is concerned entirely with the circulation of the atmosphere, and various ways in which satellite data help to indicate the patterns of large-scale atmospheric flow.

Global scale patterns of cloudiness

The earliest broad-scale maps of cloud cover compiled from meteorological satellite data were prepared by longhand methods by Clapp (1964). Fig. 7.1,

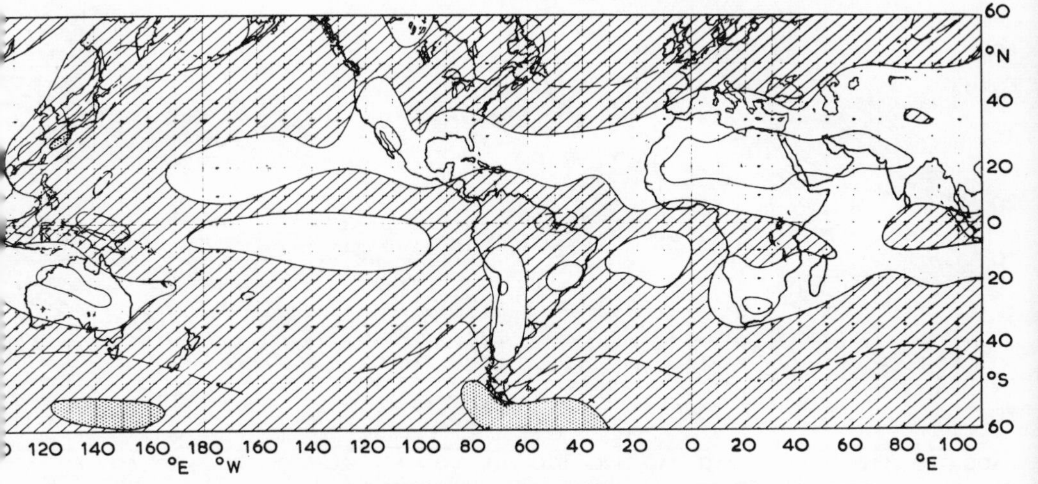

Fig. 7.1 Normal cloudiness, March–May, suggested in pre-satellite days by H. E. Landsberg (1945). (After Clapp 1964; from Winston 1970)

a pre-satellite cloud map compiled by Landsberg (1945), is included for comparison with the satellite-based maps, figs. 7.2–7.5. Although the distribution of the data available to Clapp was less than optimum, it was considerably better than that of the conventional data upon which Landsberg based his attempt, augmented where necessary by scientific guesswork. Care should be exercised in comparing the satellite and pre-satellite results, however, since Landsberg's map was one of a series designed to portray mean cloudiness whereas Clapp's cloud charts depict cloud cover during an individual year. There can be little doubt, though, that the satellite displays have much to recommend them. Two especially significant facts emerge from them:

1 The cloudiest areas – particularly those in middle to high latitudes – are more strongly cloudy than used to be thought.
2 Cloud fields within the maritime tropical anticyclones are differentiated markedly from one sector to another. Strong cloudiness (> 75%) accompanies Equatorward trade wind flows off the western shores of most subtropical continents, while belts of weaker cloudiness (< 50%) parallel the Equator at about 10–20° N and S.

More recent work has suggested that both conventional observations and the Tiros nephanalyses from which Clapp compiled his cloud maps have led to overestimates of global cloud amounts. For example, Barnes (1966) found an overestimate of 14% in the ground-based data across a period of nineteen months when he compared satellite and surface cloud data for the United States. Such overestimates are now accepted as having been widespread. They

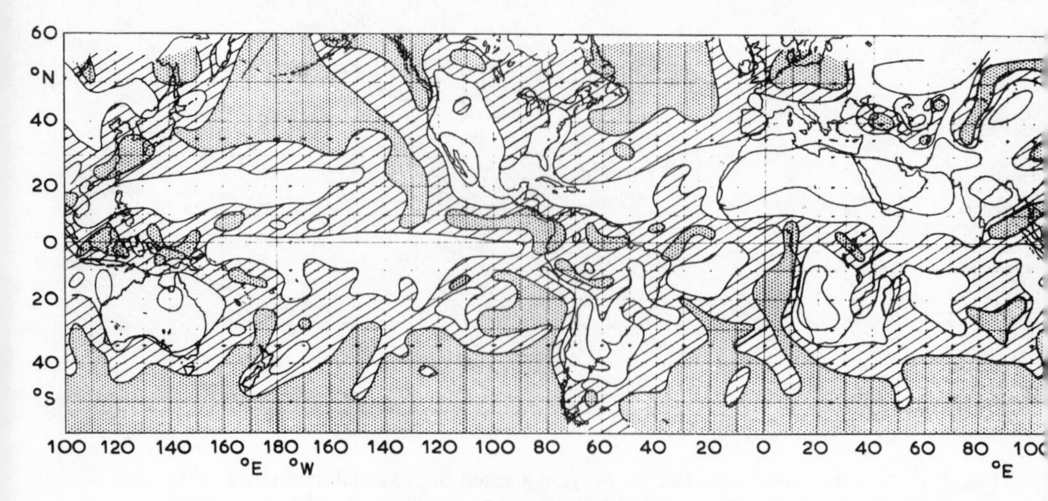

Fig. 7.2 Average cloudiness, March–May 1962, from Tiros IV photographic and nephanalysis evidence. As in figs. 7.3, 7.4 and 7.5 also, stippled shading indicates cloudiness in excess of 75% cloud cover; cross-hatching, cloudiness between 50 and 75% cloud cover; no shading, less than 50%. Isolines of 25% sky cover are indicated by solid lines within unshaded areas, and of 65% and 35% (where needed) by broken lines. (After Clapp 1964; from Winston 1970)

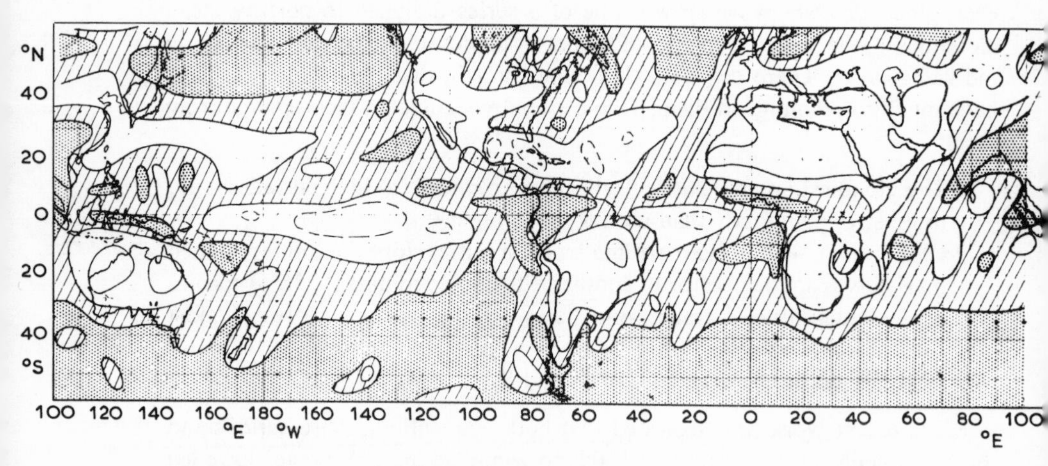

Fig. 7.3 Average cloudiness, June–August 1962, from Tiros IV evidence. (After Clapp 1964; from Winston 1970)

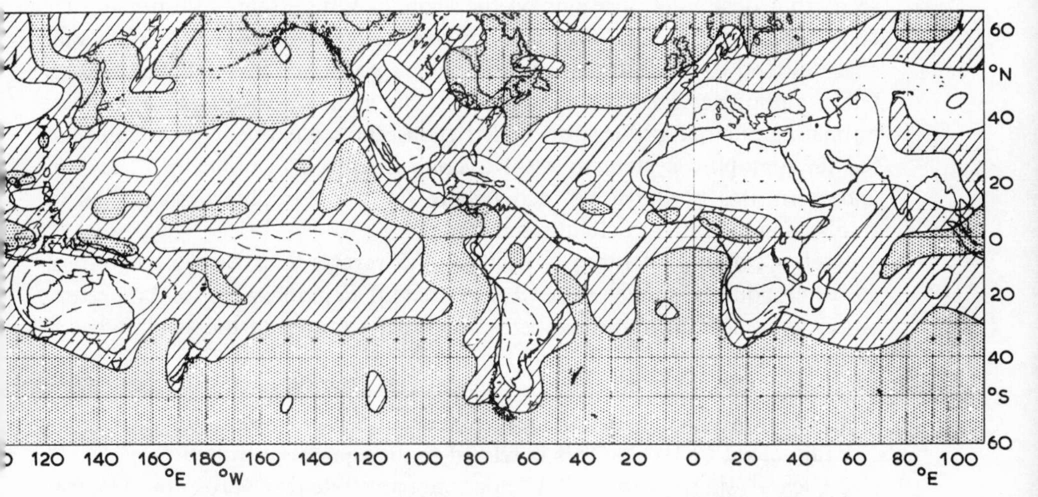

Fig. 7.4 Average cloudiness, September–November 1962, from Tiros IV evidence. (After Clapp 1964; from Winston 1970)

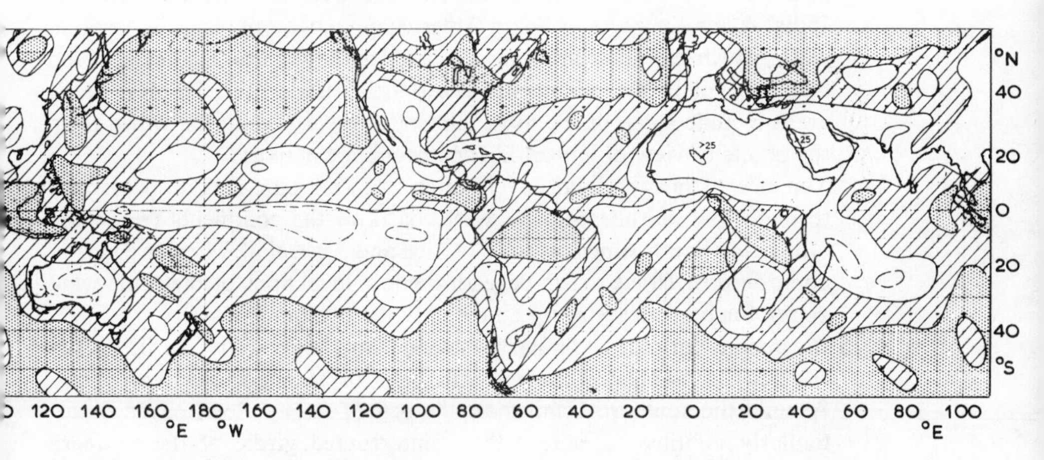

Fig. 7.5 Average cloudiness, December 1962–February 1963, from Tiros IV evidence. (After Clapp 1964; from Winston 1970)

have been explained in terms of the foreshortening effects on cloud breaks seen from observing positions on the ground. Where satellite estimates are concerned, Young (1967) fascinatingly used a 'paper cloud' test to show that analysts tend to overestimate even the cloud portrayed in a satellite picture. Other authorities hold the view that satellite analyses usually lead to overestimating of cloud cover at the higher end of the scale (due to the inability of the photographic system to resolve small breaks in a cloud field) and to underestimates at the lower end (where the photographic system fails to resolve small, scattered cloud elements).

Even a cursory reference to figs. 7.2–7.5 reveals that the substructure of global cloudiness is much better defined there than in the more solid representation compiled from surface date. The chief featurés of the satellite-based patterns include:

1 A belt of enhanced cloudiness, exceeding 75% in places, aligned generally along 0–10° N. This is related to the equatorial trough.
2 Clearer skies, with < 50% cloud cover, stretching across the tropical north and south Pacific, the north Atlantic and the north Indian Ocean oriented strongly from east to west. Similar, but smaller areas are observed across parts of the tropical south Atlantic and south Indian oceans. These are associated with 'Hadley cell' subsidences within the tropical maritime anticyclone belt.
3 Very clear skies (< 25% cloud cover) are found over the tropical/subtropical continental land masses of north Africa and the Arabian peninsula, parts of southern North America and Central America, north-east India, Australia, parts of South America and southern (especially southwestern) Africa. These are also associated with the mean Hadley cell, and help to explain the continental desert belts astride the Tropics. These areas include some of the most inhospitable surfaces in the world from the points of view of human life and economic activity.
4 Tongues of heavy cloud cover (often > 75%) extending Equatorward through low latitudes off the west coasts of the continents (especially North and South America, south Africa and Australia). These are mostly comprised of cloud sheets beneath the low inversions of the trade wind zones. Such areas are again dominantly anticyclonic.
5 In middle latitudes, similarly strong cloudiness (often > 75% sky covered) is associated mainly with belts of extratropical depressions. Around the southern hemisphere the belt of strong cloud cover is particularly continuous, across the uninterrupted girdle of the southern oceans. The vicinity of southern South America is the least cloudy at comparable latitudes throughout the year, a fact which indicates the type of influence exerted by land masses on regional climatic patterns in midlatitudes.

\n example of Essa
hotography, complete
h computed grid,
astlines and
ntification
end (Essa 7
ilar). (See p. 29)
S.S.A. photograph)

2 A rectified cloud photo-mosaic of the eastern tropical Pacific (4 July 1969) on a Mercator projection. (See p. 49) (E.S.S.A. photograph)

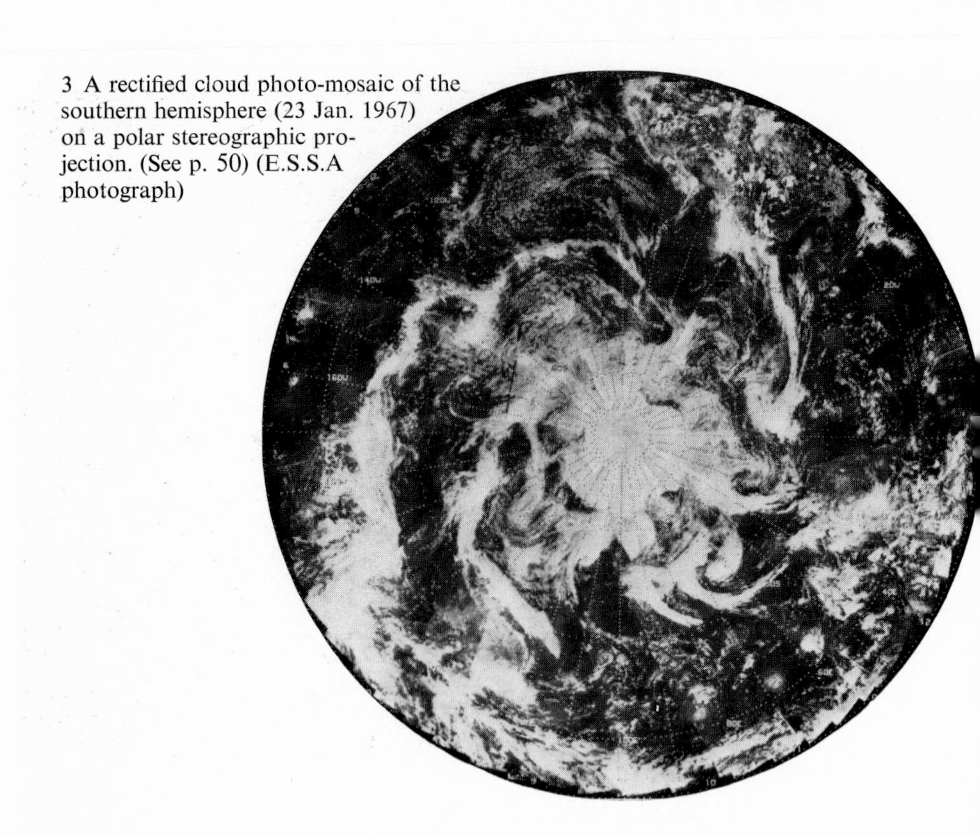

3 A rectified cloud photo-mosaic of the southern hemisphere (23 Jan. 1967) on a polar stereographic projection. (See p. 50) (E.S.S.A photograph)

4 A rectified cloud photo-mosaic of a quadrant of the southern hemisphere (23 Jan. 1967) on a polar stereographic projection. (See p. 50) (E.S.S.A. photograph)

5 A rectified cloud photomosaic of the tropical Far East (23 Jan. 1967) on a Mercator projection. (See p. 50) (E.S.S.A. photograph)

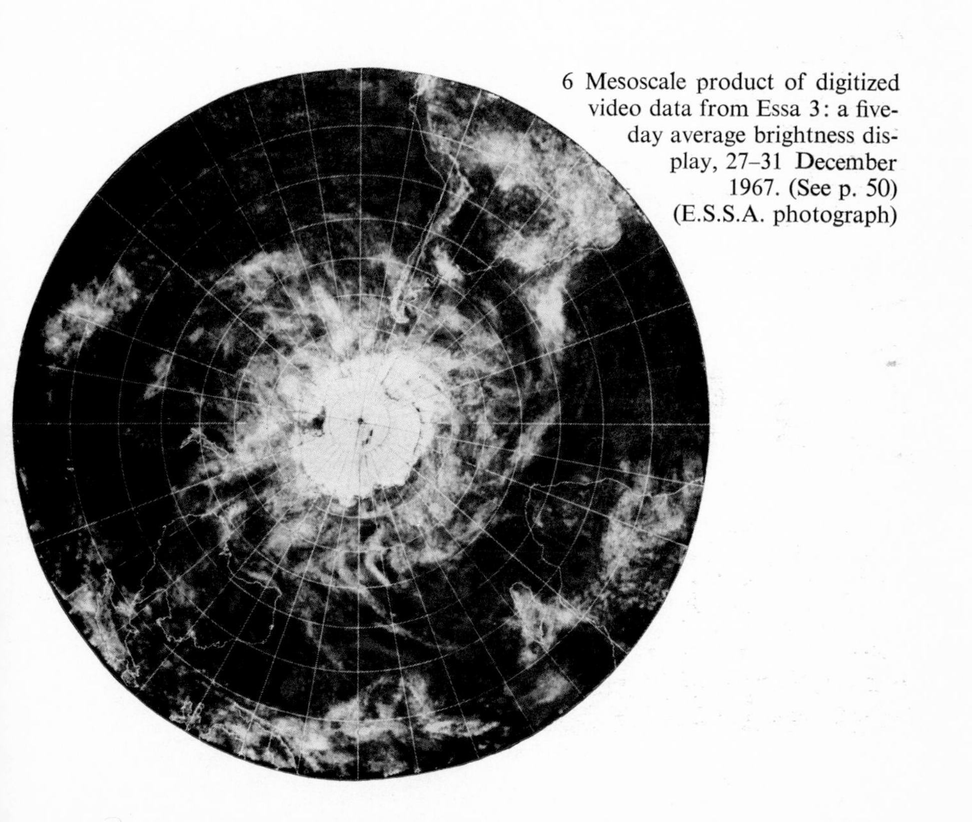

6 Mesoscale product of digitized video data from Essa 3: a five-day average brightness display, 27–31 December 1967. (See p. 50) (E.S.S.A. photograph)

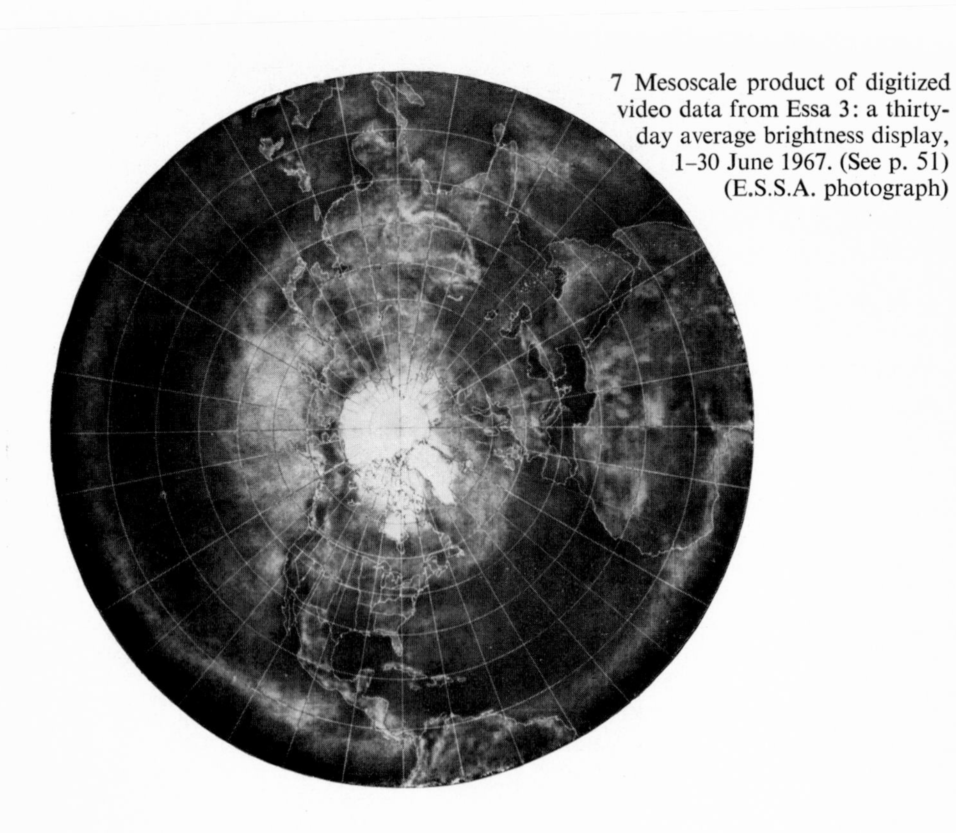

7 Mesoscale product of digitized video data from Essa 3: a thirty-day average brightness display, 1–30 June 1967. (See p. 51) (E.S.S.A. photograph)

8 Mesoscale product of digitized video data from Essa 3: an enhanced ninety-day seasonal average brightness display, June–August 1967. (See p. 51) (E.S.S.A. photograph)

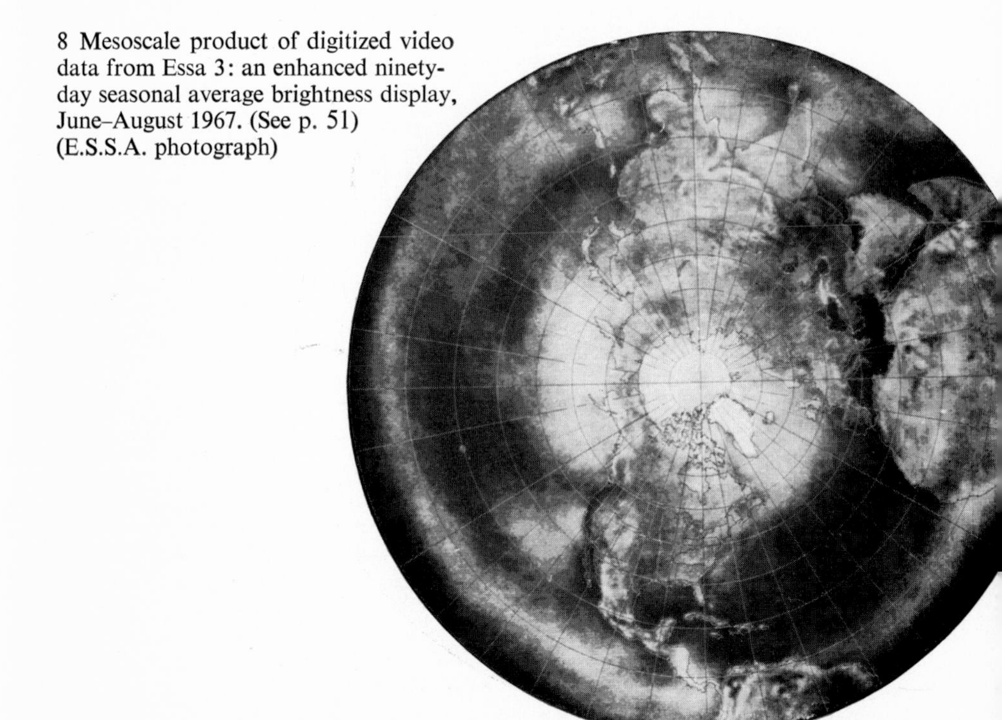

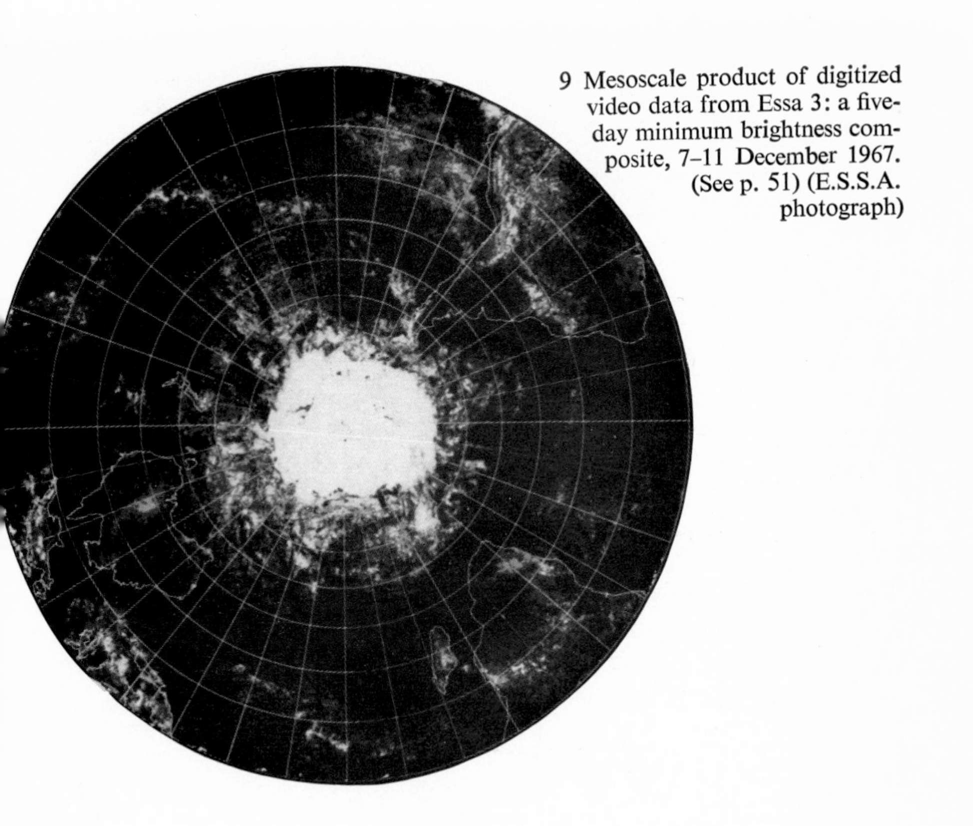

9 Mesoscale product of digitized video data from Essa 3: a five-day minimum brightness composite, 7–11 December 1967. (See p. 51) (E.S.S.A. photograph)

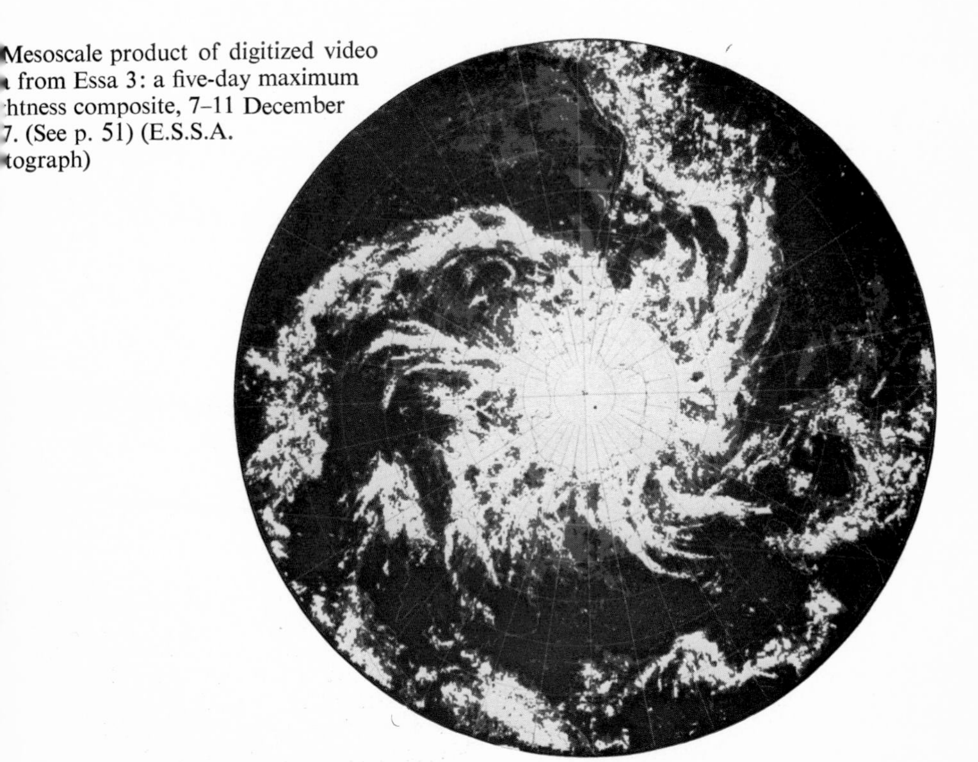

Mesoscale product of digitized video ı from Essa 3: a five-day maximum ·htness composite, 7–11 December 7. (See p. 51) (E.S.S.A. tograph)

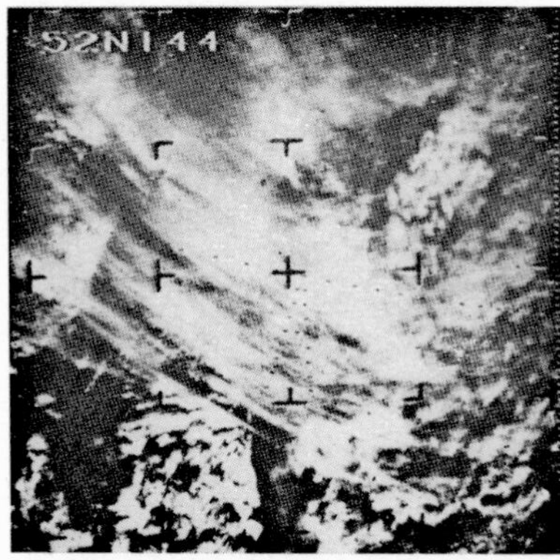

11*a* Fibrous cirrus above towering cumulus and altocumulus clouds, recorded by Nimbus II. (See p. 83) (N.A.S.A. photograph)

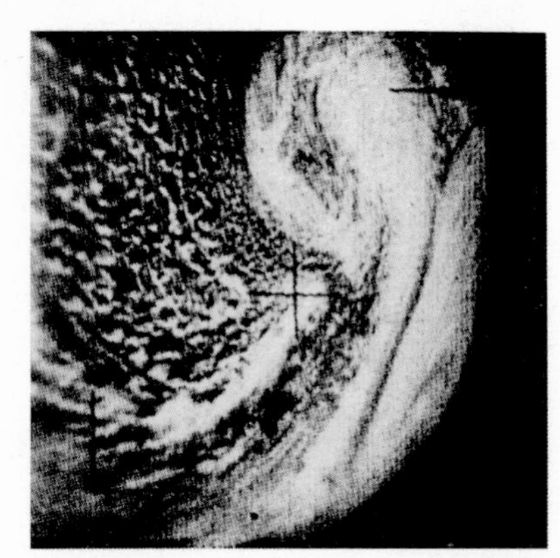

11*b* Post-frontal instability line and convective cloudiness, recorded by Nimbus II. (See. p. 83) (N.A.S.A. photograph)

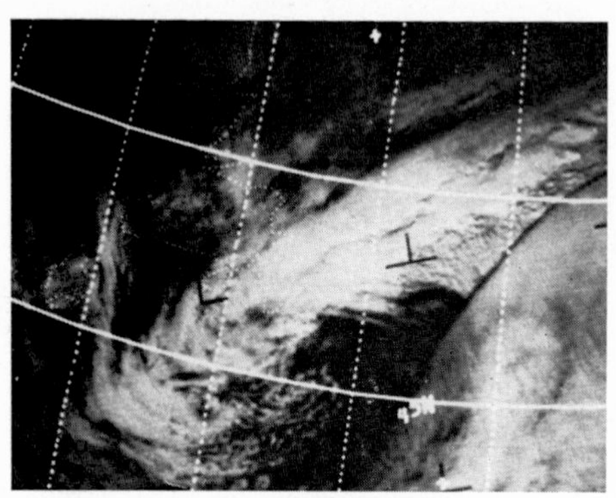

11*c* The distinctive shadow cast by the edge of a cirrus deck on a layer of lower clouds indicates the position of a jetstream. (See p. 84) (E.S.S.A. photograph)

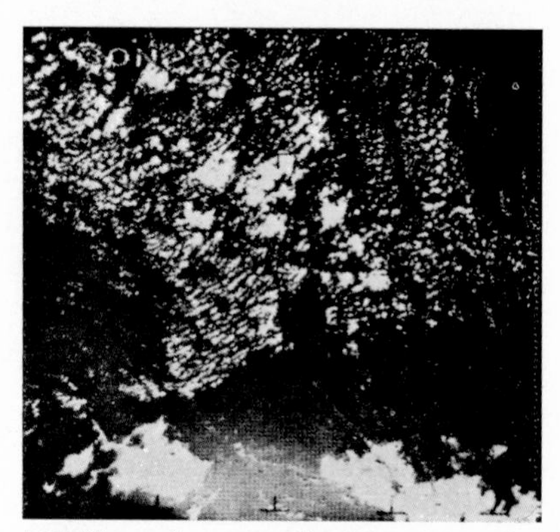

11*d* Towering cumuli with scattered cumulonimbus clusters, photographed by Nimbus II. (See p. 84) (N.A.S.A. photograph)

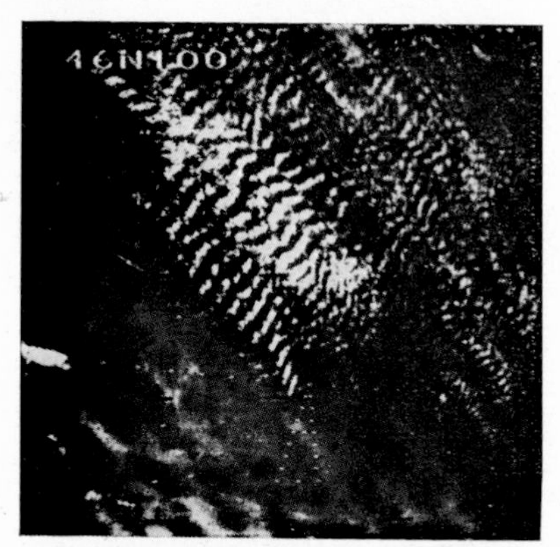

11*e* Terrain-induced wave clouds photographed by Nimbus II over Mongolia. (See p. 84) (N.A.S.A. photograph)

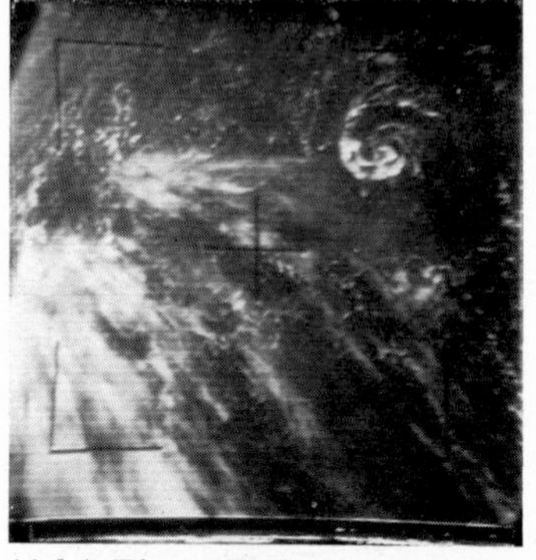

11*f* A Tiros VI photograph, 28 September 1962. Hurricane Freda occupies the south-west corner, while a smaller vortex, which did not appear on map analyses of the area, is seen in the north-east. (See p. 84) (E.S.S.A. photograph)

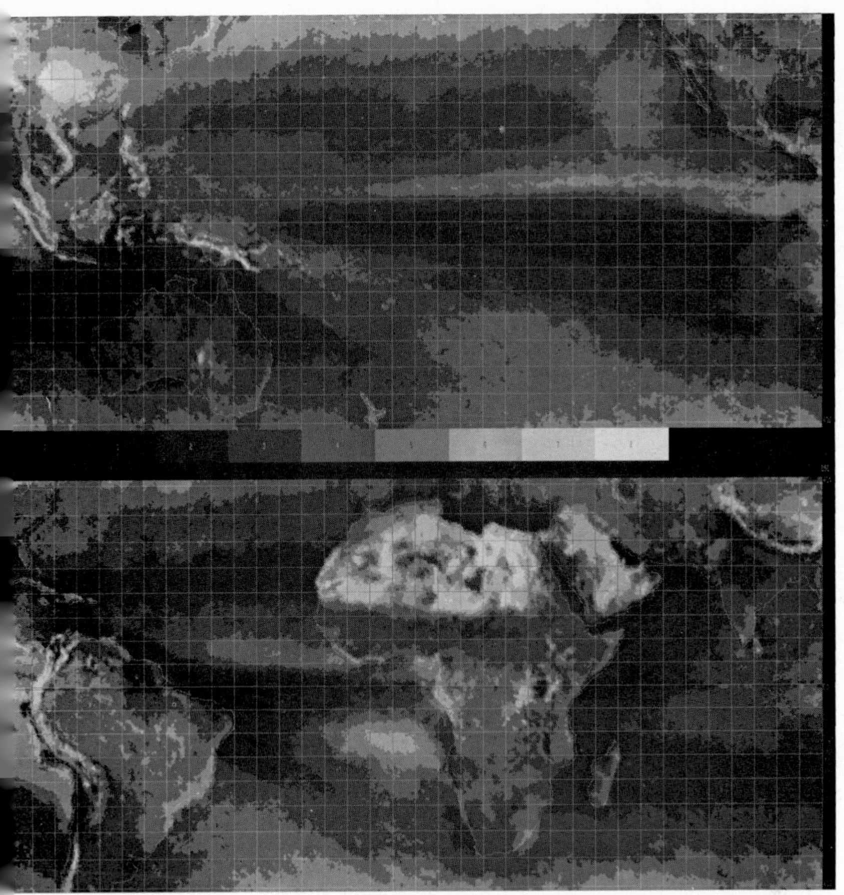

Satellite relative cloud cover (1400 h local time) for 40° N–40° S, September–ctober–November 1967–70, in mean oktas. Background brightness has not been tered off, hence the apparent strong cloud cover over the Saharan and Arabian serts. (See p. 101) (U.S.A.F./E.T.A.C. and N.O.A.A./N.E.S.S. photographs)

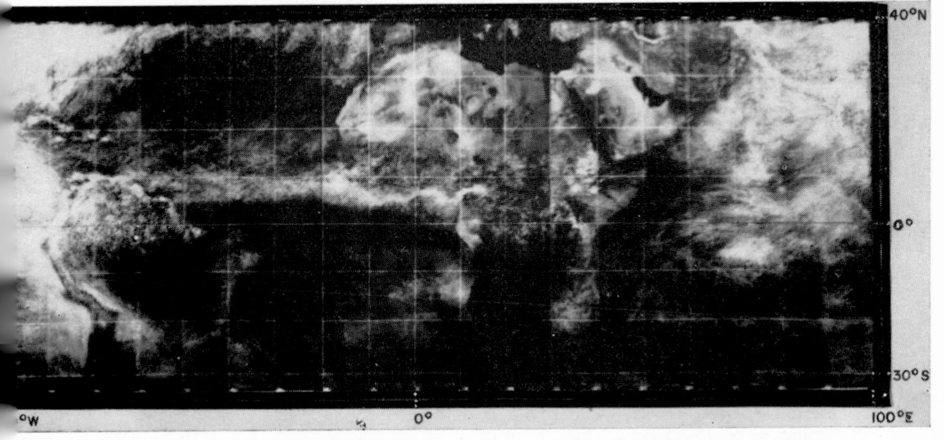

A 'multiple exposure average' of Essa cloud photographs from 90° W–100° E, 15 July 1967, on a Mercator projection. (See p. 101) (From Kornfield *et al.* 1967)

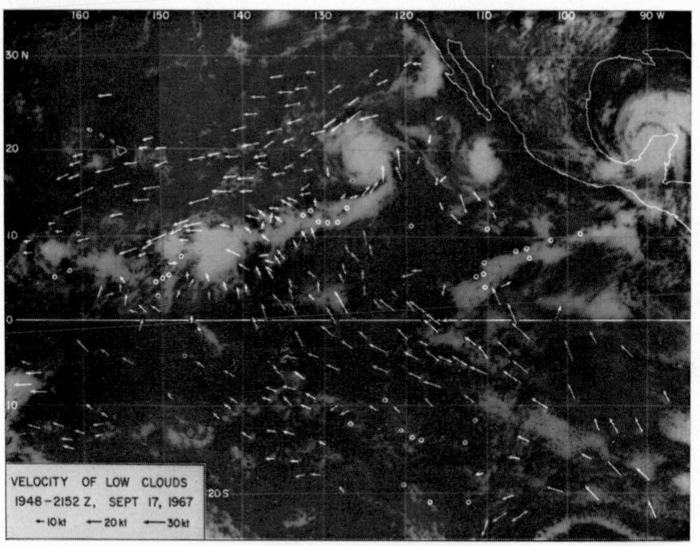

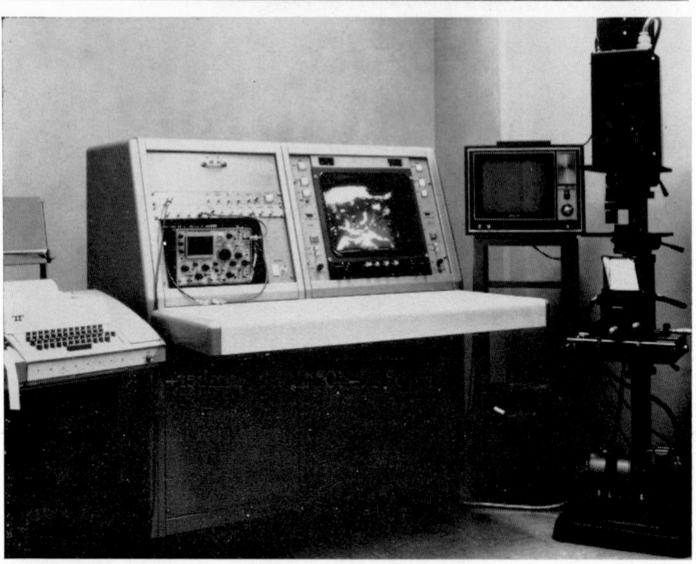

14 *top* Velocities of low clouds over the eastern tropical Pacific, plotted on an Essa Mercator projection digital mosaic. Velocities were obtained from six A.T.S.-I pictures taken at about 25-min. intervals. (See p. 138) (From Fujita *et al.* 1969)

15 *centre* A purpose-built electronic display console developed by the Stanford Research Institute, California, for ease of handling, and analysis of, geosynchronous satellite photographs. (See p. 141) (S.R.I. photograph)

16 *left* An A.T.S.-I photograph of the Pacific Ocean, 7 November 1969. (See p. 143 and fig. 5.16) (N.A.S.A. photograph)

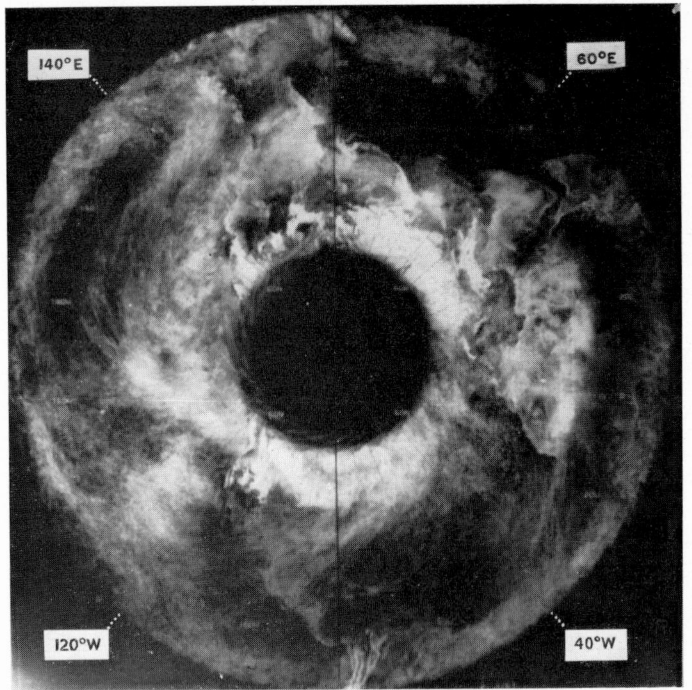

17 A 'multiple exposure average' of Essa cloud photographs, northern hemisphere, 1–15 February 1967, on a polar stereographic projection. (See p. 210) (From Kornfield *et al.* 1967)

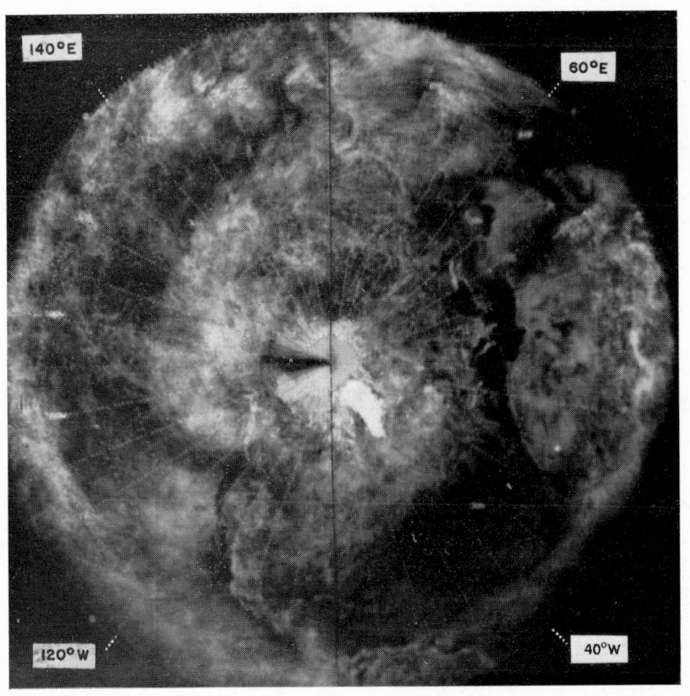

18 A 'multiple exposure average' of Essa cloud photographs, northern hemisphere, 1–15 July 1967, on a polar stereographic projection. (See p. 211) (From Kornfield *et al.* 1967)

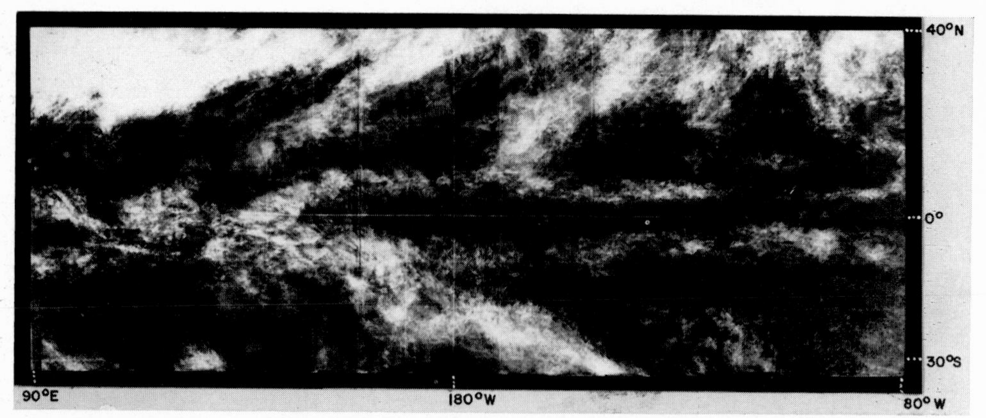

19 Tropical Pacific cloudiness portrayed by a 'multiple exposure average' of Essa 3 Mercator mosaics, 16–31 March 1967. (See pp. 242 and 244) (From Kornfield *et al.* 1967)

20 Time–longitude section of satellite photographs for the period 1 July–14 August 1967, for 5–10° N across the Pacific Ocean. Black strips indicate missing data. (See p. 252) (From C-P. Chang 1970)

21 An A.T.S.-III photograph, 1333Z hours, 19 November 1967. (See p. 263) (N.A.S.A. photograph)

22 An A.T.S.-III photograph, 2001Z hours, 19 November 1967. (See p. 263) Note particularly the more extensive convective cloudiness along the Andes, and the less dense stratocumulus cloud sheets off the west coast of South America compared with those in plate 21. (See p. 263) (N.A.S.A. photograph)

23 An Essa 3 photograph of the Arabian Sea, 9 November 1966, portraying a mature hurricane complete with eye, and a second tropical vortex over Ceylon. (See p. 274) (E.S.S.A. photograph)

24 One of the most damaging hurricanes in history, the 'Bangladesh Cyclone' of November 1970, photographed over the Bay of Bengal by Itos-1 on 11 November 1970. (See p. 268) (N.O.A.A. photograph)

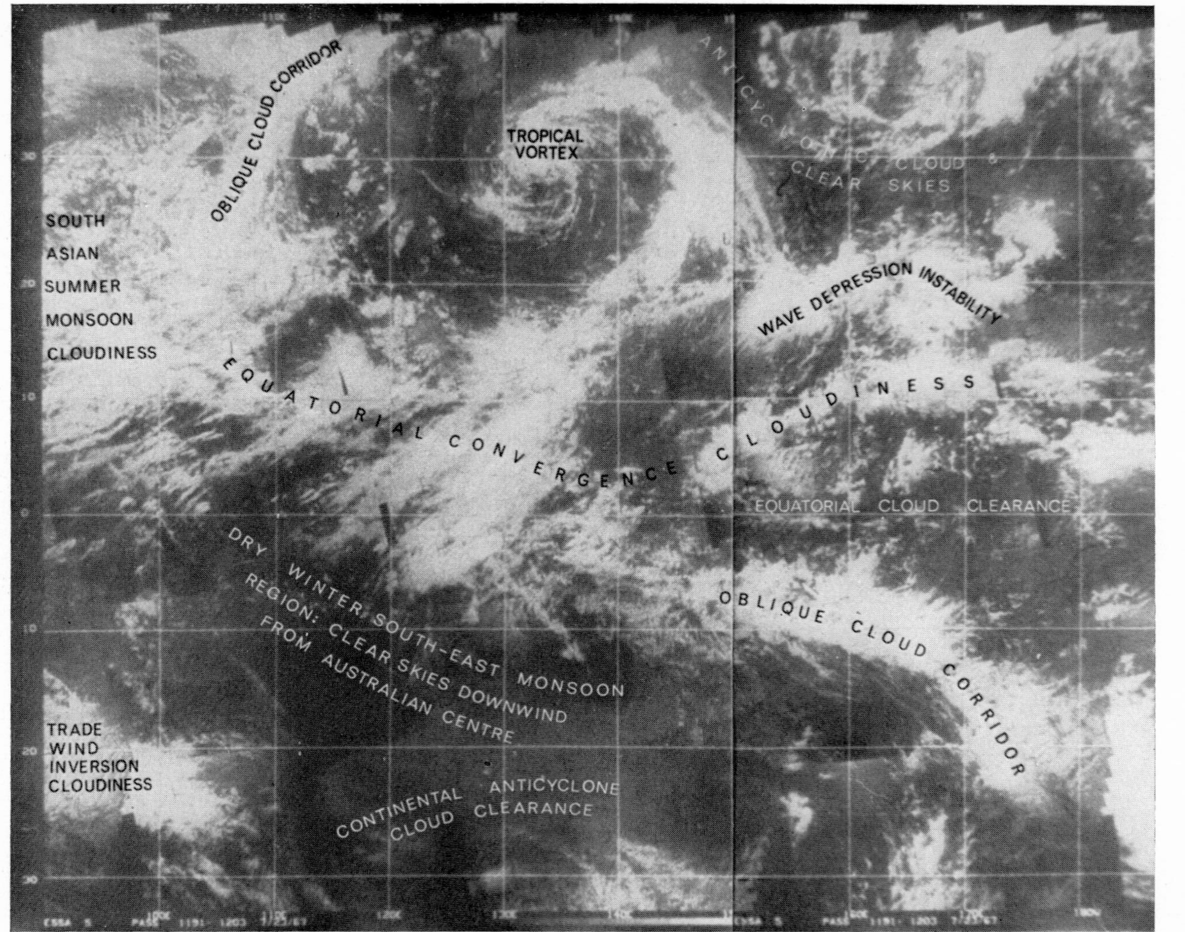

25 An annotated Essa 3 Mercator mosaic of cloud photographs, tropical Far East, 23 July 1967. (See p. 309) (From Barrett 1971)

26 An annotated Essa 3 Mercator mosaic of cloud photographs, tropical Far East, 7 January 1967. (See p. 309) (From Barrett 1971)

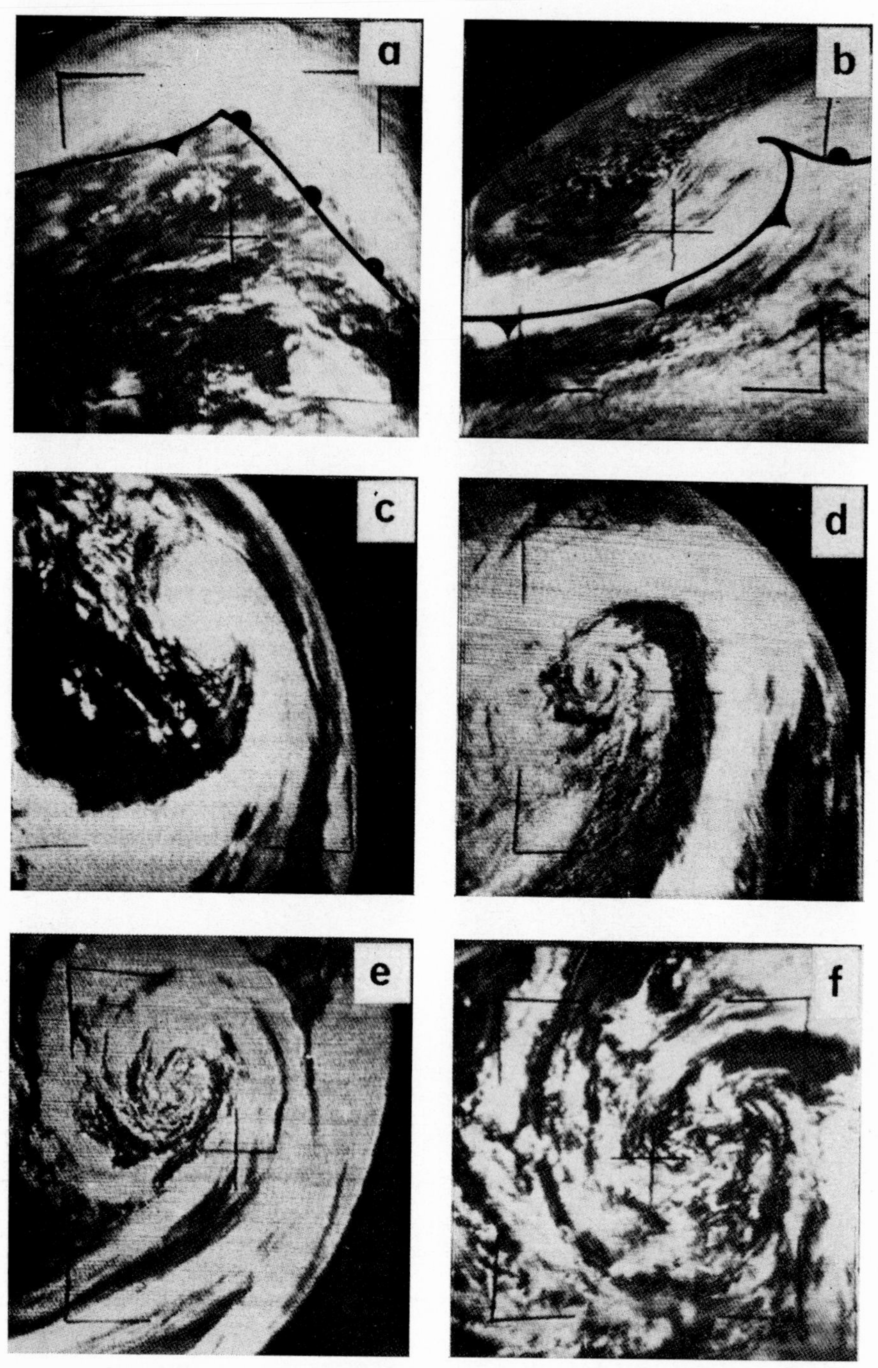

27 Stages in the lives of extratropical depressions seen by Tiros satellites. (See p. 332)
(E.S.S.A. photographs)

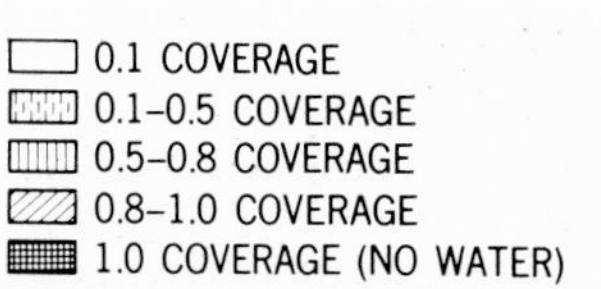

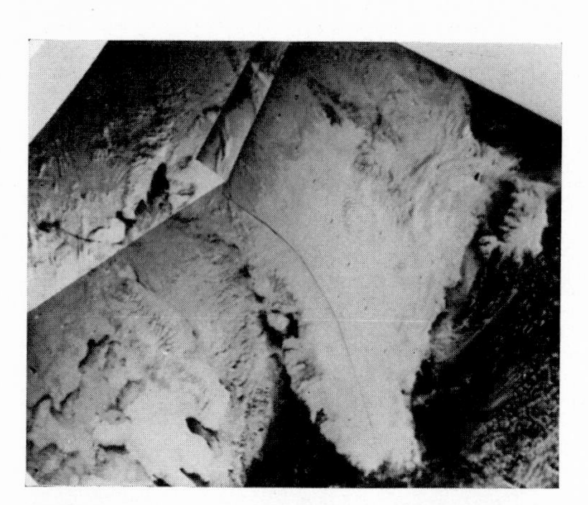

28*a* Ice and snow recorded by Nimbus III through its I.D.C.S., 15 April 1969. (See p. 354) (Plates 28 and 29 are from Allison 1971)

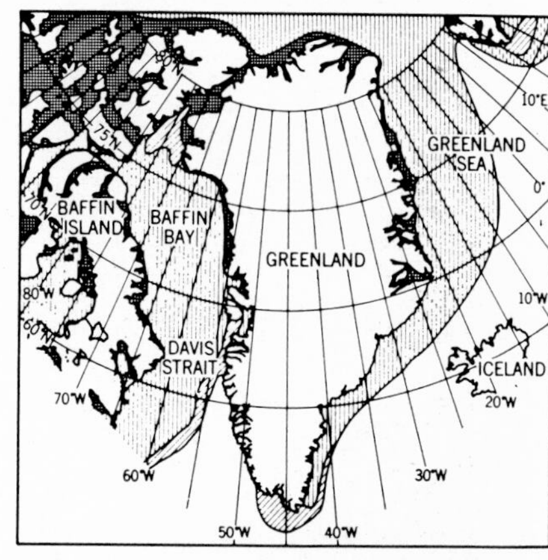

28*b* Mean Ice concentration for Greenland area in April. (See p. 354) (U.S. Navy map)

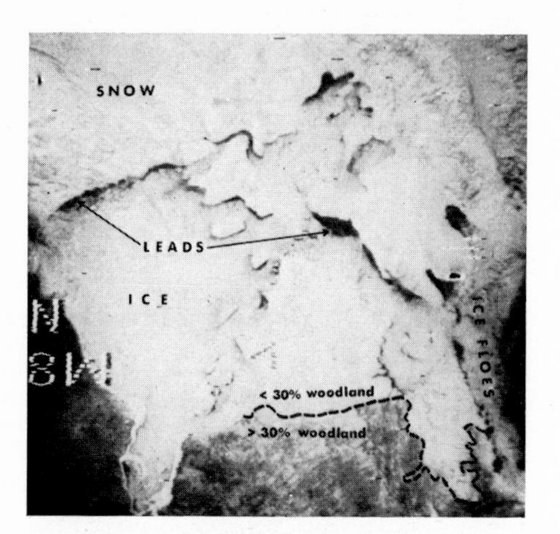

29*a* Ice, snow and vegetation boundaries, north-east Canada, as seen by Nimbus II through its I.D.C.S., 30 April 1969. (See p. 354) (ARACON photograph)

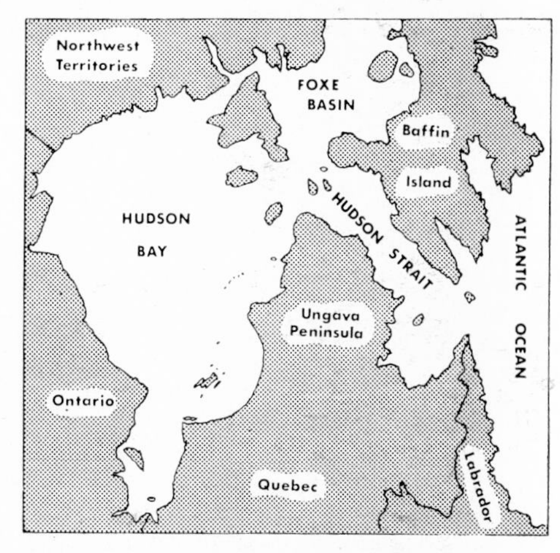

29*b* A geographical map of the Hudson Bay area for comparison with plate 29*a*. (See p. 354) (ARACON map)

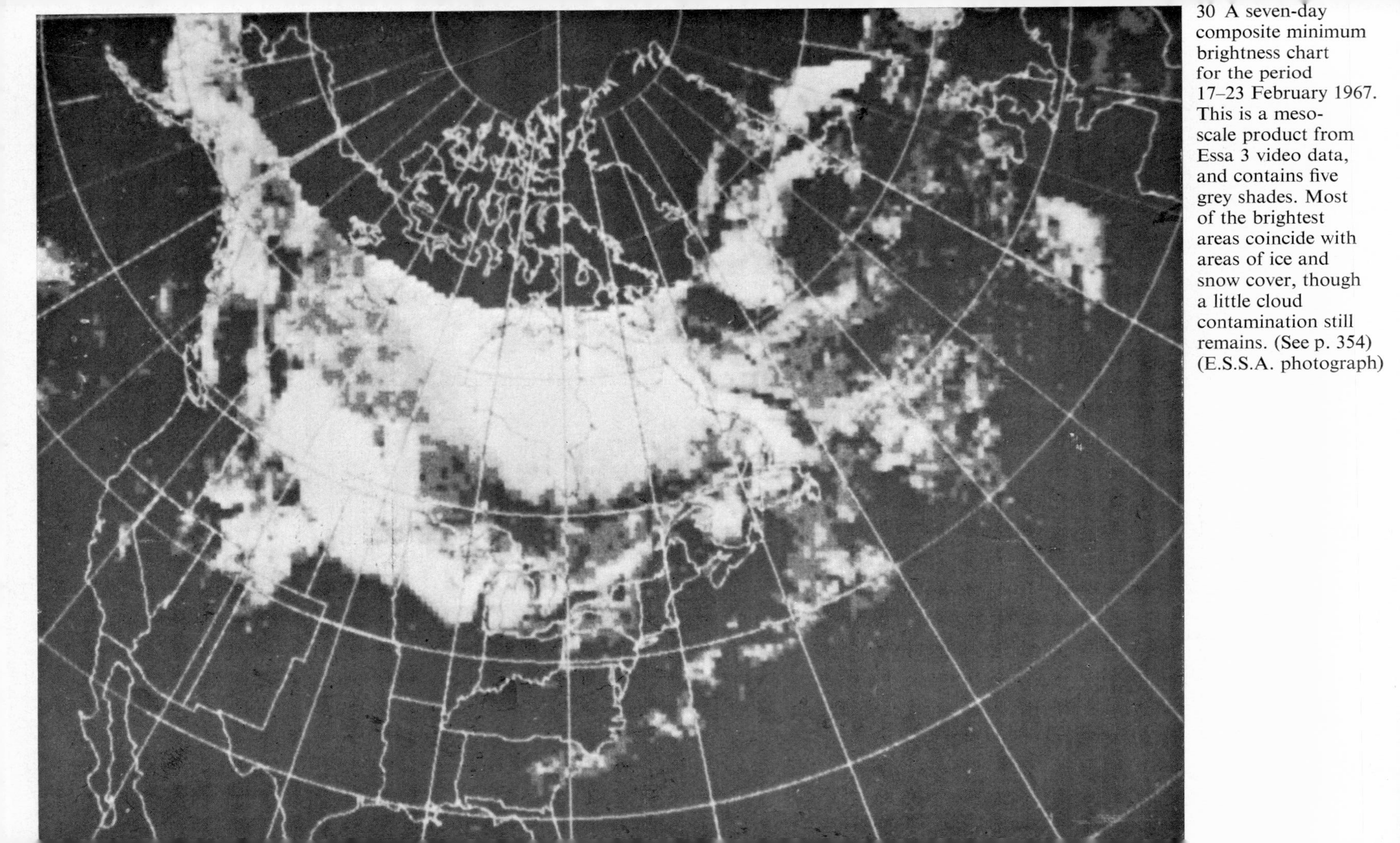

30 A seven-day composite minimum brightness chart for the period 17–23 February 1967. This is a meso-scale product from Essa 3 video data, and contains five grey shades. Most of the brightest areas coincide with areas of ice and snow cover, though a little cloud contamination still remains. (See p. 354) (E.S.S.A. photograph)

6 Particularly in the local summer seasons, relatively clear skies spread across parts of the northern continental interiors, associated with drier conditions than those prevailing across the north Atlantic and north Pacific oceans where temperate depression belts are well defined. It is possible that some of the apparently strongly clouded regions across North America and Asia in the winter months were rather less cloudy than the Tiros nephanalyses suggested; it is often difficult to distinguish between cloudiness and surface ice and snow across possibly frozen continents.

Hence, the fact that figs. 7.2–7.5 do not extend poleward of 60° N and S does not reflect on them as poorly as might be thought at first. In higher latitudes, permanent and semipermanent areas of ice and snow make cloud recognition and delimitation very difficult indeed, not only from the evidence of photographs, but from infra-red data also. The more unexpected features of the seasons represented by figs. 7.2–7.5 may be summarized and related to anomalies in the general circulation.

March–May 1962 (fig. 7.2). European weather reports confirmed that the excessive cloudiness over north-eastern Europe was associated with unusually frequent cyclonic activity and abundant rainfall. Strong temperate-zone cloudiness in the north Pacific and north Atlantic theatres (> 70% cloud cover as far south as 30° N in some places) was associated with storm tracks displaced further south than usual. March 1962 was the month of a disastrous storm along the east coast of the U.S.A. (Posey 1962). A feature in the March–May period that is more difficult to explain involved excessive cloudiness in the southern Bering Sea. Clapp suggested that this may have been composed of stratus clouds formed in moist air trapped below a low-level inversion caused by subsidence over an icy sea. Conventional analyses pointed to the existence locally of the centre of an anomalous anticyclone during that time.

June–August 1962 (fig. 7.3). An area of below normal cloudiness seemingly extended north from Spain to England and on to southern Scandinavia associated with a region of above normal anticyclonic activity and northerly air flow. Above normal cloudiness north of Japan was associated with lower than average atmospheric pressure. The monsoonal strengthenings of cloud cover across southern Asia are prominent, especially compared with the scant cloud cover there from December to May.

September–November 1962 (fig. 7.4). In this northern hemisphere autumn, below normal cloudiness near the Black and Caspian seas occurred within a region of anomalous anticyclonic activity, while above normal cloudiness in north-eastern North America lay north of a region of unusually persistent cyclones. It is difficult to explain the greater than normal cloudiness over the central north Atlantic and north Pacific oceans, since seasonal 700 mb height contours, and the patterns of their departures from normal, indicated that conditions there were more anticyclonic than usual. However, Clapp noted that in all four seasons there was a tendency for anomalous anticyclones over the sea to be associated with above normal cloud cover, while the opposite was true over land.

December 1962–February 1963 (*fig. 7.5*). The winter of 1962–3 was unusually severe around much of the mid-latitudinal belt in the northern hemisphere (see Andrews 1964). The Pacific, the eastern Atlantic and Europe were all affected by southerly displacements of the dominant storm tracks, bringing unusually strong cloudiness to quite low latitudes. There were marked reductions in the areas of low cloud amounts within the oceanic anticyclones of the subtropics. Hence it is all the more curious to note that the equatorial low pressure belt of cloudiness in both major northern oceans was further north than might have been expected (between 0° and 10° N). Not enough is known of even the major meridional teleconnections in the atmosphere to propose explanatory links between such anomalies.

Lastly, it is noteworthy that trade wind inversion cloudiness off the west coasts of tropical and subtropical continents was generally more pronounced during local winter rather than in summertime. This observation seems to be in accordance with suggestions by Crowe (1949, 1950) concerning anticipated greater strengths of flow and constancies of direction of trade winds during winter, and broader extensions of trade wind airflows in that season of the year. Together these points underline the fact that the maritime tropical anticyclones are stronger in the winter hemisphere than in the summer hemisphere.

More recently, Clapp (1968) has constructed a series of cloud maps for the late autumn season in the northern hemisphere covering 1962, 1963 and 1964, thereby illustrating the scope for comparative studies based on the accumulating archives of satellite data. Perhaps the most interesting point emerging from this later study is the evident tendency for the more prominent cloud changes to show reversals of sign in 1962–3 as compared with 1963–4. The tendency for large departures from normal to balance out from year to year is demonstrated thereby. A similar tendency is well known for parameters such as pressure and temperature, but may seem somewhat surprising for cloud cover, with its more transient character on a day-to-day scale. A mutual interdependence between cloudiness and large-scale circulation patterns may be anticipated by way of explanation. Fig. 7.6, however, depicting percentage changes in late autumn cloudiness in 1963 compared with 1962, hints at the complexity of such a link when compared with fig. 7.7, which shows corresponding changes in mean sea level pressure in millibars.

Thus satellite cloud maps are helping to establish the mean global picture with more accuracy than before, and are providing evidence of shorter-term anomalies that would otherwise have been poorly documented if documented at all. In both ways dynamic climatology stands to gain, concerned as it is

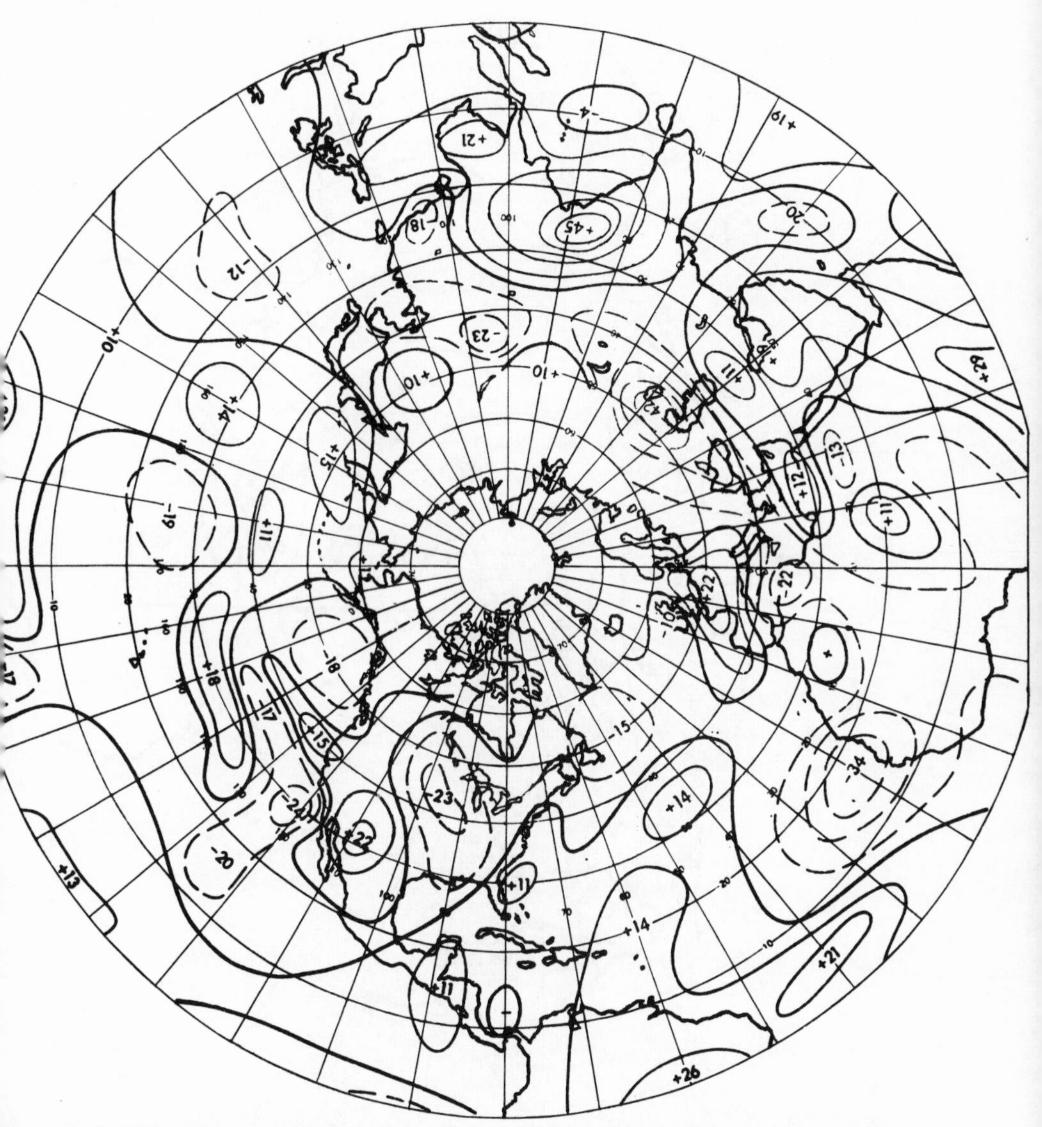

Fig. 7.6 Differences in late autumn (October–December) cloudiness, 1963 minus 1962, from Tiros cloud maps. Isolines drawn for each 10% change, with zero line heavier, positive change depicted by solid lines and negative change by broken lines. Centres labelled in per cent. (From Clapp 1968)

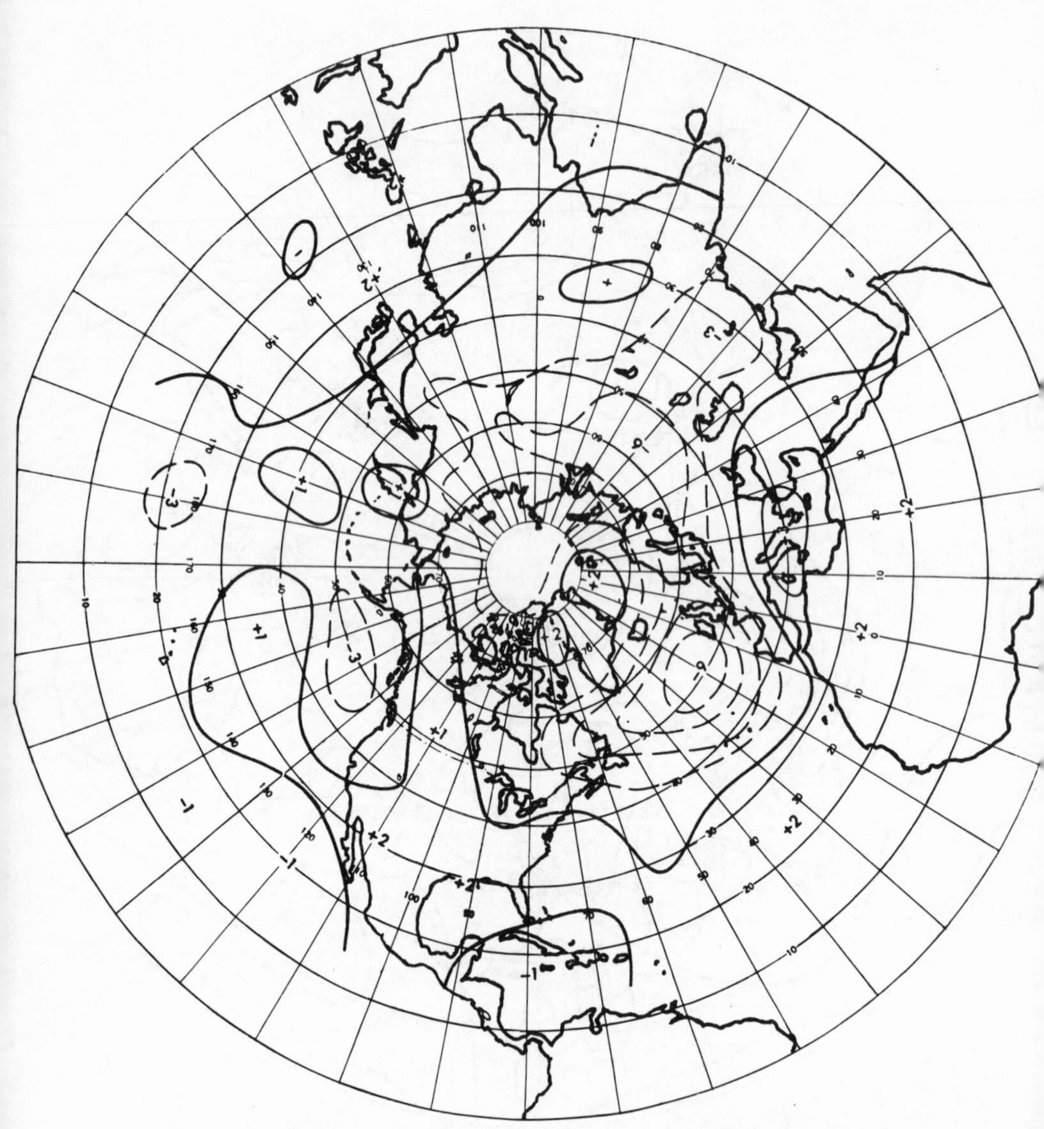

Fig. 7.7 Differences in late autumn sea level pressure, 1963 minus 1962, from conventional evidence. Isolines drawn for each 2-mb change, with zero line heavier, positive change solid, and negative broken, lines. Centres labelled in mb differences. (From Clapp 1968)

with deviations from large-scale mean conditions in the atmosphere. We await with interest and anticipation the routine cloud maps that will be produced by the application of the computer techniques introduced in chapter 4.

Meanwhile, in applied climatology, these new possibilities are most significant *a propos* the attempts by Adem and others to improve weather prediction through extended and long-range forecast periods. For such studies, however, other mean cloud analyses will be required also. Adem pointed to such a need when, in discussing his more complex prediction model, he said that it would require a more realistic appraisal of the vertical distribution of cloudiness. This will involve the type of cloud as well as the strength of cloud cover in the formulation of an appropriate descriptive framework. There is considerable scope in this area for research by satellite climatologists.

Global scale patterns of relative humidity

By the time of writing, very few published studies have involved global scale applications of the method developed by Möller (see p. 108) to determine the mean relative humidity of the troposphere from medium resolution radiation data. However, fig. 7.8 is drawn from one such study, undertaken by Bandeen *et al.* (1965) for one week, namely 11–18 April 1962, based on data from the 6·0–6·5 μm and 8–12 μm channels in Tiros IV. Clearly such a pattern must owe a great debt not only to the general circulation of the atmosphere, but also to the patterns of synoptic-scale systems during those seven days.

Notwithstanding this, the humidity map can be compared usefully with fig. 7.2, which depicts mean cloudiness through the season in which that week occurred. In this way the components associated with local anticyclonic cells or disturbances may be differentiated from those anticipated within the general circulation. So one is not surprised that a belt of generally high humidity meanders around the globe in low latitudes, nor that very low humidities are observed especially across the oceans astride the Tropics – even across areas covered by strong cloudiness, so long as that cloudiness is very shallow. In middle latitudes, where organized weather systems are more mobile than their counterparts of a similar size within the Tropics, something of the probable pattern of frontal structures becomes evident within the pattern of relative humidities. This is especially the case across the eastern seaboard of North America and over the western north Atlantic Ocean (see chapter 12). Great care is necessary, however, in such interpretation work, especially if one is interested less in relative humidity *per se* and more in the actual water vapour mass content of the atmosphere. For example, high relative humidity over the winter Arctic, where temperatures are very low, may involve much less water vapour than a low relative humidity over the Sahara Desert. Thus, for some climatological purposes, other representations of atmospheric moisture may be preferred instead (see chapter 4, p. 104).

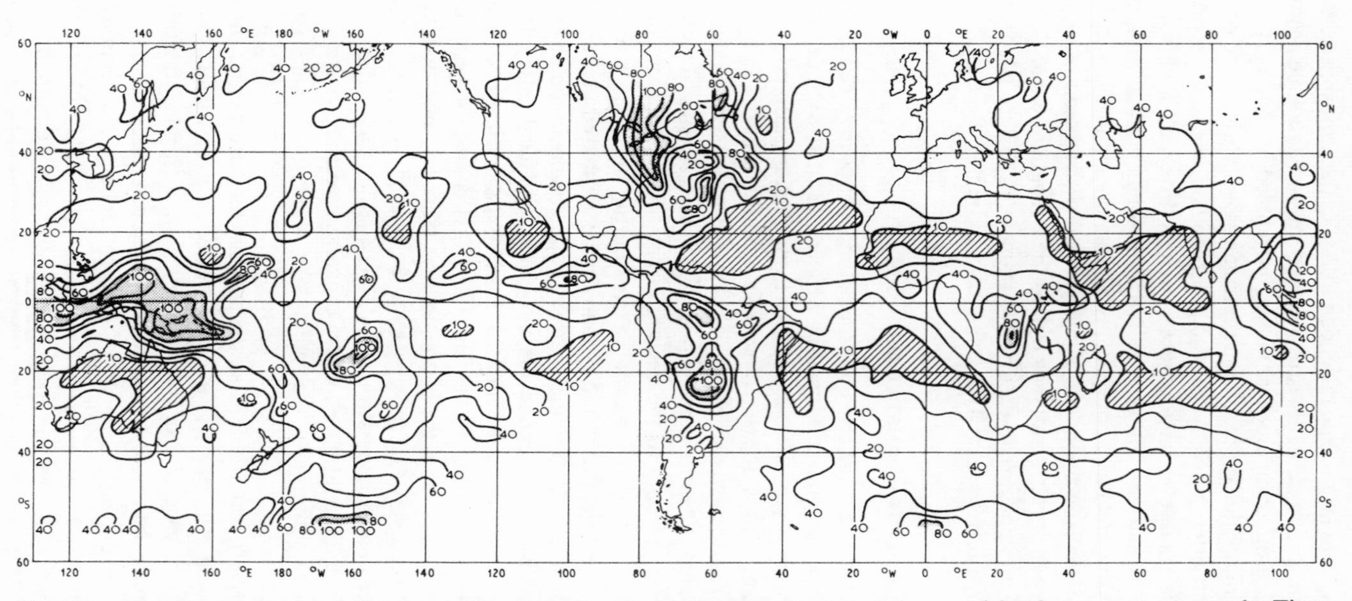

Fig. 7.8 Mean tropospheric relative humidities determined from simultaneous 6·0–6·5 μm and 8–12 μm measurements by Tiros IV during the period 11–18 April 1962, in per cent. (After Bandeen *et al.* 1965)

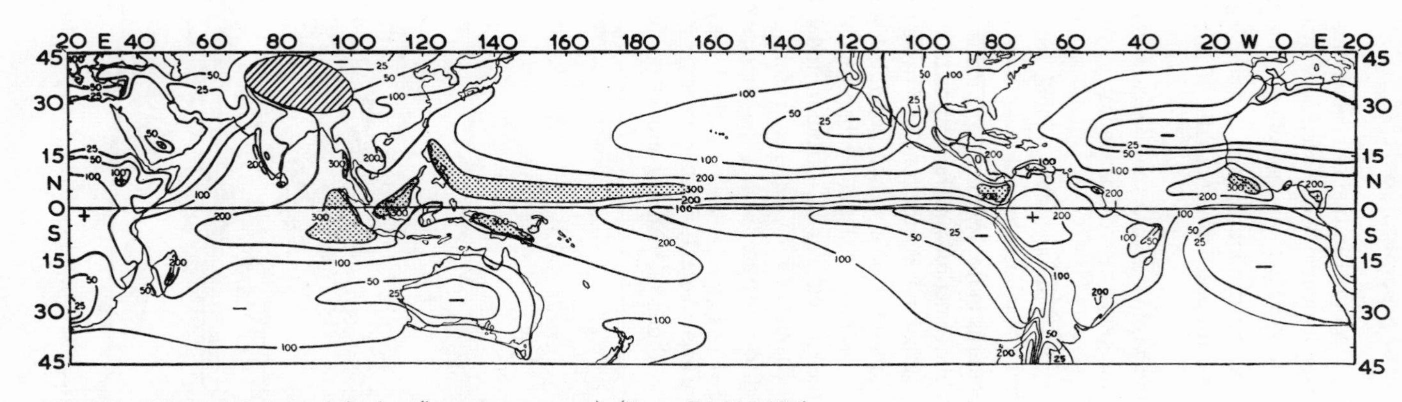

Fig. 7.9 Mean global precipitation (in cm per annum). (From Riehl 1954)

Global scale patterns of precipitation

Fig. 7.9 depicts the global distribution of mean annual rainfall on the basis of conventional data, supplemented where necessary by seemingly reasonable estimates. Three problems in particular confront the compiler of such a map:

1 The available data are produced from a wide variety of rain gauge models, whose measurements of equal falls of rain are by no means all the same.

2 Especially over the world's oceans (but over substantial land areas also) even the present rain gauge network is inadequate for global scale rainfall mapping.

3 The lengths of many of the available rainfall records are too short to give clear indications of climatological mean conditions.

Consequently the patterns in fig. 7.9 are at best tentative approximations of reality. Hopefully, satellite methods of rainfall estimation may lead to improvements in maps of this kind. By the time of writing this is still one of the few central areas in climatology for which no global scale study involving satellite observations has been completed. Perhaps the largest-scale study yet attempted dealt with the tropical Far East from latitude 30° N to 30° S, and from 90° E to 180° E (Barrett 1971). The results for July 1966 and January 1967 may be included to illustrate further the application of the technique outlined on pp. 115–19. Later, such a method may be developed for application to the globe as a whole. It would have at least two advantages over present conventional methods of mapping global rainfall: (1) the estimates

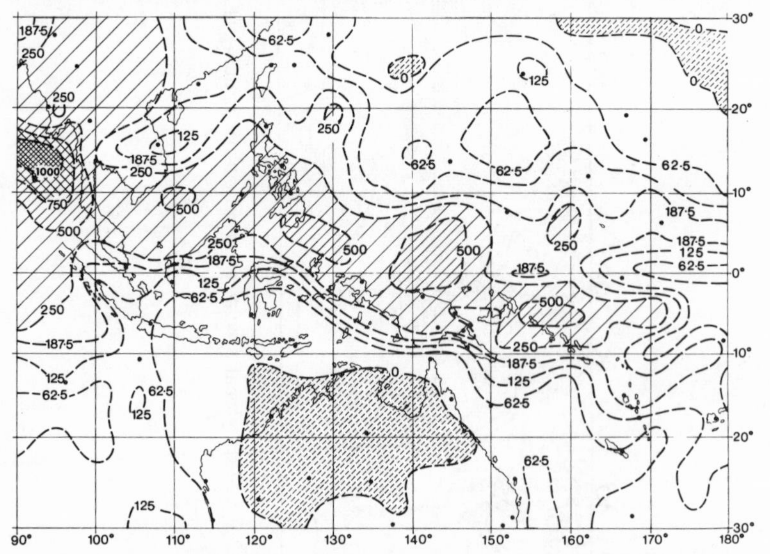

Fig. 7.10 Estimated precipitation, tropical Far East, July 1966 (in mm). (From Barrett 1971)

would be uniformly derived for all grid intersections, and (2) they would all be derived from actual observations.

Fig. 7.10 shows rainfall in July 1966 in the tropical Far East. Rainfall estimates for every 5° grid intersection within the region between 30° N and S, and 90° to 180° E form the basis for this map of an 'estimated rainfall field'. In it, strong gradients of estimated rainfall generally distinguish lower latitudes from those astride the Tropics, although in the western half of the region the belt of heaviest rainfall trended northwestward to the Bay of Bengal. It is interesting to note that centres of especially heavy rainfall (> 500 mm in July 1966) were spaced at rather regular intervals along the main east–west belt of heavy rainfall, but do not seem to have been influenced much by the distribution of land and sea, except over the large equatorial island of New Guinea.

Meanwhile, very low rainfall totals occurred in northern and central Australia, and within areas dominated at that time of the year by maritime anticyclones. However, the equatorial region in the extreme east was, less predictably, a further region of relatively little rainfall. Chapter 9 attempts an explanation of such westward-thrusting penetrations of dry weather between the mean position of the equatorial trough to the north (at about 5° N) and the oblique zone of higher rainfall to the south extending south-eastward from the Solomon Islands to Fiji.

In January 1967 the pattern of heavy rainfall within the equatorial trough adopted an even more intriguing guise than in the previous July (fig. 7.11).

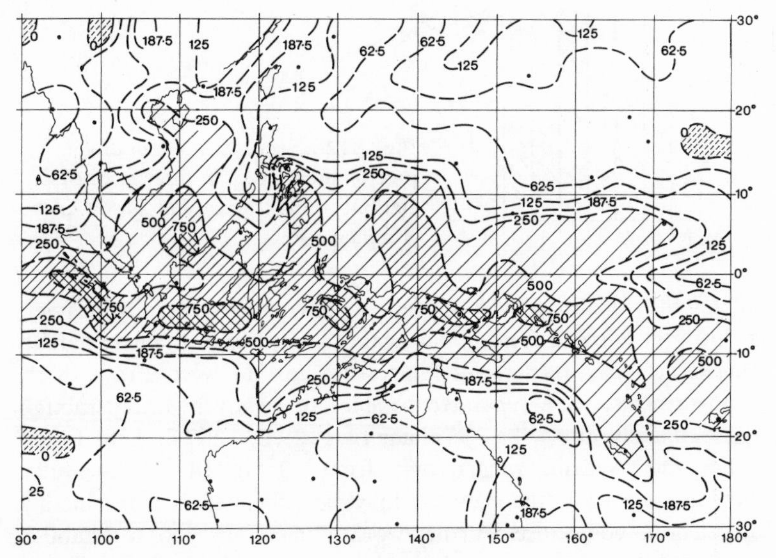

Fig. 7.11 Estimated precipitation, tropical Far East, January 1967 (in mm). (From Barrett 1971)

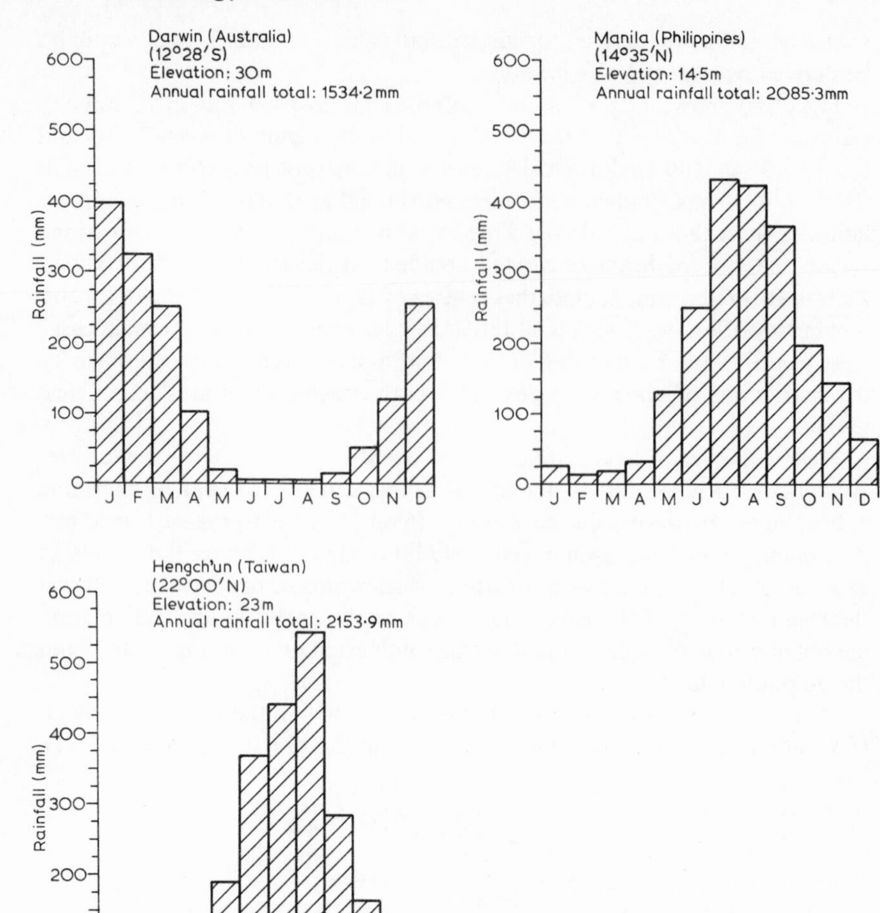

Fig. 7.12 Sequences of monthly precipitation means at three selected stations in the tropical Far East. (After Rumney 1968)

Rainfall along the equatorial belt appears to have been generally heavier, exceeding 750 mm in some areas. Once again the centres of heaviest rainfall were arranged at fairly regular intervals from east to west. From them, extending northwards, were separated tongues of relatively heavy rainfall aligned approximately along the meridians of 112° E, 126° E, 138° E and 150° E. Thus their spacing ranged only from 12° to 14° of longitude. Exactly how these unexpected patterns may be explained is a fascinating matter for scientific conjecture. Further work is necessary, too, to establish whether such patterns recur frequently in the months of January or whether they were an abnormality of the rainfall distribution in 1967.

As in July 1966, drier areas extended across much of Australia, although the wet 'north-west monsoon' affected some of its north coast. Other dry zones flanked the northern tropic on either side of an oblique corridor of higher rainfall off the coast of eastern Asia. Lastly, the vicinity of the Bay of Bengal was much drier than in June. This was anticipated in view of the fundamental circulation changes known to occur there in keeping with the south Asian monsoon (see chapter 11).

Fig. 7.12 summarizes observations of mean monthly precipitations at three Far Eastern meteorological stations particularly subject to strong seasonal rainfall fluctuations. All three stations are critical in the sense that they stand near the limits of the rain-bearing air streams that usually affect them during some part of the year. Reference to figs. 7.10 and 7.11 reveals that their January and July means are matched approximately by extrapolated rainfall estimates read from the satellite-based maps.

One other, more demanding, check was made on the acceptability of figs. 7.10–7.11. The contingency tables, figs. 7.13–7.14, relate extrapolated estimates of rainfall for the surface weather stations indicated in figs. 7.15 with their actual rainfall observations during July 1966 and January 1967. The contingency tables suggest that, although the maps of rainfall estimates were constructed from grid intersection data, not from data for the positions

RECORDED PRECIPITATION (in m.m.'s) \ ESTIMATED PRECIPITATION (in m.m.'s)	0	0 62·5	62·5 125	125 187·5	187·5 250	250 500	500 750	TOTALS
0	5	2						7
0 62·5	2	13	2		1			18
62·5 125			6	3	2			11
125 187·5			1	2		1		4
187·5 250			1	2	1	1		5
250 500				1	4	9		14
500 750					1	4		5
TOTALS	7	15	10	8	9	15	0	64

Fig. 7.13 Contingency table for satellite-estimated and conventionally-recorded monthly precipitation totals for selected stations in the Far East, July 1966. (From Barrett 1971)

of the ground observatories, the satellite patterns are by no means condemned by the available 'ground truth'. Most rainfall estimates fall either in the correct category (adjudged by the conventional measurements), or only one box removed therefrom. Most of those that are two or more boxes removed relate to rainfall stations affected by abrupt topography, such as high mountain ranges or steep volcanic islands surrounded by oceans. It can be argued that the influences of such topographic peculiarities would be well removed from any large-scale precipitation map, but this asks the broader question in climatology: 'How can an equitable selection of rainfall data be made for mapping large-scale distributions?'

In a sense, the satellite method above, doubtful though its estimates over sea surfaces may be (see p. 117), sidesteps that issue through employing cloud

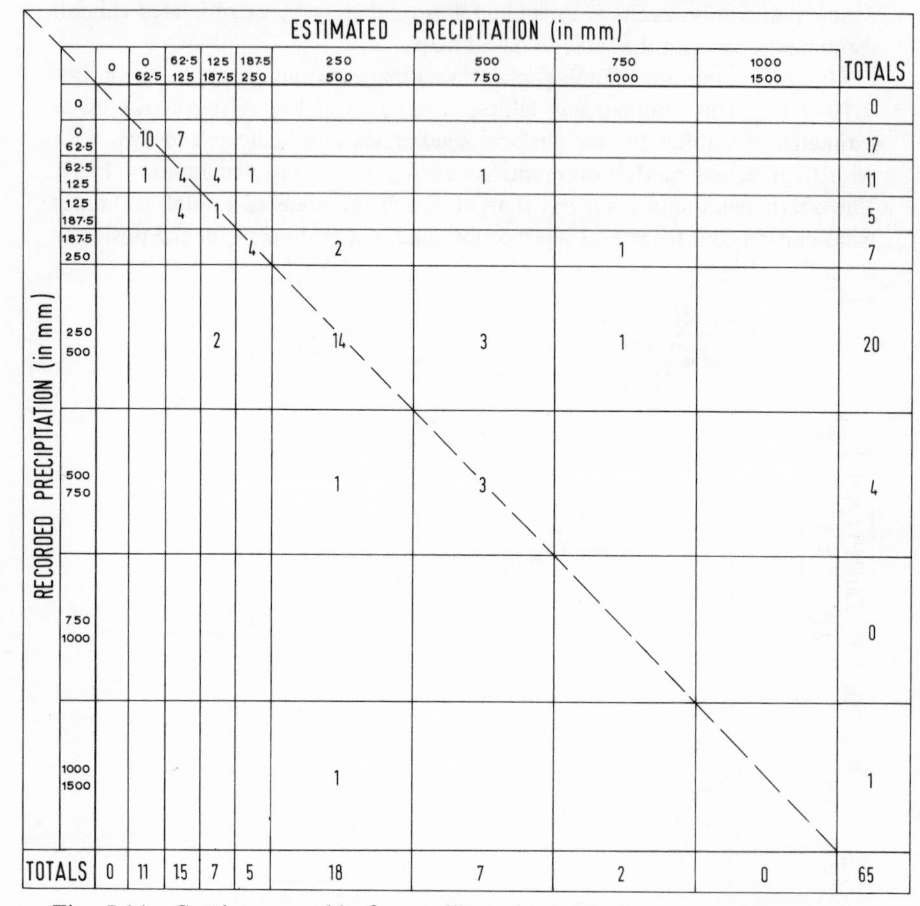

RECORDED PRECIPITATION (in mm) \ ESTIMATED PRECIPITATION (in mm)	0	0–62.5	62.5–125	125–187.5	187.5–250	250–500	500–750	750–1000	1000–1500	TOTALS
0										0
0–62.5		10	7							17
62.5–125		1	4	4	1		1			11
125–187.5			4	1						5
187.5–250					4	2		1		7
250–500				2		14	3	1		20
500–750						1	3			4
750–1000										0
1000–1500						1				1
TOTALS	0	11	15	7	5	18	7	2	0	65

Fig. 7.14 Contingency table for satellite-estimated and conventionally-recorded monthly precipitation totals for selected stations in the Far East, January 1967. (From Barrett 1971)

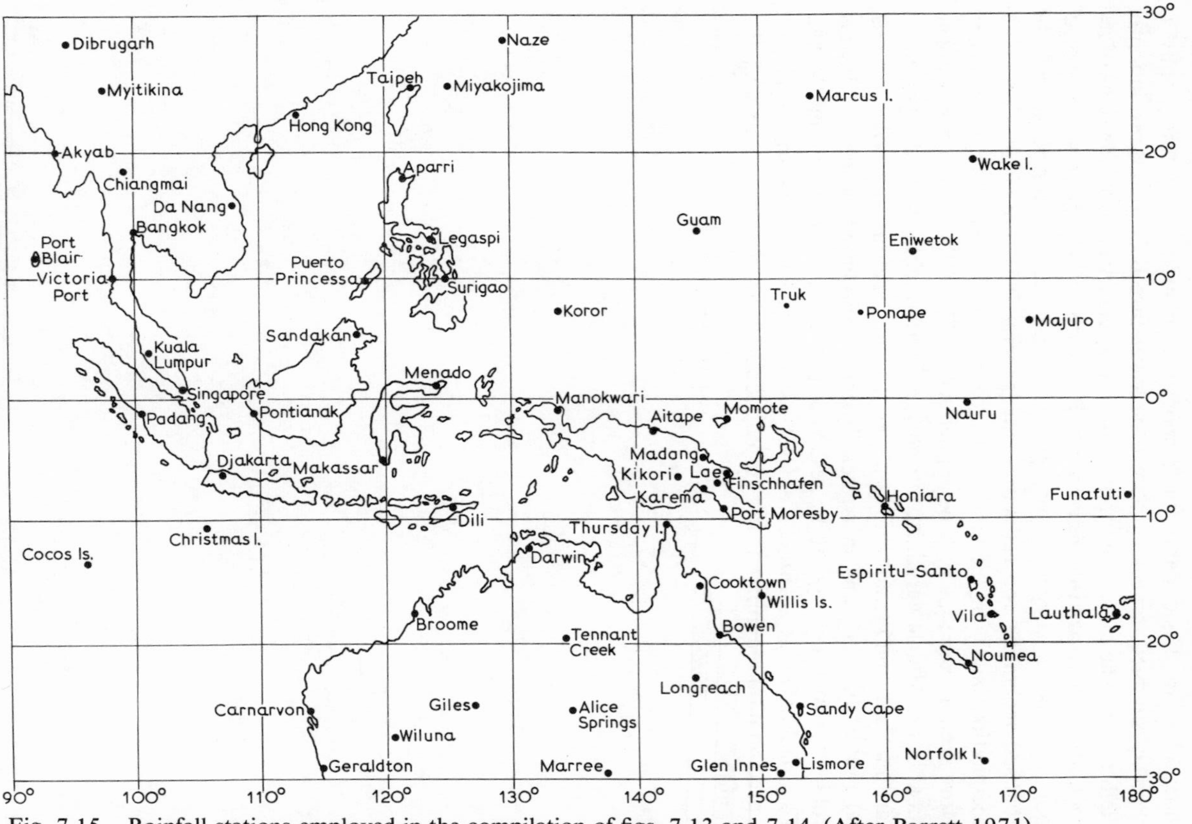

Fig. 7.15 Rainfall stations employed in the compilation of figs. 7.13 and 7.14. (After Barrett 1971)

evidence as its basic data; the response of cloudiness to topography is relatively subdued, especially above the lower threshold of the resolution of satellite data. The most important result is that very local perturbations of the precipitation field are thus filtered out. Hence, the relatively small number of poor agreements suggested by figs. 7.13 and 7.14 cannot condemn the proposed technique for estimating the general structure of a broad-scale precipitation field. On the other hand, of course, further work is necessary before such a technique may be applied more widely, for example on a quasi-global basis. Almost certainly, different evaluation tables will be required for maritime and continental mid-latitudes. In mountainous and polar regions, and mid-latitude regions in winter, frozen precipitation poses still further problems, which have yet to be resolved. However, notwithstanding the lack of a global coverage at present, at least a start has been made towards that end in precipitation mapping, as in mapping global cloud parameters and relative humidity.

8 Global patterns of atmospheric circulation

Scales of circulation

Before we proceed to discuss the climatological contributions made by satellites to circulation studies some background information is in order. With this in mind the extents – and limitations – of satellite-based contributions will be more clearly apparent. It was remarked in chapter 5 that organized airflows are conveniently divided into three classes determined by their scales.

Primary circulations: the general (or global) circulation of the atmosphere

Large-scale atmospheric circulations are so complex and variable that mean flow patterns must be compiled if the general order of the atmosphere is to be examined. Commonly such patterns are represented in two dissimilar ways, illustrated by figs. 8.1 and 8.2.

The first involves a modelling of the observed zonal wind systems, along with indications of the vertical links that must unite upper and lower tropospheric patterns in an integrated three-dimensional whole. Of course, model depictions of the general circulation differ from one authority or school of thought to another. Fig. 8.1 simply represents one of the more widely accepted at the time of writing. An excellent summary of the historical development of thinking on this subject has been prepared by Lorenz (1967).

Alternatively, or additionally, mean flow may be represented by average streamlines of airflow at the earth's surface as illustrated by fig. 8.2 (see Mintz and Dean 1952). Such representations are helpful in that they take account of longitudinal variations in flow, i.e. persistent local effects, mostly caused by differences of heating over land and sea and/or variations in surface friction.

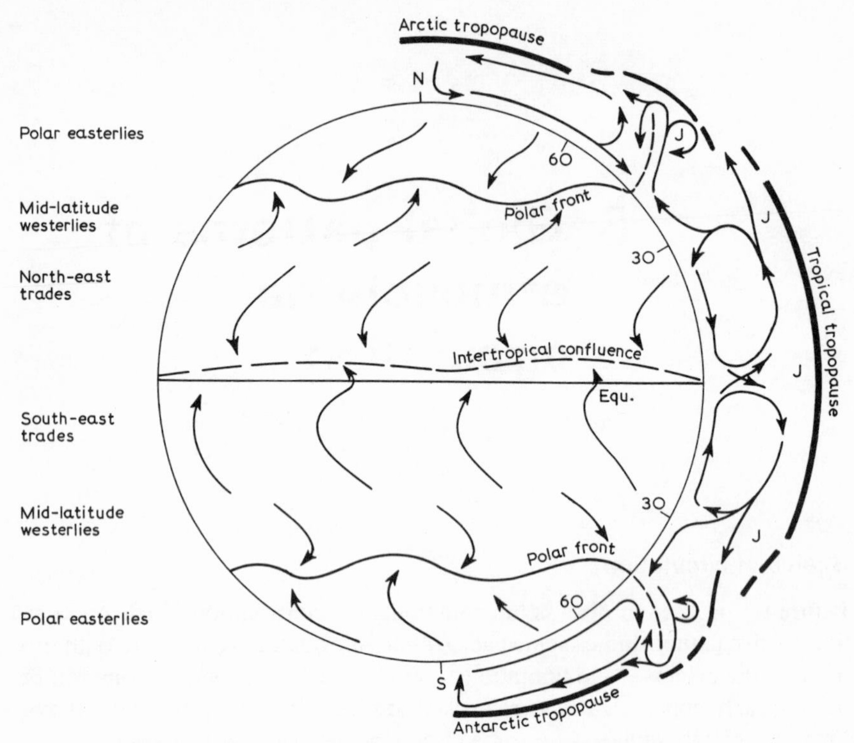

Fig. 8.1 A three-cell model of the northern hemisphere meridional circulation, and the primary wind belts in the lower troposphere. (J indicates most frequent jetstream positions) (Largely after Palmen 1951)

In both seasons the maps show large diverging swirls over the oceans (associated with warm subtropical highs) and a pronounced zone of equatorial convergence. In plan in middle latitudes the westerlies undulate meridionally in irregular stationary waves rather than exhibiting purely zonal (i.e. west–east) flow.

The very different distributions of land and sea in the northern and southern hemispheres prompt strong hemispheric contrasts in atmospheric flow especially polewards from 20° N and S. In particular, in the northern hemisphere winter a strong divergent vortex develops over Siberia (a region of intense cold and anticyclonic subsidence), while relatively persistent convergent eddies form over the Pacific and Atlantic oceans near 60° N. These more or less permanent zones of convergence and low pressure are the so-called Aleutian and Icelandic lows. In summer these features weaken, but remain identifiable in the mean wind field picture. In middle latitudes in the northern hemisphere, seasonal reversals of circulations are more or less well marked over the continental interiors, prompting the most remarkable seasonal contrasts of weather in the monsoon lands of southern Asia. In

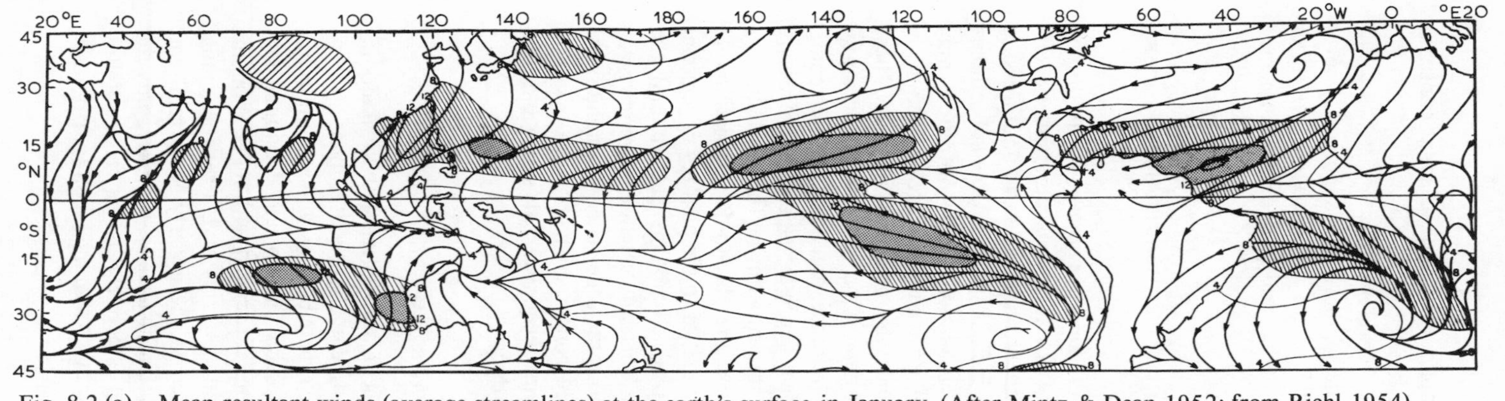

Fig. 8.2 (a) Mean resultant winds (average streamlines) at the earth's surface in January. (After Mintz & Dean 1952; from Riehl 1954)

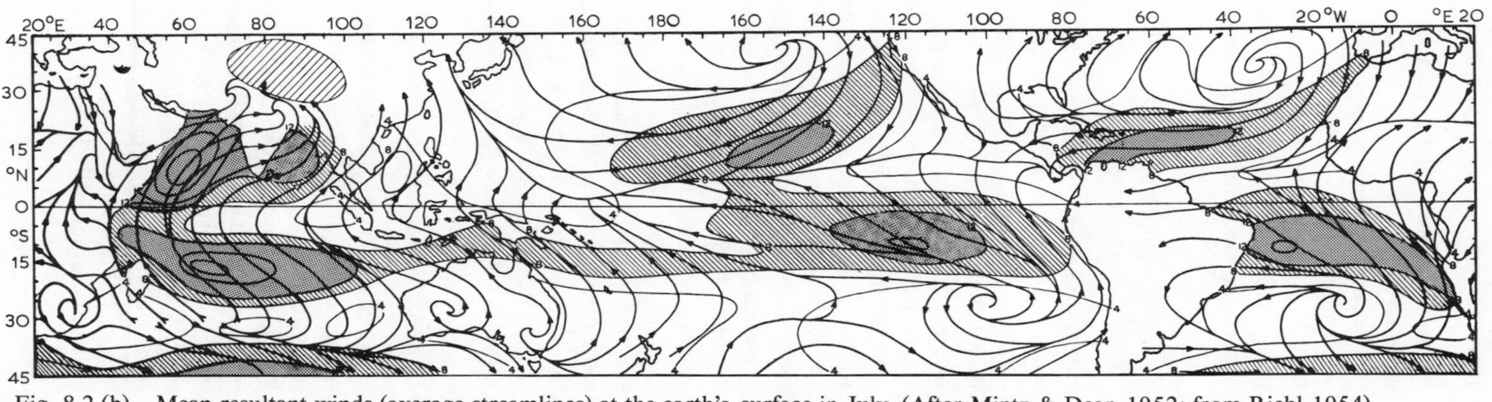

Fig. 8.2 (b) Mean resultant winds (average streamlines) at the earth's surface in July. (After Mintz & Dean 1952; from Riehl 1954)

the southern hemisphere, meanwhile, comparatively tiny seasonal contrasts appear.

The supposed mean circulations aloft are represented by figs. 8.3–8.4: the conventional data coverage, especially in the southern hemisphere, has never been satisfactory. Strong circumpolar vortices are indicated around either pole, yielding to cellular highs in lower latitudes, and to poorly documented upper-level circulations near the Equator.

Secondary circulations: short-lived synoptic scale contributions to mean global circulations

One of the dangers of the general circulation concept is that it is based upon statistical abstractions from the reality of instantaneous atmospheric circulations. In daily synoptic weather maps the generalized wind and pressure belts indicated by figs. 8.3–8.4 are seen to include very variable subordinate centres of cyclonic and/or anticyclonic circulations. Some of these may fluctuate more in shape than position, but all arise and dissipate or fluctuate in intensity through short periods of time, generally measured in days. These secondary circulations include a variety of depression types in low latitudes: anticyclonic cells in subtropical and polar latitudes, and across winter continents in mid-latitudes; maritime extratropical depressions; and low pressure organizations across arid mid-latitudinal continents during the local summer seasons. These are largely responsible for the day-to-day variations in weather, and therefore add mesoscale complexities to the pictures of regional climates etched out generally by the global circulation pattern and the seasonal swing of its principal components. The daily satellite coverage gives more direct evidence of secondary circulations than the general circulation of the atmosphere. Chapters 9–13 are mostly concerned with those synoptic scale systems. In this chapter we shall examine the global scale space and time averages constructed from runs of satellite data.

Tertiary circulations: local scale embroideries of weather and climate

Many generally small-scale circulations arise in association with local modifications of the strength and direction of wind flow. Usually prompted by local topographic factors, such as relief, land/sea distributions and city morphologies, many tertiary winds are of considerable human and economic significance. Apart from small-scale circulations associated with more spontaneous atmospheric phenomena such as thunderstorms and tornadoes, most tertiary circulations are quite tightly related to the underlying terrain, and tend to be diurnal in intensity and direction. Unfortunately, however, only the largest tertiary circulations can be studied effectively as yet from satellite evidence; paramount among these are the south Asian monsoon (partly to be

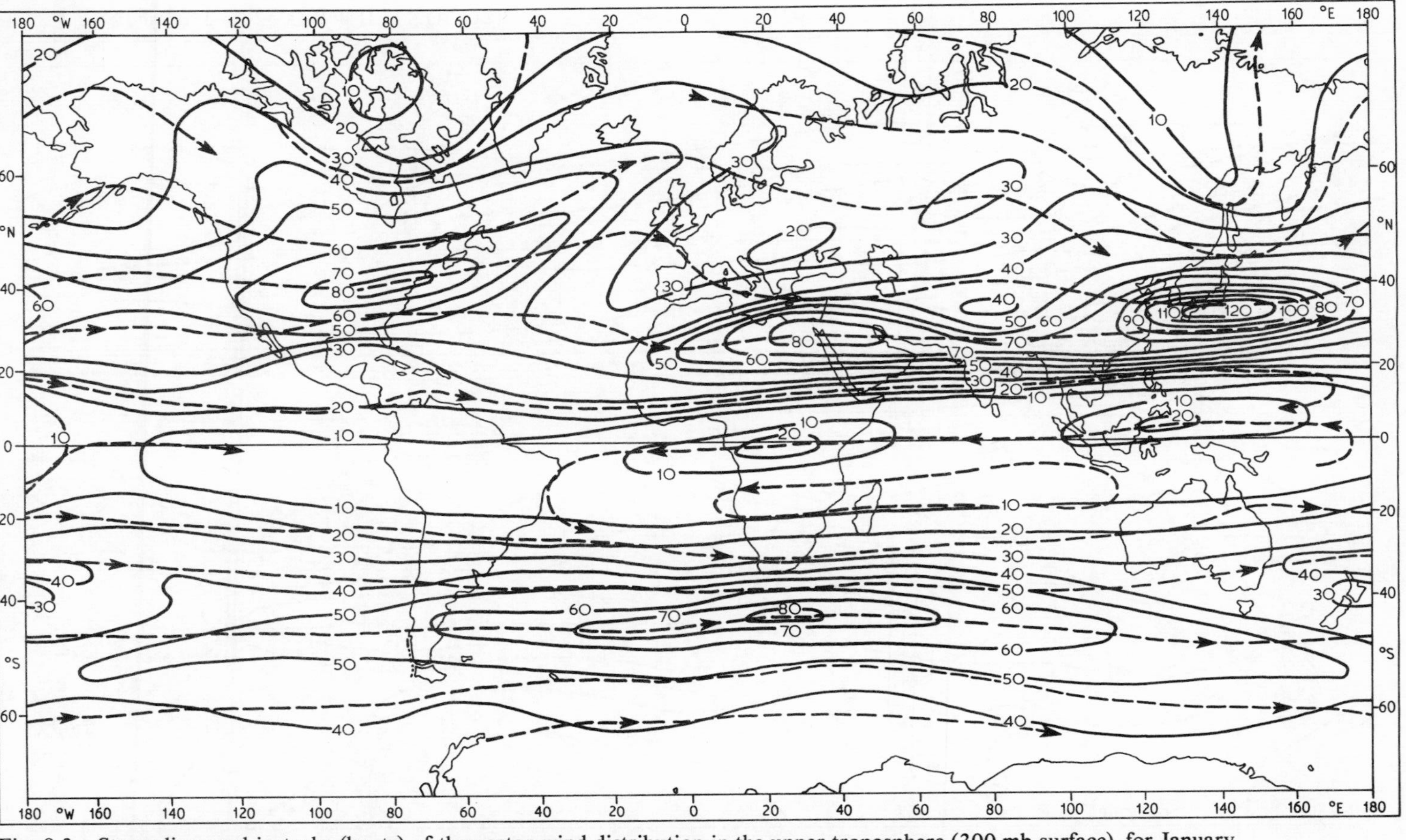

Fig. 8.3 Streamlines and isotachs (knots) of the vector wind distribution in the upper troposphere (300-mb surface), for January. (From H.M.S.O. 1971)

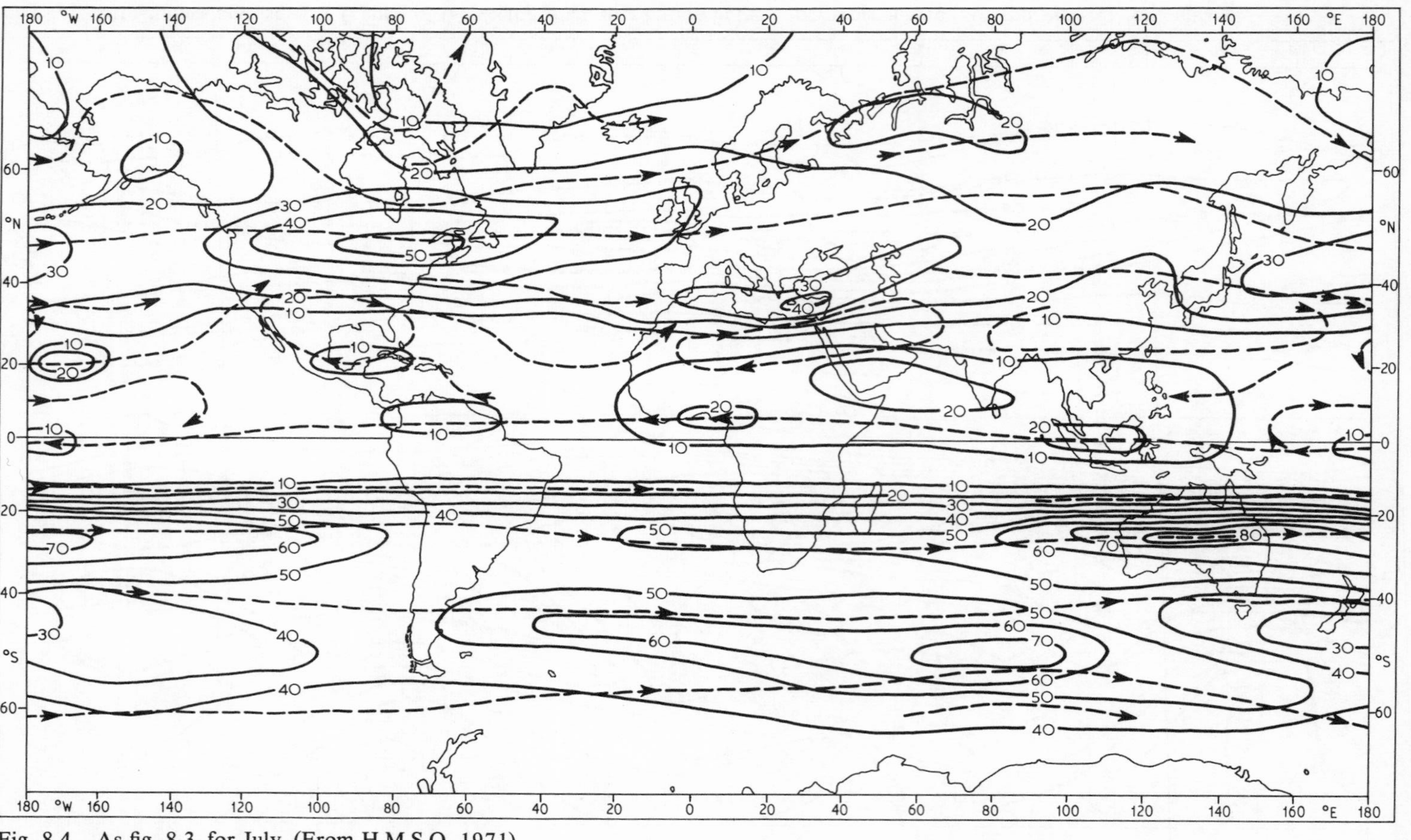

Fig. 8.4 As fig. 8.3, for July. (From H.M.S.O. 1971)

conceived as a land/sea breeze phenomenon), and the Chinook winds of North America. Hence little consideration is given to tertiary or local wind systems in the regional section of this book.

Summary of major circulation patterns

Now large-scale circulation patterns within the earth's atmosphere not only vary somewhat from time to time but also differ from one level within the atmosphere to others. Modern climatology is less preoccupied than climatology used to be with patterns at the earth/atmosphere interface alone. Much more than previously the atmosphere is being seen by climatologists as a three-dimensional continuum in which changes and variations at one level are always likely to have been influenced by, and/or may in turn have influenced, other levels above or below. In this chapter, we shall be concerned with quite highly contrasting major circulation patterns at three different levels: (1) the lower stratosphere, (2) the lower troposphere and (3) the upper troposphere.

By way of introduction, fig. 8.5 shows the mean vertical temperature

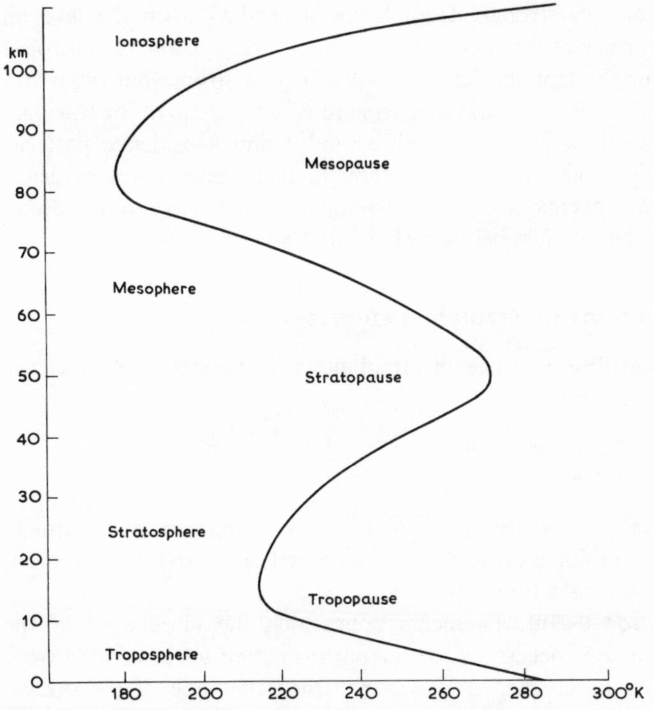

Fig. 8.5 The generalized vertical distribution of temperature through much of the earth's atmosphere. (From Petterssen 1969)

profile of the earth's atmosphere. In this chapter, as in this book as a whole, we are concerned almost exclusively with the 'lower atmosphere' – that is to say the layer between the earth's surface and the level of the commencement of the negative lapse rate (rise in temperature) within the stratosphere. Above this layer the climate is dominated almost exclusively by solar controls and owes little to earth surface geography.

In the lower stratosphere, where water vapour concentrations are very low, ozone and carbon dioxide are the chief absorbers of radiation. Ozone, which has its peak density at about 22 km, absorbs ultra-violet radiation from the sun. However, owing to its efficiency as an absorber of solar radiation, ozone filters out most of the short-wave energy in its upper layers before the lower stratosphere is reached. Carbon dioxide, on the other hand, absorbs in longer wavebands, especially terrestrial radiation that would otherwise be lost to space. Hence, the mean circulations of the lower stratosphere would seem likely to owe something to influences operating in both directions, downward from space, and upward from the surface of the earth. Both influences might be expected, of course, to differ from one latitude to another.

Meanwhile, in the troposphere, circulations may be expected to be influenced more or less strongly from below depending upon the level in question. At the surface of the earth, topography is a very powerful control of climate; just below the tropopause this control is less strong, but often still quite significant. By definition, the troposphere is the 'turbulent' or 'turning' layer of the atmosphere, being stirred by uplift and subsidence patterns prompted basically from below. Note, though, that some recent evidence suggests that some events above the tropopause may have quite direct influences on circulations just below it (Fritz and Soules 1970).

Satellite contributions to circulation studies

Let us consider satellite evidence of circulations in the three layers of the atmosphere differentiated above:

The lower stratosphere

Two major aspects of the lower stratosphere will concern us – namely seasonal contrasts in its indicated circulation patterns, and the apparent behaviour of the layer as a three-dimensional whole.

In the search for useful statements concerning the circulation of the stratosphere, recourse is necessary to medium resolution infra-red data from early Tiros flights. In chapter 2 it was recorded that one of the optical channels of Tiros VII measured infra-red radiation emitted mainly by carbon dioxide at wavelengths ranging from 14·8 to 15·5 μm. More than 96% of such radiation stems from altitudes above 10 km (Nordberg 1965). It is possible, therefore, to translate these M.R.I.R. measurements into evaluations of mean temperatures through the isothermal layer of the lower stratosphere, i.e. through the layer in which temperature varies little with height. At that level isotherms and isobars may be considered as generally parallel or near parallel to each other, a fact that permits its temperature maps to be read as pressure maps in the absence of geopotential observations. Furthermore, since it is more reasonable than at any level beneath to assume that the wind flow in the lower stratosphere is geostrophic, mean monthly wind parameters can be deduced from the temperature maps; on account of the physical links between temperature, density, pressure and wind flow such maps suffice to suggest much of the basic climatology of that atmospheric layer. In the figures that follow high temperatures indicate high pressure at some middle level in the lower stratosphere, and low temperatures indicate low pressure. An illustration of the reasoning behind this statement is included in the first of the three (northern summer, northern winter and transition) seasonal accounts that follow.

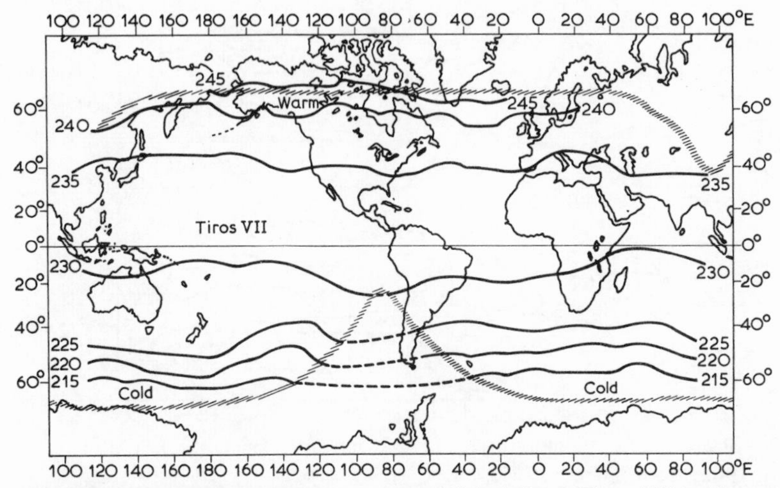

Fig. 8.6 Stratospheric temperatures established from Tiros VII evidence, 19–25 June 1963 (°K). (After Nordberg; from Barrett 1967)

Northern summer: 19–25 June 1963 (fig. 8.6)

During this period the lower stratospheric isotherms were arranged roughly along lines of latitude. An overall gradient of some 35 °K was directed from maximum temperatures in subarctic latitudes to minima over Antarctica. In order to understand the relationship between temperature and pressure it helps to consider the likely slope of some appropriate pressure surface, say the 50 mb surface whose mean altitude is at about 25 km. Heating and expansion of air in northern latitudes must raise the level of the 50 mb surface while cooling and contraction in southern latitudes must lower it below its global mean height. Thus, during the northern summer, atmospheric pressure at 25 km in the north must normally exceed 50 mb, while failing to achieve that value in the south. The result is the establishment of an enormous anticyclone of hemispheric proportions over the northern hemisphere, and an equally large cyclone south of the Equator. A global balance of winds in, and against, the sense of the rotation of the earth on its axis is a physical necessity; at this lower stratospheric level it is achieved by opposing circulations in the two hemispheres. The chief cause appears to be the length of daylight. In northern latitudes in summer the long hours of daylight promote even higher temperatures there than around the Equator. Similarly, higher temperatures occur around the Equator compared with south polar regions during their period of winter night.

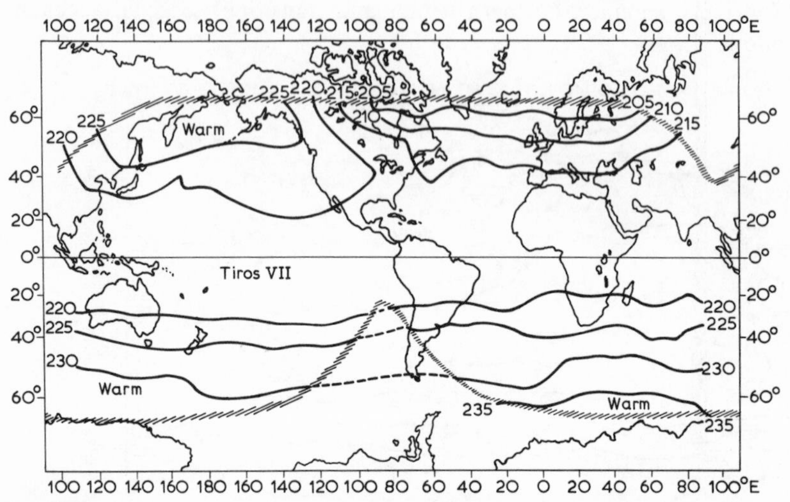

Fig. 8.7 Stratospheric temperatures, 15–22 January 1964 (°K). (After Nordberg 1965; from Barrett 1967)

Northern winter: 15–22 January 1964 (fig. 8.7)

Here, the June situation is seen almost exactly in reverse. The highest temperatures are found in the Antarctic region, and the lowest to the north of Iceland. It will be noted, however, that the pattern in the northern hemisphere is by no means as simple as that further south. In particular, temperatures were curiously high, over much of eastern Asia and the northern Pacific, finding their culmination over the Bering Straits in a pool of anomalous warmth. In circulation terms, the southern hemisphere was covered by a vast anticyclonic circulation, while a great circumpolar vortex around the northern hemisphere was interrupted by a so-called 'Aleutian anticyclone'. An unexpectedly warm stratosphere, occuring as a regular climatological feature during winter in the general area of the Aleutian Island chain, has long been recognized as the cause of a strong anticyclonic circulation in that region of the stratosphere (Wege 1957). The morphology of this warming high above the north Pacific has received much attention in the past, and various theories have been advanced to explain its origin. However, no general agreement has been reached yet on this matter (see Boville 1963; Hare 1962).

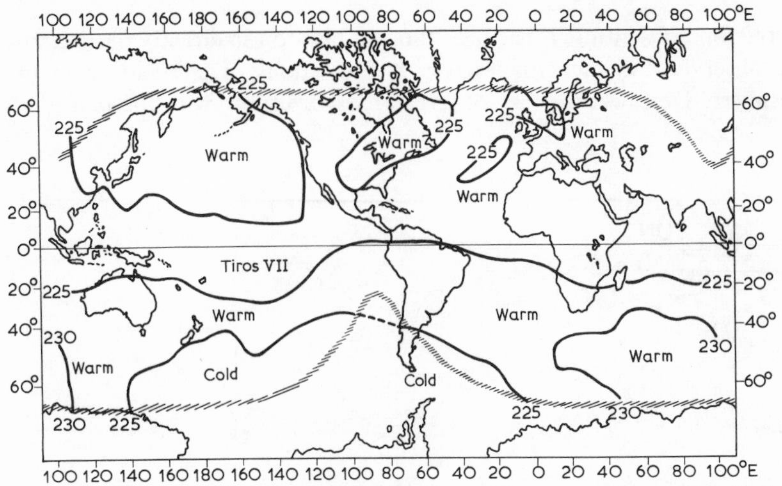

Fig. 8.8 Stratospheric temperatures, 25 September–1 October 1963 (°K). (After Nordberg 1965; from Barrett 1967)

Northern autumn: 25 September–1 October 1963 (fig. 8.8)

This period of transition may be seen as a representative of both transition season patterns, i.e. of patterns in spring or autumn. Such patterns are particularly important, since obviously most of the stratospheric year must witness some such distribution of temperatures intermediate between the more or less mirror-image distributions in January and July. Unfortunately, but not unexpectedly, such patterns are particularly complex, though the gradients of temperature, density and pressure are all quite slight. The indicated circulations must be relatively light and variable by comparison with their high season counterparts.

More recently than the Tiros VII observations, SIRS data from Nimbus III have prompted further analyses of large-scale circulation dynamics in the stratosphere. Such studies have been begun only recently, but their earliest results have already thrown some useful new light upon stratospheric climatology (Fritz 1970; Fritz and Soules 1970). The new evidence depends upon measurements from one of the eight spectral channels in that spectrometer, specifically the channel centred on 15 μm. As Fritz and Soules explained, 'changes in the radiances observed are due mainly to temperature changes somewhere in the upper 100 mb of the stratosphere'. However, two severe problems are encountered:

(*a*) The translation of the radiances into temperature evaluations necessitates a number of complex assumptions (Fritz 1969).
(*b*) In such translation procedures voluminous data processing is unavoidably involved.

To the time of writing, it has been thought best to use directly the patterns of the observed radiances themselves as indications of circulations in the stratosphere. Here we may be concerned most fruitfully with the two types

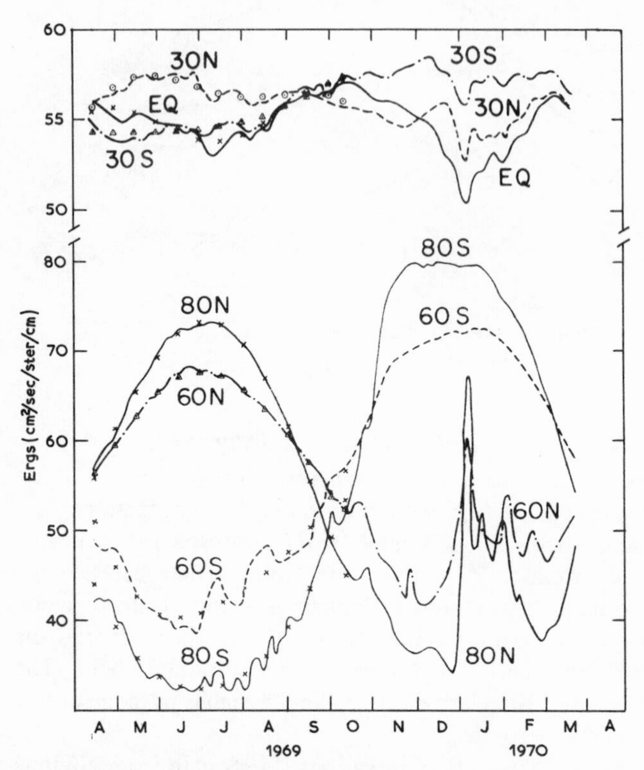

Fig. 8.9 Annual marches of radiances for selected seasons from measurements made by Nimbus III in the 15 μm–CO_2 waveband. Data at one latitude represent observations for a 4° latitude circle once daily. Crosses, circles, etc., indicate particular radiance values for the annual and semi-annual periods calculated from the data by a least-squares method. (From Fritz & Soules 1970)

of pattern changes through time that changes in those radiances have indicated.

(*a*) *Seasonal changes.* These are akin to those discussed above on Tiros evidence. However, fig. 8.9 takes the story further by illustrating seasonal marches of radiances for selected latitudes. It appears that there is a relatively smooth, rapid rise in both hemispheres from each spring to summer, followed by a similarly smooth and rapid fall from summer into autumn. In low latitudes the twin-peaked cycle was expected, since the sun traverses the Equator twice each year.

It is most interesting to note that the summer radiance at 80° S rises to a higher maximum than the comparable radiance at 80° N. The earth is closer to the sun in December than in June, and apparently heats the southern stratosphere in summer to the higher temperature. In contrast, winter and equinoctial radiances were more nearly equal everywhere.

Shorter-term fluctuations seem to be embroidered on the seasonal march of radiances especially in high latitudes in autumn and winter. In low latitudes,

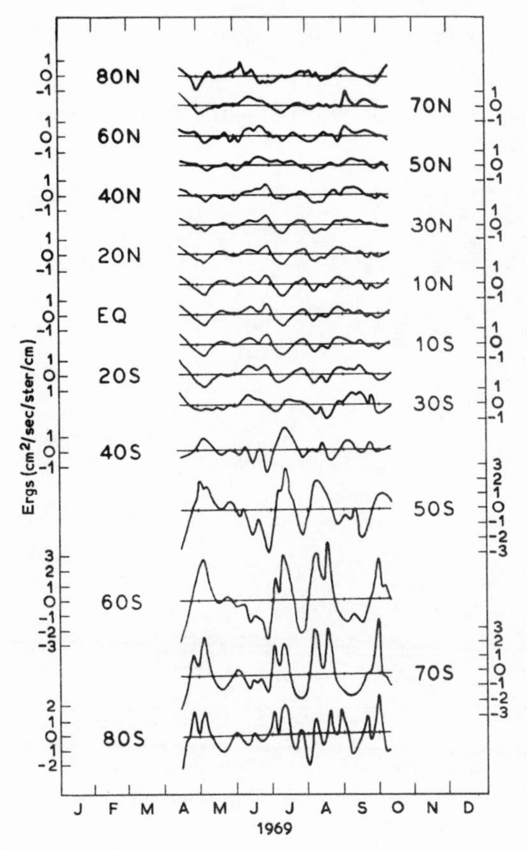

Fig. 8.10 Deviation of averaged latitudinal radiance from a least-squares fit for 80° N to 80° S, from 14 April to 10 October 1969 showing relationships between warming and cooling periods in the northern and southern hemispheres. (From Fritz & Soules 1970)

short-term fluctuations must be investigated with more caution on account of the intrinsically more complex underlying annual cycle.

(*b*) *Irregular fluctuations*. We see from fig. 8.9 that, in tropical latitudes, irregular short-term fluctuations are all in phase, while at 60° in the winter hemispheres they are exactly out of phase with those in the Tropics and the summer hemispheres. Fig. 8.10 illustrates and elucidates such relationships. This is a graphical plot of further deviations of latitudinal radiances from a least-squares fit for 80° N to 80° S from 14 April to 10 October 1969. It shows the gradual reversal of warming and cooling trends from middle latitudes in the southern (winter) hemisphere to the relatively much more subdued trends in the northern (summer) hemisphere.

Geographically, the pattern of change of radiances must be much more complex. Hence the call for maps like fig. 8.11, which depicts changes in radiance patterns between 25 June and 10 July 1969. Three zones can be differentiated:

1 A zone of variable change north of 40° N.
2 A broad zone of uniform cooling between 40° N and 20° S.
3 A zone of large-scale warming south of 20° S.

Fritz and Soules, following Manabe and Hunt (1968) invoked large-scale meridional circulations to explain the cooling observed in the Tropics. Mean-

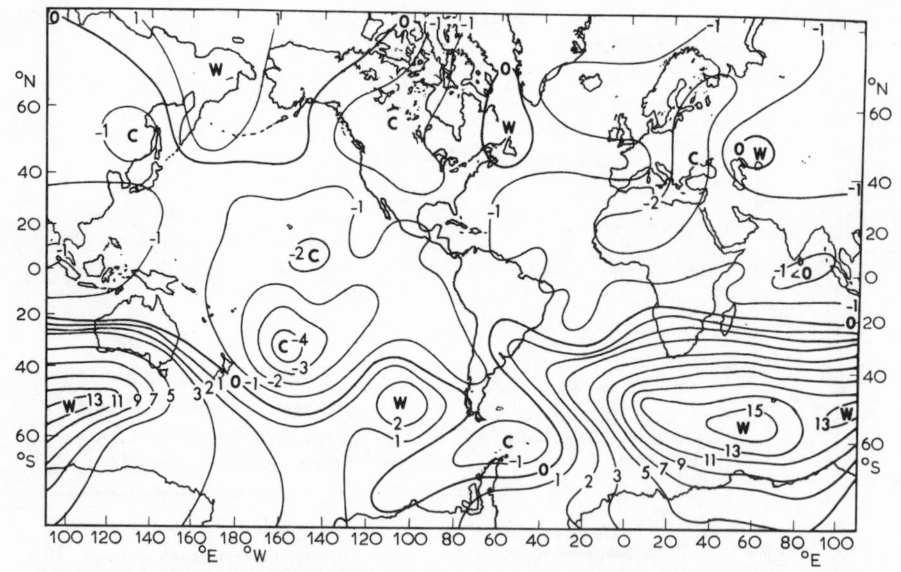

Fig. 8.11 Change in radiance (in ergs) from 25 June 1969 to 10 July 1969. Cooling occurred everywhere from 20° S to 40° N, while large-scale warming (W) took place over much of the region south of 20° S. (From Fritz & Soules 1970)

while, the winter hemisphere is warmed by large-scale heat transfer by atmospheric eddies. These are masses that retain their identities for a limited time while moving into a new fluid environment. Sinking motions, related to circulations in the vertical, possibly play a supporting role.

Lastly in this section we may make some progress towards an explanation of the circulation of the lower stratosphere with its various regional components. The neatest modern description of the system as a whole claims that the stratosphere seems to act like a huge standing wave related to the zone of maximum insolation. Within it the amplitude of change appears to be larger in middle and high latitudes of the winter hemisphere than in the Tropics or summer hemisphere, through both of which a lag effect is evident.

Fritz and Soules (1970) concluded their interesting discussion of dynamically prompted variations in the stratosphere by suggesting that, in certain circumstances, those patterns of change aloft may have significant influences on large-scale events at least in the tropical troposphere. If such a possibility becomes substantiated – perhaps through observed relationships between stratospheric temperature variations and changes in surface pressure, wind circulation or cloudiness – then these studies of stratospheric circulation will become even more significant in climatology, even the climatology of the surface of the earth.

The lower troposphere

While the observed large-scale patterns of wind and pressure in the lower troposphere (outlined briefly in chapter 5) result primarily from the imbalance of radiation between lower and higher latitudes, they are controlled importantly by the angular momentum of the earth and its atmosphere. This may be conceived as the tendency for the earth's atmosphere to move with the earth around the axis of rotation. With a uniformly rotating earth and atmosphere, the total angular momentum must remain constant, as required by the principle of the conservation of angular momentum, i.e.

$$\frac{d}{dt}(f+\zeta)=0 \tag{8.1}$$

where f, ζ and t have the usual meanings (see chapter 5). Within the general circulation of the atmosphere, poleward transports of both energy and momentum are effected by quite complex patterns of airflow. For two primary reasons, relatively few satellite contributions have been made as yet to our knowledge and understanding of the large-scale components of circulation patterns within the lower troposphere:

(*a*) Mean circulation patterns within the troposphere, particularly near the surface of the earth, have been quite well known for many years. Except in high latitudes and certain remote maritime regions, the network of surface stations, while still inadequate for much short-term forecasting, has been moderately satisfactory for large-scale climatological inquiry. Therefore less remains to be added or corrected by satellites.

(*b*) It is rather difficult to improve on conventional maps of low-level pressure and wind patterns from the indirect indicators given by satellite photographs and infra-red data. At higher levels in the atmosphere, however (e.g. the upper troposphere and lower stratosphere), comparatively few direct observations of meteorological parameters are made as a matter of routine. Hence, even indirectly drawn inferences of pressure and winds may have definite merits in studies of high-level climatologies. The same is not true of the lower troposphere, whence indirect indications of atmospheric circulation are often less definitive. The relatively simple relationships between air temperature, density and pressure that apply aloft hold much less well in the boundary layer of the atmosphere. Although cloud patterns and arrangements can sometimes be invoked as tracers of low-level airflow, especially at synoptic and smaller scales, there are many situations for which photographic analyses are of little help, especially where once-daily data are all that are available.

Little of real interest, then, can be reported as a result of satellite-based research into global scale wind circulations near the surface of the earth, but a brief discussion of representative multiple-exposure average cloud photographs may be included to introduce the global circulation as seen by satellites (see Kornfield *et al.* 1967) (plates 17 and 18). Consider, for example, the mean brightness of the northern hemisphere from 1 to 15 July 1967 (plate 18). A number of important features emerge.

(*a*) In very low latitudes, within a few degrees of the Equator, a band of variable breadth of bright cloudiness is visible. This relates to the equatorial trough of low pressure. The bright but variegated equatorial cloudiness is broken immediately east of Africa, but strengthens and become more dense over southern Asia. These are both features related to the Indian Ocean monsoon (see chapter 11). In winter the Indian region is largely cloud free, under the influence of anticyclonic outflow from Asia.

(*b*) Astride the Tropics a darker, less cloudy zone stretches around the globe, broken notably by the south Asian monsoon cloudiness, and bright, topographically stimulated, cloudiness along the isthmus of Central America. The Atlantic subtropical anticyclone appears to be the best defined of the centres of subsidence and divergence where deep cloudiness is very rare. It should be noted, however, that even across the tropical north Atlantic there is more evidence of cloud across the east of the ocean than its west. A similar contrast is even more strongly marked across the Pacific. Cloudiness there is further complicated by two central oceanic cloud corridors linking bright areas near the Equator with others in mid-latitudes. Detailed studies of photos for individual days reveal that persistent cloud fields over the eastern north

Atlantic and west of Central America were composed mostly of strato-cumuliform varieties developed under the inversions that characterize trade wind belts off western continental coasts. The north–south cloud corridors in the mid-Pacific testify to the complexity of the mean surface pressure patterns across that large ocean. A simpler situation characterizes the north Atlantic, which seems to be dominated by a single anticyclonic cell. The structure of this cell is only really complicated in July when extensions bulge eastward across the Mediterranean Sea.

(*c*) In middle latitudes bright cloudiness appears to converge upon the Arctic along two broad tracks. One of these (the stronger of the two) links low and high latitudes obliquely along the coastal zone of eastern Asia. This is the north Pacific 'polar frontal zone', along which extratropical cyclones form and run north-eastward (see chapter 12). The other track, in an analagous position in the north Atlantic, channels depressions in a north-easterly direction towards north-western Europe and the maritime passage between Greenland and Norway. The oblique cloud belt from south-west to north-east off south-eastern Asia is now known to be associated with a trough line best marked in the middle troposphere.

(*d*) Variegated cloudiness across much of North America and Eurasia is associated mainly with the low pressure that prevails in summer, and is composed of convective cloudiness and/or degenerate extratropical frontal cloudiness. In winter anticyclonic conditions become more frequent, especially over the great land mass of Asia. By contrast depressions cross northern North America more frequently.

(*e*) Around the North Pole high brightness levels may be due in part to cloudiness, though the clarity with which Greenland stands out from the rest shows that, there at least, the high brightness must be due predominantly to surface ice and snow. This suggests in turn, that the skies there must have been mostly cloud free. Thus clear, anticyclonic conditions are indicated.

Fig. 8.2 depicted the mean circulation of the lower troposphere in July, as deduced from conventional evidence. That circulation pattern should be compared with plates 17 and 18, whose greater complexities may be compared usefully with the general circulation model shown in fig. 8.1.

Similar photographs for the southern hemisphere would indicate simpler patterns than the corresponding seasonal patterns in the northern hemisphere. This fact, in large measure, is a result of a simpler geography, especially poleward from mid-latitudes. Perhaps the most interesting components of this circulation emerging from such satellite displays are the cloudy troughs east of Australia, Brazil and, less markedly, south-east of South Africa (see also p. 169). Joint analyses of photographs and conventional soundings have indicated that they are middle to upper tropospheric features, but are sufficiently well marked to have considerable influences upon surface weather (see Hill 1963; Allison *et al.* 1964).

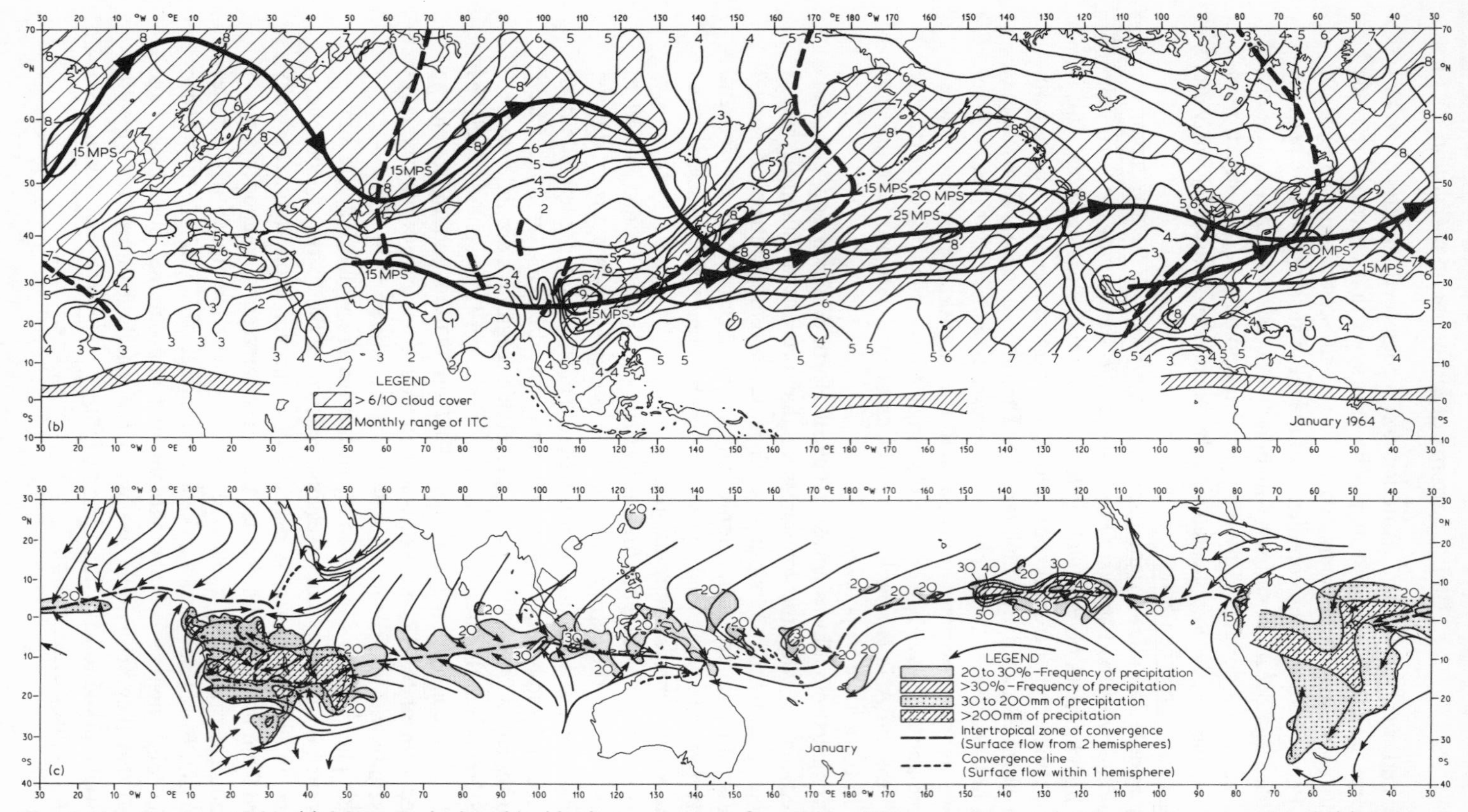

Fig. 8.12 January 1964: (a) Mean equivalent blackbody temperature from Tiros VII measurements in the 8–12 μm waveband. (b) Mean cloud cover, monthly range of the position of the I.T.C., principal axes of 700 mb maximum winds and mean trough positions. (c) Monthly mean surface streamlines in the Tropics, I.T.C. positions, and precipitation frequency and amounts. (After Allison *et al.* 1969)

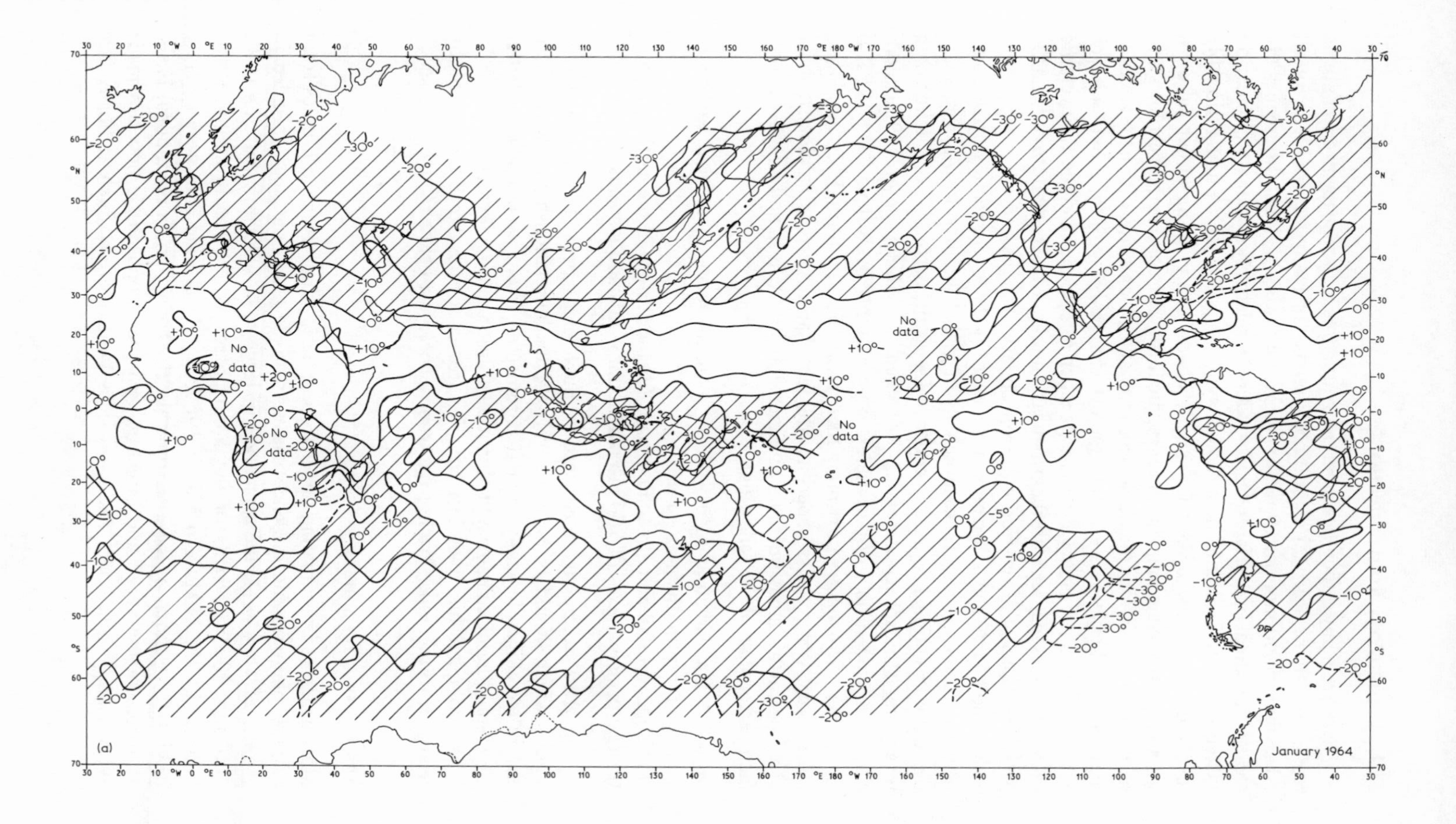
(a)
January 1964
No data
No data
No data
No data

There are many difficulties inherent in attempts to deduce mean circulations from monthly or seasonal cloud photographs, for many factors are involved other than cloud motion. However, some interesting regional differences are clearly indicated. In low latitudes, for example, relatively stationary circulation systems seem to be dominant, although synoptic processes are unquestionably involved also in the production of the characteristic mean cloud patterns, whether displayed in map or photographic form. The continental heat lows and monsoonal circulations must also wax and wane. Meanwhile, within the zones of high- and mid-latitude westerlies, the extreme mobility of the cloud patterns associated with secondary circulations (such as cyclones and the anticyclonic cells and ridges that separate them), smears the cloud data but fails to conceal the mean positions of lower- and mid-tropospheric troughs.

Recently significant attempts have been made to elucidate relationships between various types of measurements of global weather drawn from satellite and conventional sources in order to assess the usefulness of satellite data in general atmospheric research. For example, Allison *et al.* (1969) considered a wide variety of parameters on a monthly basis from June 1963 to December 1964, before presenting their results in atlas form. Even more than most recent work, theirs underlines the fact that modern climatologists must view the earth's atmosphere three-dimensionally. Allison and his co-workers included studies of the following:

(*a*) 8–12 μm radiation maps, whose patterns relate to a wide variety of heights within the atmosphere.
(*b*) Cloud, rainfall and streamline maps for the surface of the earth.
(*c*) The principal axes of 700 mb maximum winds and isotachs.

Fig. 8.12 exemplifies their results. Such maps point the way to future studies which should assess objectively the spatial relationships between these and other climatologically related parameters. From such studies a better understanding of climates and climatic variation may well emerge.

We complete our own, less complex, review of the global circulation by reference to the upper level of the troposphere, just below the tropopause.

The upper troposphere

Most progress in estimating upper tropospheric winds over the Tropics from satellite photographic evidence has been made along the lines specified in chapter 5. They assist the estimation of large-scale wind flow patterns at the 200 and 300 mb levels of the atmosphere. When that programme was initiated by the N.M.C. in March 1967, high-level wind data were severely lacking from much of the intertropical belt. Today, on an operational basis, estimates of the speed and direction of upper tropospheric flow are made every twenty-four hours and plotted as wind vectors. Such estimates are derived, as explained earlier, from considerations of the sizes, shapes and orientations of cirrus plumes extending away from active and formative cumulonimbus, cirrus spissatus (i.e. 'thickened' cirrus) and cirriform cloud shields. Jager *et al.* (1968) concluded that 'the satellite-derived winds not only "fill in" areas of sparse data, but also indicate the existence of wind anomalies which could not be detected by reference to persistence or climatology'.

Some such winds have been employed in the improvement of maps like fig. 8.13, which was based mainly on conventional climatological data. Although fig. 8.13 covers a restricted section of the Tropics, it provides a useful introduction to high tropospheric circulation patterns. Light easterly airflows are seen to dominate near the Equator, as they do on average around most of the globe. In the meantime, further north, winds strengthen from the west and even approach jet stream proportions as soon as about 15° N. Such a pattern is typical in winter over this major monsoonal region (see chapter 11).

Fig. 8.14, which places fig. 8.13 in a wider perspective, illustrates the circumglobal nature of the northern hemispheric subtropical jet stream in winter. Note especially its characteristic three wave pattern. As fig. 8.14 suggests, these jets are associated with upper tropospheric anticyclonic circulations whose centres in this season are between 5° and 10° N. Reiter (1969) rationalizes the mean subtropical jet stream in terms of the tendency of atmospheric flow to conserve its angular momentum. The northward flow of air aloft in the 'anti-trades' must, under such circumstances, give rise to strong winds from the west. Further north, in the northern winter season, the upper tropospheric flow adopts the form of a massive circulation around the polar vortex above shallow anticyclones driven mainly by cooling and subsidence across the frozen north.

To the south of the Equator, a mirror-image picture is generally obtained, though the pattern details are less well known. Fig. 8.15 illustrates the general nature of the circulation around the South Pole on 30 March 1966, at an altitude of about 12 km. This was established through tracking a constant level, super-pressure balloon of the American 'Ghost' (Global Horizontal Sounding Technique) project. The daily positions of the balloon through its thirty-three day lifetime were plotted from its radio signals (see, for example, Lally *et al.* 1966; Lally 1967). Since free-flying balloons are carried forward by the movement of the air, they act like clouds, as indicators of wind flow. Where they score over clouds is, of course, in their relative longevity. It is unfortunate, however, that technical problems are mitigating against operational usages of Ghost (or French 'Eole') balloons in tropospheric circulation studies. The chief problems involve icing at low levels, clustering, and locating each balloon's position uniquely from its radioed information.

In fig. 8.15 the circumpolar vortical circulation is seen to be the dominant flight path factor. First-order deviations from the mean latitude of flight are related to upper tropospheric troughs and ridges embedded in the large-scale

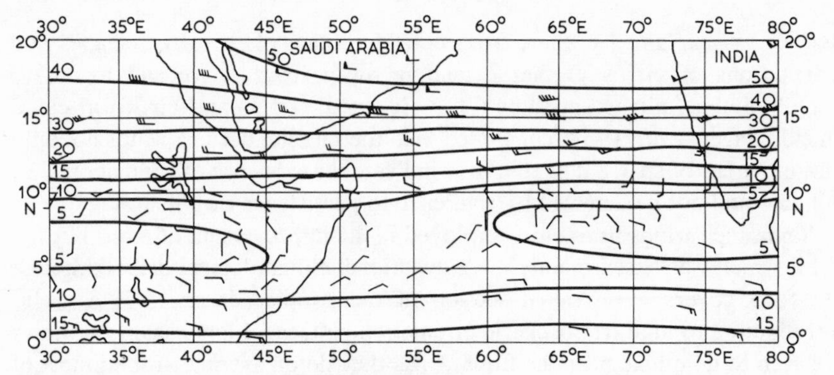

Fig. 8.13 Vector mean wind speeds and directions at the 200 mb level, December–February. Isotherms in °K. (From Jager 1968)

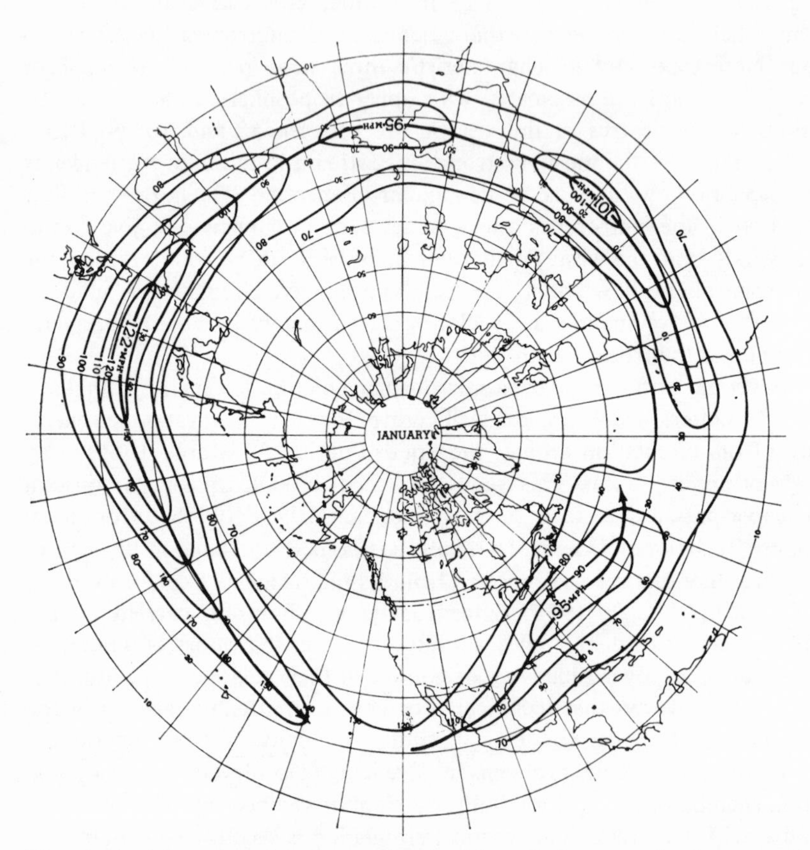

Fig. 8.14 The mean location and velocities of the westerly jetstream in the northern hemisphere in January. (In m.p.h.) (From Barry & Chorley 1968)

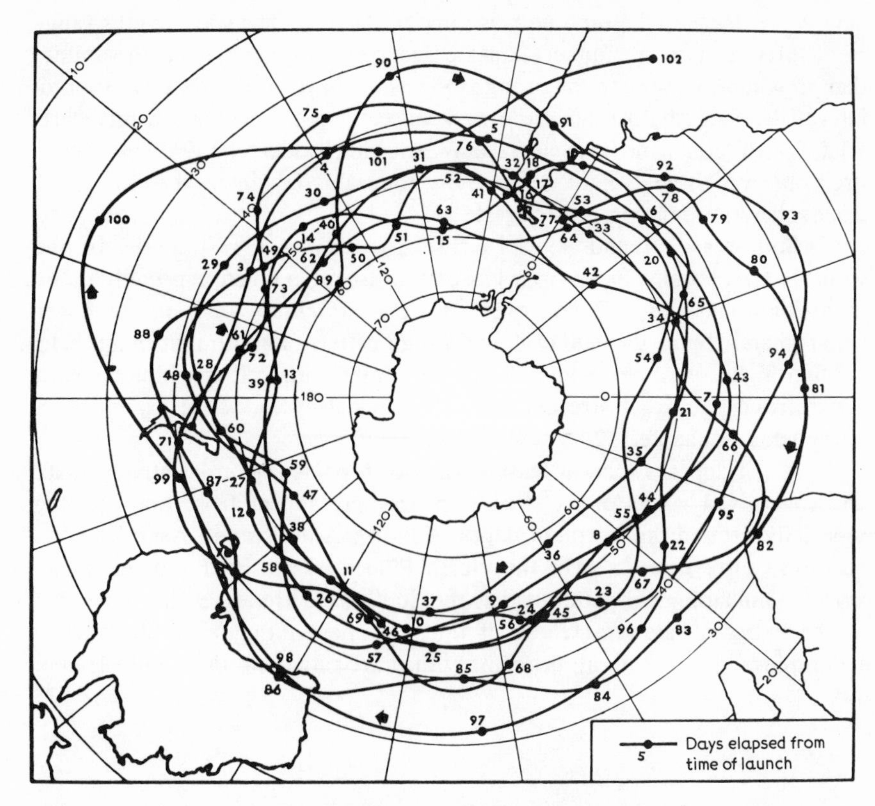

Fig. 8.15 The trajectory of a Ghost balloon released from Christchurch, New Zealand, comprised of eight circuits around the southern hemisphere in 102 days at 200 mb (approx. 13 km). (From Mason 1971)

flow. Such deviations in either hemisphere may be described in terms of long waves (or 'Rossby waves') in the upper westerlies (Rossby 1949). Cyclonic and anticyclonic curvatures of the balloon's path help to indicate the positions of ridges and troughs organized roughly radially from the pole. In a sense the shapes suggested by fig. 8.15 are misleading since the total pattern was not instantaneously observed and much development must have taken place through the month-long period. However, the pattern appears similar in kind to those regularly observed around north and south polar regions day after day. Such patterns are subject to variations of three kinds.

(1) Seasonal variations in the breadths of the latitudinal zones dominated by the upper westerlies. In general, the westerlies expand their areas of influence in winter when the local polar vortex deepens, and contract in summer.

(2) Daily variation of the numbers, and consequently the lengths, of the

long waves measured from one ridge line to the next. The wavelengths range from three to nine in number, and are generally more numerous in summer than in winter. American long-range forecasters have described a variation of intermediate length, through the American sector of the northern hemisphere, related to a 'zonal index cycle' (see Willett and Sanders 1959). This works itself out over three to eight weeks, most noticeably in winter when the upper tropospheric circulation is strongest.

(3) Both seasonal and shorter-term variations of the strengths of the westerly circulations. Polar front jet streams are often noted, especially in the more vigorous winter season over strongly baroclinic zones in the lower troposphere. Hence the rate of flow through patterns akin to that in fig. 8.15 is often highly variable. Some of the more interesting relationships between upper tropospheric jet streams and extratropical cyclones are examined in more detail in chapter 12.

Finally it appears that, at least in the north polar region, relatively weak and short-lived winds of jet stream speeds may appear from time to time especially around the northern edges of North America and Asia. The circulation centre near or over the North Pole need not be a simple, single, vortex. Climatological patterns over the South Pole are less well known due to the extreme data-remoteness of this, the most ultimate of the world's geographical corners. In all probability they are rather complex and variable too.

PART FIVE
Satellite data analyses in regional climatology

9 Linear disturbances of tropical weather

The tropical atmosphere

As we turn from more general, global scale considerations to matters of regional interest, we are justified in devoting much of the remainder of this book to tropical climatology. Four reasons can be forwarded in support of this:

1. Most satellite-based climatological research has involved the Tropics. Not unexpectedly, in the first decade of satellite studies most attention has been given to the more obvious and better defined organizations of weather appearing in satellite data presentations. Since such organizations are mostly tropical and maritime or coastal the accounts that follow reflect these early satellite preoccupations.
2. The Tropics contain the bulk of the atmosphere's source region with respect to heat, moisture and momentum. As a consequence, tropical climatic means, variations and departures from normal are not only interesting for their own sakes but also for the influences they may have on extratropical climates.
3. Many air mass characteristics and synoptic weather systems within the Tropics are unmatched by those in extratropical regions, and until recently have been generally much less well understood.
4. Fifty per cent of the area of the earth's surface, and over 35% of its population are found between 30° N and 30° S of the Equator. This belt contains most of the world's developing nations, and its chief areas of marked population growth.

Historically, the 'tropical atmosphere' seems to have achieved considerable fame as a distinctive entity more as a result of its statistical obscurity than on account of any real appreciations of its climatology. Indeed, it is sometimes argued today that the very concept of a tropical atmosphere is inappropriate, even though it is well established in atmospheric literature. Although, like any climatological boundary, that separating the Tropics as a zone marked by special characteristics must contain some element of arbitrariness, it is worthwhile attempting to define some such boundary for the purposes of climatological studies from satellites. The majority of the more traditional definitions of the tropical atmosphere fall into two general groups: geographical and meteorological.

Geographical definitions

Of these, the simplest and most obvious claims that the tropical atmosphere is bounded by the northern and southern Tropics. Unfortunately this simple definition is subject to at least one major criticism: it fails to embrace the poleward penetrations of distinctly tropical weather beyond the local tropic in the summer months. The best example involves the very warm, very moist air that extends beyond the northern tropic into subtropical areas of southern Asia during the summer monsoon season.

Meteorological definitions

These are generally more flexible than the geographical definitions. They involve such parameters as temperatures, and/or circulation patterns, and/or vertical lapse rates, etc. For example, some consider the tropical atmosphere to be that bounded polewards by the surface positions of the zonal axes of subtropical high pressure cells. The chief attendant problem here is that high pressure is normally absent from the intervening tropical land masses, and it is not easy to decide where transcontinental links should be drawn.

Recently, more complex approaches have been urged, for example by Alaka (1964). He argued that the best definition might be based on an aggregate of the common kinematic, dynamic and thermodynamic properties known to distinguish tropical from subtropical atmospheres. Unfortunately such definitions, intellectually satisfying though they might be, could prove so complex in application as to be unsuited to everyday use.

The basic characteristics of tropical weather

If anything, weather satellites have undermined still further the likelihood of the meteorological community agreeing on a simple meteorological definition of the tropical atmosphere. In removing much of the obscurity that had so long clouded the important issues, satellite data analyses have even hinted that some commonly held distinctions between 'tropical' and 'extratropical' types of weather systems were more apparent than real. For example, it is possible to find quite frequently on satellite photographs of the Tropics cloud organizations with some features typical of tropical weather systems and others more akin to those characteristic of mid-latitude patterns. The critical decision in satellite meteorology has been to present operational satellite evidence not only in the form of polar-projected images but also in Mercator map forms covering low latitudes between 30° N and 30° S (see chapters 2 and 3). Although no hard-and-fast limits need be adopted in the present account, these parallels will consequently figure more prominently than any others as the outer limits of the tropical atmosphere. Quite apart from this practical matter involving archived satellite data displays, 30° N and S of the Equator seem reasonable limits since (as chapter 6 revealed) the earth/atmosphere system as a whole is generally a heat source between these latitudes.

Table 9.1
Satellite versus pre-satellite values of radiation budget components, 0–20° N (data from Vonder Haar and Hanson 1969)

Parameter	*Mean satellite values*	*Pre-satellite estimates*
Reflected solar radiation	24%	34%
Solar radiation absorbed in atmosphere	15%	14%
Solar radiation absorbed in ocean	61%	52%

An important fact not stressed in chapter 6, however, is that the tropical atmosphere by itself is actually a net radiational heat sink (Riehl 1969). The dominant source is the surface of the tropical oceans. Satellite studies are important firstly in that they have shown the distinction between the two to be even more marked than was previously thought (see table 9.1). Satellite radiation data have indicated that earlier estimates of cloud cover, and therefore of albedo, were too high. Consequently the Tropics are known to gain more energy from the sun than was assumed earlier – but this energy excess is almost entirely oceanic (Vonder Haar and Hanson 1969). This suggests, in turn, greater transports of energy poleward by ocean currents, as well as a greater rate of air–sea energy exchange. This latter fact is most important. It is the transfer of sensible heat to, and the release of latent heat into, air overlying the oceans that sustains the warmth of the tropical atmosphere against radiational heat loss upwards, and other energy losses horizontally from the Tropics.

Many empirical studies have affirmed the particular dominance of condensation within the tropical atmosphere. They have revealed that the mean temperature lapse rate of the Tropics is nearly identical to the vertical profile obtained from parcel ascent of air of average temperature and moisture at the earth's surface: Riehl (1969) has stressed the fact, that the tropical atmosphere as a whole is not near-saturated, even along the Equator. This seeming paradox is explained largely by the observation that most tropical precipitation does not occur at random; it is concentrated mainly in synoptic scale disturbances covering perhaps only 1% or less of the area of the Tropics at any point in time. Thus, notwithstanding the unusual warmth and wetness of the tropical atmosphere, perhaps the most critical differentiating factor of its climatology in comparison with higher latitudes involves the larger number of generally small, but potent and well-defined, disturbances that characterize the Tropics. In addition to the parts they play in energy transport, they are important in that in many areas they yield some 50–75% of the mean annual precipitation. We are justified, therefore, in paying close attention to tropical disturbances in chapters 9 and 10. Much more than elsewhere around the globe, deviations from the mean frequency of such small disturbances within the Tropics may have serious effects on human activity through associated abnormalities in precipitation, through destructive winds and through other less direct environmental effects. Thus the significances of areas of deep tropical convection are twofold: (1) as active links between the general circulations of the lower and upper tropospheres, and (2) as contributors to the regional differentiation of local surface climates.

Before turning to a systematic treatment of tropical disturbances in their local contexts, however, some further remarks are necessary against the first of these two points. Consider the good approximation that the total heat (H) within the atmosphere is comprised of the following components:

$$H = gz + c_pT + Lq \tag{9.1}$$

where g is the acceleration of gravity, z is height, c_p is the specific heat of air, T is temperature, L is the latent heat of condensation and q is its specific humidity. The same equation can be written in words:

Total heat energy = potential energy + sensible heat + latent heat.

Fig. 9.1 shows that, in the Tropics, the vertical distribution of H has a marked minimum in the mid-troposphere. Riehl and Malkus (1958) suggested that, in view of such a distribution, ascent of moist air within the Tropics must take place predominantly in the central cores of cumulonimbus clouds. Such 'hot towers' are often large enough to protect their centres from mixing with drier environments, whether they are single clouds or clustered in larger disturbances. The ascent of moist air in the Tropics is concentrated in these structures, in which large quantities of latent heat are released through condensation. Fig. 9.1 indicates that H is larger aloft than even at the surface of the earth; the extra heat aloft is mostly available for transport poleward at high levels in the troposphere in the forms of sensible heat and potential energy. Most of this energy is lifted in the tall clouds of tropical disturbances, which thus comprise collectively the most active element within the at least climatically real Hadley cell circulation. Already an objective technique has been suggested for estimating the energy exchange in tropical convection systems by measuring the area change of associated cirrus outflow on successive geosynchronous cloud photographs (Sikdar and Suomi 1971). This should help to establish the spatial distributions and magnitudes of heat sources over the tropical oceans, and improve our understanding of interactions between small-scale convective regimes and large-scale atmospheric circulations.

The same workers (Sikdar and Suomi 1971) have suggested a neat set of models of deep tropical convection. Three vertical layers are involved in the largest variety of convection cells (Model I), as indicated by fig. 9.2.

(1) The layer of inflow (LIF), which includes the planetary boundary layer extending from the sea surface to the lifting condensation level (LCL). In this layer, warm moisture-laden air is drawn into the cloud region because of the local convergence which might be produced by large-scale motion.

(2) The layer of vertical motion (LVM). This has a depth from the LCL to the throat or funnel of a convective tower in which the vertical motion is maximum at the core, decreasing to a negligible value at the cloud periphery. It is assumed that the buoyant air starts condensing at the LCL and the latent heat of condensation thus released is conserved in the rising volume.

(3) The layer of outflow (LOF) between the throat and the tropopause. In

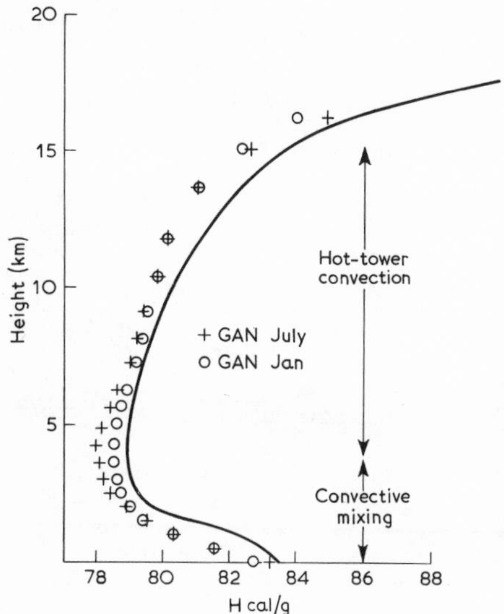

Fig. 9.1 The vertical distribution of total heat H in the mean tropical atmosphere. Mean values for Gan Island (00° 41′ S, 73° 09′ E) in January and July have been added. (From Johnson 1970)

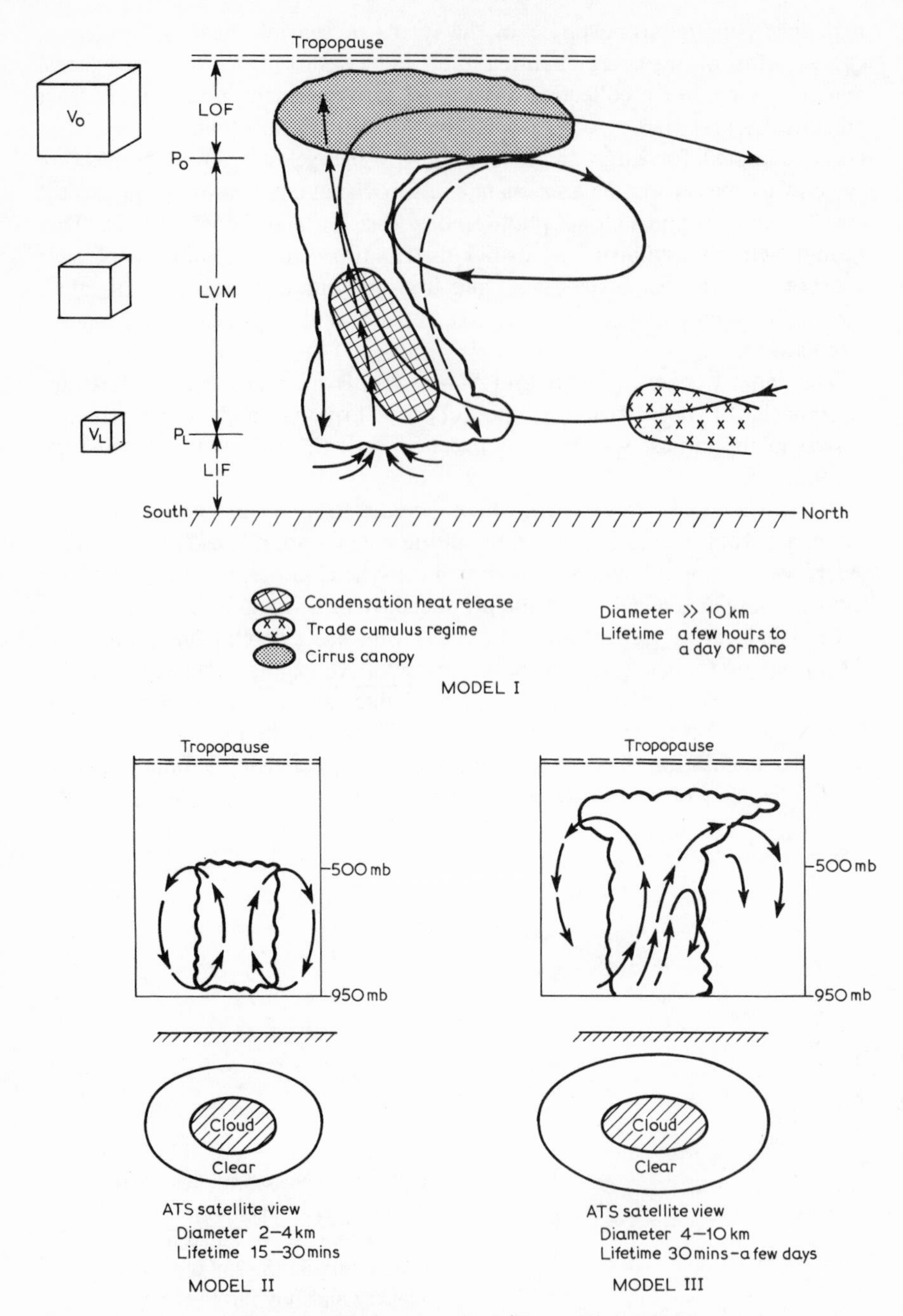

Fig. 9.2 Models of deep tropical convection. (From Sikdar & Suomi 1971)

this layer clouds diverge in the form of cirrus plumes on approaching an upper stable layer.

Sikdar and Suomi observed that deep 'wet' convection of this kind plays the main part in transporting heat from the boundary layer in the Tropics to the tropical upper troposphere, and thence as sensible heat and potential energy into higher latitudes. Thus the implications of their model square well with the earlier conclusions drawn from Riehl and Malkus.

Two other, less significant forms of convection models are illustrated also by fig. 9.2. These indicate that small-scale convective circulations (Model II: cumulocongestus cloud cells) suffer rapid mixing with the dry environment. This inhibits their growth. Lastly, deep isolated convective circulations (Model III: single cumulonimbus clouds) are organized usually on purely local scales, their associated subsidences encircling them nearby in contrast to the longer distances involved in the subsidences from Model I cloud masses.

Thus different scales of deep convection can be recognized, from the smallest tropical cumuli through cumulocongestus and cumulonimbus cells up to the largest areas of convection built of blocks like Model I opposite. Not all disturbances within the Tropics, however, are 'warm core' in the middle troposphere on account of latent heat release by condensation. Some are 'cold core', with lower central temperatures, probably due to advections of low energy air masses or, perhaps, to the effects of precipitation falling from high cloud layers and partially evaporating beneath. Even these cold core disturbances, however, are thought to assist in the export of energy to higher latitudes (Riehl 1969). In either case, it is appropriate to think not only of a heat source in and just above the tropical interface between the earth and its atmosphere, but also of a second heat source near 200 mb from which heat transport occurs outward and downward to the higher latitude mid-tropospheric energy minima. Climatological effects of such transports are seen in regions dominated by tropical maritime anticyclones. There warm, dry, subsiding air creates strong inversions often marked by sheet cloudiness on top of the cooler, moister layers of surface air.

Whereas some tropical disturbances are powerful well-organized systems, other are weak wind systems with relative vorticities approximately equal to the local Coriolis parameter. Some are basically linear in plan, while others are more nearly circular. Although no firm dividing line separates the one broad class from the other, it is convenient in the present context to deal first with the climatologies of linear disturbances within the Tropics before dealing separately with the non-linear forms.

Before reviewing these as distinct entities, however, our final note of general application is that unstable weather in the Tropics does not arise at random at the larger scales. Even the relatively random stimulus of thermal convection is influenced by some geographical factors. The three main triggers of unstable weather within the tropical zone may be summarized as follows:

1 Thermal convection. Since this is related primarily to the strength of insolation, it is strongest on average through the year as a whole within a few degrees of the Equator. Upward motion in the troposphere is enhanced where much moisture is available, and where upper level restraints on rising air currents are either weak or absent.
2 Convergence. Instability weather is encouraged especially within the thermal-convection-prone belt of equatorial low pressure where moist air streams converge, and pressure gradients are moderate to steep. Convergence, which is a characteristic of areas of cyclonic vorticity, usually leads to a building of clouds in the vertical (see chapter 8).
3 Orographic uplift. Mountains standing across the paths of moist air streams help to stimulate strong cloudiness and heavy rainfall by forcing the advected air to rise to cross them. Isolated peaks tend to warm more quickly than their surroundings early each morning, and these, too, favour the development of convective clouds.

Linear disturbances within the tropics

At the larger scales, it is convenient to distinguish between: (1) linear disturbances organized zonally, (2) linear disturbances organized meridionally and (3) linear disturbances organized obliquely across the Tropics.

Let us deal with each group in turn.

Linear disturbances organized zonally

The I.T.C.Z.

Here, convergence is the key stimulus for instability cloudiness. At the Equator, of course, f, the vorticity of the earth, is zero. Hence, the air has no additional spin imparted to it over and above any relative vorticity (ζ) which it may possess. Consequently, enclosed local circulations of air like those common at most latitudes are generally absent from the near vicinity of the Equator although wave-like eddies are frequently seen. Instead, converging air streams merge and mingle within a belt-like trough of low pressure, along an axis of moderate to very strong instability. Since the 'Norwegian School' of meteorology formulated its air mass and frontal theories in the 1920s and 1930s, the concepts of organized weather along the axis of the equatorial trough have undergone a fascinating evolution:

(*a*) At first, the enhanced convective activity noted near or astride the Equator was conceived as an intertropical or equatorial front (I.T.F.), not unlike the then newly authenticated fronts in middle latitudes.

(*b*) Later, the realization that rather little air mass contrast usually characterizes the converging air streams led to the rise in popularity of terms like equatorial or intertropical convergence zone (I.T.C.Z.) as more preferable labels in view of the near vertical air mass discontinuities. After World War II, however, it became increasingly apparent that the infrastructure of I.T.C.Z. was more complex than had been thought, so doubt was cast next on the appropriateness of even this new term. Then an explosion of observational data followed the rapid expansion of airline services across the tropical Pacific. Many of the new data indicated that, along the convergence zone, and 'within a comparatively narrow band of convergence weather ... narrow tongues of air under-run and over-run each other producing extra-ordinary cloud forms [with] clear subsident air' frequently seen on one flank or the other (Garbell 1947). Significant regional contrasts became apparent also, prompting Palmer (1951) to claim that the I.T.C.Z. is only a reality in a statistical sense, and that day-to-day weather in the equatorial trough is diversified richly by processions of atmospheric eddies.

Doubtless the meteorological complexity of low latitudes has been a major cause of the conceptual and terminological confusion that has become apparent in recent literature on the I.T.C.Z. For example, the American Meteorological Society (1959, p. 312) has defined this important feature of tropical climatology as 'the axis, or a portion thereof, of the broad trade wind current of the Tropics. This axis is the dividing line between the north-east trades and south-east trades.' On the other hand, Trewartha (1968) wrote that 'winds are usually variable and weak ... between the converging trades.... This transition zone has been given various names ... among them, intertropical convergence zone.' Such extremes have been avoided by other writers (e.g. Hidore 1969), who adopted an intermediate position by being less precise. He wrote that the I.T.C.Z. 'represents the zone in which the trade winds from the north and south of the Equator merge'. So the I.T.C.Z. itself has been conceived differently by different authorities.

(*c*) In the satellite era (post-1960) it soon became evident that arrangements of weather within the equatorial trough are even more complex and variable than the additional post-war conventional data had hinted. An ever widening phalanx of new terms seemed necessary to differentiate the various meteorologically significant aspects of the low latitude convergence. The more widely accepted of these terms include:

(i) The intertropical confluence (I.T.C.), i.e. the circulation axis of the equatorial trough. This is the principal asymptote of convergence (see Simpson *et al.* 1968), established through wind field analyses.
(ii) The intertropical vorticity zone (I.T.V.Z.), embracing the centres of maximum absolute vorticity (see Fujita *et al.* 1969).

(iii) The intertropical cloud band (I.T.C.B.), coincident with the belts of strongest convective cloudiness (Fujita *et al.* 1969).

(iv) The equatorial axis of instability (Barrett, 1970a), drawn through the geometrical centre of the I.T.C.B. This term is more specific than (iii) in that it applies, by definition, to the major east–west instability axis, and not to any other strong cloud belt within the Tropics.

In the discussion that follows the term I.T.C.Z. will be retained in its most literal sense, i.e. a *zone* of convergence more or less coterminus with the equatorial zone of positive vorticity between the divergent centres of the trade wind anticyclones. Terms (i) to (iv) will be invoked as and when appropriate.

Abundant satellite evidence confirms that linear weather disturbance patterns are frequently real features of daily weather in the equatorial zone, not just statistical phenomena as Palmer argued. As early as 1964, Fritz was able to confirm the frequent occurrence of a narrow, sometimes continuous, linear disturbance belt extending from east to west over perhaps several thousand kilometres of tropical ocean, although such cloudiness rarely appeared as a simple, single band. Let us proceed to review modern knowledge of the alignment and structure of that east–west disturbance line, before discussing its impacts on local climates, and possible reasons for its variety of forms.

A careful analysis of fig. 8.2 suggests that the equatorial belt of cyclonic wind curvature near the Equator is relatively narrow across the oceans compared with the large land masses of South America and Africa. Here it broadens under the influence of stronger surface heating and more powerful thermal convection. Since the Coriolis effect lessens as the Equator is approached, winds flow at ever greater angles across the isobars flanking the equatorial trough, until they meet along the intertropical confluence. Fig. 9.3 indicates that the I.T.C. is mostly north of the Equator in both January and July. It varies least in its position across the Atlantic, and the east and central Pacific. Across the western Pacific and the Indian Ocean it shifts seasonally in keeping with the South Asian monsoon.

Turning to satellite evidence, work published by Godshall (1968) underlined the care that must be taken to avoid assuming that the intertropical confluence, vorticity zone and cloud band are of necessity coincident in

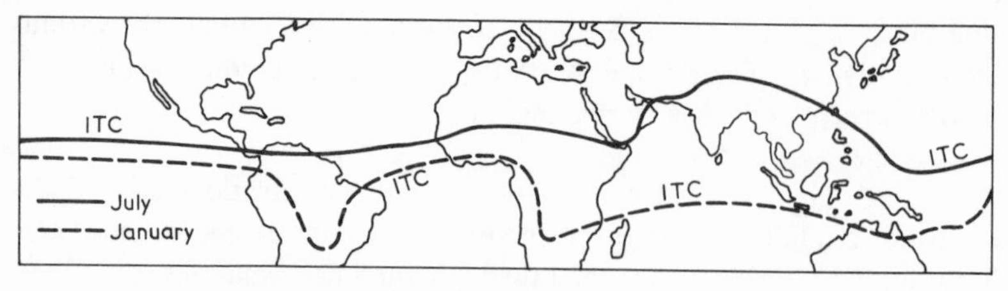

Fig. 9.3 Mean positions of the intertropical convergence in January and July, from pre-satellite evidence. (After Flohn 1969)

position. The classical view of intertropical confluence involves convergence of northern and southern hemisphere wind systems 'so that northern and southern trades reach it after complete equalization of their thermodynamic properties' (Riehl 1954). Since the constancy of the trades can be defined by:

$$\text{Constancy} = 100\ Vr/\bar{v} \qquad (9.2)$$

where Vr is the mean vector wind speed and $\bar{v}$ is the mean wind speed, obviously wind systems with large variations in speed and direction have low constancies of flow. Crowe (1951) demonstrated well that the equatorial trough is characterized by winds that are variable in both vigour and direction. However, Godshall's more recent maps (see figs. 9.4 and 9.5), indicate a sometimes considerable degree of separation between the 'I.T.C.Z.' defined in terms of low wind constancy and the axis of maximum convergence cloudiness in January and August. His maps are interesting too in that, whichever criterion is used to separate northern from southern hemisphere atmospheres, seasonal variations of the organized weather within the equatorial trough are much greater across the equatorial western Pacific than across the central or eastern regions.

Satellite studies of the equatorial Pacific

Certainly the behaviour of weather organizations east of 170° W is neither simple nor predictable on a short time scale. Recent research (Barrett 1970a) based on nephline analyses of Nimbus II photographs (see chapter 5) has documented something of the nature and extent of weather variations from day to day across the equatorial eastern Pacific and the Central American isthmus. This region, while being notorious for the complexity of its circulation patterns, has remained one for which the conventional meteorological coverage is poor. Daily weather maps are prepared at the Howard Air Force Base in the Panama Canal Zone, and by the Venezuelan Weather Bureau, but their coverage does not extend far west from the isthmus owing to the almost complete dearth of surface weather stations there. As Crowe (1951) observed so appositely, a great breadth of the equatorial Pacific, from the American coasts west to the Society and Marshall Islands at about 150° and 170° W respectively comprises, from this point of view at least, 'the most empty ocean of the world'. Here is an ideal area in which the satellite climatologist may practise his new science.

Although Nimbus II functioned efficiently for only a few weeks (from 14 May until 31 August 1966), its television coverage of the Tropics for a time was very good. Generally, Nimbus II photographs resolved sufficiently well to permit the identification of surface features down to about 3–5 km across. Such a resolution is superior even to that of the computer-rectified Essa mosaics, so, despite handling problems posed by their own non-rectification, the Nimbus pictures have proved quite useful in detailed cloud analyses.

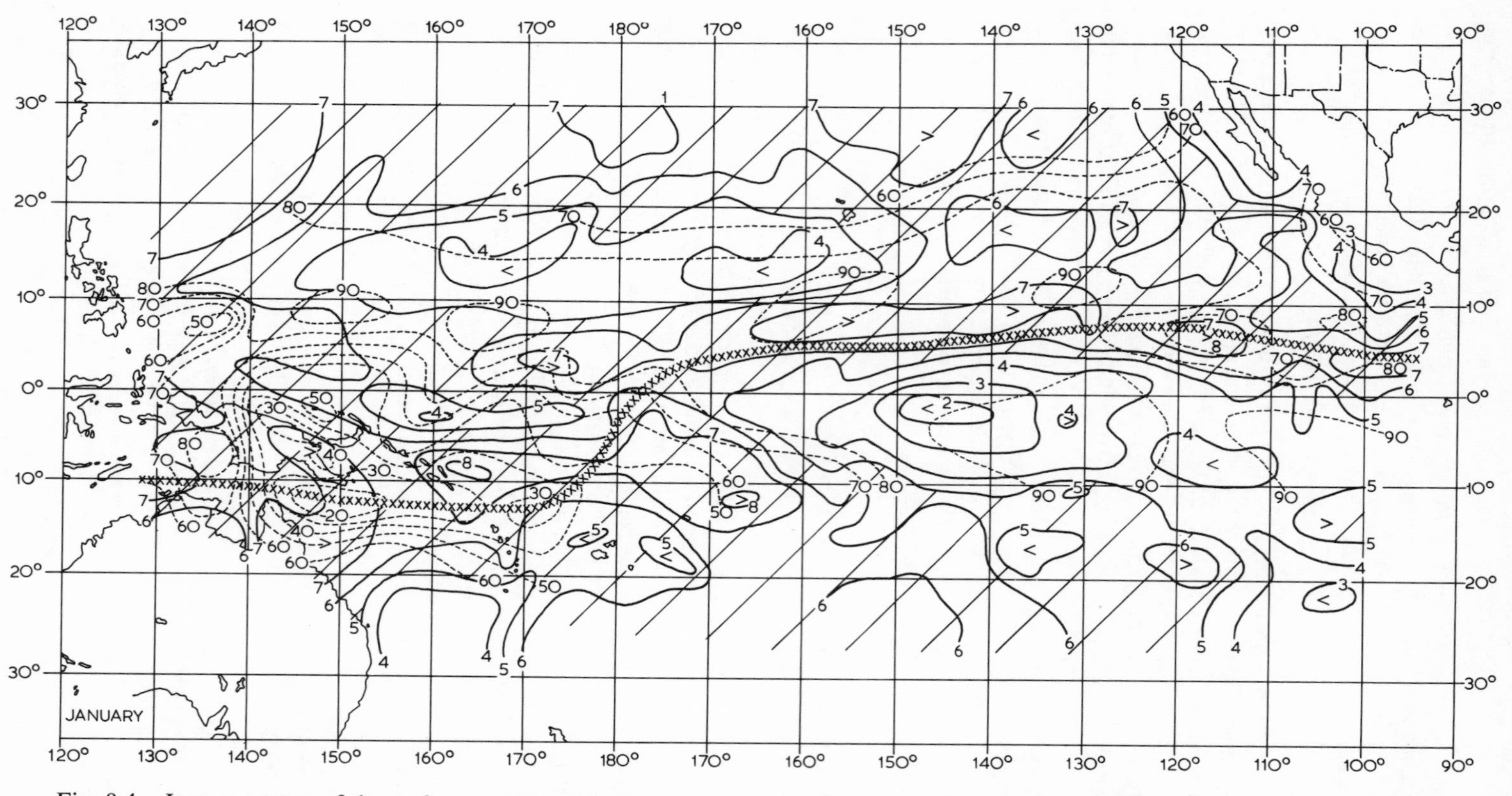

Fig. 9.4 January means of the surface position of the 'I.T.C.Z.' (line of crosses), surface wind constancy (broken isopleths, in percentages) and total cloud amount from Tiros satellite evidence (solid isopleths, labelled in tenths of sky cover). (After Godshall 1968)

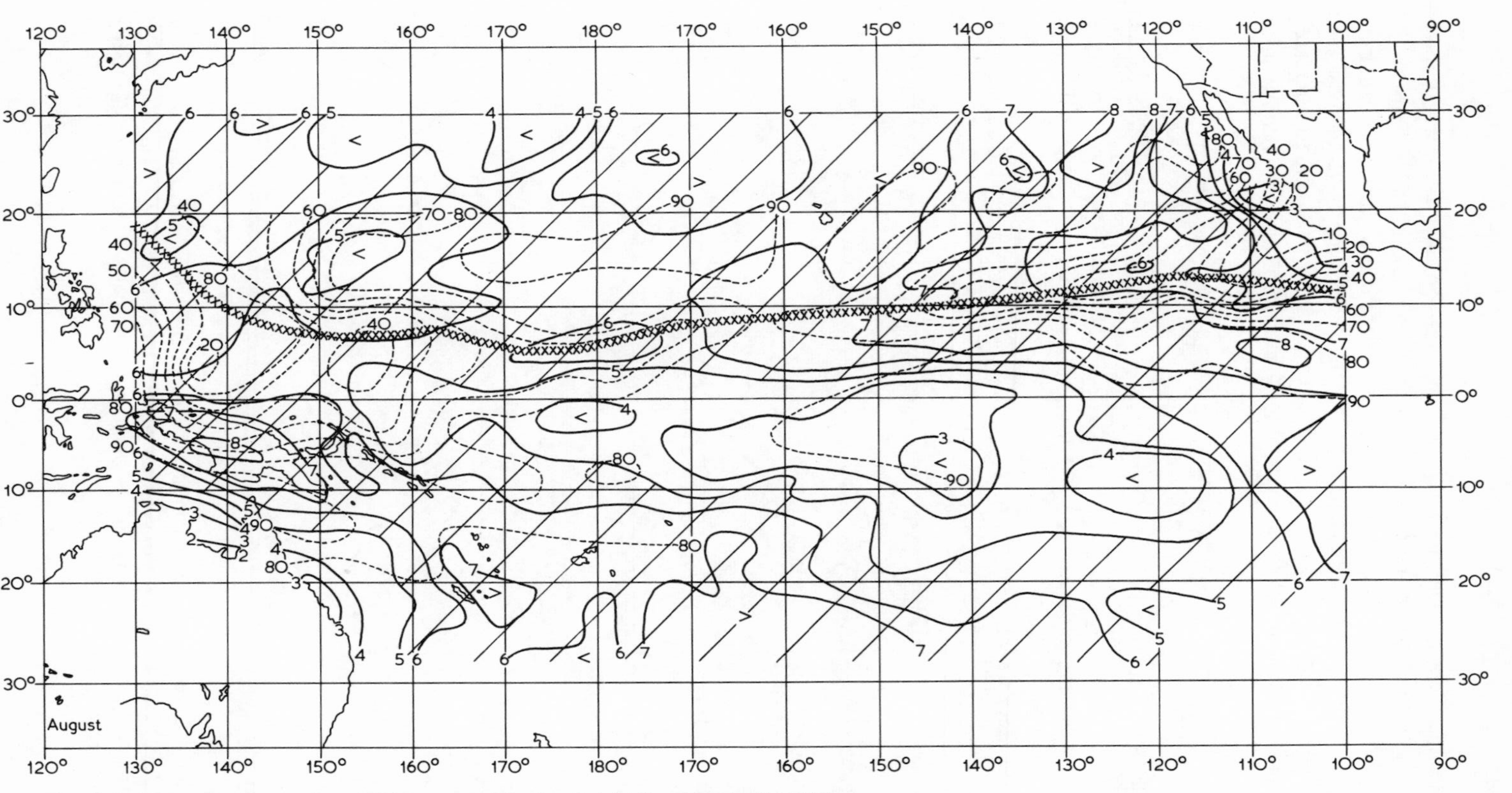

Fig. 9.5 August means of variables mapped in fig. 9.4 (After Godshall 1968)

In the Central American region, the intertropical confluence is thought generally to curve in the northern summer from about 15° N between 100° and 120° W towards the Amazon Basin, where it lies more or less along the Equator. Mean cloud maps, compiled from contemporaneous Essa nephanalyses read off across a 2° 30′ square geographical grid, help to elucidate the seasonal behaviour of the accompanying intertropical cloud band from late

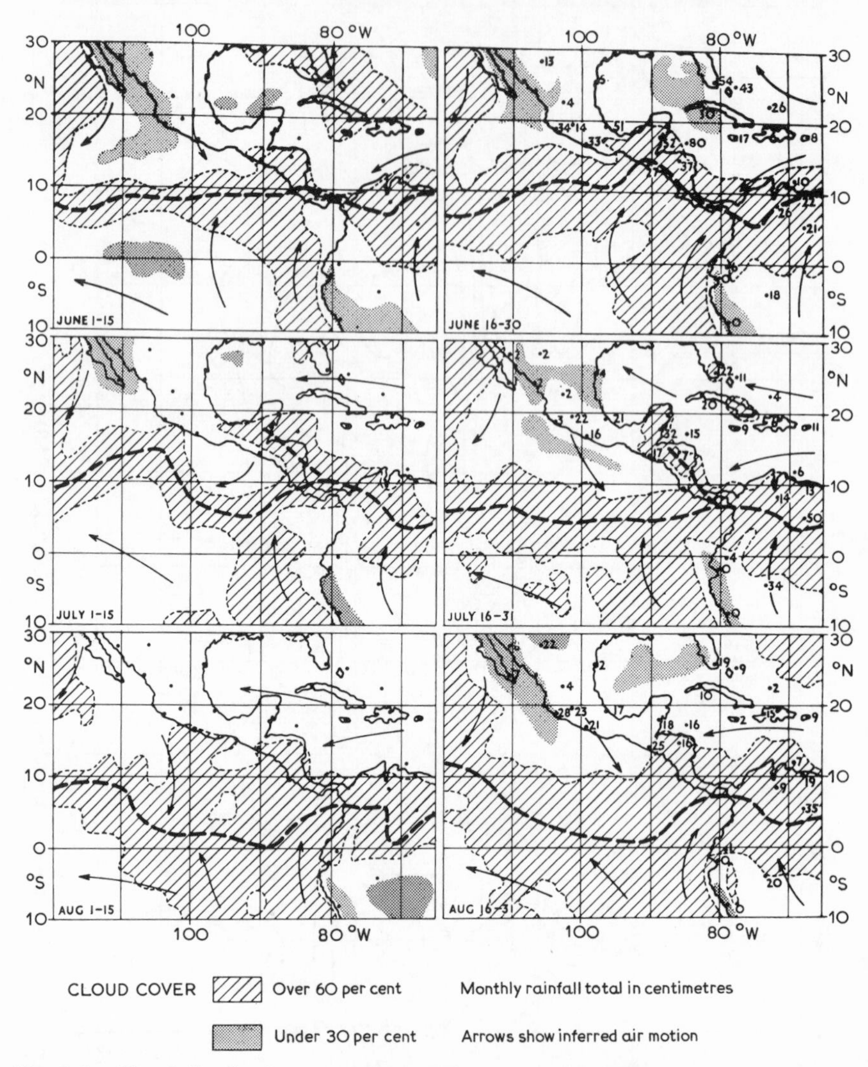

Fig. 9.6 Cloud distribution over Central America in June, July and August 1966, from Nimbus II satellite evidence. The figures in the right-hand diagrams give rainfall recorded at selected stations for the whole month in question, in cm. (After Barrett 1970b; from Crowe 1971)

May to the end of August. Even in the low resolution maps presented in fig. 9.6 some interesting sequences of cloud changes can be seen:

(*a*) In early June, the east–west belt of strongest cloud cover lay just north of the Equator west of Central America, fed by airflow down the trade wind inversion cloud corridors off the west coasts of North and South America.
(*b*) During late June, cloudiness increased north westward along the Central American isthmus, and the belt of maximum cloudiness to its west curved north, reaching nearly 15° N between 100° and 110° W.
(*c*) In early July a pool of less cloudy conditions was enclosed off the west Mexican coast between cloudier skies to the north and south.
(*d*) By late July the northern limb of heavy cloudiness has disappeared.
(*e*) Through August, the main east–west cloud axis fell away southward to approximately its late May position.

Streamline patterns deduced from daily nephline analyses (see p. 125) threw more light on short-term fluctuations hidden by the half-monthly means. Five broad synoptic arrangements of the equatorial circulation were suggested subsequently:

Type 1 The equatorial axis of instability is relatively straight and zonally aligned west of the Central American isthmus, at about 7° 30′ N.

Type 2 The equatorial axis of instability is highly sinuous west of the isthmus, suggesting a major northwards thrust of ex-southern hemisphere air a few degrees west of Central America.

Type 3 The central cloud axis lies along the isthmus itself between about 80° and 95° W, and is composed of particularly strong convective cloudiness. This may be considered as the high season situation, associated with the most northerly penetration of ex-southern hemispheric air.

Type 4 No linear belt of cloudiness appears along, or immediately westward from, the isthmus. Convective cloudiness is patchy, mostly in subcircular or amorphous blobs, over a broad area between the stratocumuliform cloud fields of the trade wind anticyclone circulations of the north-east and south-east Pacific Oceans. This, like type 5, seems to be a transition season situation.

Type 5 Instability cloud patches are fewer in number than in type 4, but rather more regular in shape, suggesting higher degrees of subsynoptic weather organization.

Thus it emerged that, even in a single region, the major organizations of deep cloudiness and their attendant weather phenomena may be both very varied and highly variable (see fig. 9.7). Gross changes, especially of latitudinal alignment, seem to be related to the shifting zone of maximum insolation, to which the intertropical cloud band is closely tied (see Gruber 1972). Day-to-day fluctuations are less predictable and more difficult to explain. Probably the relative strengths of the anticyclones in the north and south Pacific, and in the north Atlantic via the Caribbean, all have their parts to play through the strengths of their air streams which all converge upon the area in question. The influence of the isthmus itself is clearly important too, especially in summer. Nor should we forget sea surface temperature patterns (Saha 1971). Yet another definite possibility is that the alignment of the intertropical cloud band on one day may itself affect its own alignment a little later. Hewes (1968) observed that, on average, only one day out of every five at climatologically intriguing Canton Island in the west-central Pacific was

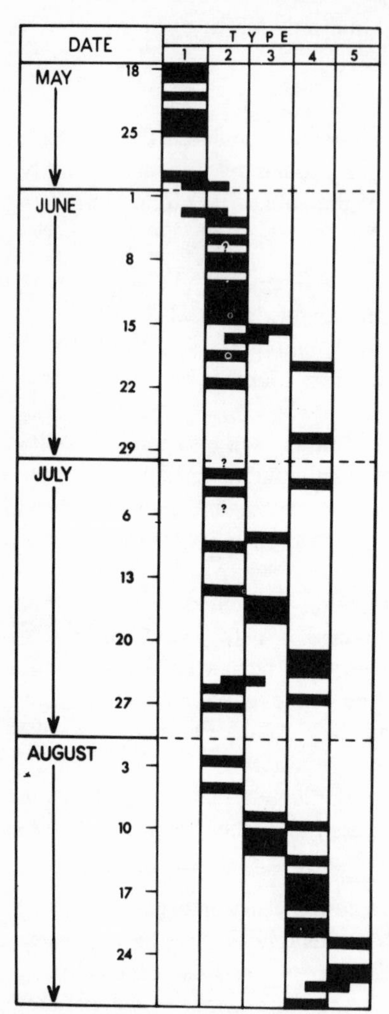

Fig. 9.7 Successive daily types of synoptic cloud patterns west of Central America, 18 May–31 August 1966. (From Barrett 1970b)

affected by disturbed weather, despite the island's proximity to the intertropical confluence. During these days of disturbed weather, the overall average transmission of solar radiation through to the earth's surface within 3° geographically of a trough axis was only 34·6%, compared with 61·3% during fair weather. If it can be assumed that a thick cloud cover near the Equator must result in some such reduction of the energy incident upon the earth's surface, then it seems reasonable to propose that, once a strong but relatively stationary cloud band has developed, it may mitigate against its own survival in that position by shielding the surface beneath from direct insolation; thus temperatures are reduced there. In other words, much of the observed short-term excitability of the instability axes in the eastern equatorial Pacific – and elsewhere in low latitudes – may result from a tendency for deep cumulonimbus clouds to be encouraged in areas recently free from cloud. This possibility deserves more research. Certainly the interactions between atmosphere, land and ocean are complex.

In conclusion, disregarding local complications caused by the Central American isthmus, three main points emerged from the Nimbus II study of weather organizations in the Central American region:

(*a*) When convergence of the trade winds from the north and south Pacific is strong, an intertropical cloud band is usually well developed over the equatorial eastern Pacific Ocean. Clearer, subsidence weather may parallel such a band on one or both sides, or cloudiness may build gradually towards the equatorial axis of instability from the anticyclonic stratocumulus cloud fields.

(*b*) Where convergence rates are moderate, cloudiness within the equatorial trough seems to be usually more subdued, often lacking abundant cirrus blow-offs aloft. Honeycombed patterns of convective cells become organized astride the axis of instability. This suggests that the horizontal wind shears are strong.

(*c*) When the convergence is weak, broad zones of clear skies dotted with amorphous patches of bright cloudiness separate the cumuliform and stratocumuliform cloud fields of the northern and southern trade wind anticyclones. No linear cloud organizations arise in such circumstances, though tropical storms or even hurricanes may develop (see chapter 10).

The range of synoptic patterns centred on the equatorial axis of instability must be broadened further to accommodate other arrangements well demonstrated by Kornfield and Hasler (1970) in the Pacific theatre through the use of multiple exposure averaging techniques applied to Essa computer-rectified mosaics (see p. 101). It seems that, at certain seasons of the year, the intertropical cloud band sometimes splits into two-well-defined belts of cirriform-topped cumulonimbus separated by relatively cloud-free (subsidence) zones. Hubert *et al.* (1969) sounded the cautionary note that such a doubling of instability belts may be more apparent than real, perhaps being something of an artefact produced by the technique of photographic amalgamation of a large number of simpler cloud organizations. There can be no doubt, however, that such arrangements do occur at certain times and in some areas. Firm evidence to this effect has been afforded by early Tiros photographs, as well as later pictures from Nimbus and Essa satellites. Recent work combining satellite and theoretical considerations suggests that both single and double structures are both within the realms of reality, but the former represents the more permanent mode (Pike 1971).

Fig. 9.8 (Hubert *et al.* 1969) suggests that doubled structures are quite well defined in the equatorial eastern Pacific during the northern hemisphere spring, while similar organizations are characteristic of the equatorial Far East between about 150° and 180° W. Here, however, the synoptic situation is rather different. The subsidence zone along the Equator is bounded by an equatorial axis of instability to its north, but by a section of the earlier described oblique cloud belt linking the Tropics with middle latitudes to its south. Others have observed that the most prominent cloud band in the eastern Pacific lies between 5 and 10° N on average throughout the year, while a second band occurs intermittently near 5° S. In 1967, this became prominent in February just west of South America. By March and through April it extended nearly to 180° W before vanishing in May. It did not appear in the Nimbus II evidence, though, and in 1968 it was present only in March and April extending much less far west than in 1967. Similar doubled structures appear occasionally on photographs of the Indian Ocean during the north-east monsoon season, but apparently are unknown across the equatorial Atlantic. Thus the dominant weather patterns differ from ocean to ocean, and vary with time. Exactly how they are related to the structure of the tropical cell within the general circulation has yet to be completely clarified.

Explaining local climatic features

The next climatological question that we should seek to answer, at least in part, is this: 'What are the local climatic effects of variations in the form and position of the intertropical cloud band?' Recent work permits the integration of some satellite and conventional observations across the equatorial Pacific, and the development of some very interesting and quite far-reaching theories of a dynamic climatological nature. Certainly it is possible to view with diminished awe the remarkable rainfall statistics from certain central Pacific island stations. Canton Island (2° 48′ S, 171° 43′ W) is a case in point.

Fig. 9.9 portrays time series analyses of air and sea temperatures for Canton Island from 1950 to 1967. Very striking variations in annual precipitation occurred during that time, observed totals ranging from only

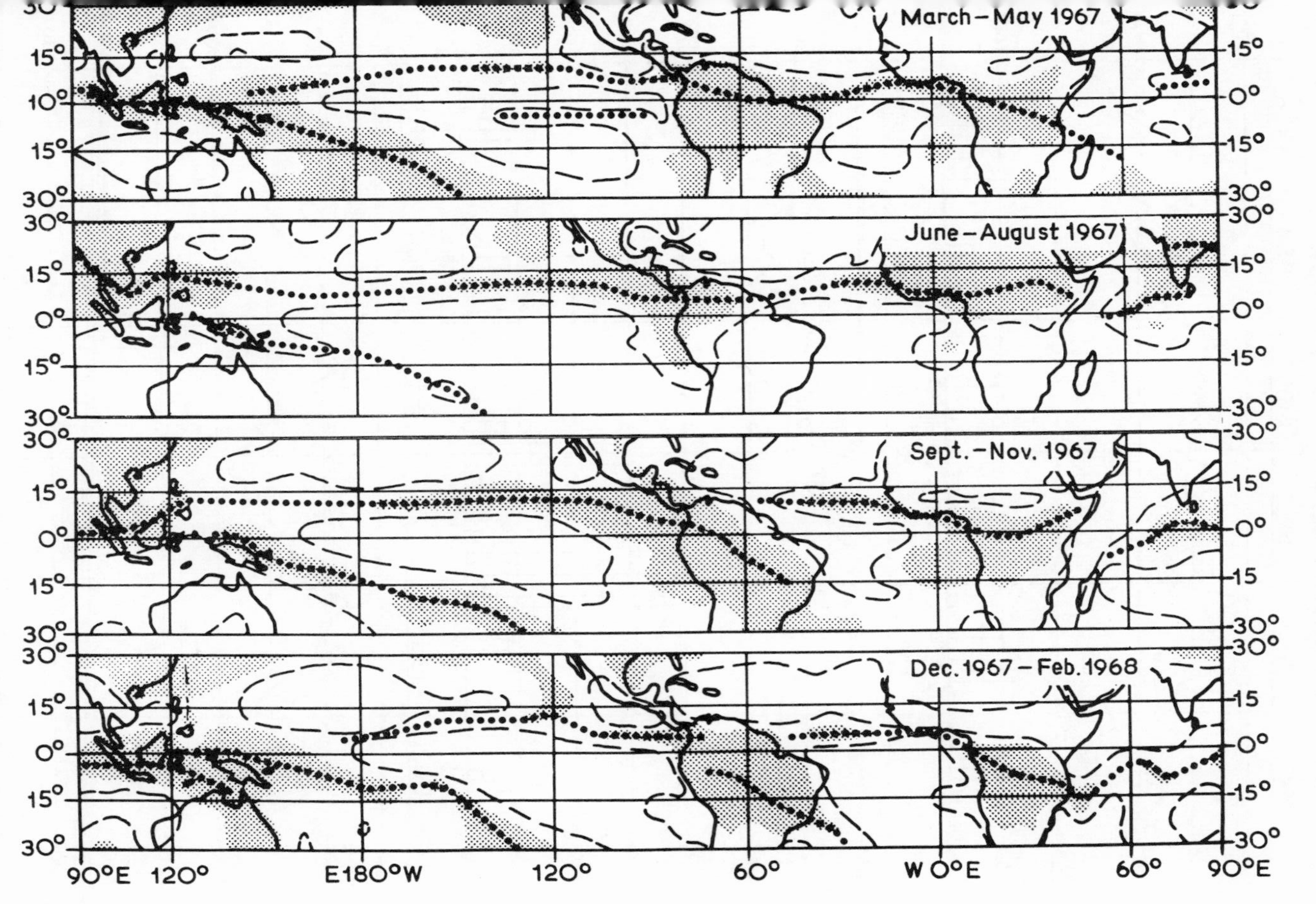

Fig. 9.8 Seasonal brightness from Essa 3 and 5 digitized pictures over the Tropics from March 1967 to February 1968. The dashed lines enclose areas of minimum brightness, while stippling indicates regions of maximum brightness. Heavy dotted lines indicate the major zonally orientated axes of maximum brightness. (From Hubert *et al.* 1969)

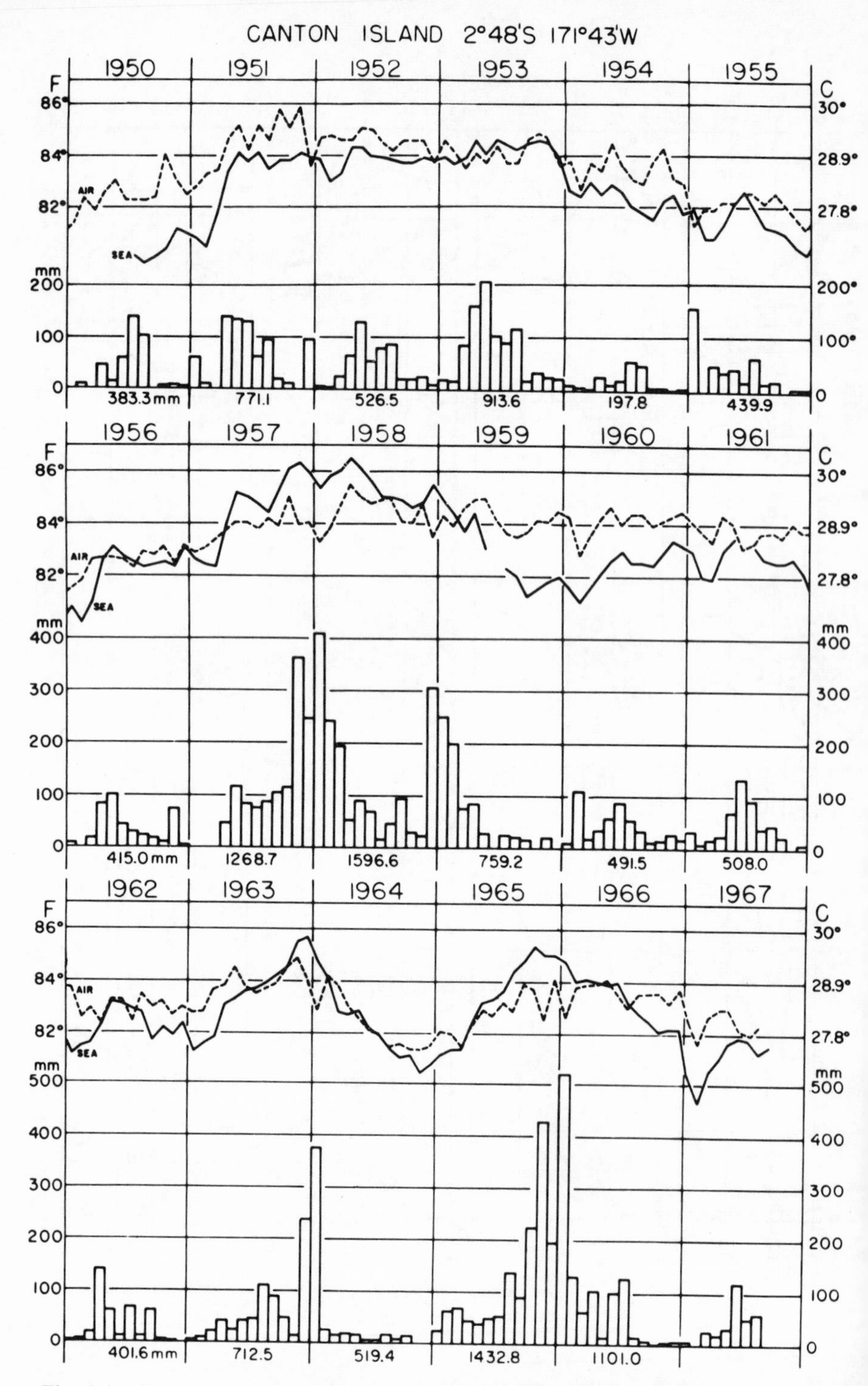

Fig. 9.9 Time series of monthly air temperatures (broken lines) and sea temperatures (solid lines) and monthly precipitation, Canton Island, 1950–67. (From Bjerknes 1969)

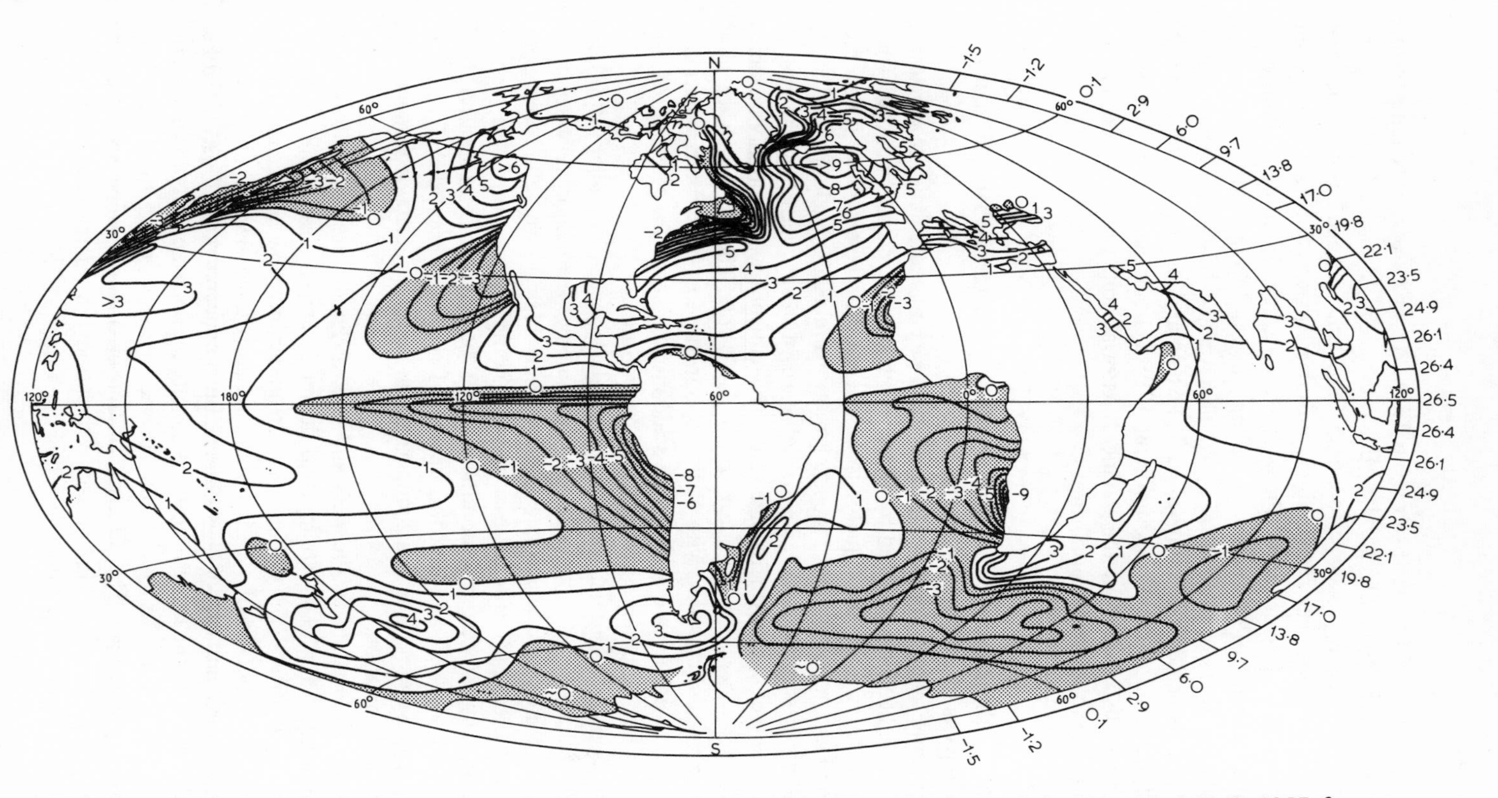

Fig. 9.10 Sea surface temperatures represented as deviations from the average at each latitude. (After Dietrich & Kalle 1957; from Bjerknes 1969)

197·8 mm in 1954 to 1596·6 mm in 1958. Air and sea surface temperature variations were generally parallel, but rather less dramatic. Satellite photographic evidence (e.g. plate 19) suggests that Canton Island lies near the mean boundary between very moist, highly unstable equatorial trough weather and the so-called 'Pacific dry zone', the wedge of widespread subsidence marked by relatively cloud-free conditions extending westward from South America just south of the Equator to about 170° W. This wedge is associated with a tongue of equatorial cold water unrivalled in extent and degree elsewhere in the world, penetrating on average to about 170° E along the Equator (fig. 9.10). The mean distribution of the Pacific equatorial cold water can be mapped through the following relationship:

$$\Delta T_s = T_s - \bar{T}_s \qquad (9.3)$$

where T_s is the sea surface temperature, and $\bar{T}_s$ its mean global value along the oceanic parts of each latitude circle.

The chief meteorological effects of these negative anomalies of equatorial water have been summarized by Bjerknes (1969). He pointed out that, when the cold water along the Equator is well developed, the air above it becomes too cold and heavy to join ascending motions of Hadley cell circulations to the north or south. Instead, the air flows westward between the Hadley-type circulations of the northern and southern hemispheres, eventually reaching the warmer waters of the western equatorial Pacific. There, through warming from below and charging with water vapour, the east–west air circulation is then able to participate in the large-scale moist adiabatic ascent that dominates weather and climate to the east of New Guinea.

The associated circulations aloft were well described by Sadler (1959). He observed that 'the upper tropospheric flow (along the Equator) reverses (from its general direction, which is east–west) near 150° E, where the west components increase towards the east and east components towards the west'. It seems that upper westerlies dominate above the Equator from about 150° E to beyond 160° W, and that this reversal from the normal direction of the upper tropospheric flow is a semipermanent feature, subject to seasonal and meridional variation.

So the circulations in the west and central Pacific are now known to be very complex. Fig. 9.11 attempts to depict supposed relationships between upper- and lower-level circulations. It underlines the significance to the regional distribution of weather and climate of the heavily clouded area generally north and east of New Guinea. Such concepts recall those of Walker (1923), who was the first to suggest a zonally organized circulation across the Pacific. His proposals have been revived recently under the title of the 'Walker Circulation'. The strength of this circulation which separates the northern and southern Hadley cell circulations, is related seemingly to the longitudinal temperature contrast along the Equator. This varies considerably

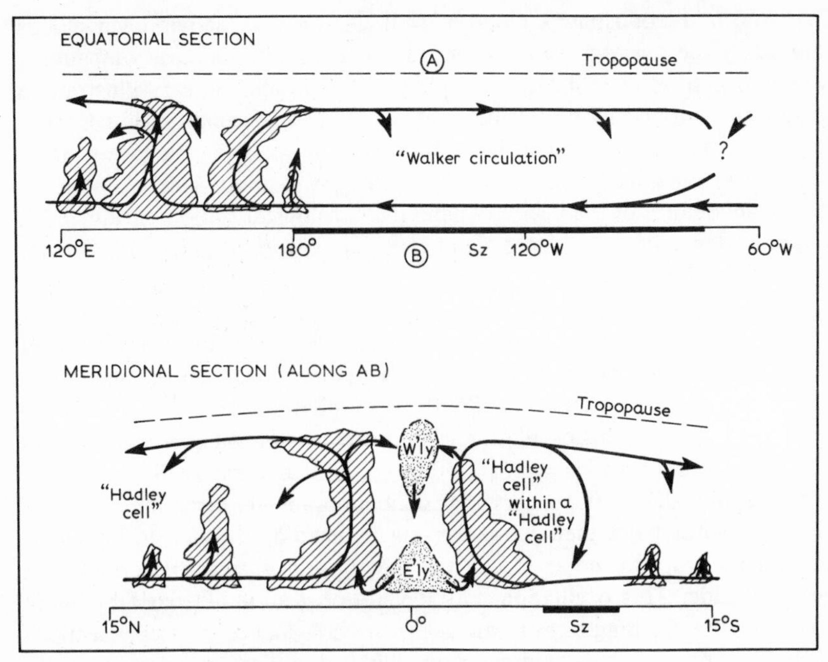

Fig. 9.11 The proposed relationships between upper and lower tropospheric circulations in the equatorial Far East during a well-established Walker Circulation. (SZ: subsidence zone; W'ly: Walker Circulation aloft; E'ly: Walker circulation near earth surface)

from year to year. The strength and persistence of each doubled intertropical cloud band seems to be influenced greatly by this variation.

The wider implications of such highly significant embroideries upon the classical circulation in the Tropics are worth investigating. They are found both within the context of the explanation of local climates, and, potentially at least, much further afield through the far-reaching effects (or 'teleconnections') of such deviations from the classical tropical model elsewhere within, and even outside, the tropical belt. Let us summarize the possible climatic implications, both locally and far afield:

Local climatological implications It has been argued by Bjerknes (1969) and by Krueger and Gray (1969) that the great variability demonstrated by the precipitation record at Canton Island reflects the variation in strength and westward penetration of the Walker Circulation. When the cold surface water extends further west than usual, subsiding air prevails over the island; when, on the other band, an eastward withdrawal of the cold anomaly occurs, the upturn of the Walker Circulation may shift to around 170° W. Rising columns of air then dominate the weather of Canton Island, giving frequent rains from mid-tropospheric cloud as well as from convective tower clouds favoured by the positive sea-minus-air-temperature difference accompanying the reduction in solar radiation reaching the sea surface. Plate 19 illustrates a period characterized by cold water and relatively arid conditions at Canton Island. The intertropical cloud band appears as a doubled structure split by an elongated subsidence zone extending well west to about 160° E. With the unfortunate demise of the Canton Island meteorological station during 1967, satellite evidence becomes more significant than before in promoting further studies of this key area.

Broader climatological implications These are more tentative than the local implications, but are possibly even more intriguing. They can be understood best by reference to the so-called 'Southern Oscillation' described by Walker in 1924. Essentially, this oscillation is an inverse statistical correlation in sea level pressure between values for Indonesia and the eastern South Pacific Ocean (see fig. 9.12). It appears that contemporaneous deviations of mean annual pressures from their long-term means have a tendency to be of one sign over the tropical Indian Ocean and the opposite sign across most of the tropical Pacific. This oscillation may deviate on a monthly basis by about $\pm 1{\cdot}5$ mb from the long-term mean, related to a sequence of teleconnections around the entire global quadrant from South America to Indonesia. Such sequences are composed of the following stages:

(i) An increase in the pressure gradient locally at the base of the Walker Circulation (i.e. off the north-west coast of South America) seems to

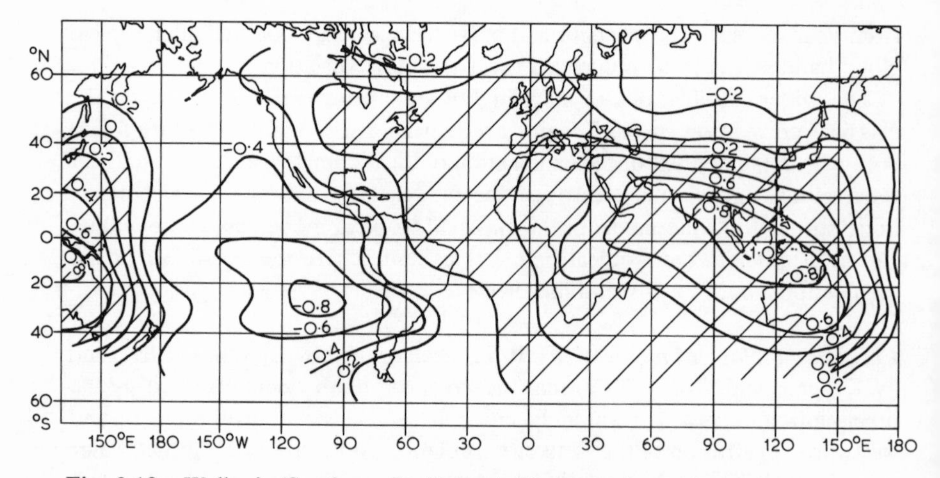

Fig. 9.12 Walker's 'Southern Oscillation'. The map shows the worldwide distribution of correlations of annual pressure anomalies with simultaneous pressure anomalies in Djakarta, Indonesia. Positive anomalies are cross-hatched. (After Bjerknes 1969)

stimulate an increase in the surface equatorial easterly air stream, and to be associated with an increased upwelling of cold water which sharpens the contrast between sea surface temperatures in the eastern and western equatorial Pacific. Put differently, it seems that an intensification of the Walker Circulation provides for an increase in the east–west temperature contrast which is itself the primary cause of the Walker Circulation. Trends of increase in the Walker Circulation, and corresponding trends in the Southern Oscillation, probably operate to reinforce one another in this way.

(ii) A case for a trend of decreasing wind speed within the Walker Circulation can be made out as follows. A decrease in the pressure gradient between South America and the equatorial eastern Pacific decreases the equatorial easterlies. The related upwellings of cold water weaken, and the sea becomes warmer and supplies more heat to the atmosphere. Thus the east–west temperature contrast within the Walker Circulation is lessened, and the zonal circulation becomes further retarded.

Thus cycles of alternating pressure trends may be envisaged, operating as a result of air–sea interactions in the equatorial zone. What is unclear as yet is how and why any trend may be reversed. Further research is necessary also on the teleconnections that operate meridionally rather than zonally, that is to say on the long-distance links that may link these tropical with some extra-tropical climatic fluctuations. It has been suggested that, when the Walker Circulation is weak, the Hadley circulations are stronger than average, and therefore export more heat and momentum than usual. The increased energy received by middle latitudes should then increase the vigour of the dominant circulations there as the imports of sensible heat and potential energy are translated in part into the energy of motion. Such seemed to be the case in January 1966. During that month, rainfall at Canton Island totalled 520 mm and sea and air temperatures were higher than average. Unusually strong westerly winds were noted in the northern hemisphere as far as 40° N. By contrast, in January 1967, only 7 mm of rain fell at Canton Island, when sea and air temperatures were relatively low. Then the contemporaneous mid-latitude westerlies were much weaker than in the previous year.

Further relationships between tropical cloudiness and mid-latitude weather features have been examined in detail, using both satellites and conventional data, by Allison *et al.* (1972).

The I.T.C.Z. in its new perspective

We may conclude our review of satellite contributions to the climatology of the intertropical convergence zone with reference to three general matters: (*a*) the structure of a preliminary model embracing the various satellite-observed structures of the equatorial cloud band, (*b*) the global distribution of such structures, and (*c*) a satellite-based definition of the I.T.C.Z.

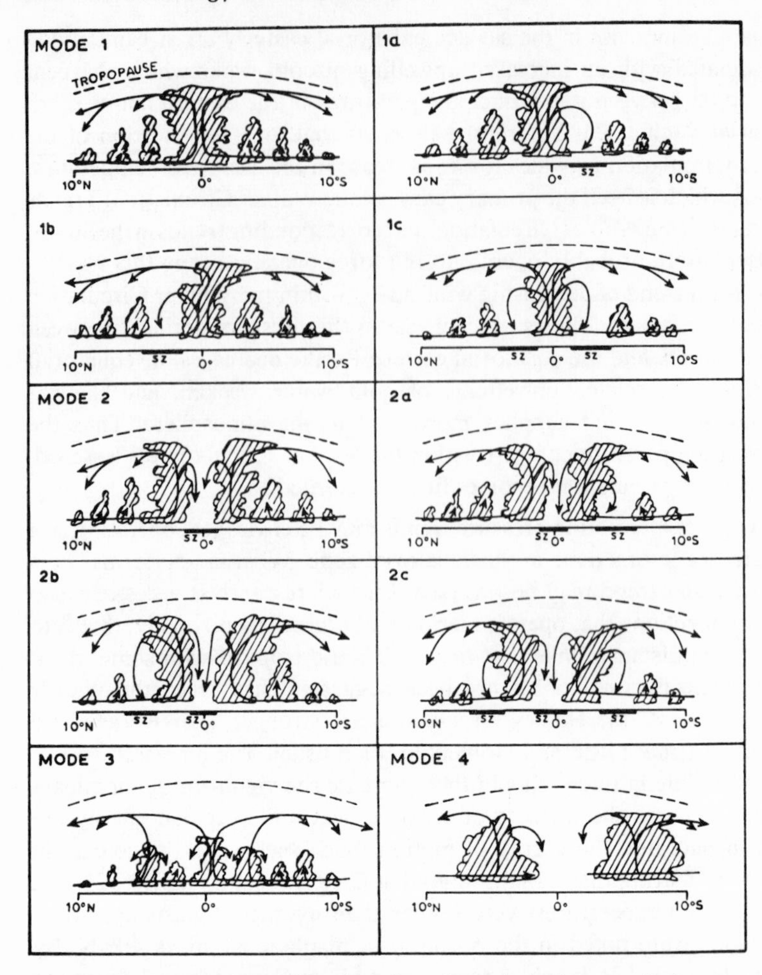

Fig. 9.13 Idealized meridional sections through the intertropical convergence zone, illustrating the range of forms that the intertropical cloud band may adopt. The classification is based jointly on the cloudiness and the intervening subsidence zones (S.Z.) that develop roughly parallel to the main instability belts. The chief distinctions are, therefore, between arrangements involving single major cloud bands (Mode 1), double major cloud bands (Mode 2), shear-lines with subdued convective cloud bands (Mode 3), and patchy convective cloudiness (Mode 4) lacking a simple linear organization. (From Barrett 1970b)

Fig. 9.13 shows a comprehensive model of intertropical cloud band structures identified across low latitude oceans by satellite studies (Barrett 1970b). Most of the ten forms depicted have been mentioned already at different points in the text above. It is obvious that several of the suggested meridional cross-sections represent considerable departures from the classical Hadley cell arrangement postulated for the Tropics. This is true especially of the double-banded structures, which involve subsidence rather than uplift along their equatorial axes.

As far as the variants of each basic type are concerned in fig. 9.13, (1 (a)–(c), 2 (a)–(c)), local overturnings of air on the poleward flanks of major instability axes were suggested first by Riehl as long ago as 1954. Riehl pointed to the possibility that 'Hadley cell within Hadley cell' arrangements may occur from time to time.

Recent theoretical studies have thrown further light on such structures. Fig. 9.14 summarizes some of the related conclusions arrived at by Asnani (1968). He demonstrated that more complex circulations than the Hadley cell

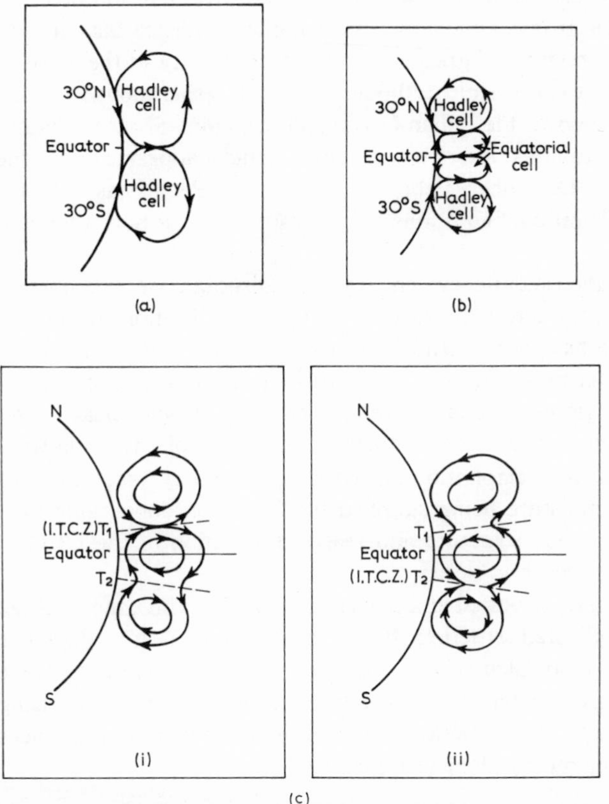

Fig. 9.14 Old and new views of the tropical meridional circulation: (a) a simple, twin-celled Hadley arrangement; (b) two equatorial cells separating two Hadley cells (Rossby 1949); (c) equatorial cells separating two Hadley cells. T_1 and T_2 are sea-level positions of maximum hydrodynamic stability: (i) I.T.C.Z. at T_1; (ii) I.T.C.Z. at T_2. (After Asnani 1968)

type are permitted mathematically by established vorticity equations: thus both single and double trough lines near the Equator can be accommodated by existing meteorological theory.

General evidence relating to the geographical distributions of basic single and double intertropical cloud bands has been afforded already by fig. 9.8, but only commented upon briefly in passing. Tentative research results for 1967 suggest further that the Atlantic pattern is the least variable pattern: usually a single cloud axis links very active convective cloudiness spread broadly over equatorial Africa and much of northern South America. Satellite infra-red observations (see e.g. Winston 1967) have indicated that the continental cloudiness is very deep. It contrasts very markedly with oceanic cloudiness along chosen parallels of a few degrees north and south of the Equator. Johnson (1970) has concluded from similiar evidence that in both January and July the regions of greatest bad weather activity in the Tropics are continental rather than oceanic, although the south Asian monsoon is an exceptional occurrence over the Indian Ocean and the China Sea (see chapter 11). Meanwhile, continuous intertropical cloud bands appear across the Indian Ocean only in December–February. Across the Pacific, as we have seen already, cloud band configurations vary greatly through both time and space.

Finally, with all the satellite evidence at our disposal, we are able to describe and redefine the intertropical convergence zone in more detail and more acceptably than has been possible before (see Holton *et al.* 1971).

In climatological terms the I.T.C.Z. may be conceived now as a narrow east–west band of vigorous cumulonimbus convection and heavy precipitation which forms along the Equatorward boundary of the trade wind regimes. Within this zone precipitation exceeds evaporation by a factor of two or more, the excess moisture being supplied by the moist, converging low-level airflows. The detailed structure and seasonal variation of the I.T.C.Z. apparently differ from ocean to ocean.

In meteorological terms, viewed at a particular instant in time, the I.T.C.Z. exhibits a more complicated structure. It may be described best as the locus of the cloud clusters associated with a train of westward-propagating wave disturbances. Certainly it is far from a steady atmospheric state, comprising rather a consequence of the dynamic properties of large-scale disturbances concentrated along the meteorological Equator.

Meridionally organized linear disturbances of tropical weather

Easterly waves

We turn next to consider linear organizations of tropical weather formed at high angles to the intertropical cloud band. Foremost among such perturbations of tropical weather are the so-called ‘waves in the easterlies’, or, more simply, the 'easterly waves'. It has been remarked that, by comparison with Pacific patterns, those of cloudiness and weather across the tropical Atlantic Ocean appear to be generally less varied and variable. Since the tropical Atlantic Ocean is relatively narrow, the subtropical anticyclones feeding the equatorial trough through the converging trades are relatively simple in structure, usually composed of single anticyclonic cells. Significant contributions are made to the cloud and precipitation climatology of the Caribbean, however, by westward-travelling troughs of low pressure embedded in the trades. The roots of these troughs are usually within the equatorial trough itself. In both form and behaviour easterly wave disturbances are very different from the intertropical cloud band:

(*a*) They extend characteristically over hundreds, rather than thousands, of kilometres
(*b*) They are often less sharply defined in terms of cloudiness and associated weather.
(*c*) They are organized meridionally rather than zonally.
(*d*) They tend to move in one direction only, from east to west.
(*e*) They often travel long distances as discrete, readily identifiable entities, i.e. they progress, rather than vacillate.

The equation of conservation of potential vorticity is:

$$\frac{f+\zeta}{\Delta p}=K \qquad (9.4)$$

where f is the Coriolis parameter, ζ the relative vorticity, and Δp the depth of the tropospheric air column involved. Thus, considering a meridionally orientated trough narrowing poleward from the equatorial low towards a trade wind anticyclone, it is clear that air overtaking the trough line from east to west must curve polewards (f increasing) towards a zone of cyclonic curvature ($+\zeta$ increasing) so that, if the left hand side of the equation is to remain constant, Δp must increase also. This vertical growth of air columns necessitates low-level horizontal convergence, and is accompanied by strong cloud growth. Thus an elongated zone of thunder shower activity often occurs in association with the maximum convergence just behind the axis of the trough (see fig. 9.15). Ahead of this axis, where f, ζ and Δp all decrease, divergent zone weather becomes progressively better marked in the descending, warmer air.

Exactly why such wave-like perturbations should occur at all is still something of a meteorological mystery; meanwhile, some authorities have suggested that the classical form never occurs in actuality. Sadler (1966) even referred to easterly waves as 'the biggest hoax in tropical meteorology'! Satellite evidence, however, leaves little doubt about the quite frequent occurrence of organized cloud clusters embedded in the trades, both in the western

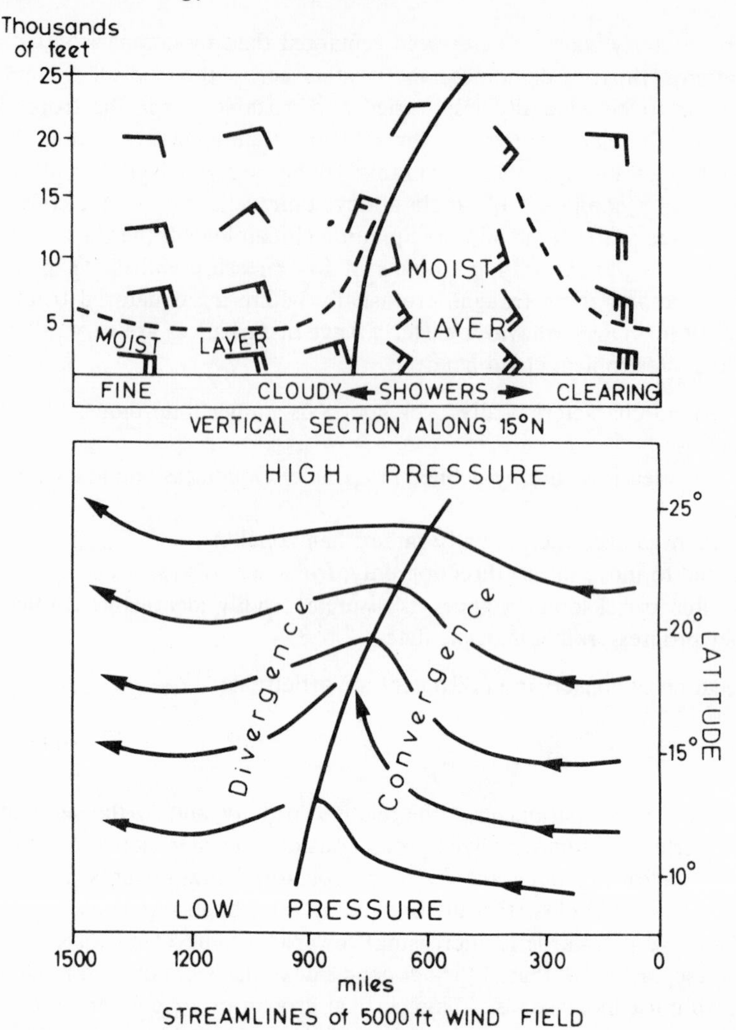

Fig. 9.15 Easterly wave model. The lower diagram shows the instantaneous wind pattern (streamlines) at about 1500 m (5000 ft); the upper one a vertical section along latitude 15° N. The wind arrows show direction and speed at different levels; the dashed line represents the trade-wind inversion; the heavy line marks the wave trough. (From Barry & Chorley 1968)

Atlantic and the western Pacific, moving along the flanks of the maritime anticyclones. On the other hand, though, it is clear that easterly wave organization is more varied than was earlier thought. For example, some such cloud arrangements are generally linear, but are built of east–west cloud bands instead of north–south bands. Others still are more amorphous or subcircular than the classical pattern. Such forms are often associated with the closed circulations of tropical disturbances or depressions: it has been reckoned for many years that linear instability within the tropical easterlies often constitutes a likely arrangement from which tropical storms or even hurricanes may arise (see chapter 10).

Two matters in particular may be elucidated through the analysis of satellite photographs of the Tropics. First, the range and nature of 'easterly wave' type organizations of cloud clusters within the trades, and, second, the frequencies and preferred paths of such organizations. Such knowledge is of vital interest to the modern climatologist; it is fortunate that several satellite-based studies have thrown important new light upon the origins of such phenomena, their ranges of form, and the contributions of their weather to local climatic statistics.

Interpretation of linear organizations of cloud clusters within the trades

The key problem in much satellite climatology concerns the correct identification of cloud patterns that are thought to be related to certain classes or types of circulation patterns. Within the Tropics, as the next chapter emphasizes, significant contributions have been made to the understanding and interpretation of tropical storms and hurricanes from satellite data, but these systems often possess the advantages of being clear-cut and well defined. Rather less, but still useful, progress has been made where cloud organizations are less definite, as in the present case.

An interesting recent survey (Frank 1969) revealed that, over the tropical north Atlantic, cloud patterns with the appearances of 'inverted V's' could be frequently identified. These patterns travelled from east to west, had a reasonably good day-to-day continuity and were associated with wave perturbations in the pressure fields in the middle troposphere. Four important points emerged from Frank's study:

(i) The 'inverted V' cloud organizations often developed near the west African coast, many growing to extend over synoptic scale areas, and extending at times from 5° to 25° N.
(ii) They moved westward at speeds approximately equal to the mean speed of the trades, ranging from 12 to 19 knots, with an average of 15–16 knots.
(iii) The 'inverted V's' were best defined in the east and central tropical north Atlantic, becoming less distinct as they moved west. During 1967, for example, almost every pattern disappeared before crossing the Antilles.
(iv) The intertropical cloud band may or may not be involved in individual patterns.

Fig. 9.16 shows that, in the 'inverted V' model, the linear organization of clouds is along, rather than across, the streamlines, in sharp contrast to the classical arrangement. It should be noted that practically no circulation

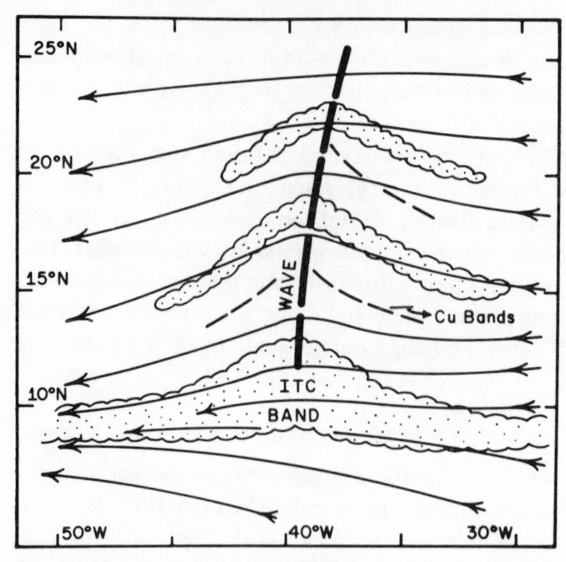

Fig. 9.16 A model of the relationship between the lower tropospheric flow and the 'inverted V' cloud pattern. (From Johnson 1970; after Frank 1969)

changes are apparent with the 'inverted V' cloud pattern. The V-shaped cloud lines usually lie nearly parallel to the low-level wind shear.

Although Fett (1966) claimed that 'the classical easterly wave model . . . is extremely well related to observations obtained by satellites', the abundant cloud evidence that is now available generally substantiates the contention of Simpson *et al.* (1968) that the classical model has been greatly overworked. Whereas on some occasions there is indeed a good resemblance between the established meteorological model in satellite-viewed cloudiness, on other occasions the resemblance may be slight. In addition to the 'inverted V' it seems that another pattern sometimes occurs too, in which the convective cloudiness is concentrated near and/or downstream from a wave axis, rather than upstream as the classical model suggests. Further organizations are also apparent from time to time in the western half of the tropical Pacific, although the degree of organization of these roughly elongated cloud clusters is generally less than in the Atlantic cases. Their rates of movement, however, seem quite similar; a GARP study group (1968) has given the speed of translation of the westward-moving clusters over the Pacific as 5° of longitude per day, which is close to the value Frank established for Atlantic 'inverted V' waves. More recently, Chang (1970) has studied the Pacific clusters through time-composite satellite photographs (see plate 20), and has shown that their movement is remarkably uniform, the phase speed being about 9 m/sec, and the period between neighbouring clusters in the range from three to six days, corresponding to a horizontal scale of 2000–5000 km. These values correspond well with those originally given by Riehl (1945) for easterly waves, and others established by Yanai *et al.* (1968) from spectral analyses of the fluctuations of lower tropospheric wind components over the tropical Pacific (four to five days). Chang further observed that the intensity of the activity of the mobile cloud clusters varied greatly from day to day. The elongated clusters were well marked on some days, almost vanishing on others. Their continuity in time was, however, readily apparent – a fact supported by the comparable time-longitude photographic section studies of Wallace (1970). In some cases it was even possible to trace the westward movement of identifiable cloud clusters continuously all the way from Africa to the western Pacific Ocean.

The contribution of easterly wave weather to local climates

Since most published work has related to the Atlantic, our regional review reflects that preoccupation. Frank (1969) concluded from his satellite-based studies that classical easterly wave patterns over the Caribbean may not be such recurrent features of local climates as previously believed, but some wave forms develop sufficiently far east to perturb weather along the west African coast, a second finding contrary to popular belief. Barry and Chorley (1968) remarked that 'It is obviously very difficult to trace the beginnings of these waves over the tropical Atlantic.' In doing so, they represented essentially the pre-satellite case. Satellite studies, almost all based on cloud photo interpretation, have done much to further our knowledge of the origins, paths and effects of this interesting family of weather disturbances.

Figs. 9.17 (a) and 9.17 (b) and table 9.2 summarize some new views proposed on the basis of specific satellite research findings. Fig. 9.17 represents schematically flow patterns and mean cloud cover models for typical west African disturbances (after Carlson 1969). In the meantime, table 9.2 lists observed frequencies of initial wave positions against longitude. This suggests that the generating mechanism, which must involve convective processes and advection of relatively warm air, may involve topography also. The Cameroun mountains seem to be one possible site favouring wave initiation, and other mountainous energy sources may exist still further east in the Sudan and Ethopia. Fig. 9.18 depicts the frequencies and fates of tropical wave systems crossing the north Atlantic from the African coast to Central America. Contrary to the emphatic statement of Thompson (1965) that 'easterly waves play no part in the weather of Africa', good evidence seems to be forthcoming from joint satellite and conventional data analyses for the formation of wave-like disturbances further east than the west African coast, indeed further east than the west African 'bulge'. Such systems seem to occur in continuous wave trains rather than singly. Some individual systems survive

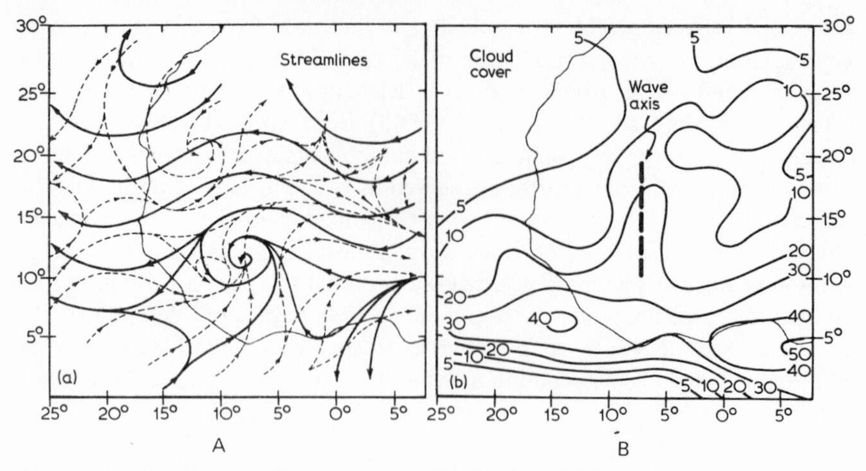

Fig. 9.17 (a) Schematic flow pattern at 10,000 ft (solid lines) and 2000 ft (dashed lines) for the typical west African disturbance. (b) Contours show a mean percentage cover of cloud with respect to the schematic flow pattern in (a). The field represents an average cover for 23 disturbances as measured near 10° W. (From Carlson 1969)

long enough to constitute perhaps even the majority of the disturbances in the eastern Caribbean region. Fig. 9.19 shows that similar sequences of rainfall and sea level pressure often occur at St Thomas (Virgin Islands) and Barbados related to the passage of 'African disturbances' in wave, vortex or amorphous forms. Thus it appears that many trade wind disturbances spawned over Africa later exercise considerable influence upon cloudiness and rainfall over the eastern Caribbean and its islands. Although many of these disturbances are probably rather weak, disintegrated wave forms, yet they must still retain sufficient vertical motion fields to contribute so significantly to the island precipitation totals. On the other hand, of course, their contribution may be little more than a heightening of locally induced island convection patterns, although this is less likely for they can be seen offshore too. Similar effects might be expected in the western Pacific theatre, but these have yet to be substantiated through joint satellite and conventionally based research.

Table 9.2

Frequency distribution of initial wave positions versus longitude (from Carlson 1969)

Longitude interval	West of 4° E	4° E – 6° E	7° E – 9° E	10° E – 12° E	13° E 15° E	16° E – 18° E	E. of data network
No. of fixes on satellite photos	4	2	1	6	3	2	14

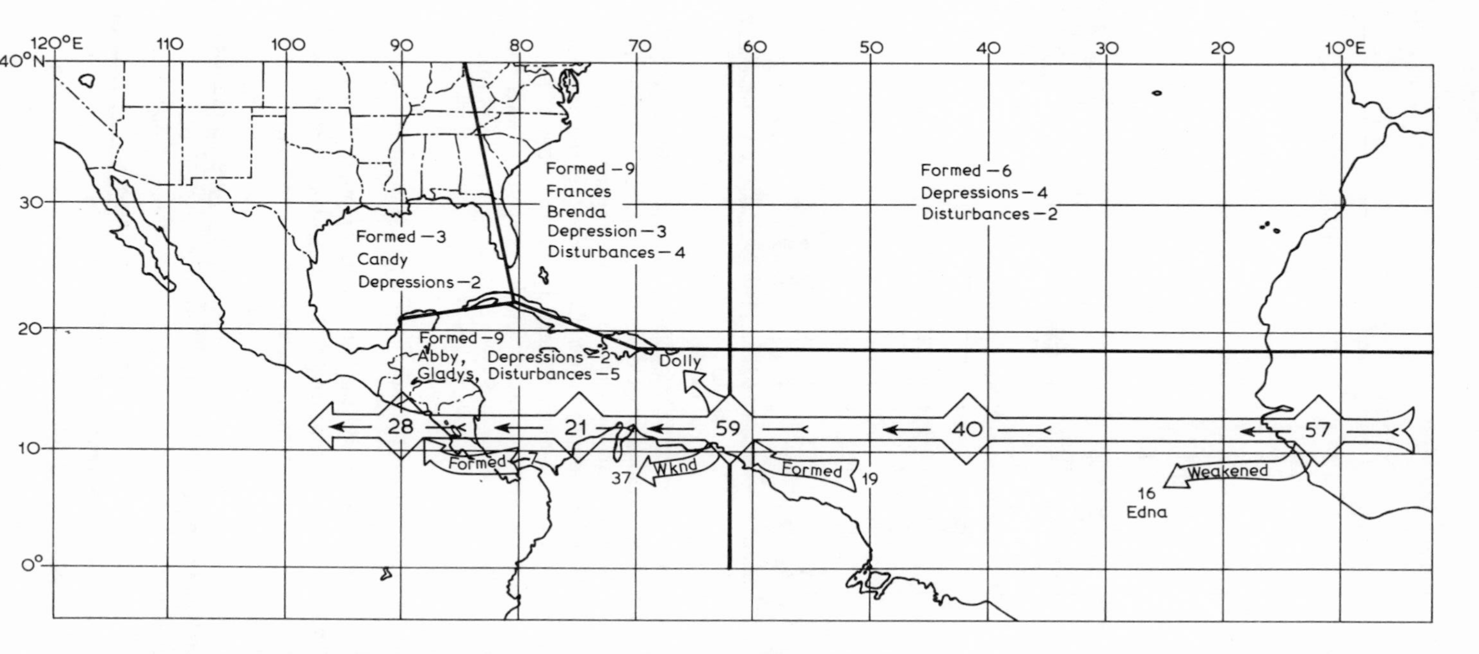

Fig. 9.18 Summary of the synoptic-scale tropical systems observed from western Africa to the eastern Pacific during the hurricane season, June–November 1968. (After Simpson *et al.* 1968)

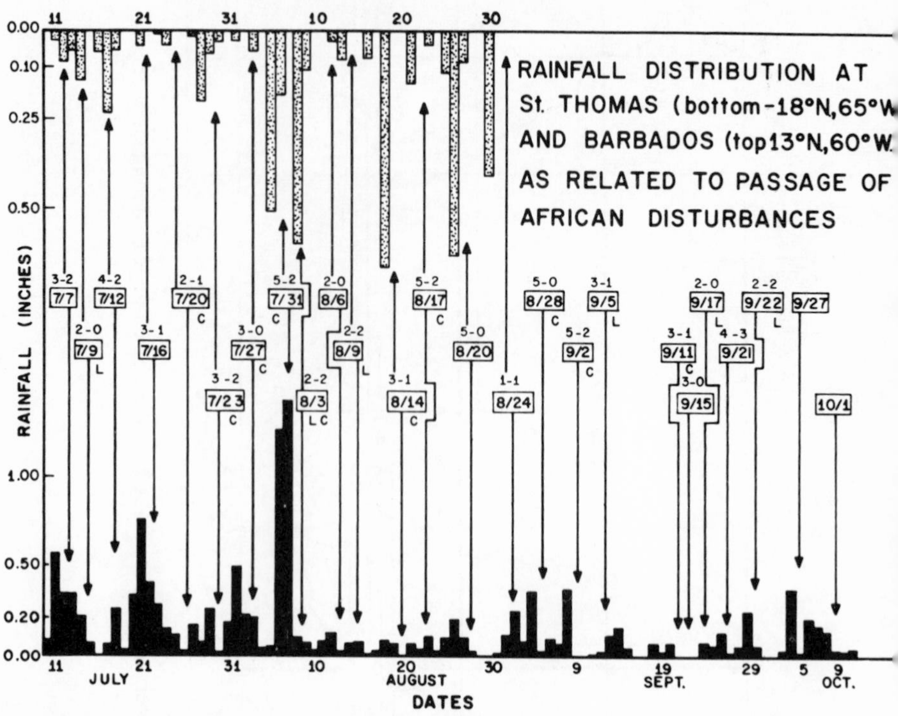

Fig. 9.19 Distribution of average daily rainfall at St Thomas (bottom: average of seven stations) and Barbados (top: average of four stations), related to the passage of African disturbances (labelled arrows). The two digits at the top of each box are the sea level pressure fluctuations (in mb) for the disturbance over the African coast at 17° W (left figure) and over the Antilles (right figure). The letter C below a box denotes a system with a circular cloud pattern with attendant low-level vortex over Africa. The letter L denotes systems at a lower latitude than usual. (From Carlson 1969)

Obliquely orientated linear organizations of tropical weather

Lastly let us turn to some of the earliest results of photographic analyses based on the data from geosynchronous A.T.S. satellites which permit short-term developments of tropical weather systems to be more closely inspected. Detailed analyses for data-scarce regions of the eastern tropical Pacific, especially by Fujita *et al.* (1969), have thrown further light on the nature and development of patterns such as those described earlier on the basis of Nimbus II evidence (pp. 231–7). It has been shown conclusively that geosynchronous satellite data can be employed very effectively in determining the scales and modes of various tropical circulations. In particular, the use of A.T.S. pictures in determining fields of cloud velocities is proving to be of value in understanding their dynamics.

Of special interest in the present context is the new model of migratory equatorial anticyclones proposed by Fujita and his team. Their A.T.S. photo analyses, along the lines listed in chapter 5 (p. 136), revealed that, just west of Central America, organized cross-equatorial flow from the southern hemisphere in winter gains relative anticyclonic vorticity in an oceanic region where sea surface temperatures are relatively low. Some such near equatorial anticyclones were noted by Barrett (1970a) west of Central America especially when the dominant intertropical cloud band curved further north than usual, leaving a relatively cloud-free pool of air between strong cloudiness over the isthmus and to its west. Fig. 9.20 presents Fujita's model of similar anticyclones, established after detailed studies of cloud patterns and wind fields over the eastern equatorial Pacific. It must be noted that mobility is an essential feature in this model. Probably some of the daily variability of the intertropical cloud band may be explained through the repetitive occurrence of such a sequence of events. The sequence may be summarized as follows:

(*a*) A 'pushing stage', in which a large-scale flow from the southern hemisphere pushes northward, thus producing a convex band of intertropical cloudiness.
(*b*) A 'recurving stage', within one to three days of the commencement of the cycle, in which the transequatorial flow begins to recurve southwards. Tropical depressions formed in the pushing stage along the zone of interaction between northern and southern hemisphere air tend to move out of the I.T.C.B. in this new stage.
(*c*) A day or so later, the equatorial anticyclone becomes characterized by an enclosed circulation in the 'cut-off stage', marked by a break in the continuity of the I.T.C.B. to the east.
(*d*) As soon as this break occurs, the northern hemisphere trades penetrate south of the anticyclonic centres, and some air becomes involved in the anticyclonic circulation which was previously of ex-southern hemisphere air alone. This is the 'mixing stage', which gives way after another day or so to the
(*e*) 'Burst stage' involving a north-westerly expansion of an I.T.C.B., the so-called 'burst band'. The joint pushing force exerted by north-east and transequatorial south-east trades aids the westward displacement of the anticyclone. This weakens alongside the very intense zone of convergence and cyclonic vorticity concentrated in the burst band. After one to two days this linear cloud organization breaks up quickly into small fragments or isolated cloud clusters.
(*f*) Finally, in an 'interacting stage' persistently intense air streams to the south-west of the travelling anticyclone centre are strong enough to prevent mid-latitude cold fronts in the summer hemisphere from moving in a south-easterly direction. A wave on the cold front is observed in this interacting stage, through which the equatorial anticyclone has been advected into middle latitudes.

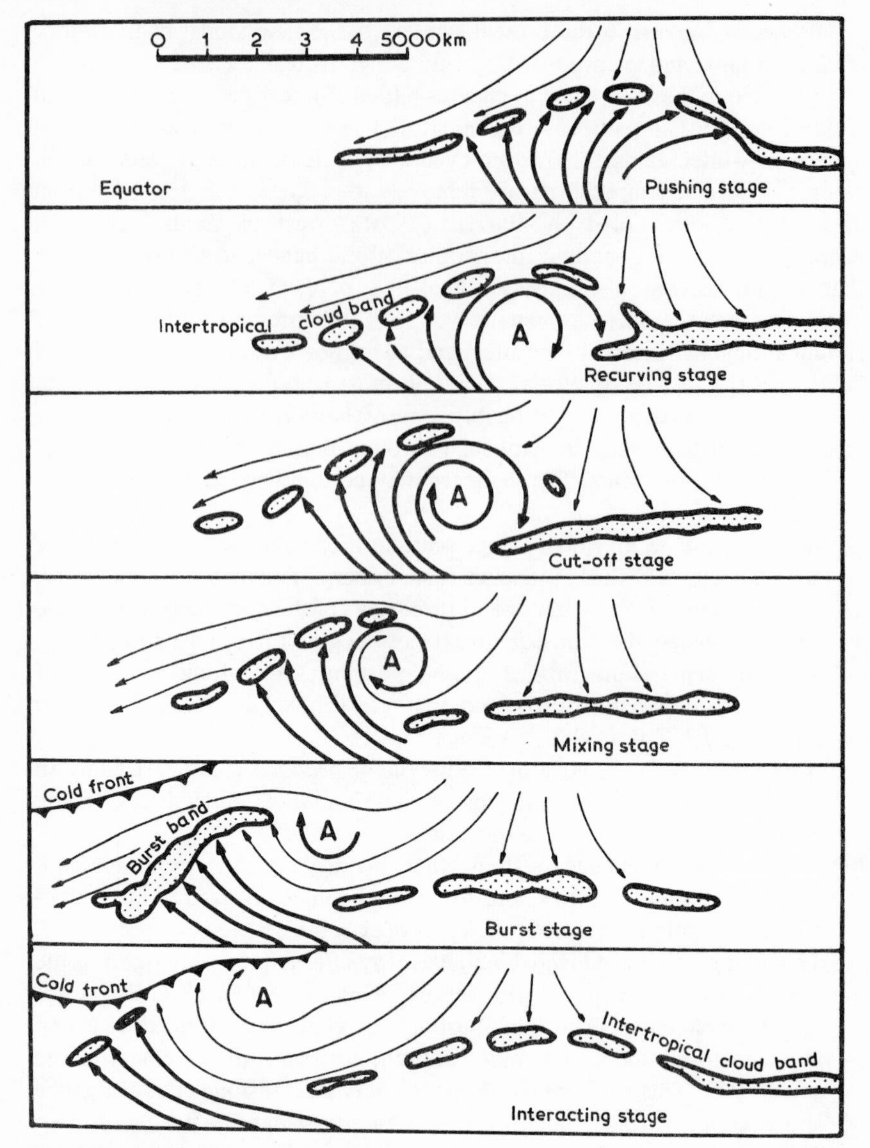

Fig. 9.20 A proposed model of an equatorial anticyclone west of Central America, in six stages. (From Fujita *et al.* 1969)

Fujita *et al.* (1969) stated that the entire life cycle of such an anticyclone requires about two weeks. Depending on the frequency of anticyclone formation, two such systems may often be seen at the same time over the equatorial northern Pacific, especially in August or September. Of particular relevance to the present discussion is the climatological impact of the linear

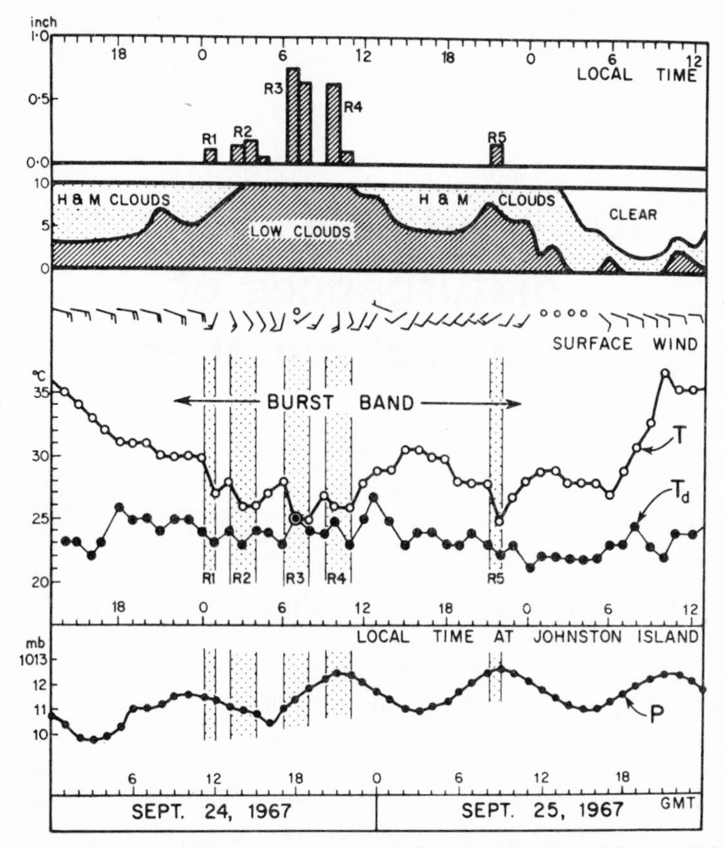

Fig. 9.21 To illustrate the passage of a burst band at Johnston Island (16·7° N, 169·5° W) on 24–25 September 1967. During the passage of the band five separate periods of rain occurred, during each of which air temperature dropped several degrees. (From Fujita *et al.* 1969)

burst bands on their north-western fronts. These previously unsuspected structures are intermediate between the normal alignment of the I.T.C.B. proper (from east to west) and the classical alignment of easterly wave weather (from south to north). Although long-term assessments of their climatic impacts have yet to be completed, fig. 9.21 hints at what may emerge. Although this time plan portrays a complex temporal distribution of associated cloudiness, rainfall and surface temperatures, it points to a considerable and significant effect of burst band weather sequences.

So burst band weather must be viewed as a significant contributory factor to the diversity of eastern equatorial Pacific climates, alongside other major linear organizations of tropical weather such as intertropical cloud bands and wave forms in the easterlies.

10 Non-linear disturbances of tropical weather

Introduction

The difficulty implicit in dividing the spectrum of tropical weather systems into two groups, 'linear' and 'non-linear', is considerable indeed. It was necessary to make passing references to vortical and amorphous cloud arrangements in chapter 9 despite its preoccupation with elongated organizations of cloud. In turn, no discussion of non-linear arrangements is possible without some mention of the linear. Not infrequently, a cloud form belonging clearly to one of the two major groups may develop from, or degenerate into, a form belonging to the other. To complicate matters further, some cloud arrangements clearly characteristic of one group at one scale of study may seem to be built of substructures characteristic of the other at a different scale. Hence the subdivision we have chosen is one of general convenience rather than absolute meteorological or even geometrical worth. Indeed, if anything, satellite studies have tended to illustrate the interdependence of linear and non-linear forms, rather than underline the separate and different contributions they may have been thought to make to tropical climatology and meteorology. Thus, for example, band structures are now known to develop frequently in association with the vortices of tropical cyclones, while, on the other hand, the intertropical cloud band often appears to be comprised of numerous smaller semi-independent centres of convection and convergence linked together (often by upper level cirrus) like beads on a chain.

Hence, it can be more helpful sometimes to classify tropical atmospheric circulations and their attendant organizations of cloud in terms of scale rather than form and appearance. We are able to differentiate usually the following scales of features depicted by satellite observations:

1 Planetary scale organizations. These include, for example, the equatorial trough and trade wind regimes, subtropical anticyclones, jet streams and the Indian monsoon.
2 Synoptic or large wave scale organizations. These include waves in the easterlies, tropical cyclones, upper tropospheric waves, etc.
3 Mesoconvective scale organizations, including cumuliform clouds and various cloud aggregates.

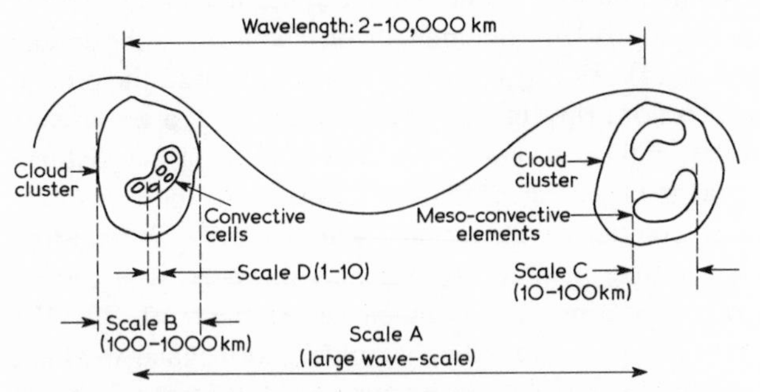

Fig. 10.1 The scales of atmospheric motion systems in the Tropics. (From Sikdar & Suomi 1971)

Bearing these different sized atmospheric building blocks in mind, a more precise range of satellite-observed cloud masses has been defined for use in tropical investigations within GARP and the W.W.W. (see Sikdar and Suomi 1971; and fig. 10.1) including:

1 Scale A disturbances (large wave scale). Wavelengths range from 2000 to 10,000 km.
2 Scale B disturbances (synoptic and subsynoptic scales). These cloud masses range from 100 to 1000 km across. They are called 'cloud clusters'.
3 Scale C disturbances (mesoconvective scale). These masses range from 10 to 100 km across. They comprise 'cumulonimbus clusters' or 'mesoconvective elements'.
4 Scale D disturbances (local convective scale). These clouds include individual cumulonimbus cells, ranging from 1 to 10 km in diameter.

The nature and purpose of this account dictate that most attention should be paid to Scale B disturbances, since these seem to make the biggest contribution to tropical weather and climate, although Simpson *et al.* (1967) argued from satellite evidence that progressive travelling waves (Scale A), or vortical perturbations in the wind field (Scale B), while often present and of undoubted importance, may not be, after all, the main or even the most common *controls* of tropical cloudiness and rainfall variability. Hence, even in this discussion some comments are in order upon the general climatic significance of lesser scale disturbances (Scales C and D), since these are certainly the most numerous and, in some regions at least, may have combined effects even greater than those of the larger systems.

Martin and Karst (1968), in a census of about 1000 cloud systems over the tropical Pacific based solely on satellite data, were to find that fully half were 'oval' in shape – meaning that they lacked strong linear or vortical organization (implying no strong wind disturbance) – and that their mean lifetime was only two days (compare this with the findings of Hamilton in the Indian Ocean, reported in chapter 11). As Zipser (1970) has pointed out, the crucial question is whether convection, through its organization on the mesoscale, can have such a controlling influence on the structure of some larger disturbances in the Tropics that it becomes meaningless to discuss these disturbances apart from their interactions with smaller scale features. Whether or not the mesoscale influence may be a controlling one, however, it is certain that the current view of tropical weather is much nearer the truth than the traditional: far from there being an overriding simplicity in tropical weather, it is now obvious that tropical weather is highly complex in terms of its sequences of events and their controls. Pursuing this theme of complexity a little further, we note that two short-term periodicities of tropical cloudiness and rainfall have been suggested recently to be added to their already well known seasonal variations.

(1) A diurnal periodicity. For many years it was suspected that tropical clouds and rain exhibit maximum amounts and frequencies during the hours of darkness, especially in continental coastal and maritime regions. That suspicion has been confirmed by various recent studies, for example by Hutchings (1964), referring to the volcanic island of Rarotonga in the south Pacific, by Lavoie (1963) concerning several tropical atolls, and by Garstang and Visvanathan (1967) and Holle (1968) with respect to observations from a research vessel at sea. Various theories have been propounded in attempts to explain such departures from the pattern previously thought to prevail throughout most of the Tropics, where a late afternoon peak of rainfall activity was involved. Possibly the nocturnal cooling of cloud tops by radiation makes the cloud layer less stable, and encourages growth by mixing of droplets at different temperatures; or the increase in the temperature contrast between sea and air rises to its maximum just before dawn and prompts the largest supply of oceanic heat to the atmosphere then, stimulating most rapid cloud development. Both factors, perhaps even operating together, would be most marked in the early morning.

(2) A semidiurnal periodicity. Here the main effects seem to be a maximum of cloudiness and rain near dawn, with a weaker maximum near sunset. Many recent observational analyses have substantiated the reality of such a pattern, including analyses reported by Kiser *et al.* (1963) for Wake Island in the

north Pacific, and by Malkus (1963) for various tropical locations. The most widely accepted explanation involves the semidiurnal pressure oscillation, related to the so-called solar atmospheric tide, which seems to encourage convergence and enhanced convective activity in the early morning and evening, but divergence and suppression of convection around midday (Brier and Simpson 1969). This semidiurnal effect, it should be noted, has a relatively small amplitude compared with diurnal variations, amounting to a range of roughly 5–15% in cloudiness, and perhaps a slightly larger percentage in rainfall.

By the time of writing, no definitive satellite studies have been completed to demonstrate from photographic or combined photographic and infra-red evidence that these short-term cloud cycles are really above the satellite levels of resolution. The difficulties of night-time photography, and of comparing cloud outlines seen through visible wavebands with others through non-visible wavebands of the electromagnetic spectrum are both technically and meteorologically real. One study of oceanic satellite radiation measurements has, however, yielded some confirmatory evidence for diminished cloudiness and a reduction of the mean height of cloud tops just after midnight (Merritt and Bowley 1966).

Satellite photographs would seem to be potentially valuable for studies of:

1 Comparative diurnal strengths of cumulonimbus cloud distributions over both land and sea.
2 Developments of individual cloud cells and cumulonimbus clusters into larger cloud clusters or even tropical storms.
3 Analyses of wind shear through considerations of the detailed shapes and forms of satellite-viewed cloud elements.

Plates 21 and 22 are portions of two A.T.S.-III photographs taken during the same day over South America. Later in the day more substantial widespread cloudiness is characteristic of tropical continental areas where sufficient moisture is available in the local atmosphere for strong cloud growth. A fascinating point which has emerged from A.T.S. photographs organized into daily cinefilm loops is that the western rim of the Amazon Basin, including the eastern slopes of the Andes, often acts as a preferred zone for the daily birth and growth of cumulonimbus clouds and cloud clusters. Many of these seem to move off in a downslope direction towards the centre of the Basin during the later hours of daylight. By morning this cycle seems to recommence with less cloudy conditions. For this to occur, the cloud cover must largely dissipate through the night. The precise climatological effects and implications of such cloud and rainfall cycles will be evaluated more fully through infra-red analyses and photographs available from low-light intensifier cameras able to function adequately through both daylight and night-time.

It is obvious already that a climatology based on pictures taken during the afternoon hours in the Tropics is much different from one based on morning pictures. This is true not only over land, but also over the oceans to a much larger extent than previously believed. For example, the large areas of stratus and stratocumulus off the west coasts of North and South America show about a 50% reduction in total cloud amount between 10 a.m. and 4 p.m. local time. Over land the change in percentage cloud cover is even more dramatic. The tropical rain forests appear to be covered by stratus and fog at dawn; the skies largely clear as the air warms in mid-morning, and then become studded with convective clouds by afternoon. This seems to be a matter on which the Goes series of satellites, giving a twenty-four hour coverage of selected tropical regions, should throw much new light.

The classification of tropical vortices

At the outset of this review of Scale B tropical vortices, careful note must be made of certain terminologies. The first outlined below was formulated in pre-satellite days. Hence its subdivisions are not quite the same as those in schemes compiled later for use in satellite photo interpretation. Some indication will be given where necessary concerning the degrees of equivalence between schemes devised for different purposes.

(1) *Tropical disturbances*. These are discrete systems of apparently organized convection. They are generally 100 to 300 nautical miles (185–555 km) in diameter, originate in the Tropics or subtropics, and have a non-frontal, migratory character. They maintain their identities for twenty-four hours or more. They may or may not be associated with detectable disturbances in the wind fields.

(2) *Tropical depressions*. These have definite closed circulations, but maximum wind speeds of less than 34 knots (17·5 m/sec).

(3) *Tropical storms*. The most significant characteristics of these systems include large size and a good degree of evident cloud and weather organization. They are warm-core vortices with sustained wind maxima of at least 34 knots, but generally lack a central 'eye' of relative calm.

(4) *Hurricanes*. These are strongly organized vortices with sustained wind maxima in excess of 65 knots (33·4 m/sec). They are very vigorous weather systems, on average somewhat smaller than many tropical storms, but having well-developed convective cloud areas usually marked in their centres by 'eyes' of clearer, calmer conditions.

Satellite-based work in the field of tropical vortices has been both rich and varied, commencing in the earliest days of satellite meteorology when it became clear that particularly destructive tropical weather organizations like tropical storms and hurricanes are especially evident in both photographic and infra-red data displays. On any one day, numerous synoptic scale cloud systems appear in the Tropics and subtropics. It is particularly important in forecasting generally and in storm warning preparation in particular to be able to: (1) identify those areas of convection which, on the evidence of their appearance, would seem to be those most likely to develop, and (2) classify existing vortices in terms of their intensities.

Tropical vortex cloud organizations

Basically, the shapes and appearances of cloud systems associated with tropical vortices seem to depend on: (1) the types of atmospheric perturbations initiating their convection, and (2) the vertical wind shear of each storm's environment.

Since the second of these factors is related closely to local patterns of winds within large-scale circulation features such as the subtropical anticyclones and high-level mid oceanic troughs, vortices developing in different regions of the world follow rather different cycles of development, although once they have reached maturity they all modify their immediate environments to such an extent that they look much the same world wide (Oliver 1969).

The first comprehensive classification scheme to come into operational use at N.E.S.S. combined two earlier approaches developed with rather different purposes in mind. One (by Fett 1968b) was an aid to the identification of pre-typhoon stage storms in the Pacific. The other (Fritz *et al.* 1966) sought to permit the determination of the maximum wind speeds of well-developed vortices from cloud parameters observed in the satellite data. The N.E.S.S. scheme (fig. 10.2) classified tropical vortex cloud systems into four groups identified as stages A, B, C and X. The basic parameters include the size and position of each overcast area, the degree of circularity of the spiral band structure characteristic of such a system, and whether the centre of the spiral band structure is located inside or outside the major cloud mass of a storm. In disturbances classified A, B or C, the centre of the pattern produced by the curved cloud bands (i.e. the apparent centre of circulation) either does not exist (as in stage A), or, if it does exist, lies outside or very near the edge of the overcast cloud area. Generally speaking, A and B represent formative stages, C applies to storms that are growing or decaying, and X (especially categories 3 and 4) applies to storms or hurricanes that have already reached maturity; stage X is subdivided into four categories based on degrees of organization. Recently a new scheme has been proposed by Dvorak (1972) but no climatological results are forthcoming yet from its application to the hurricane regions of the world. Hence we shall concentrate on the original N.E.S.S. scheme.

A NO CURVED CLOUD LINES OR BANDS	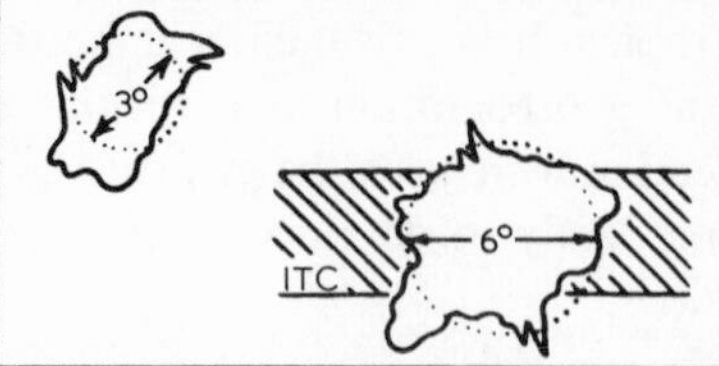 3° 6° ITC	*Stage A* is a dense amorphous cloud mass composed of cumuliform, cirriform, and layered middle cloud in any combination. Some cirrus outflow is usually present. The cloud mass must have an average diameter of 3° latitude or more. Exceptions: (1) If the cloud mass is contiguous to or within the ITC in the Atlantic, Pacific, or south Indian Ocean, it must have an average diameter of 6° latitude or more and be partially isolated by breaks from the general cloudiness. (2) In the Arabian Sea and the Bay of Bengal, the cloud mass must be 8° latitude or more in diameter.
B POORLY ORGANIZED CURVED CLOUD LINES AND BANDS ILL-DEFINED CENTER	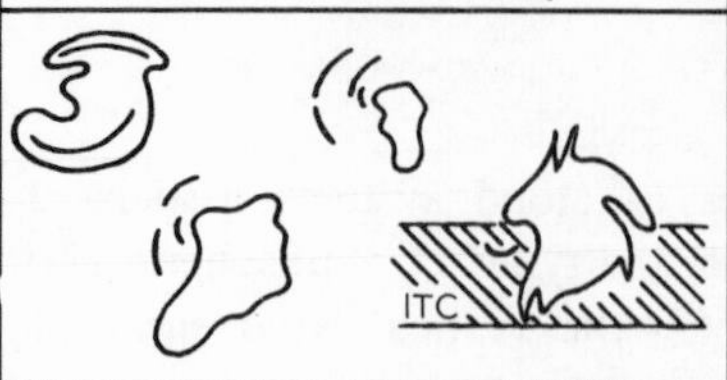 ITC	*Stage B* is a dense cloud mass with adjacent curved cumulus cloud lines and/or curved bands of middle cloud which are either detached from, or form part of, the major overcast area. The curved cloud lines and bands are often poorly organised. The pattern produced by the curved lines and bands is poorly defined – it does not appear to have one definite center. Along the ITC, the cloud mass and associated curved cumulus cloud lines and/or bands must be separated from the ITC cloudiness on at least one side and cirrus outflow must be evident.
C WELL ORGANIZED CURVED CLOUD LINES AND BANDS WELL-DEFINED CENTER OUTSIDE DENSE CLOUD MASS	 C− C C+	*Stage C* has well organized, curved cumulus cloud lines and/or broad curved bands of middle and high cloud. The pattern produced by the various curved lines and bands has a well-defined single center. The center of the pattern generally lies outside but adjacent to an associated dense cloud mass, but it can be on the edge or as much as one-half degree latitude within the cloud mass. A C̲− has no associated dense cloud mass. A C̲+ appears very well organized with a large amount of curved cirrus outflow.
X CAT. 1 POORLY ORGANISED SPIRAL BANDS ILL-DEFINED CENTER OF ORGANIZATION WITHIN CENTRAL CLOUD MASS	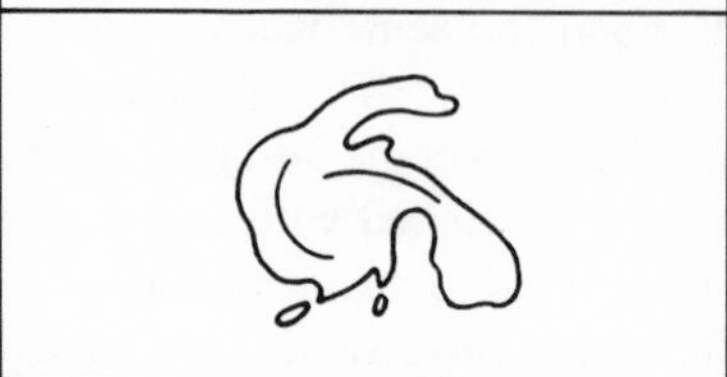	*Category 1* has a bright generally circular central overcast which is cirriform in appearance. Curved cirrus outflow is often restricted to one quadrant. Poorly organized, slightly curved cumuliform cloud bands appear near the periphery of the central overcast and cross into it at a large angle. This banding remains close to the overcast edge; away from the overcast, organized curved bands are usually absent. An eye is not visible. The center of the spiral pattern can be located approximately by extrapolating inward along the curved peripheral bands. This estimated center must be more than one-half degree latitude within the central cloud mass.
X CAT. 2 WELL ORGANISED BANDS SPIRAL BANDS DEFINE CENTER WITHIN CENTRAL CLOUD MASS		*Category 2* has a bright, often asymmetrical central overcast. Cirrus outflow is curved and more extensive. At least one long, major, well-organized band spirals at a large angle into the central cloud mass. A linear curved break accompanies this band. Within the central cloud mass, the break is covered by thin cirrus but is readily detectable. Minor peripheral bands outside the overcast are poorly organised. An eye is not visible. The central tip of the major spiral band defines the center. This center must be more than one-half degree latitude within the central cloud mass.
X CAT. 3 MODERATE DEGREE OF CONCENTRICITY TO CLOUD BANDS IRREGULARLY SHAPED EYE WITHIN CENTRAL CLOUD MASS		*Category 3* has a bright central overcast that is compact and tends to be circular. There is considerable curved cirrus outflow visible at the edge of the central overcast. Curved striations within the central cloud mass define spiral cloud bands which are moderately concentric about a visible eye. Well organised peripheral bands, some with well-developed cirrus, are present. A ragged and irregularly shaped eye is normally visible. This defines the storm center.
X CAT. 4 HIGH DEGREE OF CONCENTRICITY TO CLOUD BANDS ROUND EYE NEAR CENTER OF CENTRAL CLOUD MASS		*Category 4* has a very circular bright central overcast. The edge is often sharp and smooth over one or two quadrants, otherwise, it is striated cirrus. Highly concentric striations appear within the central overcast. Banding outside the central overcast is very well organised and circular. The entire cloud system is very symetrical in appearance. A well-defined eye appears as a small dark circular area surrounded by a bright ring. This defines the storm center.

Fig. 10.2 A tropical and subtropical disturbance classification from satellite data. (From Oliver 1969)

The distribution of weak tropical vortices

Regional contrasts are of greatest significance where stage A systems are involved – that is to say, the least intense tropical disturbances. These are worth special examination prior to our turning to a closer examination of the stage X systems, which are the most important of all on account of their extreme vigour. Each stage A system is comprised of a dense, amorphous cloud mass produced over a region where stronger than normal convection is taking place. Such may occur within, or contiguous to, the intertropical cloud band in the Atlantic, Pacific or South Indian Ocean regions, but the broadened masses of convective clouds that appear frequently in such a synoptic situation must be larger than 6° of latitude to qualify them for inclusion in this unavoidably somewhat subjective category. Smaller cloud centres generally appear too briefly as distinctive features to justify their inclusion in such a scheme. Meanwhile, in the Arabian Sea and Bay of Bengal regions, where the intertropical cloudiness is very extensive in the summer season, a minimum diameter of 8° of latitude is specified. On the other hand, when removed and separated from the intertropical cloud band, a system need have a minimum diameter of only 3° latitude for it to be designated as a stage A vortex. Generally, cumulonimbus clusters and cloud clusters within the Tropics but outside the main east–west convergence cloud band fail to attain even this diameter unless a local vortex centre has been initiated.

Analysts of N.E.S.S. classified tropical vortices routinely on the bases outlined in fig. 10.2 from 1967–1972. Table 10.1 lists all those in categories A and C observed between January 1967 and July 1968. The geographical areas are listed in order of descending frequencies of vortices. The western north Pacific is clearly the area frequented most, claiming almost half the total storms recorded. The Atlantic and eastern north Pacific oceans rank as rather poor, but almost equal, seconds. Comparatively little activity is indicated across the north and south Indian oceans and the south Pacific. It should be noted, of course, that large-scale disturbances do occur additionally from time to time across tropical land masses also, most notably across central Africa. Johnson (1970) has described what is probably the largest and most significant of such phenomena, the so-called 'Congo disturbance'. This is observed to arise from time to time over the Congo Basin before propagating eastward across the south-west Indian Ocean. Such developments seem to recur in spells, often associated with frontal cloud band extensions across Africa from middle latitudes. Within each spell disturbances occur at intervals of 3 to 4 days, stimulating enhanced cloudiness and rain in many of the affected areas. Other large, but poorly organized, tropical cloud patches more conservative by nature than those of most tropical instability cloud masses are often noted over the large islands of the south-east Asian archipelago, especially Borneo and New Guinea.

Table 10.1

Geographical occurrences of tropical and subtropical disturbances in stages A, B and C (see fig. 10.2) from January 1967–July 1968, from satellite evidence (after Oliver 1969)

Region Stage:	*A*	*B*	*C–*	*C*	*C+*	*Total*
Western North Pacific	51	128	9	80	47	315
Atlantic	40	51	9	48	18	166
Eastern North Pacific	4	26	20	62	18	130
South Indian Ocean	6	11	—	29	25	71
South Pacific	4	5	—	12	12	33
North Indian Ocean	6	11	—	7	8	32

Mature tropical storms and hurricanes

Their distribution nature and development

As we turn to the global distribution and climatology of stage X tropical vortices we note the introductory picture of their global distribution afforded by fig. 10.3, compiled by Gray (1968) from various conventional sources. It is only unfortunate that fig. 10.3 applies to periods of different lengths in different regions. Gray suggested from satellite evidence that most of the initial disturbances from which that scatter grew might have appeared slightly east of the positions shown, except in the north Atlantic. Here the Cape Verde Island area or even west-central Africa much further east might have been the main spawning grounds, as indicated in chapter 9.

Following Riehl (1954), it is generally accepted that the chief physical factors permitting and/or encouraging growth and development from an initial disturbance towards a system of mature hurricane status include the following:

1 A sufficiently strong Coriolis force to encourage a spiral airflow around an initial low pressure centre. Within about 5° of the Equator the Coriolis force seems to be too weak to be effective in this way.

2 A sufficiently high sea surface temperature to maintain the necessarily steep vertical lapse rate that accompanies the vertical circulation of a hurricane. 27 °C seems to be the lower threshold.

3 A suitable starting mechanism in the low troposphere. This is often provided by a disturbance within the intertropical cloud band, an easterly wave or an old temperate low advected into tropical latitudes.

4 Upper tropospheric conditions conducive to the deepening of a surface depression through the encouragement of more rapid divergence of air aloft than convergence just above the surface of the earth.

It is certain that the supply of heat and moisture from below must be rapid and abundant if a large warm core system is to form. This fact, coupled with

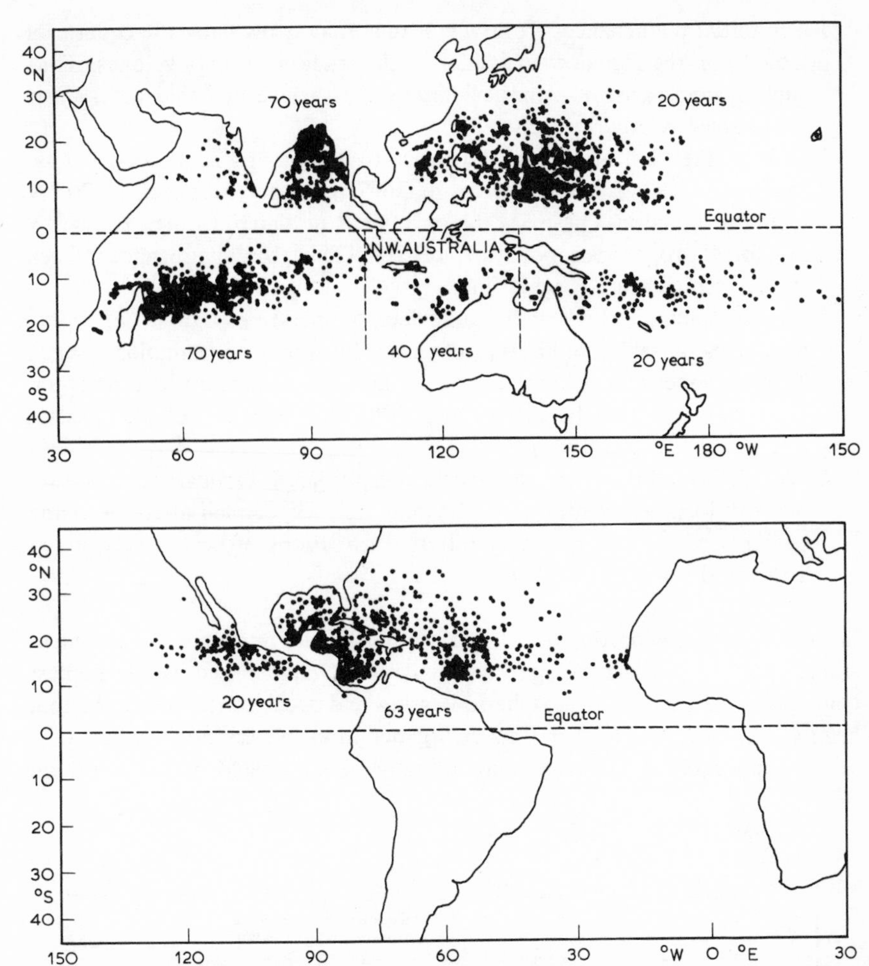

Fig. 10.3 Location points of first detection of disturbances which later became tropical storms. (From Gray 1968)

the requirement for low frictional drag at the base of an incipient system if it is not to fill too rapidly, dictates that tropical storms and hurricanes are characteristically maritime weather systems which fill and dissipate quite rapidly on moving over land. The populated regions that suffer most from their attentions are coastal regions in tropical low latitudes between about 10° and 25° N and S of the Equator, especially on the eastern seaboards of the continents. North-western Australia and western Central America are anomalous in this respect.

More recently, further light has been thrown by Gray (1968) upon the factors affecting the development and distribution of severe warm core circulations in the Tropics. He pointed out that:

1 The initial disturbances from which they may grow flank the equatorial trough, on the Equatorward sides of the trade wind anticyclones. Thus such systems arise in zones of large-scale surface cyclonic wind shear (large-scale relative vorticity).
2 Only in the Gulf of Mexico, north-west Atlantic and north-west Pacific regions do sea surface isotherms of 26·5 °C extend poleward of 20° N from the Equator. These are the only areas in which disturbances grow to storm proportions poleward of 20°. Thus the significance of sea surface temperature is underlined.
3 The key factor linking high sea surface temperatures and strong storm development seems to be the 'potential buoyancy' of cumulus clouds. This is conceived as the difference between the equivalent potential temperature (θ_e) at the surface and 500 mb (a measure of the possible rate of vertical cloud growth).
4 'Ventilation effects' are significant where large vertical wind shears prevent local concentrations of latent heat of condensation in rising currents of air by promoting different directions of heat advections at upper and lower tropospheric levels.

Consequently, according to Gray's version of the hurricane development story, cyclogenesis is most marked in regions characterized by the highest combined potential for tropospheric heating and areal concentration of that heating. He suggests further that divergence aloft, an essential feature of a system approaching maturity, may actually be stimulated by the optimum

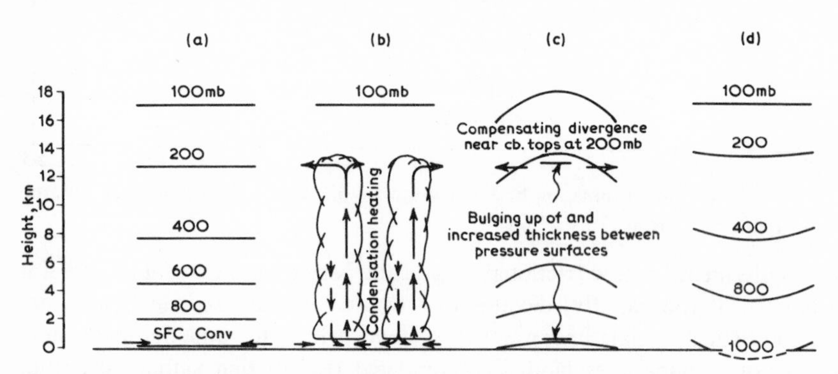

Fig. 10.4 Model growth cycle applicable to the genesis or intensification of tropical storms: (a) Initial pressure surface at beginning of convergence; (b) Generation of cumulonimbus by surface convergence, and release of condensation heating; (c) Initial effect of mass convergence and condensation heating: the bulging-up of pressure surfaces (greatly exaggerated); (d) the resulting slope of pressure surfaces after compensatory upper level accelerations and outflow have dissipated. Pressure surfaces are concave below due to increased pressure thickness from condensation heating. (From Gray 1968)

growth conditions of cumulonimbi themselves in the lower troposphere. This view lessens the requirement proposed by Riehl and others for the arrival of an upper tropospheric high before cyclonic maturity may be achieved; perhaps a developing storm may promote its own upper level centre of divergence (see fig. 10.4).

Satellite photographic data have contributed much information useful to our understanding of tropical storm initiation. For example, despite occasional protestations to the contrary (see Sadler 1964), some hurricanes have been seen to develop from wave patterns in lower tropospheric equatorial easterly flows (Fett 1966). Fig. 10.5 illustrates a suggested cycle of

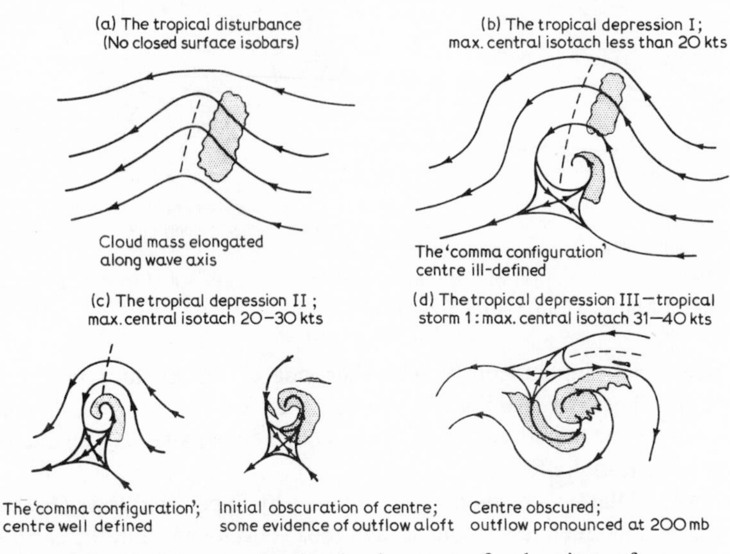

Fig. 10.5 Early stages in the development of a hurricane from an easterly wave streamline pattern. (After Fett 1964; from Barrett 1970)

development from such a beginning. In this fashion, from time to time, the earliest cloud masses from which Atlantic storms have been seen to form became visible first over the African continent (Carlson 1969). On other occasions, satellites have confirmed the potential of the intertropical cloud band as a zone of vortex generation and intensification. Fig. 10.6 portrays the model suggested by Fett (1968a) after a careful satellite study of 'Marie' in the western Pacific in October 1966. At that time, the intertropical cloud band was very active. Mass convergence fed the strong cumulonimbus cloudiness in which release of latent heat encouraged further buoyancy and a lowering of surface pressure through outflow aloft.

Leigh (1969) and Cox and Jager (1969) have documented and explained occasional fascinating occurences in the western Pacific in which two mature

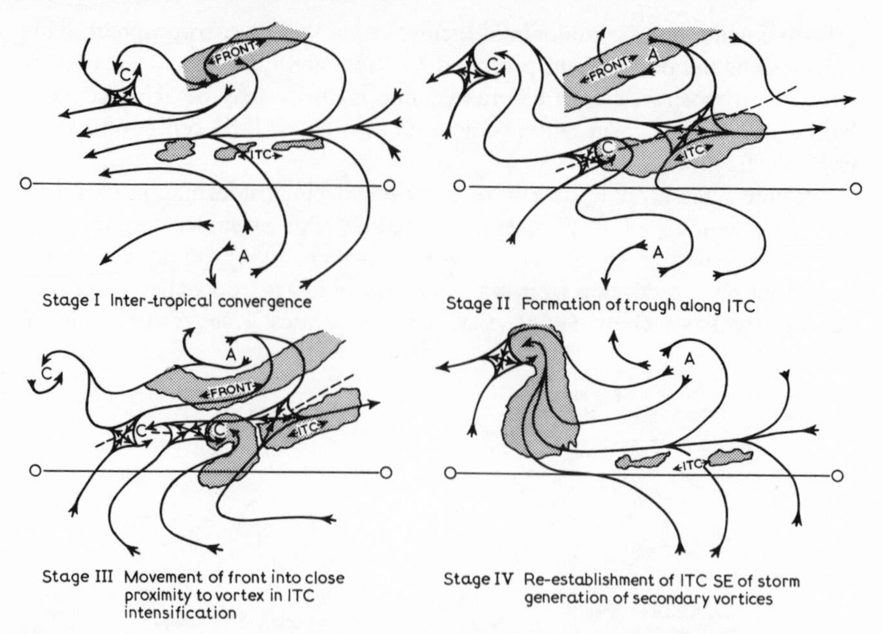

Fig. 10.6 Schematic depiction of the main sequence of events noted in the development of typhoon Marie, 1966. (After Fett 1968a)

cyclones developed simultaneously on either side of the Equator near the same meridian. These are similar to others noted earlier over the Indian Ocean ('Sumatran' or paired vortices). It seems that such pairs may be due in part to cross-equatorial effects.

Occasionally, Atlantic storms may be initiated by upper cold lows (Frank 1971), but these are much less frequent than those developed from the earth's surface upward.

Mature hurricane structure

Before we turn finally to more purely geographical aspects of tropical storm and hurricane development, let us review the structures characteristic of mature hurricanes. Apart from the intrinsic interest and importance of such a study it will help to explain ways in which the satellite classification outlined earlier becomes useful operationally in storm forecast preparation. Time and space do not permit detailed comparisons between the model outlined below and those proposed in pre-satellite days. The interested reader could usefully have a pre-satellite meteorology text open by his side for private purposes of comparison.

Fig. 10.7 incorporates many features seen by satellites to be characteristic of mature hurricanes. These include the dense overcast clouds of the central

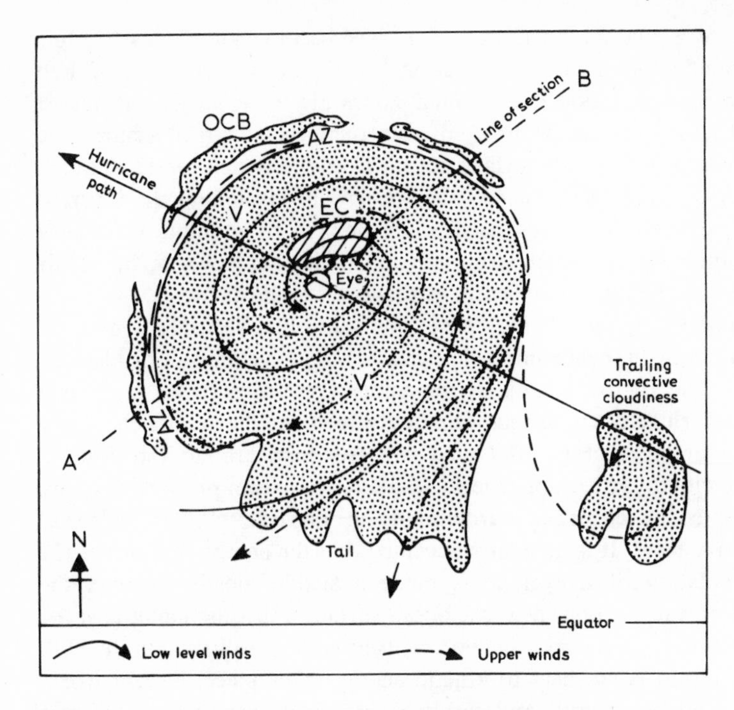

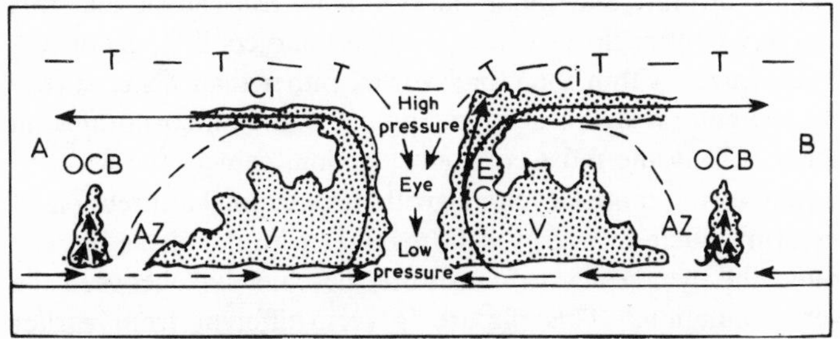

Fig. 10.7 A model of a mature northern hemisphere hurricane incorporating features discovered and/or confirmed by satellite studies. T: tropopause; Ci: cirrus; V: hurricane vortex; OCB: outer convective band; AZ: annular zone; EC: primary energy cell. (From Barrett 1970)

('vortex') area, a central clearance (the 'eye') appearing as a small, dark, circular area surrounded by a bright ring of especially deep tower clouds (the 'eye wall' clouds), a wedge of broken convective cloudiness (the 'tail') on the Equatorward side of the vortex, and an advance band of bright clouds (the 'outer convective band') separated from the main cloud disc by a belt of clearer skies (the 'annular zone'). Sometimes an apparently less well-organized patch of rather amorphous convective cloud is observed travelling in the wake of a mature hurricane. This has become known as 'trailing convective cloudiness'. Time cross-sectional analyses of hurricanes by Fett (1964) led to his suggestion that annular zones are zones of subsidence; it seems that at least some of the air uplifted through the vortex of a hurricane sinks back towards the earth's surface not far from the vortex centre as much warmer, drier air. This air possibly plays a part in fashioning the inverted bowl of the vortical cloudiness itself while helping to promote instability along the outer convective band through its mixing with warm, moist air drawn in towards the hurricane centre in the lower troposphere.

Aloft, of course, as fig. 10.7 suggests, the sense of air circulation is the reverse of that beneath, being anticyclonic rather than cyclonic aloft. Plate 23 indicates that one of the results of the upper level divergence is often a circumferential fringe of cirrus feathering outward from the area of the lower level vortex cloudiness. Fig. 10.7 also suggests that, around the circumference of the cloud vortex, upper level winds of jet stream proportions may occur, stimulated by enhanced baroclinicity around the annular zone and outer convective band. It is generally accepted that the energy that drives the powerful circulation of a cyclone is made available mostly through the condensation of water vapour near the ocean surface, releasing latent heat for consumption as kinetic energy. The vortex itself is the area in which the bulk of the change from latent heat to kinetic energy takes place. Recent work, however, assisted by aircraft and radar observations, has shown that this energy release, largely contingent as it is upon uplift and cooling, is concentrated in localized centres within the cloud vortex rather than being spread uniformly across the central overcast area. Fig. 10.7 thus incorporates an appropriate 'energy cell' in the most common location, namely the forward right hand quadrant of the model cyclone (with respect to the direction of the path of the storm). Such an energy cell, together with the circle of deep wall clouds around the eye, seems to constitute the chief link between the upper and lower circulations. This picture is very different from earlier ones that portrayed hurricanes as areas of almost uniform convectional instability.

Hurricanes are, therefore, neither simple systems, not yet as simple internally as once was thought. Last of all, it must be remarked that no two individual systems are ever quite the same, either in structure or behaviour. The detailed differences that may occur, however, are matters chiefly of meteorological, rather than climatological, significance, and need not be reviewed here. An excellent modern meteorological review of hurricane characteristics has been prepared by Miller (1967). Numerous articles and papers since 1960 have described the individual characteristics of particular storms through analyses of satellite and conventional observations (see, for example, Warnecke *et al.* 1968, and Erickson 1967).

Climatologically, hurricanes have two primary significances: (1) as intense, highly powerful components of tropical weather, and (2) as important heat transport mechanisms within the general circulation of the atmosphere.

Some general remarks are in order under these two headings before we turn to examine in more detail regional patterns of hurricane behaviour, and the scales of their local economic and climatological impacts.

Hurricanes as features of tropical climatology

Hurricanes are not very frequent phenomena, even in the areas where they occur most often. Hence their contributions to climatological statistics other than rainfall are generally slight. However, the rain-bearing capacities of mature storms can be quite enormous. In extreme cases as much as 2000 mm may be yielded in a single locality over a few days. Not infrequently rainfalls of 200 mm or more are recorded from hurricanes in the course of a single day. In some regions, especially coastal strips most exposed to hurricanes, rainfalls from such systems may comprise more than half the annual rainfall totals.

Hurricanes are of principal significance, however, in that they constitute the most destructive group of weather systems in the world. Hurricane circulations are less vigorous than those of mid-latitude tornadoes, but tornadoes are comparatively tiny, and blaze trails of destruction on average only a few hundred metres broad; at the other end of the range of scales hurricanes (on average about 400 km across) are smaller than extratropical depressions, but are much more intense. Much of their impact results from associated phenomena. Great losses of life may be caused by storm surges inundating low lying coastal areas with wind-driven sea water in which all floating objects act as gigantic battering rams. Additional flooding is often caused by heavy rains. Substantial damage may be caused by hurricane winds which frequently exceed 70 m/sec. The hurricane that struck the head of the Bay of Bengal in November 1970 was directly responsible, mostly through flooding, for over 300,000 deaths; indirectly, its political repercussions through the Bangla Desh affair were long and far-reaching (Frank and Husain 1971). Hurricane Betsy caused damage of about $1·5 billion in Florida and the Gulf States of the U.S.A. in 1965 (Goudeau and Connor 1968). Small wonder that a constant watch on hurricanes is maintained by satellites and conventional means; or that continuing efforts are being made to develop feasible methods whereby the energy of hurricanes might be controlled (see, for example, Malkus and Simpson 1964).

Hurricanes as features within the general circulation

The regional studies at the end of this chapter indicate that most hurricanes move poleward as they mature and dissipate. Hurricanes have been described

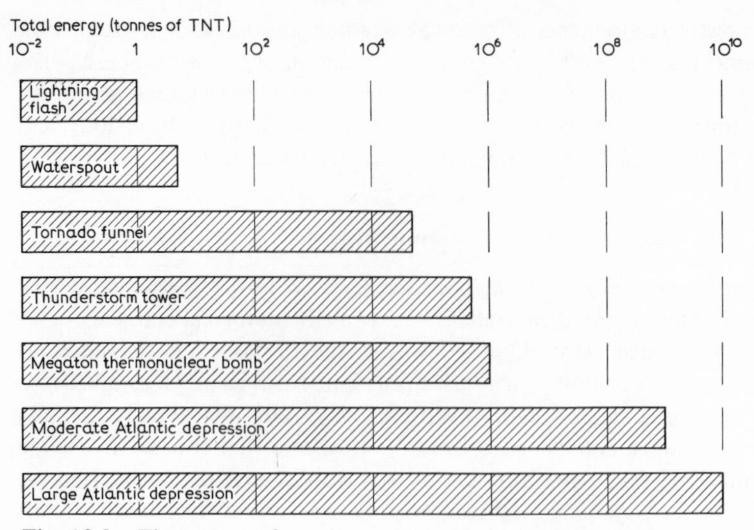

Fig. 10.8 The energy of weather systems. This is so large that all but the smallest systems are much more energetic than man's thermonuclear weapons. (After Macdonald 1968)

as simple heat engines (Riehl 1954), but, as such, they are most inefficient. Only about 3% or less of the heat released by condensation within a hurricane is converted into kinetic energy. Much of the remainder is converted to potential energy and exported through the outflow layer aloft. (Riehl and Malkus 1961). In the lower troposphere, some energy is carried into higher latitudes by hurricanes as they travel poleward, gradually losing their unique characteristics as they go. A measure of the energy involved was estimated by Dunn and Miller (1964) (see also fig. 10.8). Their calculations indicated that from $2{\cdot}0$ to $6{\cdot}0 \times 10^{26}$ ergs are liberated daily in the average hurricane in the form of latent heat. Clearly, hurricanes must transport enormous quantities of heat toward the poles, thereby contributing not a little to the heat balance of the earth/atmosphere system. There are good reasons for arguing, therefore, that it would be premature at this moment to contemplate frequent, large-scale hurricane control; some benefits might be accrued through reductions in loss of life and damage to property within the Tropics, but decreased rainfall totals in immediately affected coastal areas, and potentially enormous and complex climatic consequences outside the Tropics might be expected to weigh heavily on the debit side.

Hurricanes in regional climatologies

It has been noted that every tropical ocean except the south Atlantic has its share of the annual crop of hurricanes. In the remainder of this chapter we shall review the distribution of hurricanes region by region in circumglobal fashion. We shall begin with the better documented regions (the north Atlantic and eastern north Pacific), and conclude with those that are less well documented (the western Pacific and the north and south Indian oceans).

The tropical north Atlantic

It has been recognized for a long time that the usual trajectories of hurricanes lead from the eastern north Atlantic toward the Caribbean, whence they curve parabolically north and north-east along the eastern seaboard of the U.S.A. The detailed paths followed by individual storms can be established with much greater degrees of accuracy from satellite photographs than from conventional data. Fig. 10.9 portrays the storms tracked by the U.S.W.B. in the 1966 hurricane season. It is clear that, although the most common track followed in that season of unusually frequent hurricane activity was indeed parabolic, two systems in particular behaved less normally; one arose in the area of Nicaragua and Honduras before moving east to cross Cuba and the Gulf coast of Florida; the other developed south-east of Florida before tracking west to Mexico. Thus, although a standard track may be recognized, wide departures from such a track occur from time to time. In 1966, hurricane Faith was notable for the depth of the penetration of its degraded form towards the Arctic Ocean, following a not uncommon path down which poleward transports of energy are effected by hurricanes and hurricane-derived mid-latitude depressions. The role of Africa as an initial breeding ground for Atlantic hurricanes seems to be substantiated at least in part by fig. 9.18, since several storms originated in the African region.

Table 10.2 lists the number of hurricane days in the north Atlantic from 1954 to 1968. Considerable variation from year to year is indicated by the large range, from eleven to fifty-six days. August, September and October are the months affected most, while the season from December to May is practically hurricane-free. Carlson (1971) analysed sea surface temperature over the tropical Atlantic for a five-year period, and demonstrated that a correlation exists between the number of tropical storms formed within the July–September season and ocean temperatures over a wide area centred near 10° N and 35° W. Namias (1969) suggested that such seasonal variations of sea surface temperatures in the Tropics may be influenced in turn by anomalies in the circulation at higher latitudes and their effects upon the amounts of cloudiness and insolation receipts over wide areas. It has been noted from satellite evidence that more cloud was spread over the eastern tropical Atlantic during August in the relatively inactive year of 1968 compared with the same month in 1967. Oddly enough, the strength of the trades did not seem to be affected, although the strength of the trade wind inversion was. This spread south and restricted the area of warm surface water still more when southward intrusions of cold water occurred along the coast of

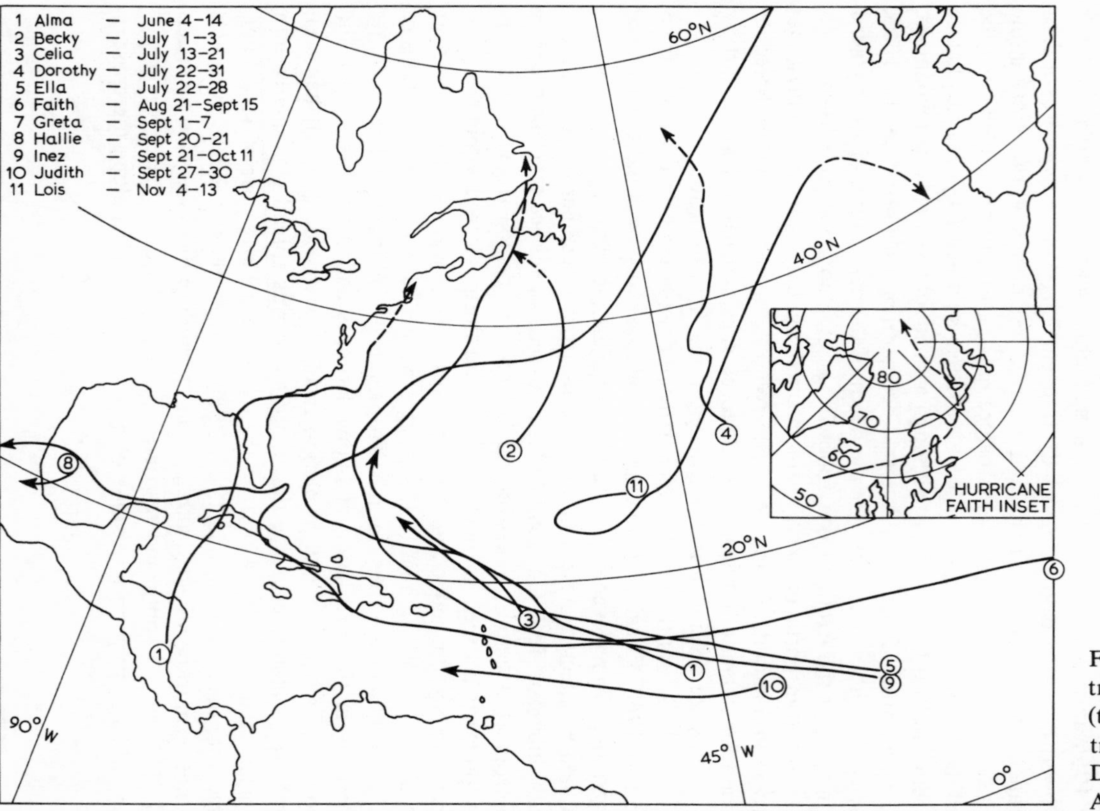

Fig. 10.9 Tracks of hurricanes and tropical storms, north Atlantic, 1966 season (tropical stages, continuous lines; extra-tropical stages, broken lines). (After U.S. Dept. of Commerce, Weather Bureau North Atlantic Hurricane Tracking Chart)

Africa. So the observed year-to-year variations in hurricane tracks and frequencies seem to be involved in widespread and very complex cause-and-effect relationships. The contributions of hurricanes to local climatic statistics are also very varied.

Table 10.2
North Atlantic hurricane days, 1954–68 (from Sugg and Hebert 1969)

Year	*J*	*F*	*M*	*A*	*M*	*J*	*J*	*A*	*S*	*O*	*N*	*D*	*Year*
1954	—	—	—	—	—	1	—	5	8	16	—	1	31
55	4	—	—	—	—	—	—	22	28	2	—	—	56
56	—	—	—	—	—	—	1	9	2	—	3	—	15
57	—	—	—	—	—	3	—	—	19	—	—	—	22
58	—	—	—	—	—	—	—	14	16	5	—	—	35
59	—	—	—	—	—	1	2	—	10	11	—	—	24
60	—	—	—	—	—	—	4	2	13	—	—	—	19
61	—	—	—	—	—	—	4	—	35	9	1	—	49
62	—	—	—	—	—	—	—	1	—	10	—	—	11
63	—	—	—	—	—	—	—	11	7	23	—	—	41
64	—	—	—	—	—	—	—	7	33	6	—	—	46
65	—	—	—	—	—	—	—	6	21	3	—	—	30
66	—	—	—	—	—	7	8	9	11	10	5	—	50
67	—	—	—	—	—	—	—	—	33	11	—	—	44
68	—	—	—	—	—	3	—	5	—	5	—	—	13
Total	4	0	0	0	0	15	19	91	236	111	9	1	486

Note: If two hurricanes are in existence on one day, this is counted as two hurricane days.

The eastern tropical north Pacific

Here, severe tropical vortices occur frequently especially along the west coast of Mexico. Their season of maximum activity is much the same as that in the Caribbean and Gulf of Mexico, but their frequency is greater. Fig. 10.10 compares the occurrence of hurricanes and tropical storms along both coasts from 1921 to 1969, showing the monthly frequencies of storms whose geometric centres passed within a distance of 200 nautical miles (370 km) from the shore. Many such storms move inland. The amounts of the associated rainfalls are so important for irrigation purposes that the success or failure of crops in a given area may well depend upon such rainfall. Serra (1971) observed that, on the basis of observed hurricane frequencies, the probability that at least one storm would affect or move inland each year attains 0·97 over the Baja California peninsula, exceeds 0·90 in five of the ten west Mexican coastal states, and is greater than 0·74 in nine of these ten. The gross climatological trajectories of the west Mexican storms are suggested by fig. 10.11. Although contributions from satellite evidence to the knowledge of storm tracks and intensities along the west Mexican coast have been consider-

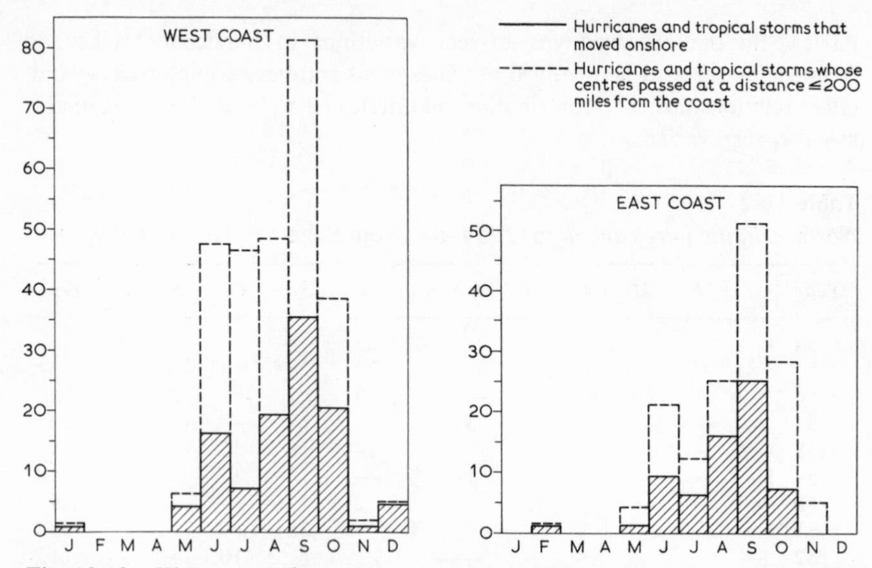

Fig. 10.10 Histograms of monthly occurrences of tropical storms and hurricanes, east and west coasts of Mexico, 1921–69. Storms whose centres moved inland are shown by shaded columns, those that passed at a distance of 200 n.mi. are shown by unshaded columns. (After Serra 1971)

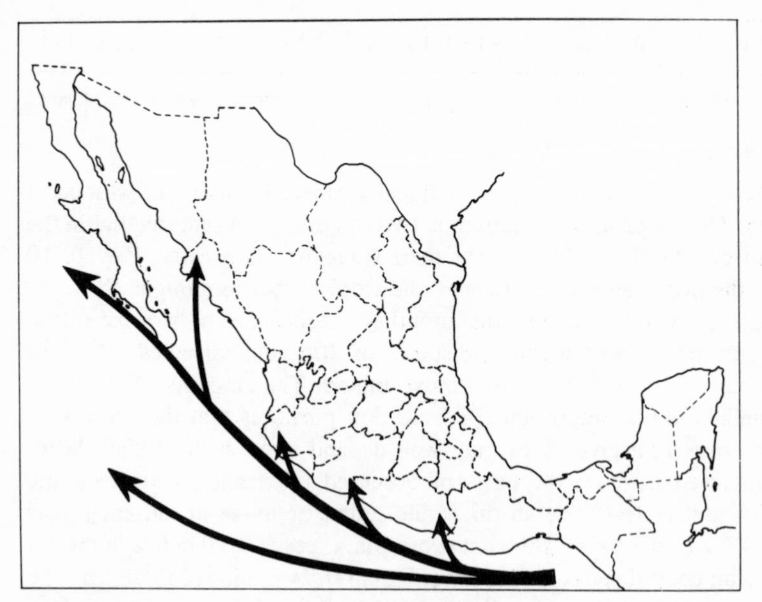

Fig. 10.11 Gross climatological trajectories of west Mexican coastal hurricanes. (From Serra 1971)

able, such contributions have been even greater across the more remote stretches of the central Pacific whence very few ship reports are usually forthcoming. Those tropical storms and hurricanes that were observed from May to November 1970 are plotted in fig. 10.12. This portrays a characteristic picture for one season in this extensive area. Table 10.3, summarizing a tentative climatology based on five years of operational satellite data coverage, suggests that the annual variability is considerably less than in the north Atlantic region, despite the debt that this area owes to ex-Atlantic and even ex-African disturbances. Frank (1971), after tracking all the synoptic scale disturbances in the wind and pressure fields across the tropical north Atlantic, concluded that systems originating on the Atlantic side of Central America play a very important part in producing east Pacific storms and hurricanes. In 1970 eleven of these were initiated by systems of African origin, and two by waves that formed over the Caribbean. Only five actually arose within the eastern Pacific I.T.C.Z. itself.

Table 10.3

Tabulation of hurricanes (Hu) *and tropical storms* (TS) *by month and year in which they began in the Eastern Pacific*

	May		*June*		*July*		*Aug.*		*Sept.*		*Oct.*		*Nov.*		*Totals*		
Year	*Hu*	*TS*	*Hu*	*TS*	*Hu*	*TS*	*Hu*	*TS*	*Hu*	*TS*	*Hu*	*TS*	*Hu*	*TS*	*Hu*	*TS*	*Combined*
1966			1				4		2	4		2			7	6	13
1967			1	2		4	2	2	1	2	2	1			6	11	17
1968				1		4	3	5	2	1	1	2			6	13	19
1969					1	2	1	1	1	3	1				4	6	10
1970	1			3	1	5	1	3		1	1	1		1	4	14	18
5-year total	1		2	6	2	15	11	11	6	11	5	6		1	27	50	77
Annual averages															5·4	10·0	15·4

A rather curious fact of the western Pacific hurricane area is its almost total independence of its counterpart over the eastern north Pacific. Denney (1969) remarked that the latter is located uniquely near dissipating influences; it is thought that the dissipation of many an east Pacific hurricane in mid ocean is caused by upper level shearing associated with a subtropical jet stream. Such a jet usually lies to the north of a mean ridge line aloft stretching from about 10° N, 160° W to the American coast around 25° N, 100° W. Earlier, Sadler (1963) pointed to the existence of a mid north Pacific trough line lying across the westward path of the storm systems nearly 180° W. This, too, must constitute a significant obstacle not only to the movement of intense systems across the mid Pacific, but also the development of similar systems in that area. Satellite evidence suggests that, just occasionally, systems may traverse the entire north Pacific Ocean, but hurricane organization is very rarely retained throughout. The pattern of cold water upwellings in the

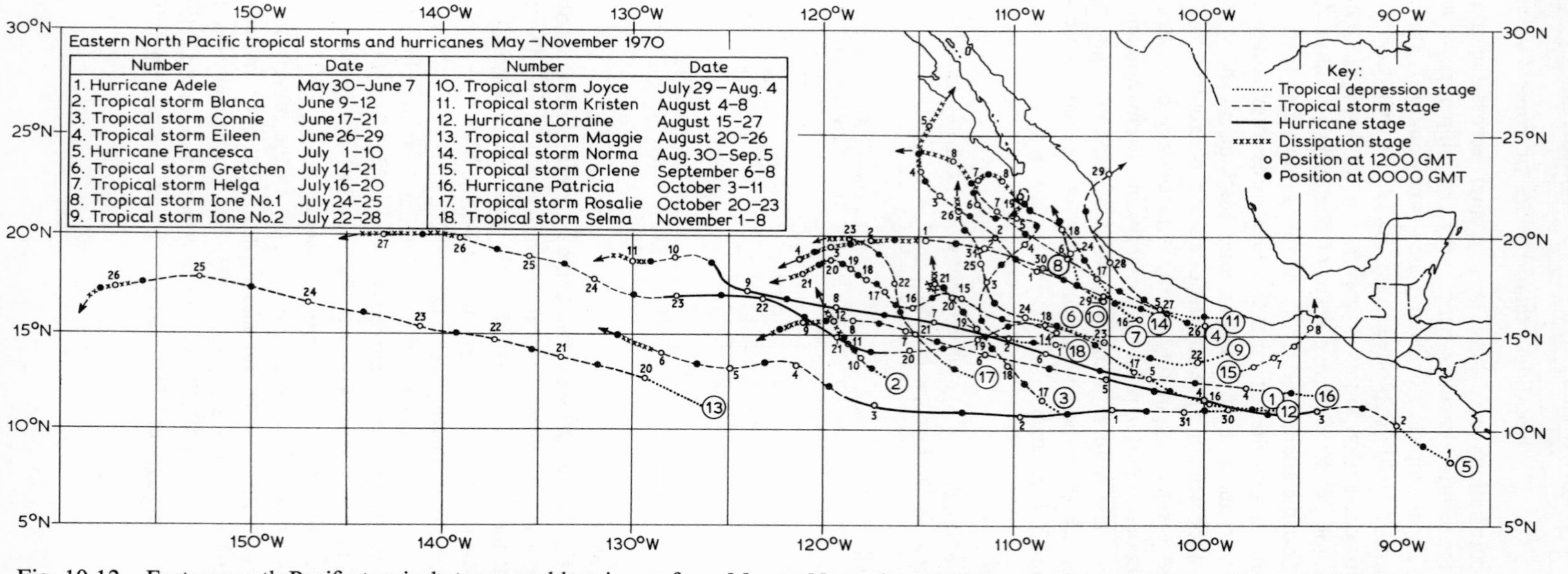

Fig. 10.12 Eastern north Pacific tropical storms and hurricanes from May to November 1970, largely from satellite evidence. (After Denney 1971)

central north Pacific is thought to play a part also in restricting the extent of the area frequented by tropical storms and hurricanes in the eastern north Pacific (Denney 1969). Across the centre of the ocean, only tiny island populations are affected by the few trans-Pacific systems.

The western tropical north Pacific

Records indicate that the largest and most powerful tropical cyclones are the typhoons of the western Pacific. These develop most frequently during the summer season when the north-east trades weaken and withdraw from a broad area stretching from New Guinea to Japan. The chief spawning grounds seem, from satellite evidence, to cover the region between 130° and 160° E, and from 5° to 25° N. In this oceanic region, many vigorous storms and hurricanes arise, especially in late summer as the intertropical cloud band begins to fall back southward after the height of the south Asian monsoon. Usually the storms set off on quite classical paths, first travelling westward, then curving north and north-eastward around the north Pacific tropical anticyclone. Heavy rains and high winds often inflict much damage on the Philippines and nearby island chains, as well as along the mainland coasts of the China Sea, across South Korea, and the more southerly islands of Japan. Along the southern and south-eastern coasts of China typhoons account for as much as 20% of the mean annual rainfall, playing no small part in determining that the local summer is the wetter half of the year (see fig. 10.13). Such climatic effects, however, are quite strongly concentrated along the coasts; hurricanes soon dissipate on passage overland, although regeneration may occur if the parabolic trajectories turn storms back over the sea before the active links between the upper and lower circulations have been irrevocably broken. Monsoon depressions play a larger part in the precipitation climatology of peninsular south-east Asia whereas frontal depressions do the same at least in winter across Japan and Korea. However, especially in these more northerly regions, the occasional late summer and early autumn typhoons may have disastrous side effects. These contribute significantly towards the common September maxima of precipitation (see fig. 10.13), occasionally yielding quite exceptional falls of rain. Typhoon rains at Tanabe in southern Honshu on 29 August 1889 totalled nearly 900 mm; at Nagasaki in Kyushu on 5 September 1923 nearly 90 mm fell in a single hour!

The Australian region

Two areas off the coasts of Australia are particularly prone to hurricane activity: (1) the Timor Sea, along the north-western coasts of Australia, and (2) from about 170° W across the south Pacific and the Coral Sea towards the eastern coasts of Queensland and northern New South Wales.

As in other southern hemispheric regions, however, the frequencies of hurricanes are low compared with those in the northern hemisphere. More-

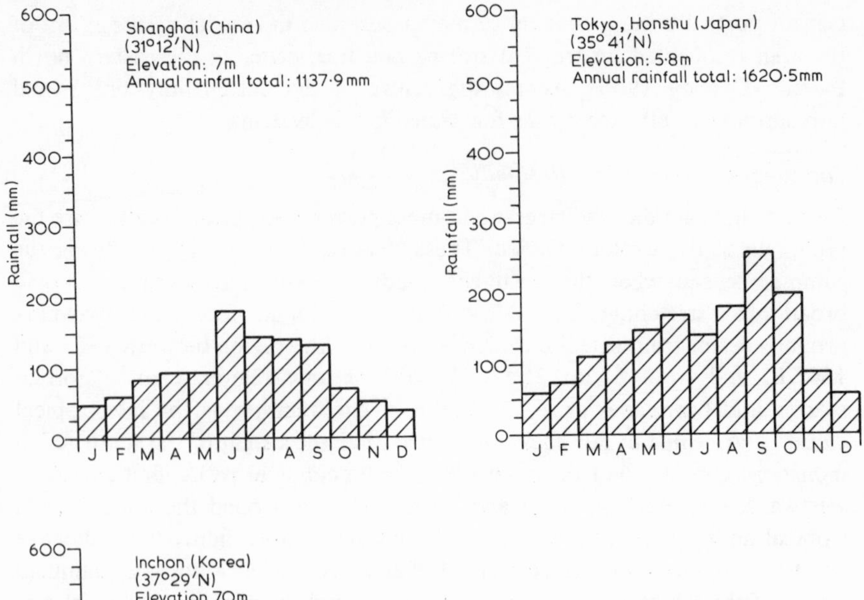

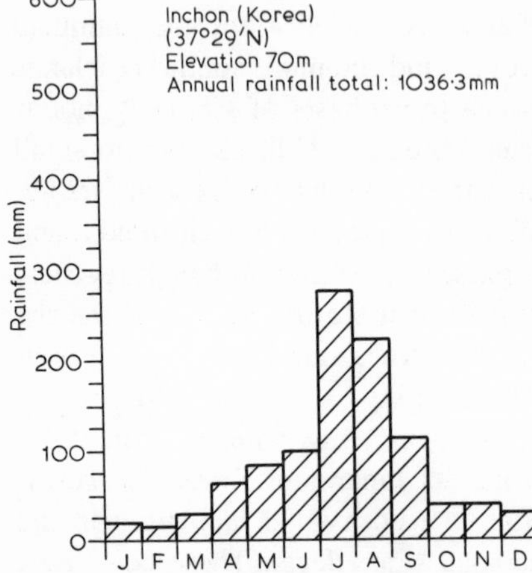

Fig. 10.13 Mean monthly rainfall graphs for selected stations in eastern Asia. (Statistics from Rumney 1968)

over, the systems that occur are, on average, less intense than those north of the Equator. Yet the Australian storms still contribute significantly to local climates. December to April is the season of greatest activity, but most storms appear in the months of January, February and March. Representative tracks are illustrated by fig. 10.14. Some are characteristically parabolic, but many are highly irregular. Hence the movements of storms in these regions are especially difficult to forecast from day to day.

Although the frequencies in Australian waters are low compared with those in most other hurricane regions (see table 10.4), their economic consequences may be quite considerable. Their seasonal distribution in the better known and more densely populated east Australian region attains a high plateau

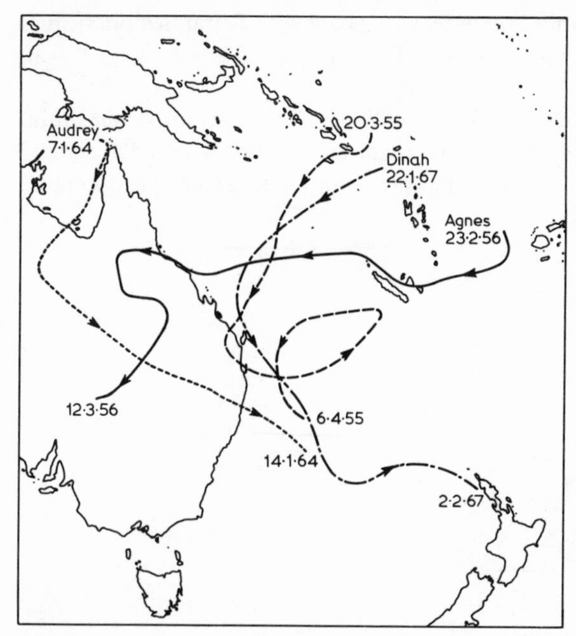

Fig. 10.14 Representative tracks for hurricanes in the Eastern Australian region.

rather than a peak from January to March. Then the chances of hurricanes arising are between two and three times the chances in December and April.

Table 10.4
Areas where tropical storms develop. (*Following the W.M.O. definition*) (after Gray 1968)

Area	*Average % of global total*	*Average number of storms per annum*
Northern hemisphere	76	47
N.W. Pacific	36	22
N.E. Pacific	16	10
N.W. Atlantic	11	7
Bay of Bengal	10	6
Arabian Sea	3	2
Southern hemisphere	24	15
S. Pacific	11	7
S. Indian Ocean	10	6
off N.W. Australia	3	2
Totals	100	62

Detailed analyses of rainfall patterns associated with Coral Sea hurricanes have been carried out by Brunt (1966). Two interesting climatological conclusions emerged:

(1) Coral sea hurricanes do not constitute the most important rain-producing mechanism over the open ocean, as stratiform clouds seem to dominate their structure there, and high precipitation totals are the exception

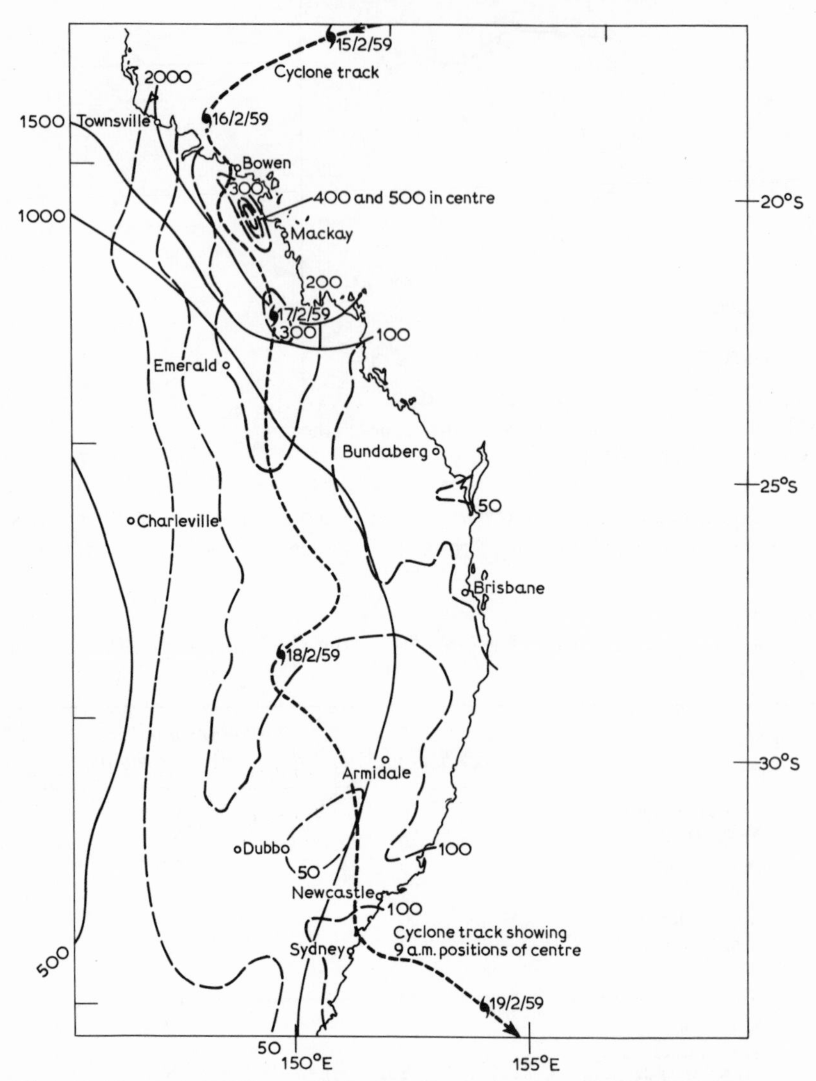

Fig. 10.15 Isohyets for a single hurricane over eastern Australia (the 'Bowen cyclone', February 1959) (broken lines) and mean annual isohyets (solid lines). The cyclone track is also shown. (After Brunt 1968)

rather than the rule. The density of convective cloud in these mature tropical cyclones seems relatively low, even within the overcast of the vortex. At Willis Island (a small, low coral cay 16° S, 150° E from 1926 to 1965 hurricane days only accounted for 49 out of 187 days of heavy rainfall (50 mm). Three non-hurricane days yielded more rain than the wettest with a hurricane within 300 miles (483 km) of Willis Island. Thus, statistically, the total hurricane contribution to rainfall at sea is probably rather small.

(2) The rainfall picture changes dramatically as a storm moves across a coast. Increased low-level convergence results from a frictional drag on the wind circulation and a related weakening of the Coriolis deflection. The most immediate effects are along the forward right-hand quadrant of such a storm, causing it to elongate with its axis along the direction of the coast. Lapse rates are usually steepened by passage overland, and, coupled with orographic effects, help stimulate outstanding falls of rain in regions of abrupt relief, for example along the eastern flanks of the Great Dividing Range. At Springbrook, Queensland, in 1954 one hurricane yielded over 1000 mm of rain in thirty-six hours; areal depths of precipitated water have exceeded 170 mm in twenty-four hours and 280 mm in forty-eight hours over about 130,000 km^2. Clearly, the frequency of such heavy rain-bearing systems must play a key part in shortening or prolonging those droughts that can be so significant only just over the north–south mountain crests beyond the normal reach of coastal rainfall from the north-east trades. Fig. 10.15 shows the pattern of the percentage contribution made by a single storm, the 'Bowen cyclone' of February 1959, to that year's rainfall in north-east Australia.

In north-west Australia, the annual frequency of somewhat less than one storm each year along the coast implies comparatively little impact on the climate, or on man's economy, in that very thinly populated region.

The Indian Ocean

The distribution of hurricanes in the Indian Ocean is, in many ways, the most complex and variable of all. Apart from north-west Australia, other areas affected (mainly in local late summer seasons) include the Bay of Bengal, the Arabian Sea and the vicinity of Malagasy south of the Equator. Especially in the north Indian Ocean, the local period of storm activity is at least potentially longer than in most other hurricane regions, despite the intervention of the southern monsoon, splitting the hurricane season into two. From April to June, well-developed storms often accompany the onset of the monsoon, while the monsoon rains from July to August are associated more with amorphous, short-lived patches of bright cloudiness within the southerly monsoon flow (see chapter 11). As the southern monsoon falls back in autumn, leaving a broad area of warm sea surfaces, shallow pressure gradients and light winds across the Bay of Bengal and the Arabian Sea, quite severe hurricanes may develop in September, October and November (see plate 24). Weather satellites are now the kingpins of early warning systems for the densely populated lands that may be affected by these large destructive vortices, especially in the Ganges delta area where the land shelves gently and the dangers of flooding are intense. Unfortunately, the problem of devising an adequate defence or evacuation policy in the face of happenings like those in late 1970 is much more difficult.

Long-term averages of cyclones indicate an annual frequency of about six in the Bay of Bengal, but only one in the Arabian Sea. Pedgley (1969) employed weather satellite data in support of coastal and shipping reports, and earlier cyclone observations, in the preparation of a climatology of

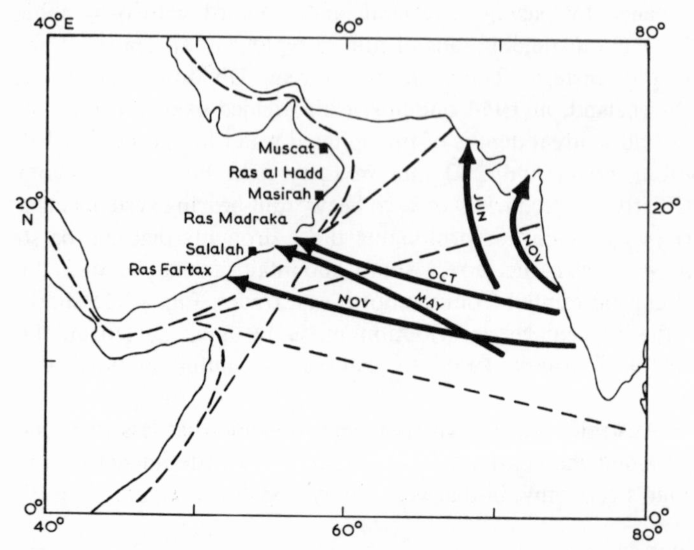

Fig. 10.16 The usual tracks of Arabian Sea hurricanes (solid lines) and main shipping routes (broken lines). (From Pedgley 1969)

storms in the Arabian Sea region. The usual storm tracks are depicted by fig. 10.16, which indicates additionally the months in which they usually occur. It seems that hurricanes develop preferentially over the south-eastern quadrant of the Arabian Sea and then move west-north-west towards Arabia, occasionally recurving north or even north-east towards the border region of India and East Pakistan. Seasonally, they are about equally distributed between May–June and October–November, the monsoon transition months.

Although they affect the coasts of Arabia very infrequently, about once every three years on average, their local contributions to rainfall totals and long-term rainfall means are quite significant. In such regions, where rainfall is generally very slight, occasional intense weather systems can easily play disproportionately large parts in local climatologies. Thus particular care is required in the interpretation of climatic mean statistics. The unshaded portions of the mean monthly rainfall columns for Salahah and Masirah (fig. 10.17) indicate the size of the hurricane contribution. Pedgley (1969)

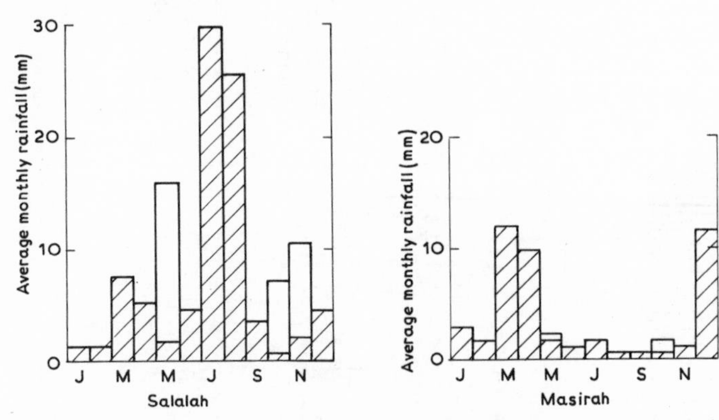

Fig. 10.17 Seasonal distribution of rainfall at Salalah and Masirah (see fig. 10.16). Contributions by cyclones are unshaded. (From Pedgley 1969)

observed that, at Salahah, the cyclonic rains are usually confined to a relatively narrow corridor along the coast. Even in that corridor, however, they might be expected to exceed 10 mm in one day only one year in ten. Between 1943 and 1967 the great bulk of rain observed at Salahah in May, October and November (three of the four most frequented months) fell from a mere four storms, two in May (24 May 1959, 117·2 mm, and 26 May 1963, 236·1 mm) and one each in October (25 October 1948, 156·8 mm) and November (13 November 1966, 202·7).

By comparison little can be said of hurricanes over the south Indian Ocean, since ship reports are few and few climatological studies have been based on evidence from satellites. Such storms are known to occur most frequently in January and February, but have been noted on Essa mosaics in every month except August and September (fig. 10.18). Apparently they often develop from wave-like perturbations in the easterly flow across the south Indian Ocean, especially when the equatorial trough is furthest south. Perhaps two storms of hurricane intensity may be expected in the course of an average year in the vicinity of Malagasy, where the abrupt topography has helped to prompt some very intense and heavy falls of rain. These have included some of the most dramatic in the whole history of meteorological observation.

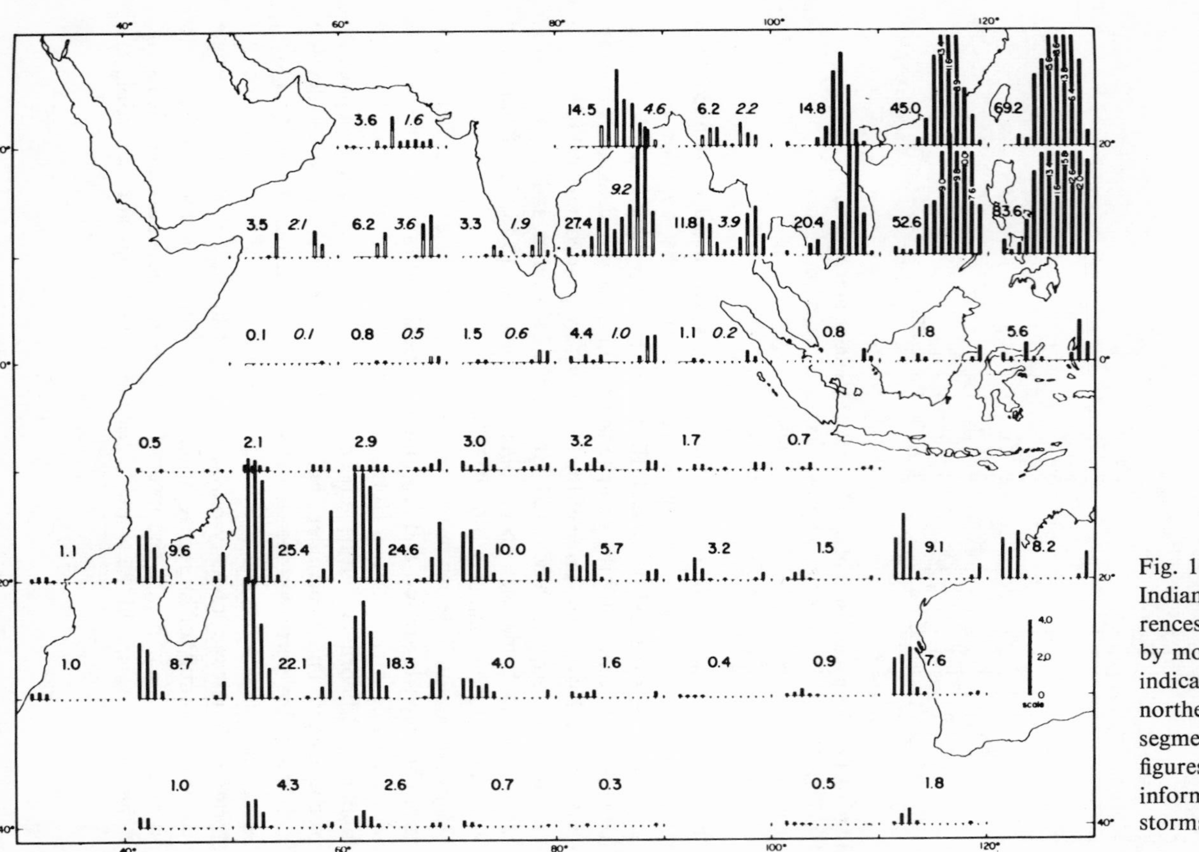

Fig. 10.18 Tropical cyclones in the Indian Ocean region. Average occurrences per 10 yr for each 10° square by months. The upright figures indicate average 10 yr totals. Over the northern Indian Ocean the unblacked segments of the histograms and the figures in italics give corresponding information for severe tropical storms. (From Ramage 1971)

We may conclude by noting the cogent observation made by Alaka (1964) to the effect that 'among the unsolved problems of tropical meteorology, none, perhaps, are more urgent than those related to hurricanes'. It is fortunate that these systems, clearly defined as they are, rank as those that can be studied best from satellite photographs and infra-red displays.

11 The south Asian monsoon

Its nature and importance

Meteorologically, the concept of a monsoon contains two key ideas – regularity and reversal. As Sutton (1962) has remarked, a monsoon is 'a system of winds blowing regularly in different seasons of the year, but with an alternation in direction from one season to another'. Implicit in this definition is the fact that strong seasonal contrasts of weather associated with the mean seasonal circulations are important features of monsoon climates. While, strictly speaking, any climate reasonably summarized in these terms may be referred to justifiably as a monsoon climate (fig. 11.1) the south Asian monsoon is the only one that warrants special study in this volume on the grounds of scale and human influence. Suffice it to say that various other monsoons have been described, especially elsewhere in the Tropics (though Miller (1966) even claims that the North American continent experiences a monsoon circulation, albeit somewhat obscured by migratory cyclones and fronts). The monsoonal character of central Africa has yet to be explored in detail from the satellite viewpoint.

The yearly cycle of weather events across southern Asia, the Indian Ocean and other coastal zones bordering upon its waters is significant within the dynamic climatology of the general circulation. It is also highly significant by reason of its impact upon the lives of hundreds of millions of people especially in the Indian subcontinent. Before discussing the more purely atmospheric aspects, some comments are in order to underline the significance of the south Asia monsoon to some of the most thickly populated countries in the world.

The greatest effect on human life and economy is exercised through the annual cycle of precipitation, which can be conveniently expressed through

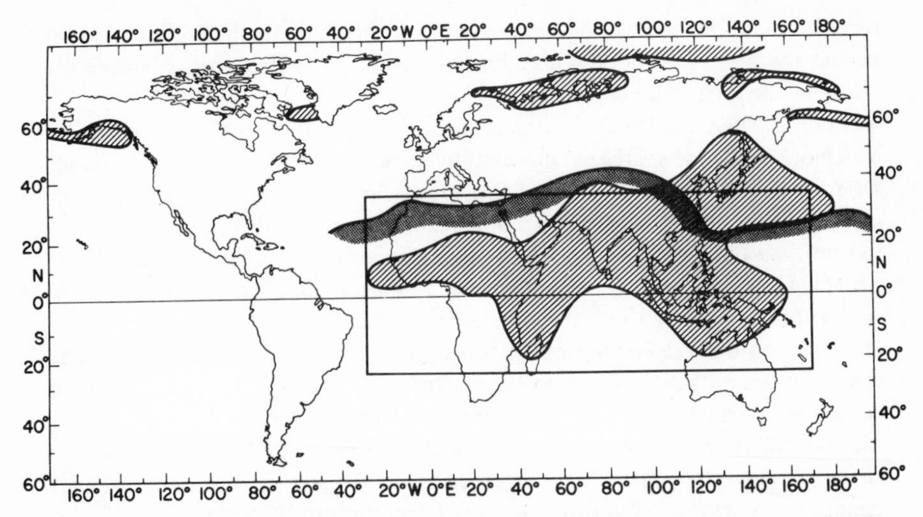

Fig. 11.1 Geographical extent of the monsoons according to Khromov (1957). Average frequency of predominant wind directions shown in three categories. Unshaded areas are non-monsoonal. (From Ramage 1971)

rainfall means and rainfall variability. Fig. 11.2 illustrates the average advance of the summer monsoonal rains in the Indian subcontinental region. The earliest onset is in southern Burma, so that the rainy season shortens thence towards West Pakistan. In this last region the monsoon rains are delayed by some seven weeks compared with the eastern coastlands of the Bay of Bengal. The variability of the rains is, in part, a function of the length of the monsoon rainy season. Along the Burmese coast the variability of annual rainfall is as low as 15% with respect to the long-term mean, but the figure rises to 40% and above in western India and West Pakistan. It is reckoned that over 90% of the water of the subcontinent as a whole is

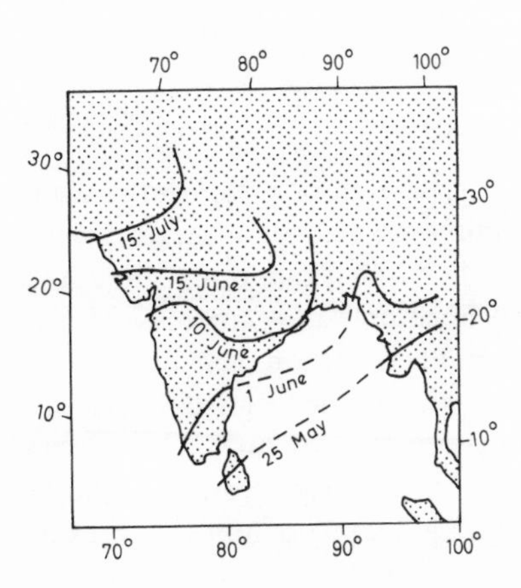

Fig. 11.2 The advance of the summer monsoon, indicated by average dates for the burst of the monsoon. (From Barry & Chorley 1968)

received from the wet season in summer. Too much or too little precipitation during the monsoon months may have almost equally dire consequences for much of the local population.

Traditionally, the south Asian monsoon has been explained in terms of an enormous 'land–sea breeze' circulation system. Modern views are rather different. They hold that the Asiatic monsoon regime is a consequence of complex interactions of planetary and regional factors both at the surface of the earth and in the upper troposphere (see, for example, Lockwood 1965). In particular, the Himalayan mountain chain and the high Tibetan plateaux are said to play key roles, both through their influences on low level circulations, and through their deflecting effects on upper tropospheric circulations. The precise mechanisms have yet to be worked out, particularly with respect to sequences of cause and effect. It is clear, however, that the final, abrupt arrival (or 'burst') of the Indian monsoon does not take place until the winter subtropical westerly jet that curves cyclonically around the high Tibetan plateaux has weakened and has been displaced north to curve anticyclonically around the plateaux, as much as 20° nearer the North Pole (Thompson 1951). An upper tropospheric easterly jet then becomes established across southern Asia at about 15° N.

Koteswaram (1958) has described relationships that seem to exist between the upper level jet stream and monthly average rainfall distributions over southern Asia and central Africa. These essentially involve reversals of the sign of divergence with height through the troposphere such that divergence aloft is generally associated with convergence below and vice versa. It is well known that surface convergence in the Tropics usually leads to a strong development of convective cloudiness especially if the converging air streams are reasonably moist. The air streams advected across southern Asia in summer are moist indeed, and the weather associated with their period of dominance is characteristically as cloudy, humid and wet as the weather associated with the divergent anticyclonic airstreams from the continent during winter is clear, non-humid and dry.

Satellite studies

Although the major controls of the south Asian monsoon as a large-scale seasonal perturbation of the general circulation seem to be northern hemispheric and continental rather than southern hemispheric and maritime, the advent of weather satellite observing systems must presage an increase in our knowledge and understanding of this important phenomenon (Ramage 1971). Analyses of satellite photographs in particular can be carried out more usefully in general for oceanic than for continental regions. These can supplement the conventional observations that have always been more readily forthcoming from the land mass of Asia than the largely empty expanses of the Indian Ocean. Since it seems possible that a high percentage of the annual monsoon rains may be brought by more or less well-defined depressions within the onshore flow, Essa mosaics would seem to have much to offer in this respect, especially the Mercator montages of the Tropics between 30° N and 30° S. Before we proceed to review current research based on such photographs, however, let us examine satellite infra-red maps to become acquainted better with the regional climatology of the whole area involved directly or indirectly in the south Asian monsoon system.

Infra-red investigations

During 1963–4, extensive meteorological observations were made in the Indian Ocean region in support of the International Indian Ocean Expedition (I.I.O.E.). For part of this period (from 19 June 1963 to the end of December 1964) Tiros VII relayed medium resolution infra-red radiation measurements through five spectral channels, of which one was in the 8–12 μm portion of

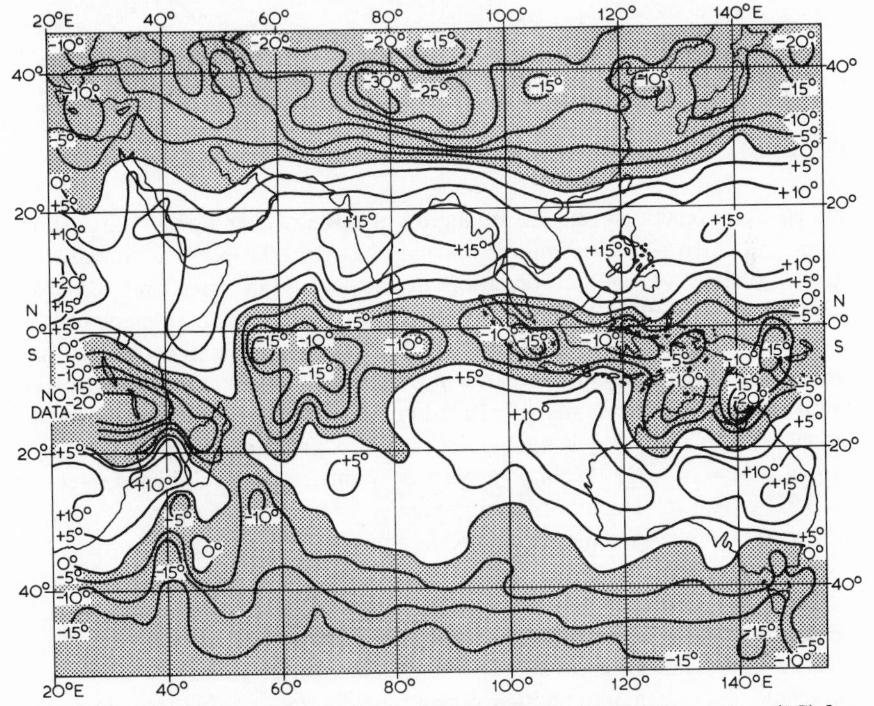

Fig. 11.3 A Tiros VII monthly map of equivalent blackbody temperature (°C) for January 1964, derived from measurements of emitted radiation in the 8–12μm atmospheric window over the Indian Ocean. (Areas below 0 °C stippled). (After Kreins & Allison 1969)

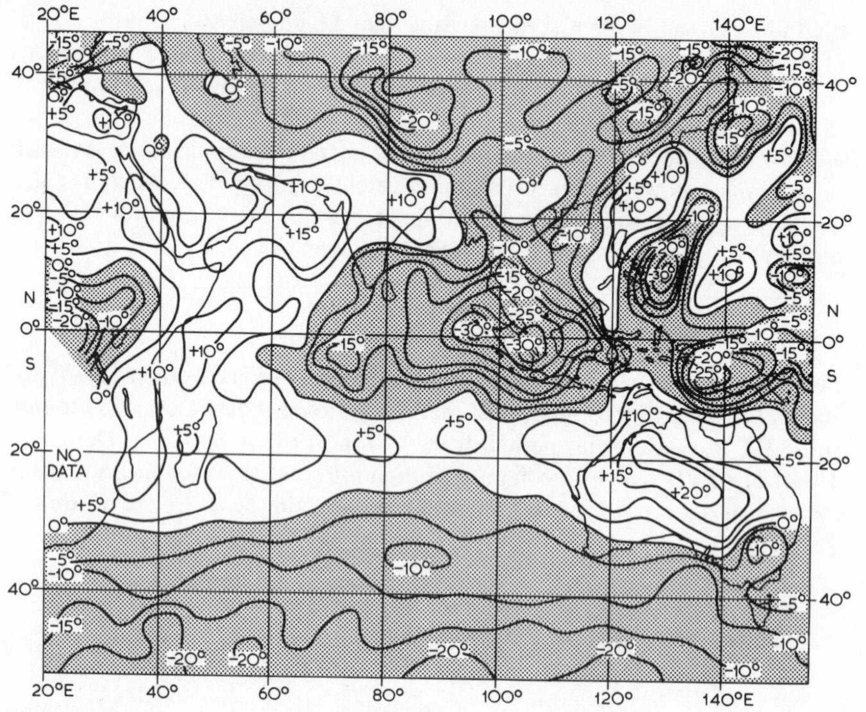

Fig. 11.4 As fig. 11.3, for April 1964. (After Kreins & Allison 1969)

the electromagnetic spectrum. Radiation in this waveband originates mainly from the earth's surface and/or its cloud cover (see p. 42). An atlas of monthly maps of emitted radiation in the 8–12 μm waveband has been prepared by Kreins and Allison (1969). A selection from these maps may be used to illustrate gross seasonal weather changes with a continuity and uniformity in space and time that is unrivalled by conventional data displays. In the preparation of the maps, selected radiation data were compiled in terms of mean equivalent black body temperatures, and then computer-averaged across a 1:40 m scale Mercator projection. Isotherms were drawn at intervals of 5 °C except in regions of steep gradients, where the values of the extreme T_{BB} temperatures were plotted and some intermediate isotherms omitted. The number of orbits per month which crossed the Indian Ocean region varied considerably, from as few as 45 (September 1964) to as many as 170 (October 1963). Since more numerical averaging was possible for those months that yielded a good supply of data, some of the variations in amounts of detail from one map to another reflect the different crossing frequencies. In addition to this interpretation problem, three other factors must be borne in mind before the salient features of figs. 11.3–11.6 can be discussed meaningfully:

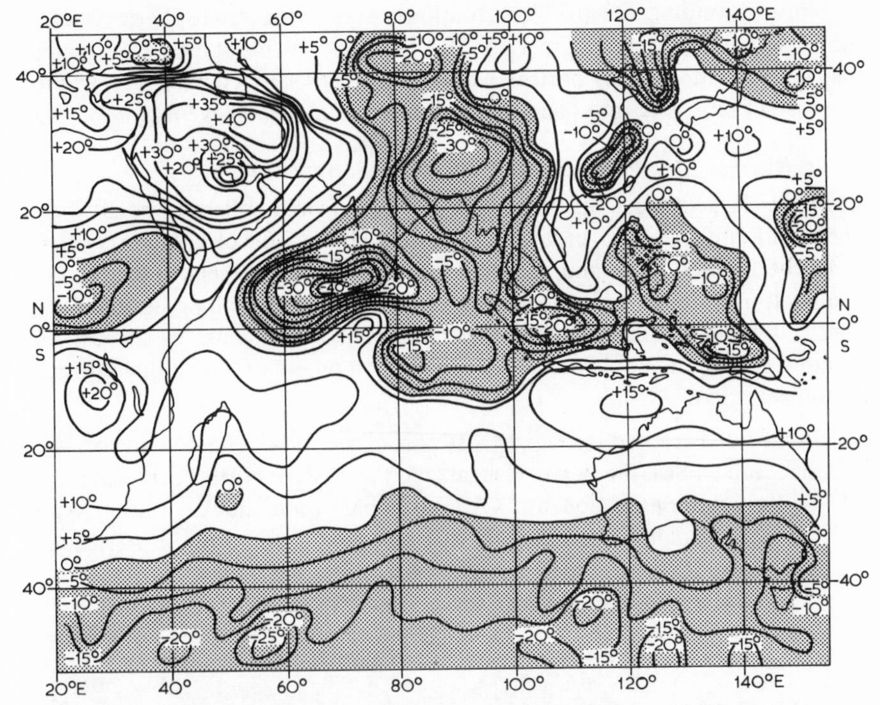

Fig. 11.5 As fig. 11.3, for July 1964. (After Kreins & Allison 1969)

1 Thin cirrus clouds act as if they are partially transparent to radiation and may permit some terrestrial and atmospheric radiation from lower levels to be measured by a satellite sensor.
2 Where broken cloud decks are involved, a satellite combines the radiances from the cloud tops and from between the clouds when observing in the 8–12 μm region.
3 Although the 8–12 μm waveband is mainly transparent to electromagnetic radiation, some absorption of infra-red energy by water vapour and ozone may cause the isothermal patterns to be colder than they would be in the absence of an atmosphere. The worst effect of this kind is associated with a very moist atmosphere where the skies are clear of clouds (Wark *et al.* 1962). Then a departure by as much as 10 °C from the real surface temperature may be produced.

As outlined in chapter 3, the optimum conditions for temperature mapping through this atmospheric window waveband involve homogeneous overcasts of cloud, or clear skies, filling the radiometer fields of view. Under these conditions T_{BB} temperatures can be interpreted uniformly in terms of surface or cloud-top temperatures, with progressively lower cloud-top temperatures indicating progressively higher cloud-top altitudes.

Notwithstanding the problems outlined above, the atmospheric window waveband maps derived from Tiros VII data are valuable at least by reason of the general patterns they portray when these are interpreted with care, and for the insight they give into one years' sequence of monsoon events.

A northern winter pattern: January 1964 (see fig. 11.3)

Here the large-scale temperature pattern has a predominantly zonal form, indicating an east–west organization of cloudiness and circulation systems over the Indian Ocean and its adjoining continents. The central feature is a temperature minimum from east Africa to New Guinea along an axis lying just south of the Equator. On either flank, higher temperatures suggest less cloudy conditions in two distinct zones composed of the following geographical areas:

(*a*) In the northern hemisphere, across part of the Sahara, the Indian coasts of the Arabian Sea, the Bay of Bengal, and the western north Pacific Ocean.

(*b*) To the south of the Equator, across the trade wind anticyclonic region of the south Indian Ocean, and the summer continents of South Africa and Australia.

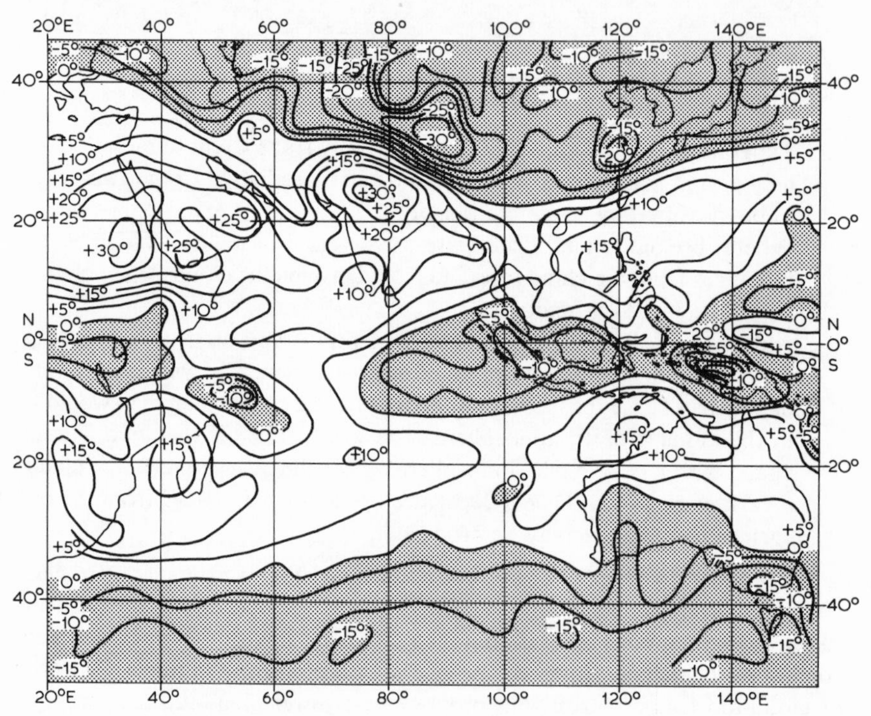

Fig. 11.6 As fig. 11.3, for October 1964. (After Kreins & Allison 1969)

Perhaps the most striking feature of the map is, however, the very steep northward temperature gradient over northern India to the temperature minimum of below −30 °C in the Tibetan region. In winter, a shallow layer of cold, high pressure air is centred over Asia, replaced over eastern Asia by a trough at the 700 mb level. The combined effect of the edge of the high plateaux and the strong latitudinal temperature gradient helps to anchor a westerly jet stream across northern India in winter. This jet curves cyclonically eastward until it crosses the Chinese coast above the position of the 700 mb trough. In fig. 11.3 the alignment of the jet is indicated quite well by the more tightly knit isotherms from West Pakistan east to northern India and the south China coast. Air subsiding beneath this upper westerly current feeds dry northerly winds across India and Pakistan. Low incidences of cloudiness and rainfall prevail during January and February when the anticyclonic outflow is at its strongest.

The northern spring: April 1964 (fig. 11.4)

The chief developments involve the equatorial trough and the Indian subcontinent. Higher temperatures associated with less persistent deep and/or high cloudiness across the western equatorial Indian Ocean suggest a circulation change between the African coast and 70° E. Similar breaks in the low temperature belt along the equatorial trough were noted in Tiros IV data in the northern spring of 1962 (Winston 1969). These may indicate the initiation of a transequatorial airflow from south to north.

Further north, the strong east–west temperature gradient across the north of the Indian subcontinent appears little different from its January alignment, although a relatively rapid warming over the Middle East seems to have been associated with a northward displacement of the strongest temperature gradient there. Unfortunately, the maps do not extend their coverage sufficiently far north to suggest the contemporaneous strengthening of a second westerly jet north of the Tibetan plateaux. Fig. 11.4 does, however, demonstrate the development of a thermal low pressure area over northern India. Tiros VII evidence showed that, in that region, T_{BB} temperatures rose from about 7 °C in March to over 30 °C in April, thereafter decreasing until the end of the summer monsoon. During April the weather is usually hot, dry and squally before the rain-bearing winds arrive from the south, depressing temperatures and raising humidity levels to an uncomfortable peak. As fig. 11.4 suggests, April–May is the hottest season across much of India, East Pakistan and Burma.

The northern summer: July 1964 (fig. 11.5)

By July, moist oceanic air streams dominate the weather of much of southern Asia. By comparison with fig. 11.4, this map for the month of July shows few differences south of 10° S, but substantial differences north of the Equator. The zonal belt of high temperatures centred on 10–20° N has largely disappeared, broken from north to south by several patches of much lower temperatures. The indications are that most of peninsular India, and the eastern coastlands of the Bay of Bengal as far as Malaysia are covered by the deep, persistent cloud of the wet summer monsoon. During this season, India is crossed by an easterly equatorial jet, as outlined earlier, but there is little direct evidence of its alignment in fig. 11.5.

During this season of the year, the transequatorial flow of air down the continuous pressure gradient from the south Indian Ocean to the northern interior of India is thought to be most pronounced. Findlater (1969) has shown that the mean flow reaches jet stream proportions at 1000 m in July from the northern tip of Malagasy, along the coasts of east Africa and across much of the Bay of Bengal as shown in figs. 11.7 and 11.8. This broad

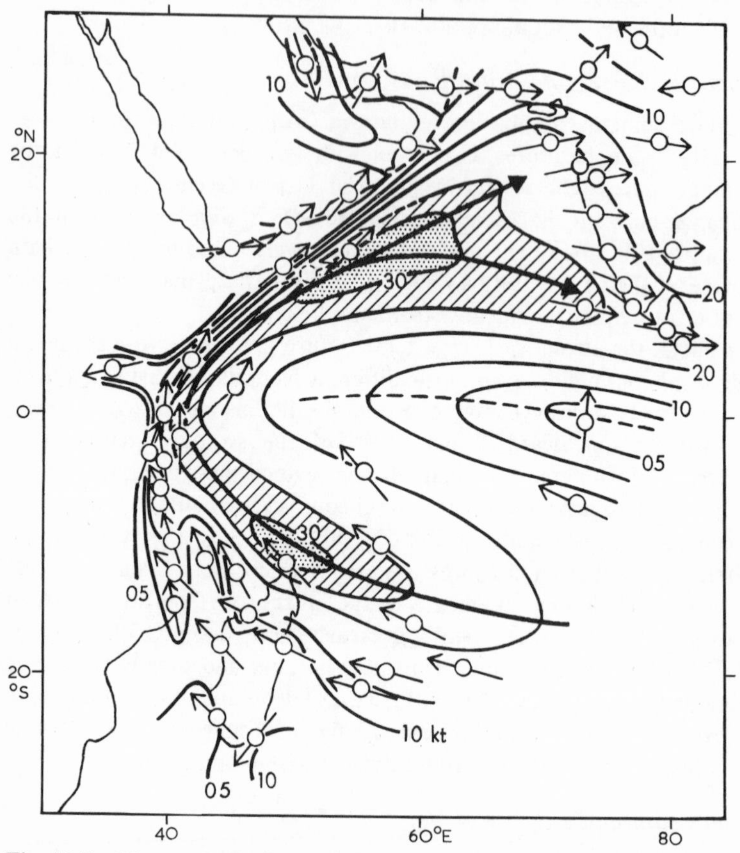

Fig. 11.7 The mean low-level jet stream over eastern Africa and the Indian Ocean at 1000 m in July. The thick arrow-headed curves are streamlines at the core of the flow. Continuous lines are isotachs at 5 kt intervals. The directions of vector mean winds at stations from which observations were available are shown by arrows through station circles. (From Johnson 1970; after Findlater 1969)

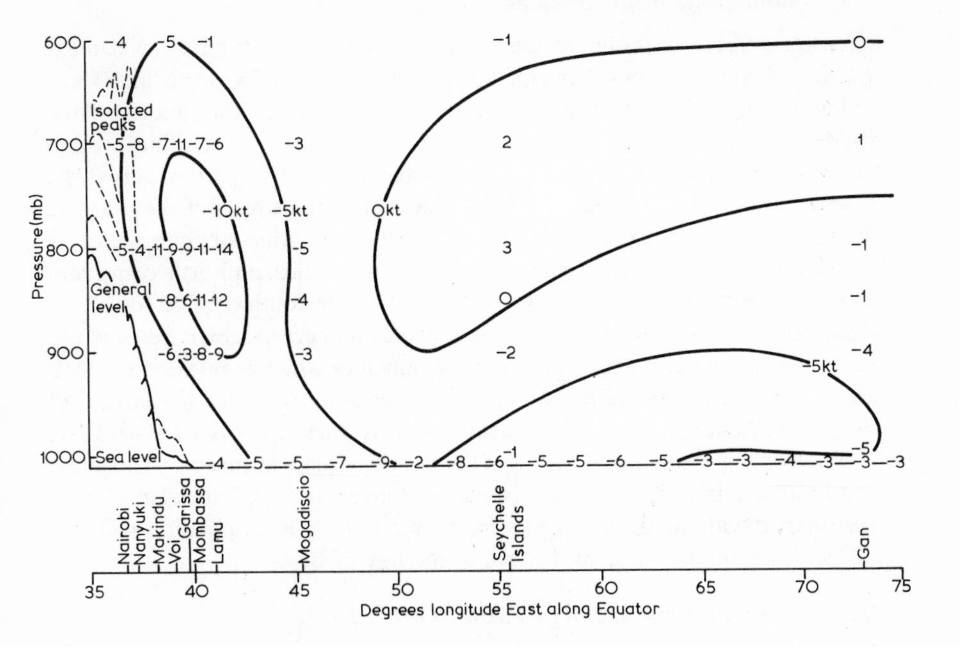

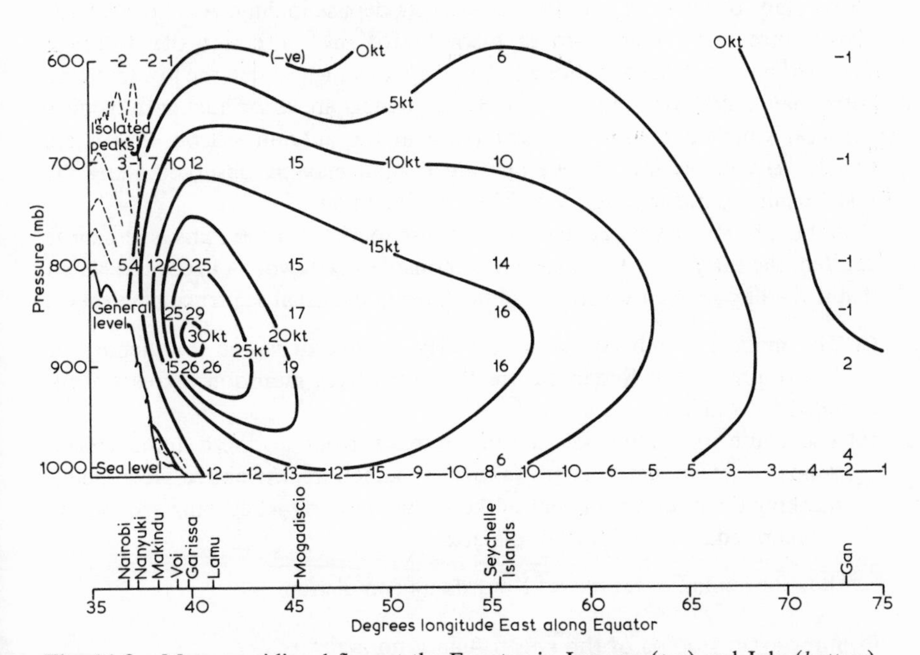

Fig. 11.8 Mean meridional flow at the Equator in January (*top*) and July (*bottom*) indicating a much stronger south-north transport of air across the Equator in July. (Positive values indicate flow from the south). (From Findlater 1969)

monsoon current is probably the most important agent for low-level mass transfer of air from one hemisphere to the other. To its north and west, airflow towards the Indian low pressure centre is across the deserts of the Middle East. The distinction between the moister, transequatorial airflow and the drier winds from the west is reflected in the temperature pattern across the Arabian Sea in fig. 11.5. Sawyer (1947) described the effects of the wedging out north-westward of moist air masses by warmer drier air; convection is inhibited around the northern shores of the Arabian Sea, and little or no rain falls in the summer months in the north-west of the continent (see fig. 11.2). Meanwhile, the air streams flowing across the southern Arabian Sea and the Bay of Bengal acquire abundant moisture, much of which is released through strong convective instability stimulated by convergence and the effects of orography. Essa photographs frequently portray a particularly persistent line of clouds along the exposed slopes of the western Ghats in summer, and a less cloudy area inland to their lees (see also Bunker 1967). Somewhat similar mesoscale cloud variations appear from time to time in association with the hills and mountain ranges to the east of the Bay of Bengal.

The northern autumn: October 1964 (fig. 11.6)

The zonal temperature pattern, so strongly marked in January and April, shows signs of returning in October after its demise in July. As expected, the temperature belts show signs of a southward shift, although the distances involved are small. Conventionally based studies suggest that the low temperatures maintained over much of the Bay of Bengal are associated with tropical vortices, which are most frequent there in the autumn season. These are largely responsible for the late autumn rainfall maxima observed along the east Indian coast, for example at Madras (fig. 11.9).

Although the autumn retreat of the moist southerly airstreams is less clear cut than the early summer onset of the monsoon, Lockwood (1965) has stated that it is still possible to discover some order in the usual sequence of events:

(*a*) The main 200 mb equatorial easterly jet stream leaves the vicinity of Khartoum in the Sudan before the south-west monsoon retreats from northern India.

(*b*) The south-west monsoon usually retreats from northern India before winds of jet stream strength (above 60 knots) appear above New Delhi, marking the re-establishment of the winter westerly jet curving around the southern edge of the Tibetan plateux.

Thus the annual cycle moves towards its completion.

Photographic studies of the south Asian monsoon

As yet few studies have been made of the sequence of monsoon events through careful analyses of cloud fields portrayed by satellite photographs.

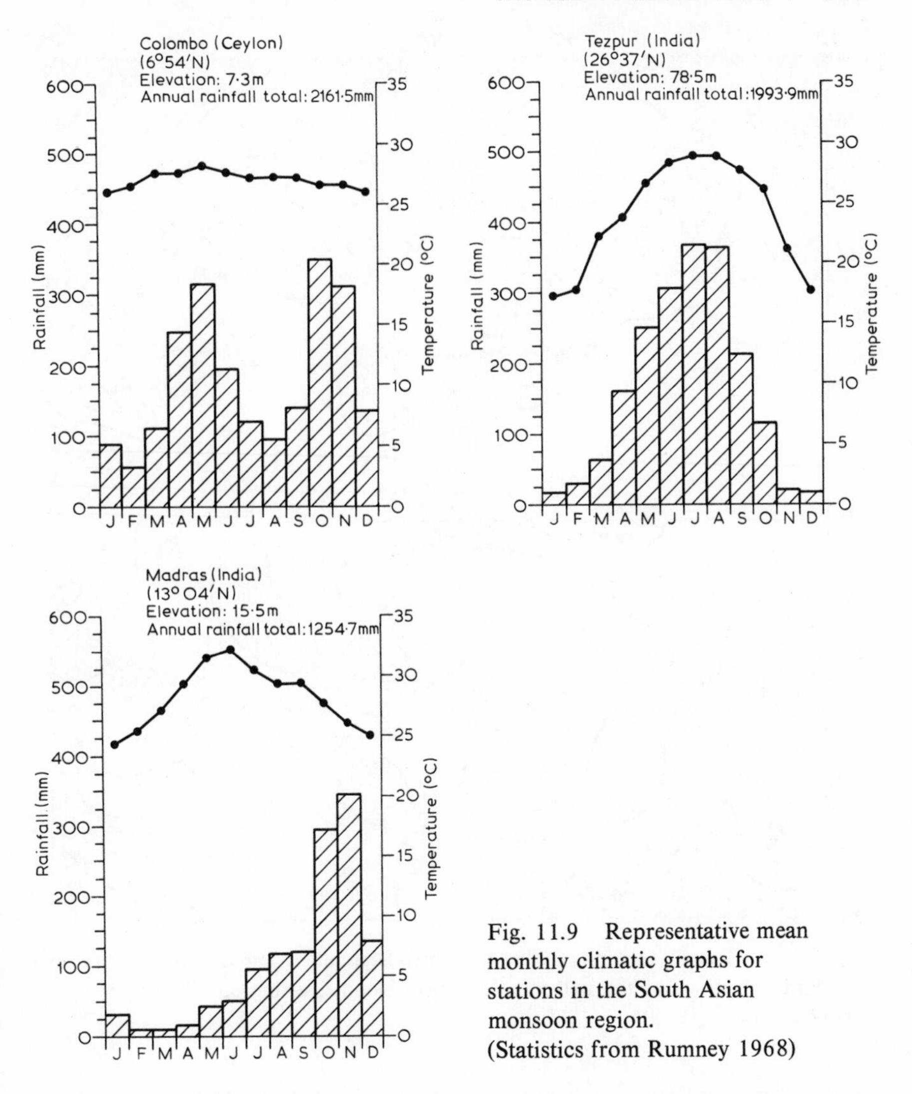

Fig. 11.9 Representative mean monthly climatic graphs for stations in the South Asian monsoon region.
(Statistics from Rumney 1968)

This is a pity, since much remains to be learnt that cannot be derived from the more equivocal evidence of infra-red radiation patterns. However, a start has been made in the right direction (Hamilton 1973), and some initial findings can be reported here.

Fig. 11.10 illustrates the Indian Ocean cloudiness during April–September 1967, through its half-monthly standard deviation pattern. Since the standard deviation of a population is one of the principal statistical measures of its variance, fig. 11.10 separates areas of little cloud cover percentage variation from those of much variation. The areas of low σ_c indicate relatively constant

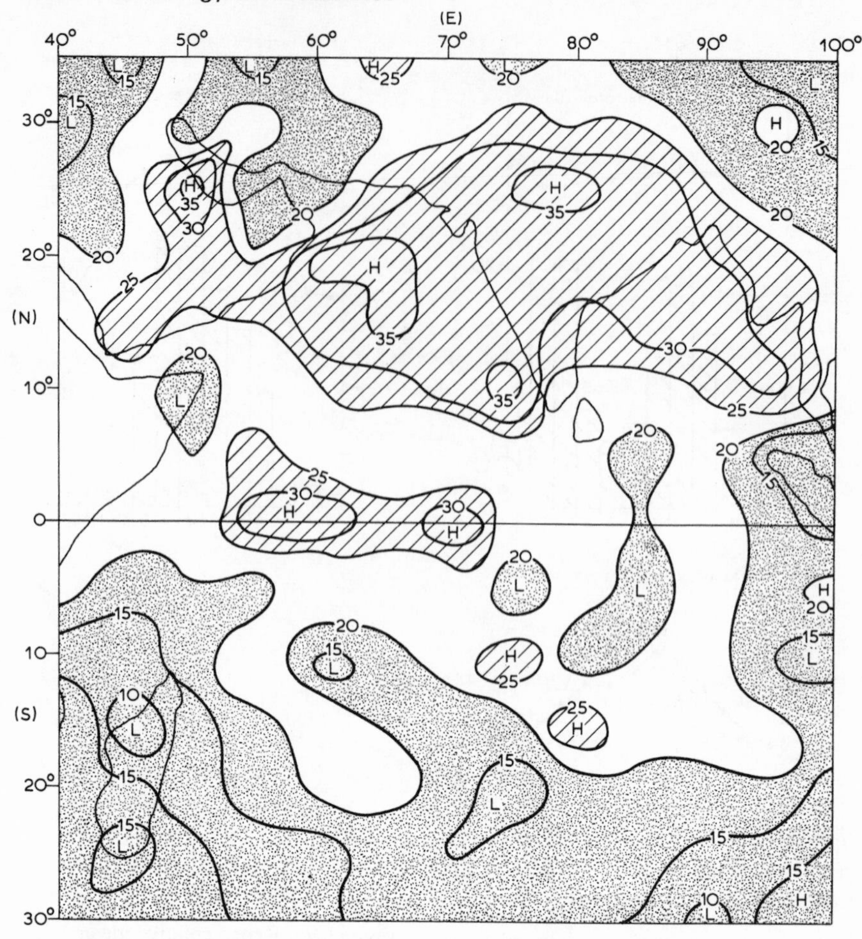

Fig. 11.10 Indian Ocean cloudiness, April–September 1967, expressed by the standard deviations of the half-monthly mean cloud distributions (in percentages). (After Hamilton 1973)

sky conditions, and the areas of high σ_c indicate areas affected by large variations of cloud cover.

In fig. 11.10, the largest area of high σ_c (>25%) indicates quite well the spread of the summer monsoon across southern Asia and its neighbouring seas. The centre of high σ_c in the centre of the Arabian peninsula is much more unexpected and difficult to explain. Perhaps it might be attributed to a variable 'subtropical cyclone' in the middle troposphere, deriving its moisture from the Bay of Bengal via easterlies crossing peninsular India (cf. Ràmage, 1971). The second largest area of high σ_c extends along the Equator between 50° and 75° E. Whereas progressive seasonal changes of cloudiness may be proposed as the chief factor prompting high standard deviations in the main monsoon areas, recourse to the photographic record indicates that short-period fluctuations from cloud to no cloud cover are responsible for the high values along the Equator.

At the other end of the scale, low standard deviations are noted over the interiors of north Africa and Asia to the east and west of the high Tibetan plateaux, and in association with southern hemisphere circulations to the south of 15° S. As earlier maps have indicated, both short-term and seasonal variations of the chief weather belts in the mid-latitudes of the southern hemisphere are rather small. The most unexpected area of low σ_c at first sight may seem to be east of 95° E just north of the Equator. Evidence discussed in more detail later, however, suggests that the intertropical cloud band varies relatively little in the region of Sumatra, which helps to anchor its position. Even during the dramatic cloud changes of the monsoon season across the north Indian Ocean cloud cover remains almost invariably strong in this area at the eastern end of the ocean.

Further light is being thrown upon the structure of the southern monsoon as a bearer of bad weather by careful day-to-day analyses of the sizes and movements of cloud masses embedded in the monsoon flow. While there is no doubt that considerable rain falls often accompany mid-tropospheric cyclonic disturbances at the onset of the monsoon (see, for example, Ramamurthi and Jambunathan 1965), the rainy season as a whole includes considerable weather variations. The traditional view of almost uniformly cloudy and wet weather for many days on end does not bear critical scrutiny when substantiation is sought even from conventional records. Not only are the monsoon rains not continuous during any one monsoon season, but also they vary considerably from year to year in intensity and distribution. Thus, the breaks that interrupt the rainy weather vary in their lengths and frequencies

Table 11.1

Occurrences of bright cloud clusters (nephmasses), 40–100° E, 30 N–30° S, from Essa photographic evidence, 1 April–30 September 1967. In detail the figures may have been affected somewhat by the fluctuation of the dividing line between the photo-coverage for successive days across the Indian Ocean, but this is not thought to have had a significant influence on the overall numerical sequence, which closely resembles a geometric progression (from Hamilton 1973)

Number of days through which specific individual nephmasses could be identified:	1	2	3	4	5	6	7	⩾8
Number of nephmasses (⩾ $\frac{1}{5}$th of 10° grid square [$\simeq 0{\cdot}25 \times 10^6$ km^2]	488	243	110	65	26	13	12	16

from year to year. On satellite photographs, the mid monsoon season masses of deep cloudiness, probably with embedded rainfall areas, appear as bright, amorphous patches. It is too early to say yet whether these have marked effects upon the wind and pressure fields, but fresh satellite evidence is forthcoming as to their lengths of life and geographical distributions (Hamilton 1973).

Table 11.1 summarizes the lives of all those bright cloud masses that

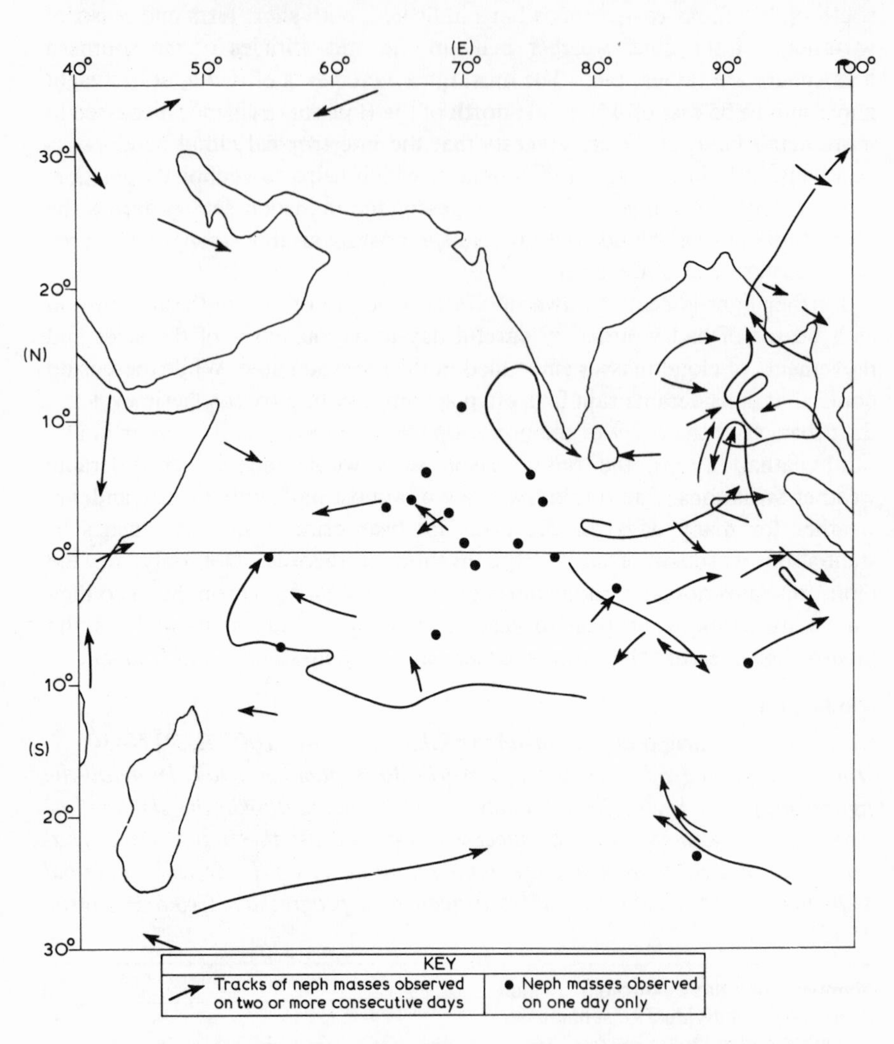

Fig. 11.11 (a) The lives and paths of discrete cloud masses photographed by Essa satellites in May 1967. Circles indicate masses identifiable on one day only. The pattern is partly contaminated by the effects of photo-overlaps 24-hours apart. (After Hamilton 1973)

occurred within the size range indicated across the area from 30° N to 30° S and from 40° to 100° E during April–September 1967. It is especially noteworthy that very few survived as recognizable entities for more than three days. Indeed, 50% did not retain their identities beyond a single day. July and August were the most active months in both hemispheres.

Figs. 11.11 (a) and (b) portray the regional distributions and paths of motion of the cloud masses photographed by Essa in May and July 1967 respectively. In May the south-west monsoon usually extends no further north and west than a diagonal line across the Bay of Bengal from the tip of India to the coastal border of Burma with East Pakistan. Hence, fig. 11.11 (a) illustrates an early stage in the spread of the moonsoon, which extends its influence from south-east to north-west as the transequatorial flow strengthens.

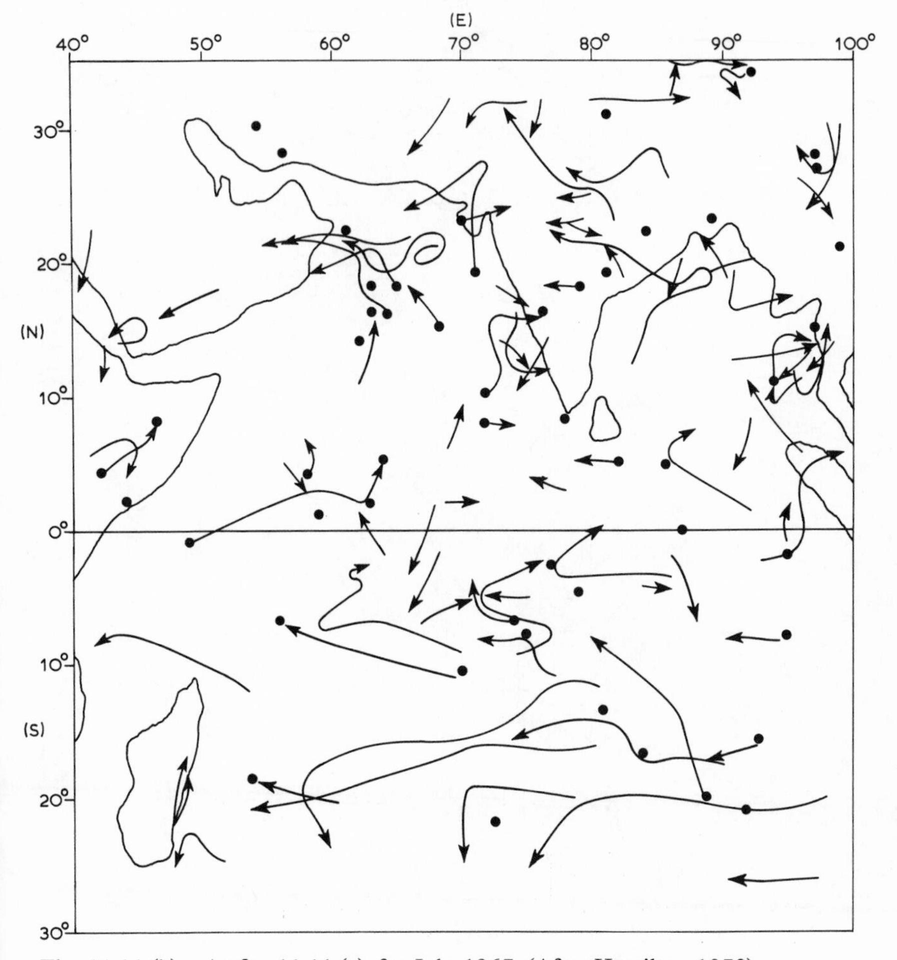

Fig. 11.11 (b) As fig. 11.11 (a), for July 1967. (After Hamilton 1973)

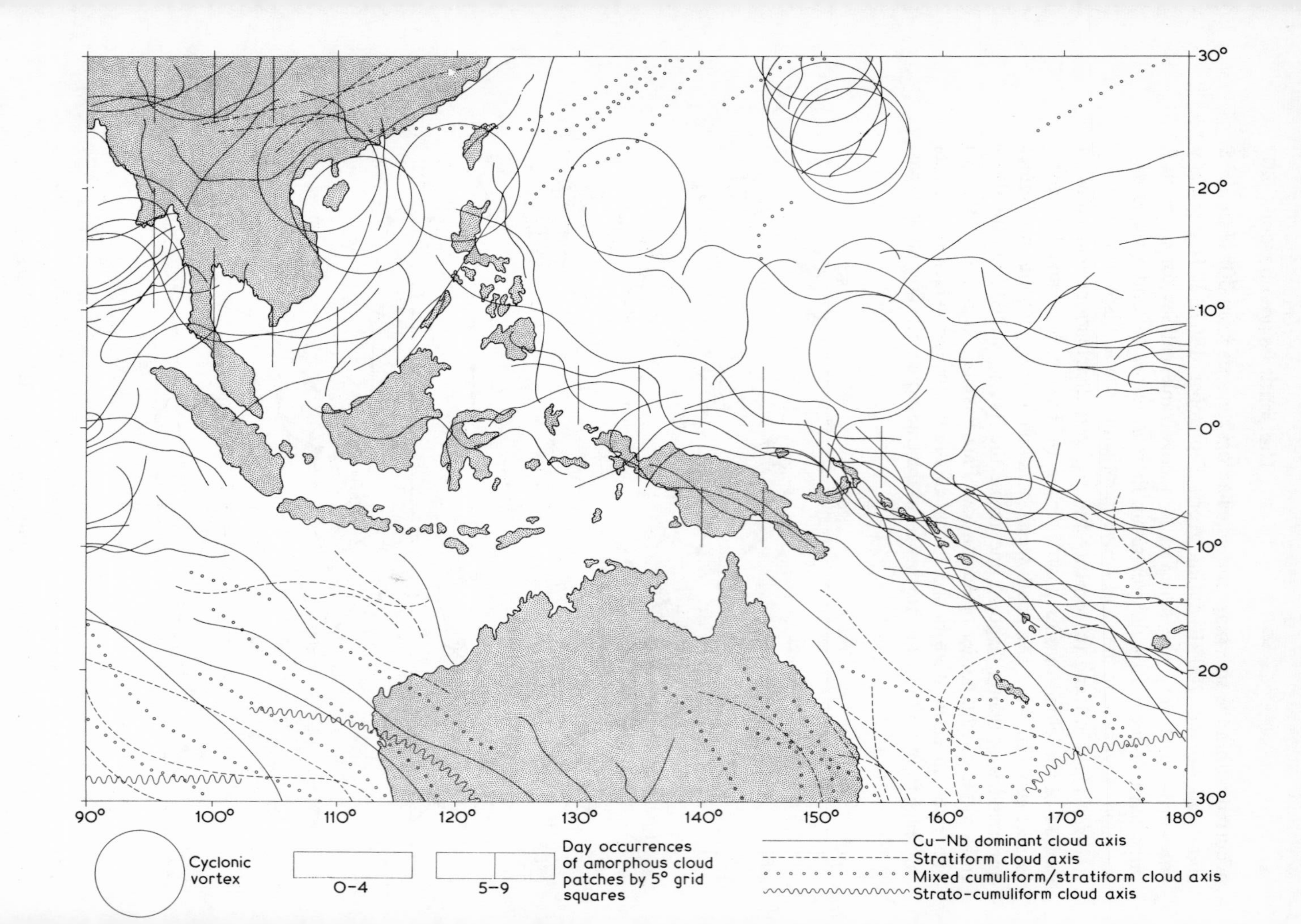
30°
20°
10°
0°
10°
20°
30°
90°
100°
110°
120°
130°
140°
150°
160°
170°
180°
Cyclonic vortex
0–4
5–9
Day occurrences of amorphous cloud patches by 5° grid squares
Cu–Nb dominant cloud axis
Stratiform cloud axis
Mixed cumuliform/stratiform cloud axis
Strato-cumuliform cloud axis

Fig. 11.11 (b), meanwhile, illustrates the month during which the monsoon usually reaches its ultimate in both extent and weather intensity. Neither map contains a single cloud mass that could be traced as an identifiable, independent entity for more than three days at a time. Of equal interest and significance is the fact that in neither month is there much evidence of discrete cloud masses crossing the Equator. This last fact applied to all six months considered during 1967, sadly enough for those who had hoped that the controversy concerning the volume, extent and corridor of cross-equatorial flow across the Indian Ocean might be resolved by the evidence afforded by long-lived tracers of satellite-viewed cloudiness. The simpler and more substantial cloud mass tracks, especially in July 1967, were around the northern flanks of the tropical anticyclone over the south Indian Ocean. The true monsoon region, north of the Equator, was characterized in both months by considerable subsynoptic confusion. Especially in July, the chief weather-producing cloud centres seemed to follow random paths through their brief lifetimes. Many of those identified at least on two successive days around 15–30° N, however, transferred generally westward. Perhaps this is evidence of steering by the summer jet stream aloft. The final, overall impression imparted by fig. 11.11 (b), is of a convectively bubbling southerly monsoon air stream following a smoothed Z-shaped course from about 10° S, turning across the Equator under the rotational influences of the earth, and recurving north-westward about 15° N.

South-east Asia and northern Australia region

We complete our review of the climatology of the south Asian monsoon, and of tropical climatology as a whole, by reference to high season weather distributions in the area adjacent to the Indian Ocean to the east, namely from 30° N to 30° S and from 90° to 180° E (plates 25 and 26). In the search for ways to increase the climatological usage of satellite photographs, Barrett (1971) employed a simple genetico-generic cloud classification scheme in order to assist the dynamical interpretation of mean monthly cloud maps. Table 11.2 explains his three-fold categorization of the larger nephanalysis cloud masses. Generally speaking, as the sixth column of the table indicates, circulariform vortex systems are indicative of tropical storms or hurricanes; elongated cloud masses are associated with intertropical convergences, appropriate topographic influences or extratropical troughs or fronts;

Fig. 11.12 *Opposite* Significant nephsystems, tropical Far East, July 1966. Elongated nephsystems are represented by their long axes, differentiated on the grounds of their dominant types of cloud components. Amorphous nephsystems are indicated per 5° grid boxes by vertical hatching; and circulariform nephsystems by circles of 5° radii, focused on the cyclonic vortex centres. (From Barrett 1971)

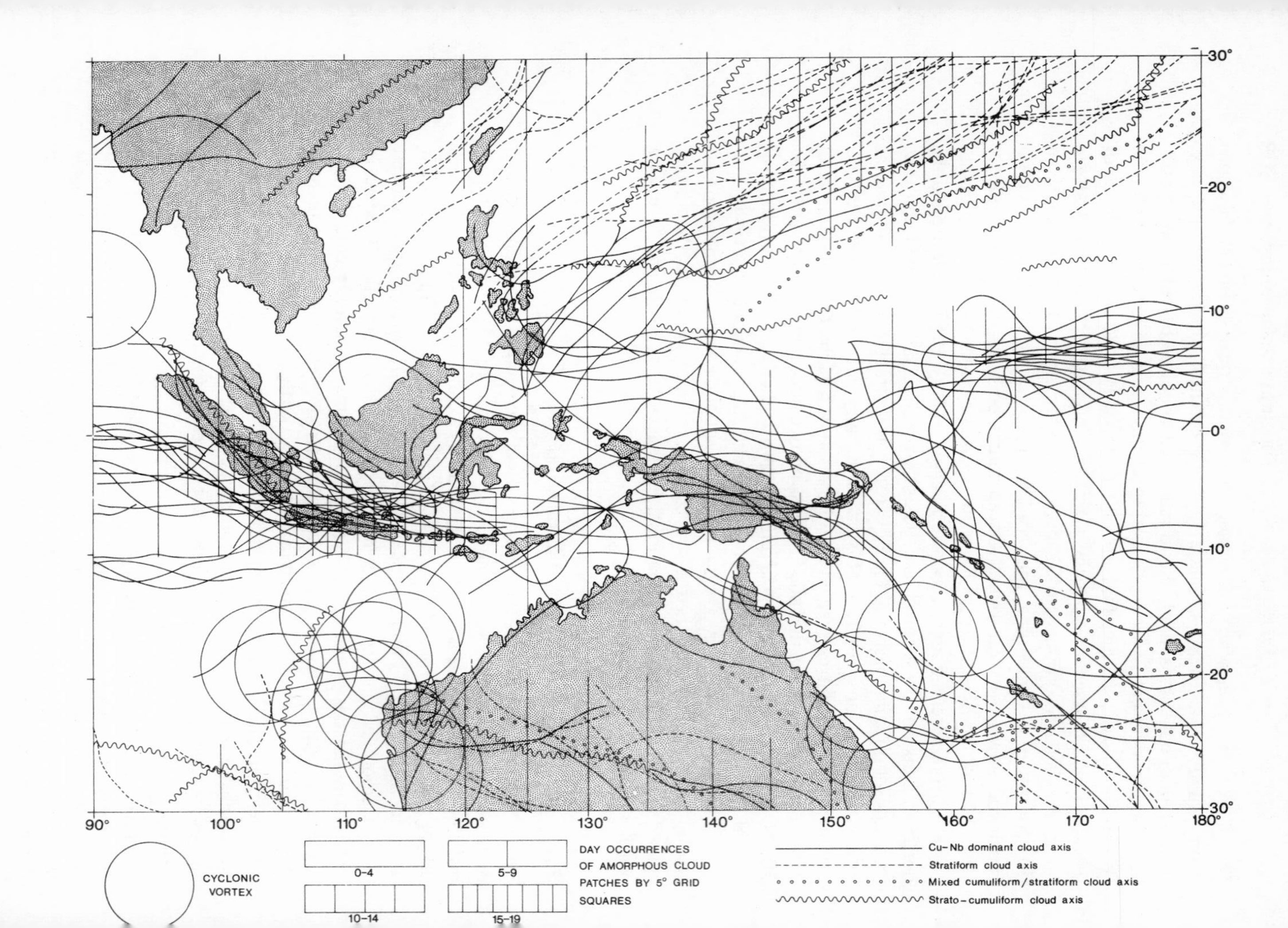
30°
20°
10°
0°
10°
20°
30°
90°
100°
110°
120°
130°
140°
150°
160°
170°
180°
CYCLONIC VORTEX
0-4
5-9
10-14
15-19
DAY OCCURRENCES OF AMORPHOUS CLOUD PATCHES BY 5° GRID SQUARES
Cu-Nb dominant cloud axis
Stratiform cloud axis
Mixed cumuliform/stratiform cloud axis
Strato-cumuliform cloud axis

and amorphous masses have a variety of origins. It should be noted that no reference is made in figs. 11.12 and 11.13 to the scatters of smaller cumulonimbus cloud clusters and other cloud clusters that appear day after day. Their distributions tend to be relatively uniform within the equatorial trough of low pressure. The significance of each individual cluster is mostly short term and local, as in the case of many Indian Ocean monsoon cloud masses.

Table 11.2
A simple classification of nephanalysis synoptically significant cloud masses (from Barrett 1971)

Type No.	*Form*	*Description*	*Dimensions*	*Length/breadth ratio*	*Synoptic indications*
1	Circulariform	Circular or sub-circular vortical cloudiness (vortex centre indicated on nephanalysis)	Vortical overcast circles ⩾5° lat. × 5° long.	Between 1 : 1 and 1·5 : 1	Well-defined vortices of hurricanes or tropical storms
2	Amorphous	Cloud masses lacking simple geometrical shape	Central overcast areas covering grid squares ⩾5° lat. × 5° long.	No simple relationship	Complex and/or various, e.g. poorly organized widespread instability and/or degenerations from types 1 or 3
3	Elongated	Relatively simple linear cloud masses, i.e. masses stretched along one straight or nearly straight axis	Lengths ⩾5° lat. or long, depending on their general orientation	⩾2 : 1	Various types of instability axes, caused most frequently by convergence and/or topography

In contrast to the Indian Ocean monsoon, that of south-eastern Asia seems more simply attributable to a seasonal fluctuation in the alignment of the intertropical cloud band. Essentially, it seems that the zone of maximum weather activity shifts little from January to July to the east of 150° E, but swings markedly on that hinge to the west. In January 1967 the intertropical cloud band lay just north of northern Australia, crossing the Indian Ocean between 0° and 10° S. Fig. 7.12 (p. 190) depicts the related rainfall activity in that month, the peak month of the wet monsoon along the northern coastlands of Queensland, Northern Territory and Western Australia (see fig. 11.13). In July, however, the convergence zone weather systems congregate more along a zone lying obliquely across the tropical western north Pacific from the east of New Guinea towards Indo-China and the Bay of Bengal (see

Fig. 11.13 *Opposite* Significant nephsystems, tropical Far East, January 1967. See fig. 11.12 for key and explanation of map symbols. (From Barrett 1971)

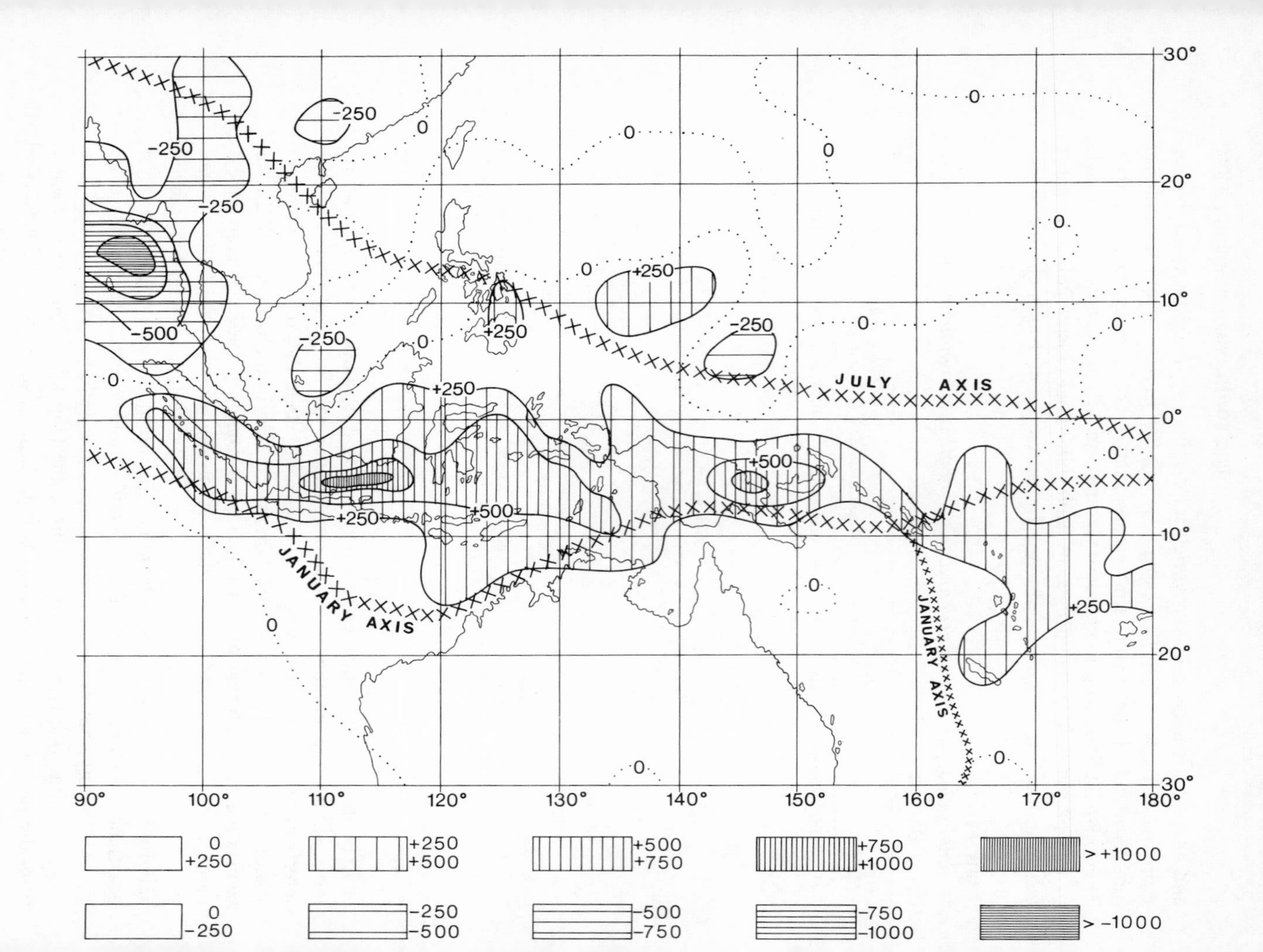
JULY AXIS
JANUARY AXIS
JANUARY AXIS
90°
100°
110°
120°
130°
140°
150°
160°
170°
180°
30°
20°
10°
0°
10°
20°
30°
0 +250
+250 +500
+500 +750
+750 +1000
> +1000
0 -250
-250 -500
-500 -750
-750 -1000
> -1000

fig. 11.12); in July 1966, a lesser zone of disturbed weather extended from Java and Sumatra in a westerly direction.

An interesting fact related to these high season contrasts in weather activity emerges from fig. 11.14, which depicts rainfall differences between January 1967 and July 1966, along with the axes of the equatorial troughs in those two months as evidenced by conventional observations. It is clear that the heavier rains associated with the trough tend to occur in the summer hemispheres, on the Equatorward sides of the equatorial low pressure axes. In other words, it appears that the monsoon rains in the tropical Far East fall from transequatorial air streams before they are diluted by admixture with converging flows on the poleward flanks of the intertropical confluence. These latter domiciled airflows must be generally cooler and drier, and consequently less sensitive to convective stimuli from below. In forecasting summer monsoon rains in affected areas such as south-east Asia and northern Australia, or twice yearly monsoon rains astride the Equator in the East Indies, the likely behaviour of the equatorial trough is obviously of paramount importance.

Further climatological studies should help to clarify the chains of interrelated causes and effects that influence year-to-year variations from the mean climatological patterns within the Tropics, and, through even longer teleconnections, into higher latitudes. For many years, the best statistical indicator of the date of the burst of the Indian monsoon has been mean surface pressure over South America (see Atkinson 1968). If, as now seems likely, the pressure gradient at the base of the south Pacific trades affects the strength of the zonal Walker Circulation and the extent of its penetration westward, pressure over South America may be expected to affect also the distribution and intensity of weather across Ramage's 'island continent' of south-east Asia. Continuing westward, the seasonal pattern and frequency of monsoon cloud masses may be affected variously from year to year by this already long sequence of cause and effect, perhaps mainly through influences on the upper easterly jet. Still further west, this jet is thought to influence rainfall across equatorial Africa. From there only a short sector of the earth's circumference must be traversed before South America is reached once more and the possibility of a novel form of 'feedback' must be considered. More than at any time in the history of the science, climatology is being forced to examine local phenomena within a circumglobal context. Weather satellites are providing many of the data that will be needed for studies of this kind.

Fig. 11.14 *Opposite* Precipitation differences between January 1967 and July 1966 in the tropical Far East, evaluated from figs. 7.10 and 7.11. The apparent axes of equatorial low pressure in January and July are indicated by lines of crosses, drawn from conventional evidence. (From Barrett 1971)

12 Temperate mid-latitudes

General considerations

In sharp contrast to the Tropics, the middle and high latitudes as a whole are characterized by strong rotational effects (the Coriolis force reaches its maximum at the poles), large temperature contrasts and, generally speaking, consumption of heat. Middle latitudes (here taken as the zone between 30° N and S of the Equator and the Arctic and Antarctic Circles) are especially interesting and important meteorologically in that the atmosphere there is strongly baroclinic. Geographically their importance stems from the fact that their populations comprise the bulk of that of the world as a whole. Hence their climates have certain distinctive features, and play key roles in man's global economy and civilization.

Fig. 12.1 illustrates the difference between barotropic and baroclinic states

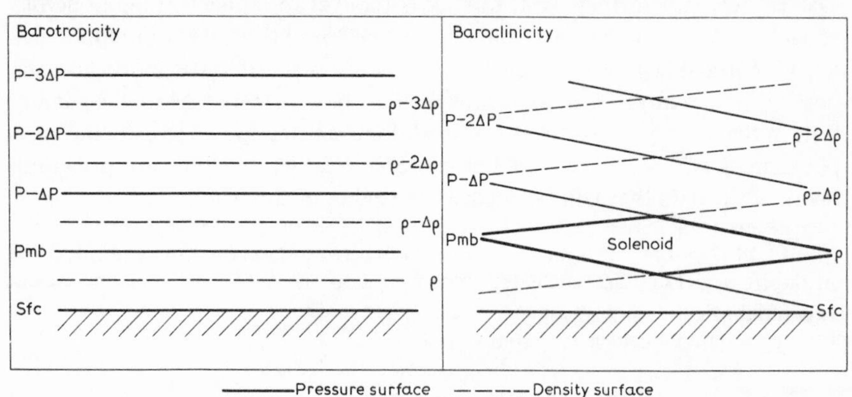

Fig. 12.1 Barotropic and baroclinic situations (left and right, respectively).

in the atmosphere. It will be seen that the barotropic case involves a hypothetical atmosphere in which surfaces of pressure and density coincide at all levels. Although barotropicity is an ideal concept, it serves none the less as a useful first approximation to reality in some areas under certain circumstances. The more usual baroclinic case, on the other hand, is one in which surfaces of pressure and density intersect at some level or levels. The degree of baroclinicity in any cross-section of the atmosphere is proportional to the number of quadrilateral intersections (solenoids) per unit area. Meteorologically, there are important consequences of the contrasts between these two arrangements. Along any isobaric surface in a baroclinic case there is a tendency for denser, heavier air to sink and lighter air to rise. Thus there is a tendency for a circulation to develop in the vertical. Such a tendency is absent in a barotropic atmosphere. The vertical circulations normally produced in baroclinic atmospheres are accompanied by transformations of potential energy into kinetic energy, often organized horizontally into cyclonic disturbances.

Within the observed pattern of the general circulation of the atmosphere, the strongest meridional temperature gradients are found in middle latitudes, concentrated markedly along the so-called polar fronts, separating ex-tropical from ex-polar air streams. Across the polar frontal zones in both hemispheres, strong temperature gradients are found in conjunction with much less strong pressure gradients; these zones are the most strongly baroclinic of all within the general circulation. Differential vertical movements of the warm and cold air masses on either side of each polar front are accompanied by the release of enormous amounts of potential energy which are converted into kinetic energy and stimulate strong wind flows.

Climatologically, the most significant feature of the mid-latitudinal zone is that it is characterized by mobile weather systems, some of which are cyclonic and others anticyclonic. It was seen in chapter 8 that the general circulation pattern aloft in middle to high latitudes is simple by comparison with that at the surface. The dominant direction of movement of the mobile pressure organizations is however, generally the same at all levels up to the troposphere, namely from west to east. Indeed many of the cyclonic disturbances in the lower troposphere form within the polar frontal zone under optimum upper tropospheric conditions, and, as we shall see later, generally travel around the globe in quite well-defined cyclic relationships with ridge and trough lines just below the tropopause. Embedded within the upper westerlies, intermittent winds of jet stream strength are associated with the stronger baroclinic zones, and play their own parts in the complex four-dimensional life cycle of the model extratropical cyclone.

It must be noted, however, that the deep frontal depressions that develop from typically wave-like perturbations along the zone of maximum baroclinicity are not the only type of cyclonically organized weather disturbances of large size within middle latitudes. Put in another way, not all mid-latitude depressions originate as polar frontal waves. Four other types that develop in distinctly different situations play their own important parts in the climatologies of those latitudes:

1 Lee depressions, comprising wave-like troughs on the downwind sides of mountain ranges, with or without closed isobars in their centres. A dynamical explanation is sometimes offered in terms of the equation:

$$\frac{\zeta + f}{\Delta p} = K \qquad (12.1)$$

(see also p. 249) where ζ and f have their usual meanings, K is a constant, Δp is the pressure difference between the top and bottom of the air column in question, and adiabatic motion is assumed. In these terms in a uniform westerly air stream for example ($\zeta = 0$), air speeds up on ascent to cross a mountain range, and slows down on descent. Δp first decreases, then increases across the crest. Therefore $(\zeta + f)$ must decrease on ascent, and, since it was previously zero, must become anticyclonic. Beyond the summit, both Δp and $(\zeta + f)$ increase, but, at first, f is decreasing on account of the southward motion. So, ζ must become less anticyclonic, and then cyclonic, if the balance is to be maintained, and a lee trough is formed. Such troughs tend to be anchored by the ranges in whose lees they form, e.g. the Rockies, especially in summer.

2 Thermal lows. These are essentially summer phenomena, resulting from strong daytime heating of continental areas. These lows may persist even at night when the heat sources weaken. Good examples are found in the south-west centre of America, over Iberia, and in West Pakistan in the northern hemisphere summer months.

3 Polar lows. These are most frequent in winter in synoptic situations involving a southward advection of very unstable maritime air from high latitudes. The wakes of strong frontal lows, and the eastern flanks of north–south orientated ridges of high pressure are especially favourable situations.

4 Cold lows. These are best marked in the middle troposphere, especially over north-eastern North America and Siberia. Their chief climatological significance results from their thick medium and high cloud sheets which offset radiational cooling from the land surfaces beneath during the long nights of the arctic winter.

The anticyclonic cells and ridges that separate the various cyclonic centres evident on daily synoptic charts likewise include mobile and more or less stationary types, related to the prevalent upper tropospheric pressure patterns, to topographic features, and to seasonal solar influences. Note that, in general, at a given latitude, mean sea level pressure is higher over land in winter, and lower in summer, a fact related to the differing thermal capacities of land and sea.

Finally, before examining satellite evidence of mid-latitude climatic structures and distributions, a word about the basic contrast between the general circulation within the Tropics and that observed in middle latitudes. Newton (1970) summarizes the situation in these terms:

(1) In low latitudes, where the component of earth rotation about the vertical is small, it has been established firmly that the dominant circulations are the Hadley cells of each hemisphere. These are driven by thermal forces and friction. They are important within the general circulation especially in that they help export on balance, energy, moisture and momentum to higher latitudes.

(2) At the poleward bounds of the Tropics, where the meridional circulations become nil, it has been established that the meridional transfer of westerly momentum, heat and water vapour into extratropical latitudes is effected by atmospheric eddies. These discrete masses operate both horizontally and vertically, but the contribution of the horizontal eddies (operating around the quasi-stationary highs and travelling cyclones and anticyclones) is the more important of the two. Thus middle latitudes are characterized by intrinsically more complex and variable circulations, and include the mean boundaries between the global zones of net export and import of energy, moisture and momentum.

It is in this kind of dynamic climatological context that most of the discussion that follows has been cast, bearing in mind the fact that, since land masses intervene more in middle latitudes in the northern hemisphere, the satellite view of reality north of the Equator is rather less clear than to the south. Thus, although relatively little work of a basically climatological nature has been attempted for the little-populated mid-latitude zone in the southern hemisphere, at least what has been done appears to be more complete. For this reason, we shall review southern hemisphere patterns first before turning to consider the more piecemeal evidence that is forth-coming to promote a better description and understanding of the comparatively complex climatology of the northern hemisphere.

Mid-latitudes of the southern hemisphere

Extratropical cyclones

Southward of 30° S, the southern hemisphere possesses a simple geography. By far the greater part of the surface area is covered by the southern waters of the Indian, Pacific and Atlantic oceans and the Southern Ocean, which unites the others in a circumglobal belt of almost uninterrupted sea between 40° S and the coasts and ice shelves of Antarctica. Only the southern quarter of South America, a similar fraction of Australia, a minute tip of South Africa and the islands of New Zealand act as significant surface complications south of 30° S so far as the development of weather and climate is concerned. It has been noted already (p. 211) that cursory analyses of satellite photographic and infra-red displays generally reveal three northward extensions of relatively cloudy conditions away from the circumglobal maximum around 50–60° S. Further light has been thrown on these cloudy extensions by Streten (1968a). Working with Essa 3 multiple exposure average mosaics for the six months of best illumination locally (January–March and October–December 1967), Streten found that two conspicuous meridional cloud bands were particularly evident:

1 Across the south Pacific, where a distinct band extended from the Equator north of Australia in a great arc to increasingly higher latitudes further east, merging with the circumglobal cloud belt at around 135° W and 45° S.
2 Across the western south Atlantic, where a dense cloud band extended on average from northern Argentina south-eastward to merge with the major cloud belt around 45° W and 45° S.

Streten averaged the areas of strong cloudiness depicted by the Essa mosaics through two 'seasons' suggested earlier by Taljaard (1967) (see table 12.1). Thus the six months in 1967 that were studied by Streten were

Table 12.1

A suggested seasonal subdivision of the year in the southern hemisphere, assuming that temperature is the most important element which determines the atmospheric circulation (after Taljaard 1967)

Months	*Sea surface temperatures*	*Season*
December–March	Warm	Summer
April–May	Intermediate	Autumn
June–September	Cold	Winter
October–November	Intermediate	Spring

divisible into two groups, constituting a complete 'summer' (January–March and December 1967) and an 'intermediate' ('spring') season (October–November 1967). Despite the subjectivity implicit in such an analysis, a number of significant points emerged:

1 Low cloudiness persisted at the latitudes of the oceanic highs, extending further south on the eastern sides of the oceans than on their western sides. Travelling cells, detached from the more permanent subtropical anticyclones are probably involved in the development of such a pattern.

These cells are recurrent features of the climate of southern Australia in particular (see Gentilli 1971).

2 Cloud amounts increase very rapidly south of 40° S in both seasons, building up a continuous ring of strong cloudiness around Antarctica.

3 A considerable reduction of the Antarctic pack ice characterized the southern spring and summer. In October 1967 the edges of the frozen seas seemed to be well north of 60° S in the central south Atlantic, though often obscured by cloud. In January the outline of the continent was much more clearly discernible following an apparent mean retreat of the ice by some 10° of latitude. The circumglobal belt of strong cloudiness, however, changed little from spring to summer in mean breadth, strength or position.

More detailed analyses of the two conspicuous meridional cloud bands detailed above showed them to be associated with high frequencies of occurrence of cyclonic vortices, and/or with frontal or quasi-frontal zones. Except, perhaps, for the tropical end of the Pacific band, which may be of a more stable nature due to geographical reasons (see p. 228), the strong cloudiness eastward of each southern continent seemed to be accompanied by relatively frequent extratropical cyclogenesis, and preferred paths of cyclones and their accompanying frontal systems. These generally moved away towards the south-east.

In a complementary study Streten (1968b) tracked numerous southern hemisphere vortices around the Southern Ocean, and established the key regions of extratropical formation and decay.

Fig. 12.2 shows that four main regions of vortex formation appeared, two in the broad south Pacific and one each in the south Atlantic and south Indian oceans. These cyclogenetic regions seem to be largely responsible for the many depressions that contribute towards the strong cloudiness around the South Ocean and the coasts of Antarctica.

Streten's evidence squares more with the view that most southern hemispheric depressions form in association with a 'polar front' generally located between 40° and 50° S (van Loon 1965) than with the rival hypothesis that both 'polar' and 'antarctic' fronts may be identified as zones of preferred cyclogenesis (Astapenko 1964). In analysing the mosaics, Streten found that in almost all cases where there were clearly identifiable cloud vortices close to the Antarctic coast, these could be retraced in the sequence of mosaics to origins at lower latitudes. Only occasionally were wave-like disturbances of local origin observed between 60° and the continent (such were described by Astapenko as 'annular vortices'). These more local systems stimulate characteristic sequences of pressure change at Antarctic coastal stations but lack significant organizations of weather such as those associated with mid-latitude cyclones (see chapter 13).

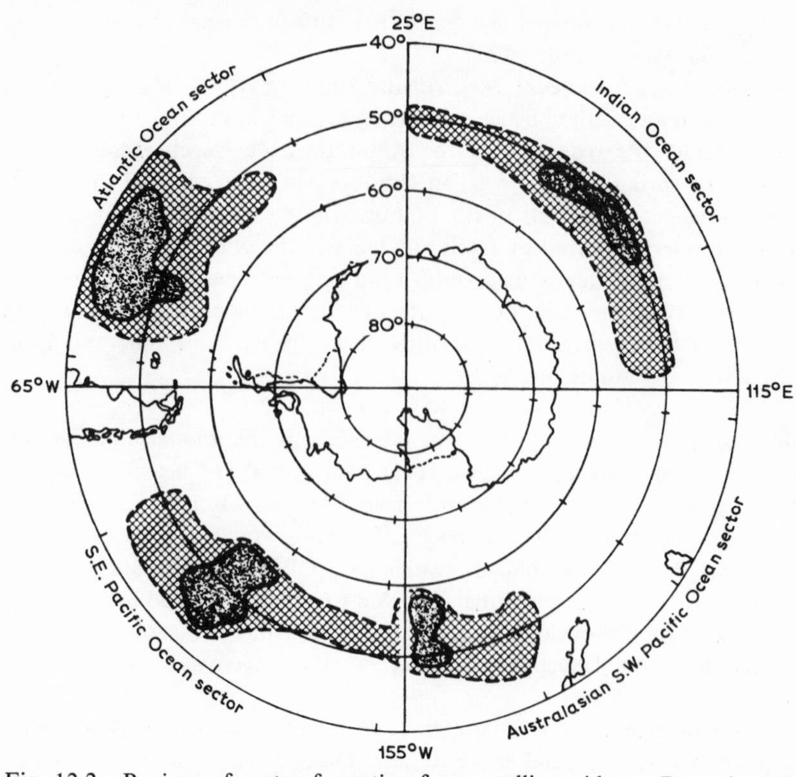

Fig. 12.2 Regions of vortex formation from satellite evidence, December 1966–February 1967, for each sector of the southern hemisphere. Areas within which one-third and two-thirds of the vortices formed during 80 days are shown respectively by stippling and cross-hatching. (From Streten 1968b)

One additional fact of considerable fascination and importance becomes obvious when fig. 12.2 is compared with Streten's map of vortex decay. Fig. 12.3 displays those regions in which vortices either became no longer readily detectable, or could be identified no longer as their cloud organizations moved over continental Antarctica. It is clear that decay is concentrated also in four areas, more or less evenly spaced around Antarctica, but much closer to its coasts than the polar frontal areas of vortex formation. Thus fig. 12.4 confirms a fact implicit in much of what has gone before—that depressions arising mostly at the southern ends of the meridional corridors of net energy, moisture and momentum export from the Tropics, generally approach Antarctica as they track around the Southern Ocean. Note also that most of the day-to-day positions of the centres of the vortices indicate a retardation south of 60° S compared with movement through 40° S and 50° S (cf. van Loon 1967).

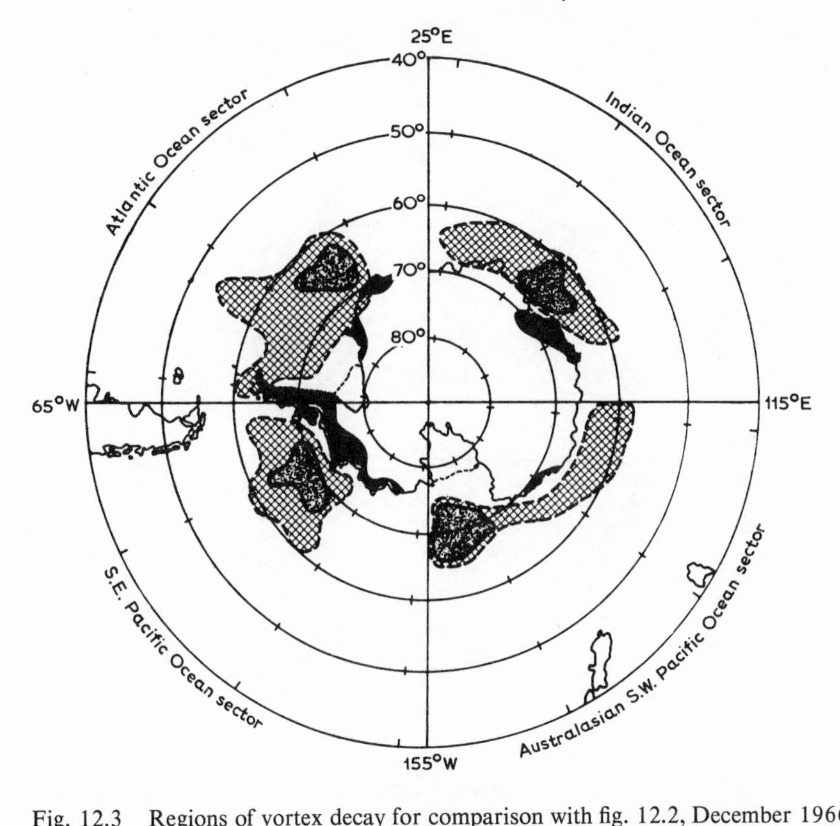

Fig. 12.3 Regions of vortex decay for comparison with fig. 12.2, December 1966–February 1967. Areas in which annual precipitation exceeds 300 mm of water equivalent are shown by dark cross-hatching. Areas within which one-third and two-thirds of the vortices decayed are shown respectively by stippling and cross-hatching. (From Streten 1968b)

Such behaviour seems curious in that anticyclonic conditions dominate the continent of Antarctica, maintaining a divergent drainage of cold air outward from the centre, though depressions are known to cross western Antarctica at least from time to time. We can only conclude that the convergence of depressions on the continental coast must be influenced by upper tropospheric flow patterns. Unfortunately the sparseness of upper air data surrounds the details of such relationships with obscurity.

Recent work by A. J. Troup and N. A. Streten (1972) has resulted in the first comprehensive classification of southern hemispheric vortices based on their appearance as seen by satellites. From this it has become clear that there are considerable differences between extratropical cyclones in northern and southern hemispheres, with respect to their structures and life cycles (see also Streten and Troup 1973). Troup and Streten have distinguished between:

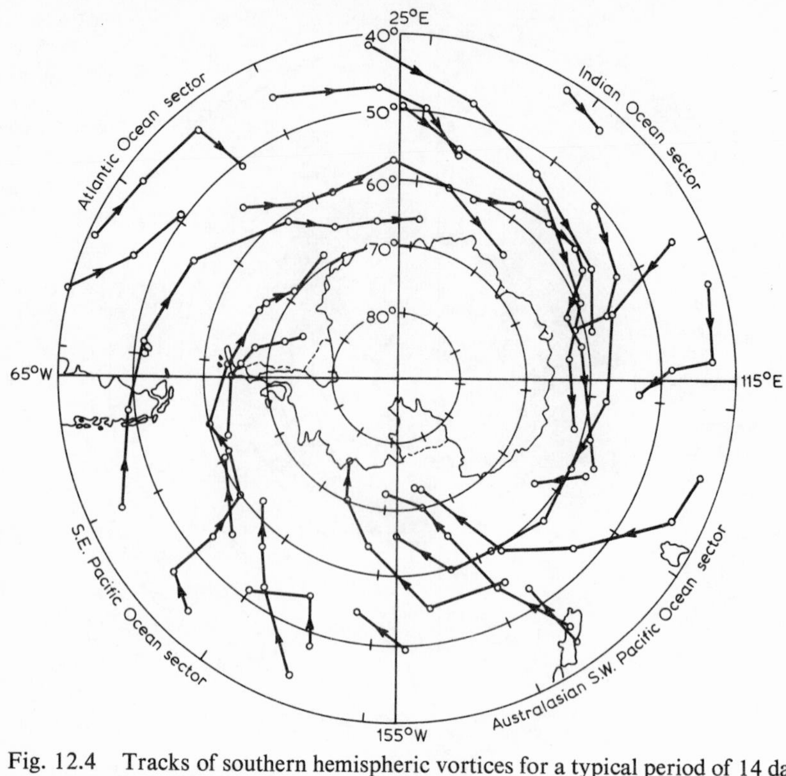

Fig. 12.4 Tracks of southern hemispheric vortices for a typical period of 14 days in January 1967. Circles indicate the daily positions of the vortex centres. (From Streten 1968b)

1 Vortex types associated with 'frontal' cloud bands to the east. They are the chief extratropical types.
2 Other vortex types of lesser frequencies of occurrence. These may or may not have associated frontal cloud bands.

The two groups may be subdivided as follows:

Type 1 Vortices (fig. 12.5)

Wave stage (W)	This is indicated either by a bulge on a major cloud band without a change in the angle of the band, or by a deformation of a major cloud band which changes angle at the deformation.
Formation stage (A)	This appears as an inverted comma-shaped cloud without a distinct slot of clear air.
Late formation stage (B)	Here hook-shaped cloudiness has an associated clear slot, but there is no evidence of clear air spiralling around a rotation centre.

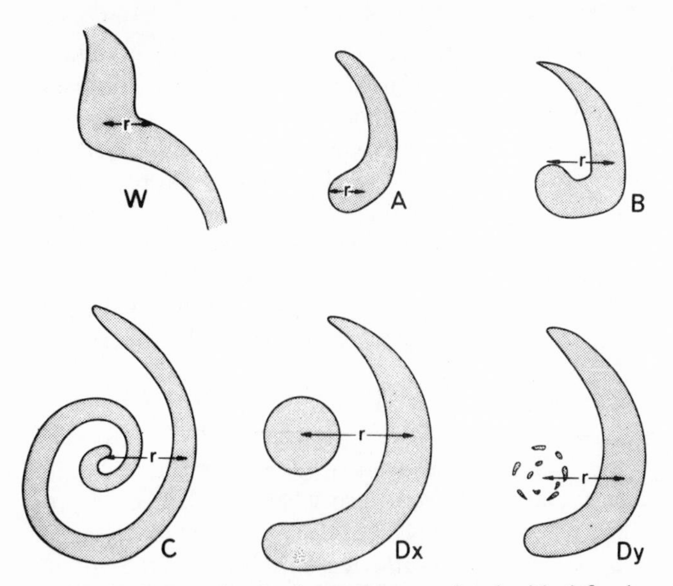

Fig. 12.5 Schematic cloud patterns associated with defined vortex types. Cloudy areas stippled. r indicates distance taken as radius from vortex centre. For non-frontal vortices, r is distance to edge of cloudimass. (After Streten & Troup 1972)

Mature stage (C) In the mature vortex clear air spirals around a definite centre of rotation.

Dissipating stage (Dx) Little or no clear air is evident in this dissipating vortex. There is a central cloud mass with banding or streaks, usually larger than A or B, and evolving from C.

Dissipating stage (Dy) This type of dissipating vortex has little cloud in its centre. Cloudiness is mainly organized in circular bands, or is spirally banded towards the west. Vertical structure is often apparent.

Type 2 Vortices

E This is a tropical cyclone in extratropical latitudes, with no 'frontal' type cloud band.

F This vortex is similar to *Dx*, excepting in its lack of an associated 'frontal' cloud band.

G This is similar to *Dy* excepting in its lack of an associated 'frontal' cloud band.

H A mesoscale vortex.

J This is a terrain-induced vortex type.

K In this case a tropical cyclone in extratropical latitudes possesses a 'frontal' cloud band.

Further research is likely to amplify the following points:

1 It seems that far more southern hemispheric vortices develop in isolation than from apparent waves on well-defined frontal cloud bands, the classic context in the northern hemisphere. Satellite photographs frequently portray new vortices with frontal bands where no frontal history has been apparent previously (see Gibbs 1949, 1960, for similar pre-satellite findings). Streten (1968b) did remark upon some classical cyclone developments from frontal waves, but noted that these were in a small minority during his period of interest.

2 As noted earlier by Langford (1957), there is little satellite evidence of cyclone families in the southern hemisphere forming from series of frontal waves.

3 Many Southern Ocean cyclones seem to be composed of eastward-moving vortices, each trailing a single, characteristically curved cloud band orientated usually from south-east to north-west. In classical northern hemisphere terms, most southern hemisphere mid-latitude depressions appear to be occluded. Possibly 'occluded frontogenesis', described by Anderson *et al.* (1969) in a northern hemispheric context, occurs more frequently south of the Equator. Certainly the mature stage of development illustrated by fig. 12.6. is very similar to the appearance of many southern hemisphere cloud arrangements. Anderson *et al.* referred to cases of apparently 'instant occlusion' over the north Pacific, in each of which an occluded-like front formed by frontogenesis (i.e. associated with an intensification of the horizontal temperature gradient in a restricted zone) rather than by a process of occlusion ('shutting off' warm air from the earth's surface by mass lifting). Often when a comma-shaped cloud, representing a maximum of positive vorticity advection aloft, approaches a frontal band, wave formation begins in the usual way. Sometimes, however, the merging of a comma-shaped cloud with a frontal band takes place in such a way that the system seems to 'jump' from the wave stage to mature occlusion without passing through the common intermediate stages. The resulting cloud pattern produced by the merging of a vorticity comma and a frontal wave gives the appearance of an occluded system, and is usually analysed in the forecasting procedure as such. In brief, satellite pictures indicate that such systems actually result from frontogenesis in a synoptic situation which involves the merging of a trough with a wave as an east–west front. The frequency, geographical distribution and characteristic weather of such a happening are currently unclear. It might, however, be particularly profitable to see whether some southern hemisphere depressions are of this kind.

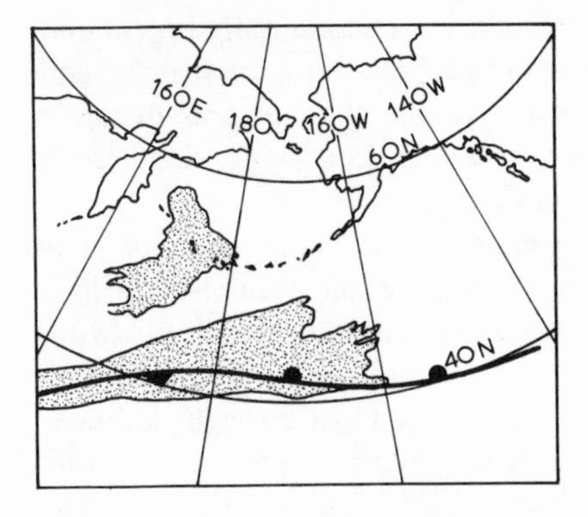

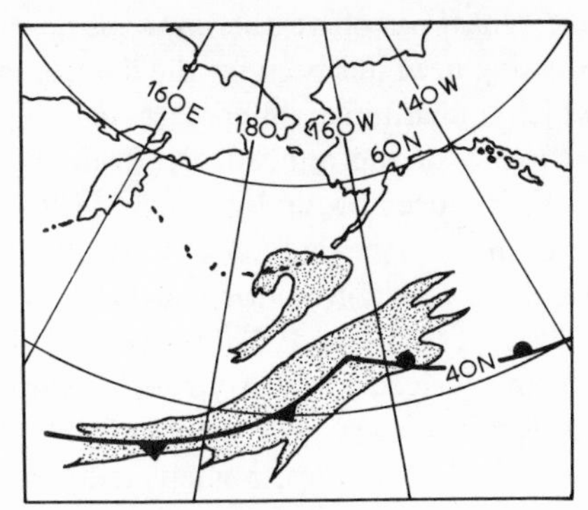

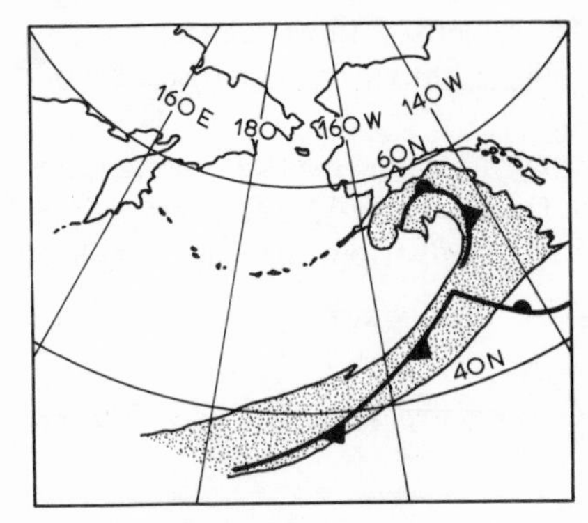

Fig. 12.6 Occluded frontogenesis. A model showing frontal positions with relation to the cloud band. The occluded front is placed near the trailing edge of the cloud band and the slow-moving portion of the cold front is located near the leading edge. (After Anderson *et al.* 1969)

Other satellite indications

Traditionally, the pre-eminent problem in studies of the meteorology of Southern Oceanic regions has been a dire deficiency of meteorological data of any kind. Zillman (1969) asserted that 'those unfamiliar with southern hemisphere meteorology may not fully appreciate the magnitude of the problem with which the Southern Ocean analyst has been faced. Over an almost unbroken expanse of some 20 million square miles [approx. 50 million km^2] of ocean poleward of the main shipping lanes vast weather systems can generate, evolve and decay undetected by conventional observation techniques, and, until recently, this observational deficiency had effectively precluded confident analysis in the high southern latitudes. The advent of the meteorological satellite has completely changed the nature of the problem.'

Similar sentiments must be true of much southern hemispheric climatology, since dynamic, synoptic and physical climatology depend so much upon routine meteorological observation and analysis. In view of the relatively little pre-existing climatological knowledge of southern hemispheric mid-latitudes it is hardly surprising that little progress has been made yet in assessing the extents of the contributions to local climatic statistics of the different types of weather systems described above. An even more urgent and pressing need has been for the development of interpretative expertise through which quantitative inferences may be made of the thermal and dynamic structure of the atmosphere from the new satellite data alone. Some such work has been begun by Martin (1968a, b, c).

Using a once-daily A.P.T. satellite coverage of the ocean south of New Zealand, and conventional data from the Macquarie and Campbell Islands (55° S, 158° E and 52° S, 168° E, respectively) for the winter of 1966, Martin derived average values of various surface and free atmospheric parameters for each of a number of cloud types. The most strongly latitude-dependent quantities, namely temperature and pressure, were processed as departures from their seasonal mean values. Fig. 12.7 shows the basis for Martin's data categorization. His scheme recognized three scales of atmospheric organization: (1) synoptic scale (500–2000 km), (2) mesoscale (50–500 km), (3) mesomicroscale (10–50 km).

Martin classified the most frequently occuring and synoptically important cloud patterns observed over the Southern Ocean. Table 12.2 summarizes the statistical inferences that can be drawn seemingly from the photographed

Table 12.2

Means of selected parameters for various cloud types (surface and 500 mb parameters) (from Martin 1968b)

	Surface			*500 mb*		
Cloud types	*MSL pressure anomaly (mb)*	*Surface temp. anomaly (°C)*	*Vector mean surface wind and standard deviation (knots)*	*500 mb Geopotential anomaly (dkm) and standard deviation*	*500 mb temp. anomaly (°C) and standard deviation*	*500 mb Vector mean wind and standard deviation (knots)*
Core of Spiral 111, 112	−17	+1	260/11 (8)	−17 (8)	−4 (2)	340/21 (12)
Within Frontal band-well organized 211, 215	−2	+2	290/17 (3)	+4 (8)	+2 (3)	300/56 (8)
– poorly organized 212, 213, 214	−6	+1	290/13 (3)	−5 (13)	−1 (3)	305/43 (5)
Pre frontal 221, 222	+3	+1	305/15 (3)	+5 (14)	+2 (3)	280/42 (6)
Post frontal 231, 232	−5	0	275/13 (3)	−8 (14)	−3 (3)	290/41 (7)
Convective 311 (open polygonal cells, mostly clear)	−8	−1	260/20 (4)	−21 (6)	−5 (4)	245/25 (9)
312 (open polygonal cells, mostly cloudy)	−4	−2	240/17 (4)	−9 (13)	0 (3)	235/54 (10)
313 (closed polygonal cells)	+16	−1	235/12 (4)	+7 (11)	−1 (2)	230/24 (16)
Layer clouds – solid un-banded, 411, 412	+11	0	255/12 (2)	+18 (7)	+2 (2)	245/23 (4)
– solid banded or broken, 413, 414, 422, 423	+5	+1	265/17 (2)	+11 (9)	+2 (3)	260/36 (7)

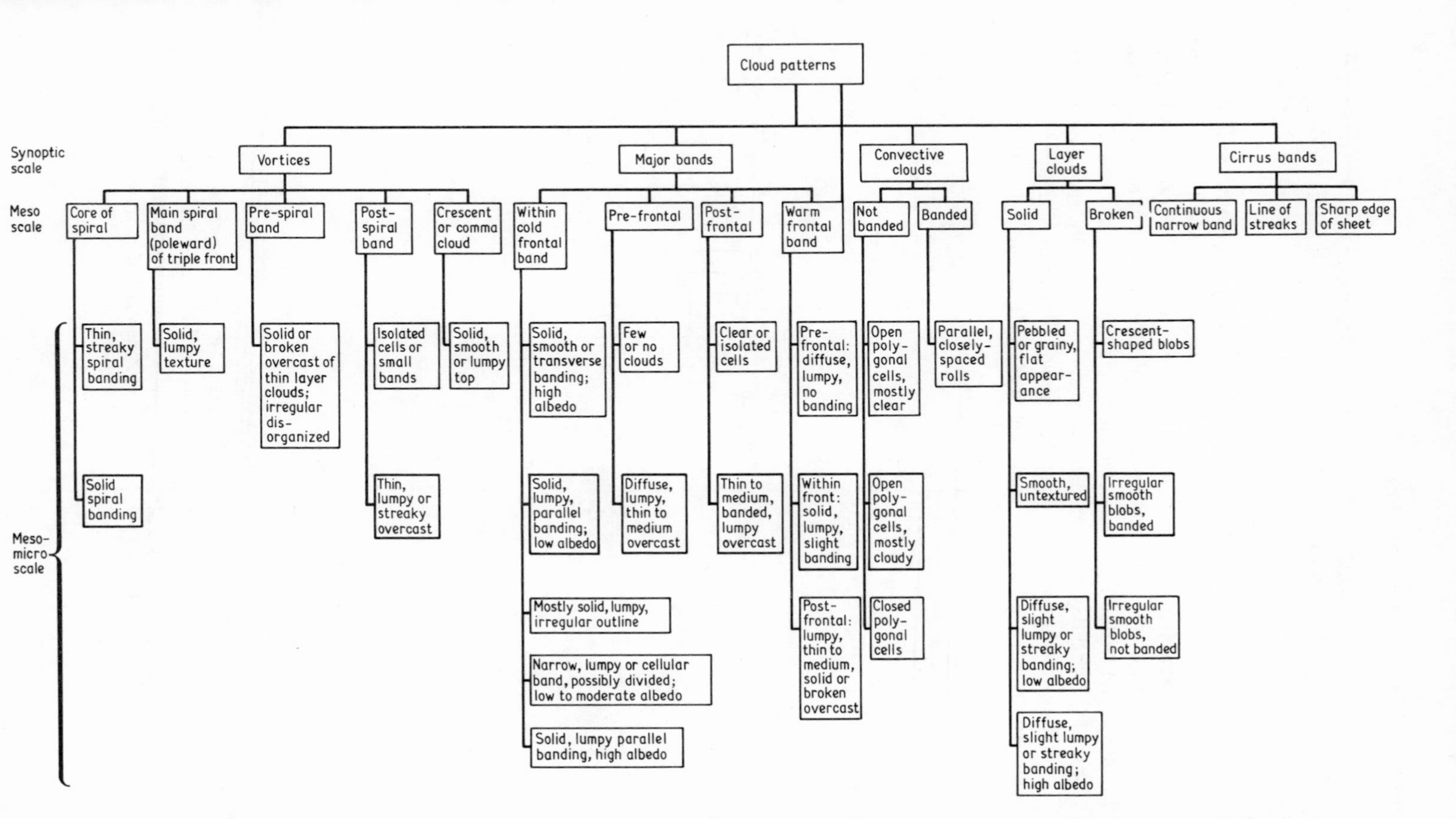

Fig. 12.7 A satellite cloud classification designed to isolate and describe the most frequently occurring and synoptically important patterns over the Southern Ocean according to geometry, texture, brightness and position relative to other patterns. (After Martin 1968b; Zillman 1969)

cloud arrangements he studied. Later work by Guymer (1969) extended the range of quantifiable meteorological parameters to include the 1000–500 mb thickness pattern. The reference numbers in table 12.2, column 1 relate to boxes in fig. 12.7.

It is through these and similar means that many more satellite-based advances in the climatological understanding of the mid-latitudes in the southern hemisphere may be made in the foreseeable future. Although representative temperature and rainfall graphs for stations in South America, Tasmania and within the oceanic girdle display little climatic seasonality (fig. 12.8), the day-to-day weather picture is one of unusually rapid variability. The mobile weather systems discussed above contribute significantly to the poleward transfer of energy from the low latitude energy sources to the

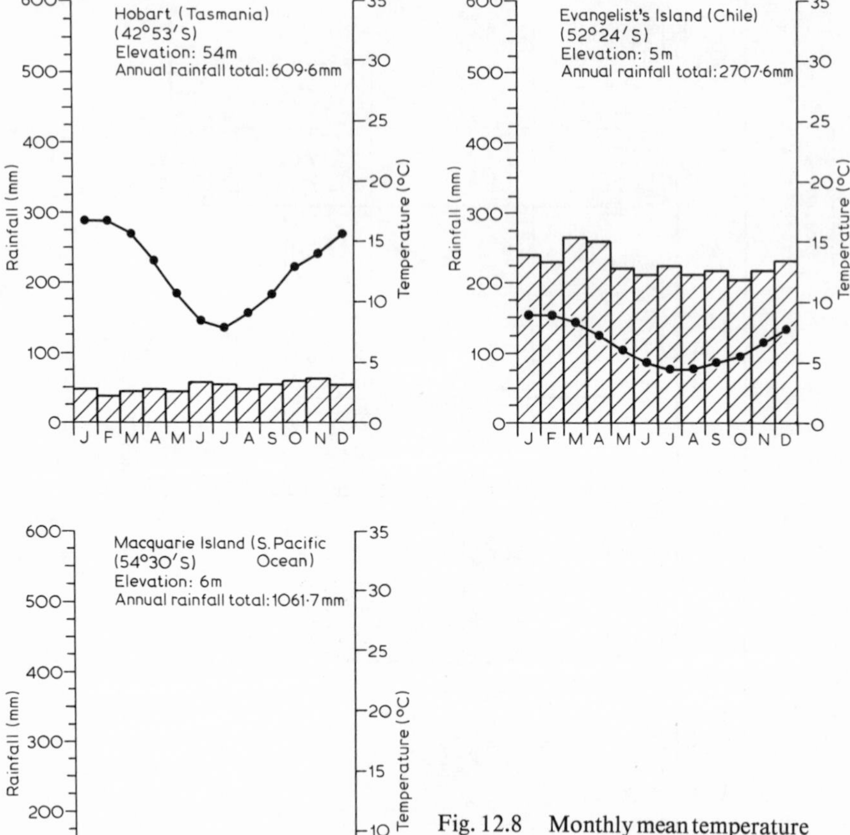

Fig. 12.8 Monthly mean temperature and rainfall graphs for representative stations in mid-latitudes in the southern hemisphere. (Statistics from Rumney 1968)

ultimate sinks around and across Antarctica. Unfortunately little is known as yet of the year-to-year differences in the frequencies, paths and intensities of such synoptic scale circulations. Even less is known of their mean variations through longer periods of time.

Past attempts to view this circumglobal zone as a whole have been based, of necessity, upon short-term concerted observational efforts, preeminent among which was that involved in the International Geophysical Year (I.G.Y.) from 1957 to 1958. The establishment of the fully operational American meteorological satellite system in February 1966 has opened the door more widely to both larger-scale and longer-term investigations.

Mid-latitudes in the northern hemisphere

Extratropical cyclones

Although the polar front in the northern hemisphere is more strongly defined than its southern hemisphere counterpart, its most definite expression occurs in two separated fragments stretching obliquely south-west to north-east off the eastern coasts of North America and Eurasia. Frontal manifestations across the major land masses themselves are often much more difficult to discern. Many distinctive arrangements of clouds have been associated by different workers with specific types of northern hemisphere cyclones and their subordinate features. For a fuller description the reader is referred to Anderson *et al.* (1966, 1969).

As in the southern hemispheric case, it is unfortunate from the point of view of the regional climatologist that most satellite-based work has been essentially meteorological, not climatological, in nature. Most published satellite studies have been concerned with the structures and developments of individual weather systems, of which many have been singled out for study not on account of their normality but rather on account of some unusual feature. Here we must concern ourselves mostly with the general conclusions that can be drawn therefrom.

The most striking and obvious cloud structures portrayed by satellite photographs in middle latitudes of the northern hemisphere are, like those in the southern hemisphere, associated with extratropical cyclones. Unlike their southern counterparts, however, the frontal cloud bands curving away from the centres of northern hemisphere vortices have been so carefully studied that several distinctive types can be differentiated from one another on the grounds of their detailed cloud appearances. Let us review the principal characteristics of the most readily recognizable types, in order of the ease with which they may be identified.

Cold fronts

Active cold fronts appear as continuous, well-developed cloud bands. These are associated with strong baroclinicity, considerable thermal advection, and powerful vertical wind shears. Whereas lower-level winds are usually at high angles to the plane of a cold front, the upper-level winds are parallel or nearly parallel to it in its active stages; this arrangement, coupled with the strong baroclinicity, helps promote the well developed band of cloud. Lower-level stratiform and cumuliform clouds are seen to be overlain in part by upper-level cirrus.

Inactive cold fronts, meanwhile, are often visible as narrow, fragmented, discontinuous cloud bands. Such bands indicate weak baroclinicity and cold air advection, and slight vertical wind shear. Generally both upper- and lower-level winds tend to be perpendicular, or nearly perpendicular, to the planes of the fronts. The constituent clouds are mainly low level cumuliform and stratiform, though some cirriform may be present. The light winds and weak vertical shear are the chief factors leading to the fragmentation of the bands of cloud. Inactive cold fronts over land are sometimes difficult to identify from satellite photographic evidence, since they may have few or even no associated clouds.

When cold frontal cloud bands are particularly weak and difficult to orientate, it may be helpful to examine the character of the cloud more broadly in the region of the suspected front. Associated with most cold fronts, especially over water, cellular cumuliform clouds characterize the cold sector air behind the fronts, while ahead of them there is often a mixture of cumuliform and stratiform cloudiness.

Occluded fronts

These spiral away from the vortex centres of mature and decaying depressions, often as powerful cloud bands with mottled, irregular or 'lumpy' appearances. Equatorwards from the vortex centre of a storm, the point of occlusion and the peak of the warm sector can be located usually by a change in cloud texture from mottled to comparatively smooth. The smoothness is due to copious cirrus in this area of maximum upglide of moist warm sector air. Polewards, the occluded frontal clouds curve around the northern edges of their associated centres of vorticity, often merging with central discs of variegated clouds. Present practice in meteorology ends the 'occluded front' in the north or north-west quadrant of a storm; it does not carry the front completely round the vortex as it might.

Warm fronts

Active warm fronts are, at best, hard to locate on satellite cloud pictures, while inactive warm fronts – evidenced by conventional observations, cannot be located at all. Well-organized cloud bands may indicate the strong baroclinicity of a warm front, but even such cloud bands are usually short and merge quickly with quasi-banded clouds ahead of the approaching cyclone. Commonly warm frontal clouds are combinations of stratiform and cumuliform beneath a well-developed cirrus canopy. Unlike cold frontal cloud bands, which may have well-defined leading edges, the leading edges of warm fronts are often very diffuse, indicative of the thin upper-level streamers that can sometimes be seen from the ground as a young to mature depression approaches. Fig. 12.8 is a model of surface warm, cold and occluded fronts in their relations to satellite-viewed cloud bands.

Pre-frontal squall lines

Sometimes thin lines of cumuliform clouds and/or cumulonimbus appear ahead of, and roughly parallel to, a cold frontal cloud band. Alternatively, groups and clusters of convective clouds may adopt a scalloped appearance in plan on a satellite photograph. These have been named 'pre-frontal squall (or instability) lines'. A counterpart, recognized by some, is the 'post-frontal instability line' which may be observed within trailing cold sector air behind a cold front and separated from it by a zone of clearer skies.

Non-frontal cloud bands

Some of the long cloud bands festooning satellite photographs are not associated with fronts, despite giving the appearance of frontal bands. In many cases these cloud bands are prompted by surface confluence, especially where warm, moist air is advected poleward. Such bands may appear generally broken, or more continuous, often depending upon the presence or absence of upper-level cirriform cloud tops. North–south oriented bands often occur in the areas of convergent flow to the rear of high pressure cells as lows approach from the west.

Problems in the recognition of frontal types

Especially in the northern hemisphere the recognition of even these basic frontal types is complicated by their different behaviours over land and sea. Not surprisingly, frontal cloud bands usually change character as they move inland at the end of an over-water trajectory. Even a well-developed, continuous band may disintegrate into a fragmented collection of clouds as its front moves over land. Similarly, frontal cloud bands moving offshore may metamorphose as they begin to travel over water, becoming broader and/or more continuous. These changes are due to changes in the amounts of moisture available to the mobile atmospheric systems.

Where fronts lie partly over land and partly over water, the former sections may seem so poorly developed by comparison that inaccurate satellite photographic analyses may result. For the climatologist contemplating studies of frontal movements across selected regions through time such complications must be borne in mind. Perhaps these problems have deterred workers from undertaking studies of vortex tracks in the northern hemisphere comparable to those outlined earlier for the southern hemisphere. On the other hand, of course, conventional observing networks are better in the northern hemisphere and less of climatological significance can be added from the satellite coverage, especially in those regions where most people live.

The satellite evidence, however, has been significant in two major respects. First, it has led to an improved knowledge of the 'classical' extratropical cyclone in the northern hemisphere, and, second, it has helped our understanding of the contributions made by such systems to the climates of affected regions.

The life-cycle of a northern hemisphere depression

A common sequence of pattern changes through the lifetime of an oceanic polar frontal low is depicted by fig. 12.9 (Barrett 1970). Not only are surface isobars and winds indicated (sold lines) but also 500 mb contours and winds (broken lines). Jet streams are indicated where appropriate, and the patterns of satellite-viewed cloud bands. Less organized cloudiness (e.g. scattered cumuliform and cumulonimbus clouds in the trailing cold sectors of stages (e) and (d)) have been omitted for the sake of simplicity. So too have upper-level clouds.

It is obvious that a general and progressive development takes place from the first stage (a) to the last stage (f), involving a change from a simple, initially radial, arrangement of the cloud bands in fig. 12.9 (a) to a more complex, finally concentric, pattern in fig. 12.9 (f). Let us examine this development in more detail from stage to stage.

Figs. 12.9 (a) and (b) represent frontal wave stages in vortex development along an initially baroclinic frontal zone. The cloud wave evident in fig. 12.9 (b) arises with the approach of an upper-level vorticity maximum associated with a trough in the 500 mb flow pattern. The upper-level changes promote an isallobaric low (i.e. a centre of reduced pressure) at the earth's surface. Where this coincides with the frontal cloud band the front broadens and forms a convex bulge towards the colder air. Pressure begins to fall at the surface, and a closed circulation is initiated.

As the wave develops, but before the onset of occlusion, the curvature of the bulge becomes more pronounced. A 'dry slot' appears between the centre of the vortex and the cold frontal band, and rapid deepening of the circulation follows (fig. 12.9 (c)). The warm frontal cloudiness is distinguished by longitudinal stripes of various degrees of brightness, while, behind the front, bands of cloud form at right angles to its long axis in the upglide zone where warm moist air is forced to rise over the cold. The cold front appears characteristically mottled on satellite photographs due to shadows cast from the tall cumulonimbus cloud tops that form along this zone of extreme instability.

Fig. 12.9 (d) illustrates a mature stage following the initiation of occlusion.

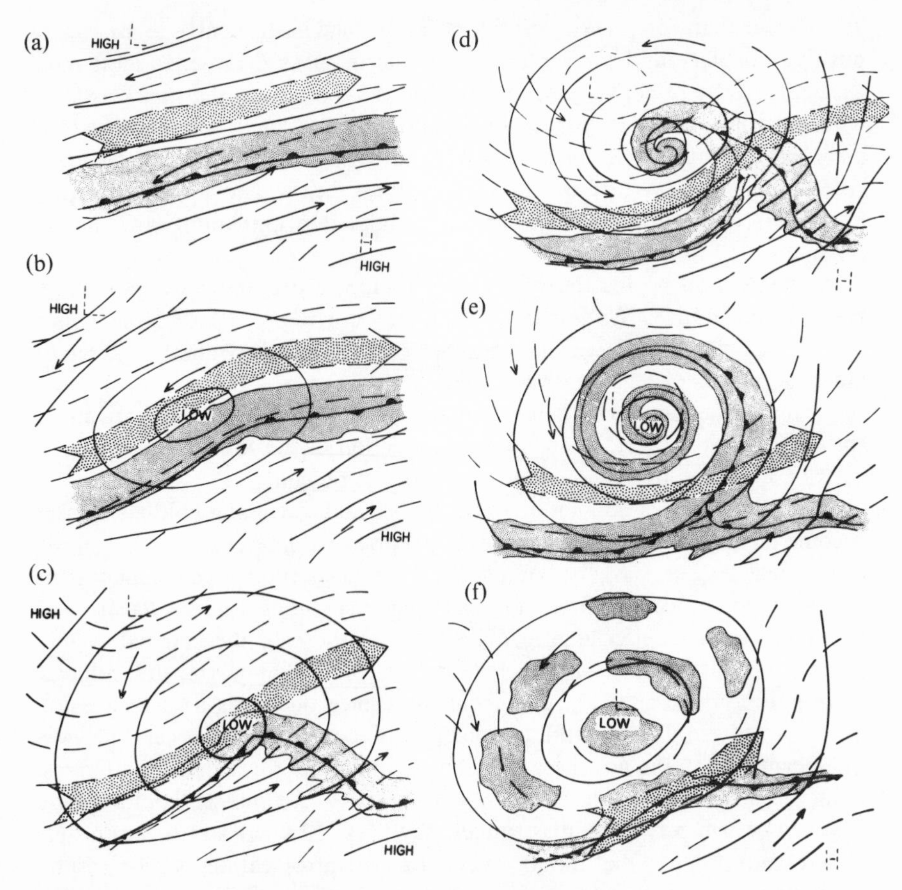

Fig. 12.9 A life-cycle model of a mobile, frontal, northern hemispheric, extra-tropical cyclone. Finely-stippled areas are major bands of sheet clouds. Broken lines and letters relate to 500 mb contour patterns and winds, and continuous lines to sea-level isobars and winds. Coarsely-stippled bands indicate jetstream cores. (From Barrett 1970)

The dry slot curls around the southern side of the cloud disc of the vorticity centre, behind the single band of the occluded front. Aloft the 500 mb low has advanced relative to the 1000 mb low pressure centre, due to changes in the storm's temperature field. The storm now approaches both its peak of organization and its maximum intensity. As the dry slot penetrates to the very centre of the vortex, the supply of moisture available to the cyclone is much reduced, and less latent heat of condensation is available for transfer to kinetic energy. More or less simultaneously the mid-tropospheric pressure centre arrives over the surface vortex centre, and convergence through considerable depth results in a rise of pressure at the surface beneath. It should be self-evident that, for pressure to fall at the surface in the centre of an area of positive vorticity and convergence, the rate of mass divergence aloft must exceed that rate of convergence below. Fig. 12.9 indicates that the jet stream embedded in the mid-tropospheric flow around the trough wave at that level plays an important part in transporting air away from the area of surface vortex formation. When the pressure centres below and aloft are superimposed (see fig. 12.9 (e)) the jet stream stimulated by maximum baroclinicity is usually well to the south.

As dissipation begins the vortex bands adopt a progressively more fragmented appearance. Well-defined cloud-free corridors separate the coiled bands of cloud. Convective cloud patches characterize the trailing cold sector areas as cold, moist air is drawn south over warmer surfaces and becomes subjected to heating from below. Finally, in fig. 12.9 (f), poorly organized spiral cloud patterns accompany the dissipation of the low pressure centre, where no vestiges of frontal cloud organization remain.

Thus the planes of contact between the original warm and cold air masses become manifestly more extensive as the depression life-cycle proceeds, until, in the final stage, a largely homogenized air mass rotates gently about the declining vortex centre. This air mass possesses temperature, density and humidity characteristics intermediate between those of the original, highly contrasting, air masses. In effect, therefore, the initial warm and cold air streams become stirred together from the centre outwards as a depression matures. Within the general circulation, of course, this process plays a significant role in the net poleward transfer of energy towards high latitudes. If the cloud areas in fig. 12.9 are considered to be mostly contained by relatively warm air, a comparison of, say, figs. 12.9 (a) and (b) with figs. 12.9 (e) and (f) hints that the life-cycle of an extratropical depression results effectively in a transfer of energy across the polar front. Bearing in mind also the earlier established fact that, in the northern as in the southern hemisphere, depressions generally progress into higher latitudes as they mature, it is concluded that the meridional energy transport effected by such systems is very considerable.

Although a range of other types of vortices has been established from satellite evidence in mid-latitudes of the northern hemisphere little need be said of them in the present account. Few genuinely climatological studies have yet been undertaken in each case. For further details of various lower tropospheric depression types, including both primary and secondary frontal cyclones, non-frontal cyclones and upper tropospheric vortices, the reader may consult the accounts by Boucher and Newcomb (1962), Boucher (1963), Widger (1964) and Anderson *et al.* (1966, 1969) prepared for use in weather analysis and forecasting.

The contributions of northern hemisphere depressions to regional climates

Generally speaking, the chief frontal cloud bands along the eastern seaboards of North America and Asia are most prominent in winter, when they may be several thousand kilometres in length and exceed 500 km in breadth. It is within these two zones of confrontation of extratropical and polar air streams that 'classical' frontal depressions are spawned most frequently. These appear either singly, or in families. As they mature they generally travel north-eastward across the north Atlantic and north Pacific oceans towards the north-western coasts of Eurasia and North America. These mobile storms (although typically less intense than their Southern Ocean counterparts) are often quite vigorous and prompt rather more complex sequences of weather than the southern hemispheric lows, on account of their characteristically more complex frontal organizations. Separated from one another by weak ridges of high pressure, they play a large part in the daily changeability of weather over the northern oceans and the land areas most open to their influences. The most notable of these is north-west Europe, which, unlike North America, is given little shelter from westerly weather by mountain ranges lying across the dominantly zonal flow of wind.

Climatologically, of course, details of the day-to-day variations in temperature, rainfall, humidity and wind speed and direction that are prompted by these mobile weather sequences are largely lost by averaging over monthly, seasonal and annual periods. Hence, the chief areas influenced by mobile lows are evidenced better by maps like fig. 11.10, based on daily standard deviations of cloudiness.

Seasonally, the land areas influenced most by the predominantly maritime air streams involved in mature extratropical cyclone circulations are characterized by distinctly cool, moist summers and mild, wet winters compared with less exposed regions. Thus there is a moderation of the more extreme high season contrasts referred to broadly as 'continentality'. Numerous indices have been suggested to express this notion quantitatively. For example, Gorczynski (1920) based his index (K) on the annual range of temperature:

$$K = 1{\cdot}7\frac{A}{\sin \phi} - 20{\cdot}4 \qquad (12.2)$$

More recently, Berg (1953) has related the frequency of continental air masses (C) to that of all air masses (N) in a comparable index, which, however, depends more on daily weather differences:

$$K = \frac{C}{N} \qquad (12.3)$$

Across Europe, both authorities established trend lines orientated generally from south-west to north-east associated with rising values inland.

It is evident from synoptic weather charts, daily weather records for selected west coast stations, and satellite data displays that the main tracks of northern hemisphere depressions, and their intensities, differ more from summer to winter than in the southern hemisphere. The chief contrasts are as follows:

(1) In winter the northern hemisphere depressions are, on average, fewer and deeper. In this way they reflect the relative simplicity of the upper westerly circulation described in chapter 8. Their paths across the north Atlantic and north Pacific are about 10° further south than in summer. Few penetrate Europe on account of frequent 'blocking anticyclones' (see, for example, Namias 1964). These divert some depressions north-eastward off the Norwegian coast, while other lows travel further south towards the Mediterranean. Not uncommonly depressions slow down, stagnate and fill in the British region bringing spells of rather gloomy, overcast and relatively quiet weather. Some of the north Pacific systems retain much of their organizations across the Rockies and are then partially rejuvenated by the warmer surfaces of the Great Lakes. Later still these storms may become involved in the active cyclogenetic processes off the east coast of North America.

(2) In summer the cyclogenetic areas and the main transoceanic tracks are further north, and the low pressure systems are generally relatively small and shallow. With the replacement of winter highs by thermal lows over the continents, however, these barriers to their movement across the land masses are removed. Thus the summer depressions often move faster and penetrate further inland than those in winter. In coastal regions like western Europe for example, classical frontal sequences of weather therefore tend to be more frequent and better defined in summer.

Lastly we may note one other broad significance of extratropical depressions. It was remarked earlier that such circulation systems transfer large quantities of energy towards the poles. Their momentum transfers from low latitudes to high latitudes are also important to the general circulation of the atmosphere. Fundamentally, the earth needs to conserve angular momentum. As it rotates upon its polar axis, so it imparts some westerly angular momentum to winds blowing in the opposite direction, i.e. from the east. Simultaneously surface westerlies steadily lose angular momentum due to friction with the earth. Strongly dominant westerly winds in middle latitudes are maintained by angular momentum derived from the easterly winds of the Tropics and subtropics. As Sutton (1965) observed, the exchange is effected by many individual circulation cells in middle latitudes. These cells, including polar frontal lows, transport momentum exactly as do the eddies in a turbulent fluid. In this way, such systems have played a vital part within the operation of the atmosphere's 'heat engine'. Put in Sutton's particularly felicitous way, the 'furnace' (located mainly in the Tropics) drives the 'flywheels' of the trade wind belts; the westerlies represent the gear-boxes where the direction is reversed, while individual cyclones complicating the daily weather patterns of mid-latitudes may be conceived as 'cogwheels'. Finally, at the poles we find two minor 'flywheels' of easterly motion.

Mid-latitude anticyclones

Generally less attention has been given both to anticyclonic structure evidenced by satellite data, and to the roles played by anticyclonic cells in mid-latitude climatologies, except in the cases of the large polar continental highs that sometimes occur in winter. However, useful rules for classifying anticyclonic axes or 'ridge lines' from cloud photographs have been formulated for forecasting use. More comprehensive classifications, treating anticyclonic cloud patterns and organizations as total entities, have yet to be attempted. Anderson *et al.* (1969) recognized four types of northern hemisphere ridges or 'ridge lines' from satellite cloud evidence, and suggested that similar rules for determining surface ridge lines also apply to southern hemisphere systems:

1 Type A ridge line. On the foreward edge of the frontal band (fig. 12.10 (a) cloud fingers are often tied in a nearly continuous fashion to the frontal cloud itself. These fingers often extend in an Equatorward direction until their points reach to the axis of the surface ridge itself.
2 Type B ridge line. On the western sides of subtropical high pressure cells changes in the characters of the clouds from cumuliform to stratiform often occur where the low level winds change direction from south-easterly or easterly to south-westerly or southerly. Points on the surface ridge line are positioned where the greatest change occurs in the cloud character. The change is caused by the difference in heating and stability; this air is unstable with heating from below in the south-easterly or easterly flow, and more stable with cooling or less heating from below in the south-westerly or southerly flow (fig. 12.10 (b)).
3 Type C ridge line. Not uncommonly in middle latitudes two extra-tropical cyclones come in close proximity to each other, separated only by a sharp polar ridge. On the northern side of the passing polar high there is a wind shift from north-westerly or northerly to south-westerly or southerly. The surface ridge is positioned along a line where the low-level cumuliform clouds first develop in the low-level cold air having a northerly component over cold water (fig. 12.10 (c)). This line is usually coincident with the forward edge of the overcast from the western cyclonic low in the belt of strong south-westerlies.
4 Type C (variation) ridge line. In some cases, the cumuliform clouds will not develop where the wind swings to north or north-west because the contrast in the air and sea temperatures is small. Thus, especially in the warmer months, a clear corridor in advance of a frontal band alone indicates the location of the surface ridge line.

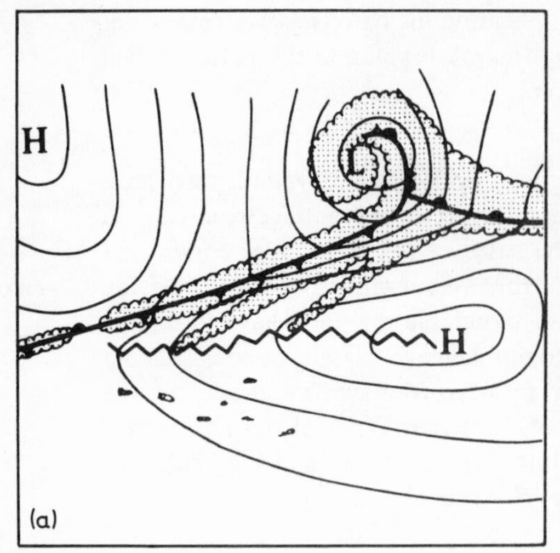

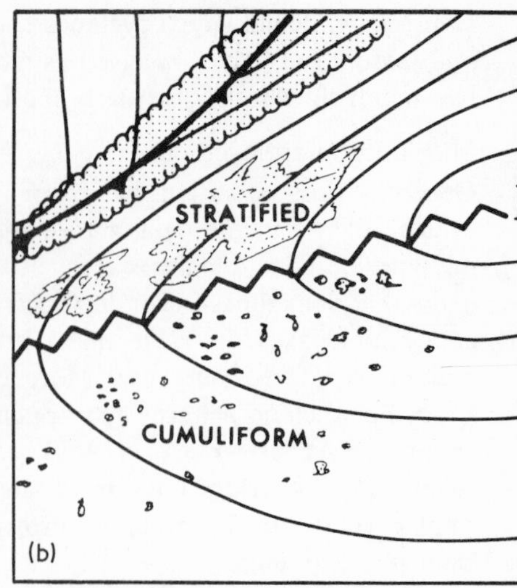

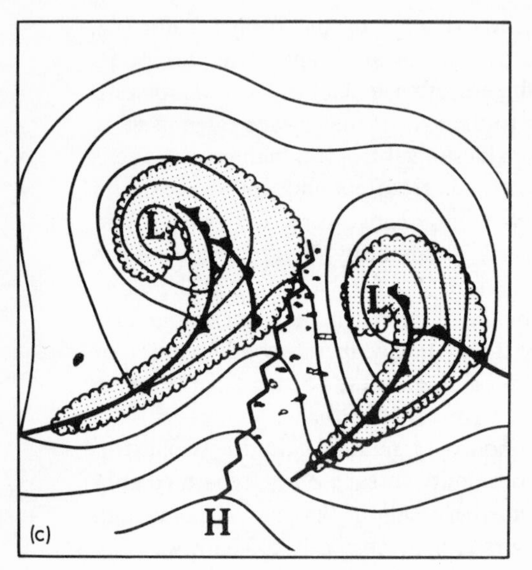

Fig. 12.10 (a) Model of a Type A ridgeline. The surface high extends west from the centre of the subtropical high, being located at the end of the cloud fingers extending from the frontal band.

Fig. 12.10 (b) Model of a Type B ridgeline. The surface ridgeline is located where clouds change from a cumuliform character to a more stratified nature in the western side of a subtropical high.

Fig. 12.10 (c) Model of a Type C ridgeline. This is located at the surface where cumulus clouds first develop, and where the solid overcast ends in the northern portion of a polar high wedged between two cyclones. (From Anderson *et al.* 1969)

Smith (1968) attempted correlations between sun glint areas over sea surfaces and the location and orientation of surface ridge lines. The most significant conclusion he reached was that, when the sun and satellite are in the correct relationship to each other, sun glint can be a useful indicator of the position of a high pressure centre. When little cloud occurs, the very light winds associated with anticyclonic subsidence and divergence cause smooth ocean surfaces with highly reflective surfaces resulting in a bright sun glint.

Over land, of course, topographic inequalities stimulate innumerable cloud pattern variations, many of which have a certain intrinsic meteorological interest but relatively little climatological importance. Many examples of topographically induced cloud patterns or pattern variations have been included in the 'Picture of the Month' series of the *Monthly Weather Review*. These have included a particularly rich variety of anticyclonic patterns, since the stronger, more continuous cloud areas associated with cyclonic circulations are less prone to visible influence by mountain ranges, broad deep valleys, dendritic drainage patterns and the like. Subparallel cloud bands comprising lee wave clouds have been described by Fritz (1965); standing cloud waves prompted by high mountain ranges were described by Conover (1964); some topographically induced cloud lines and radiation fog patterns photographed by the Russian satellite Cosmos 122 were discussed by Vetlov (1966); dense fog in the Great Californian Valley has been explained by Parmenter (1967). It could be that particularly persistent local occurrences of some such features have been invoked insufficiently in earlier mesoscale climatological descriptions and discussions, but little progress has been made in investigating such a possibility. In all, we must conclude that the relatively adequate network of conventional recording stations in middle latitudes has lessened the excitement of the satellite adventure into serendipity across the continents.

Mid-latitude jet streams

Turning next to jet stream questions, it is evident that these are being pursued with equal interest over land and sea. Most work to date has been concerned with the identification of jet-induced cloud features, and the verification of meteorological analyses of jet stream forms and strengths by conventional evidence. Particularly useful beginnings have been made by Whitney *et al.* (1966), and Viezee *et al.* (1966) who concluded that jet analyses can benefit from satellite pictures even in data-rich areas. The most definitive cloud characteristics include an extensive cirrus shield having a sharply defined poleward edge (often outlined by a shadow cast on lower cloud surfaces or on the earth), and/or transverse banding in the shield itself. In most cases the jet axes are located along the poleward cloud edges. Numerous case studies of polar frontal jets over the North American continent have suggested that cirrus streaks, first recognized as useful jet stream indicators by Conover (1962, 1963), are often undependable detectors by themselves. The greatest danger of all seems to be in confusing frontal with jet stream cloudiness.

Something of the influence of mid-latitude jets on surface weather and climate can be gathered from the theoretical model proposed by Riehl *et al.*

(1952) (see fig. 12.11). Here, horizontal divergence is required on the tropical flank of the jet in the entrance area and on the polar side in the exit (cf. the arrangement required in connection with the equatorial easterly jet, p. 294). Conversely, convergence is required in the remaining areas of wind speed maxima, although curvature can alter such a divergence pattern. Divergence

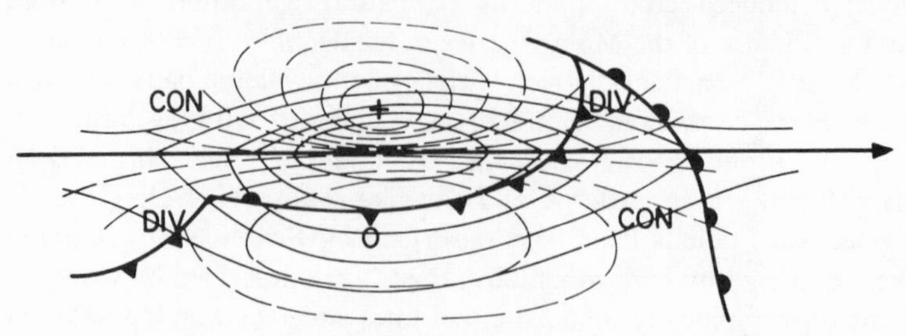

Fig. 12.11 Model of surface fronts, absolute vorticity distributions (broken lines) and divergence at the 300-mb level in the vicinity of a jet maximum. Isotachs are shown as solid lines. Note the convergence in the left-hand entry zone and the right-hand exit zone, and divergence in the right-hand entry zone and the left-hand exit zone. (From Reiter 1969)

at the jet stream level favours convergence and ascent in the layers beneath, for one of two possible reasons:

(*a*) The immediate proximity of the tropopause above the jet stream serves as a boundary or lid so that the divergent areas around the jet stream generate upward motion immediately below the jet.
(*b*) The horizontal divergence is strongest at the jet level, so that the decrease of divergence downward in the layers beneath permits upward motion.

There is general agreement among satellite workers that extensive cloudiness characterizes the tropical side of the jet stream (e.g. Oliver *et al.* 1964) and that cloud on the poleward side, if present, tends to be at a lower altitude. This is consistent with the concept of a tropopause 'break' (Reiter 1969) along the jet maximum separating a lower polar tropopause from a higher mid-latitude and/or tropical tropopause.

Whitney and his co-workers (1966) concluded that satellite evidence supports Riehl's model further through the propensity for the Equatorward cloud shield to favour the entrance area of the jet maximum. On the other hand, however, substantial cloudiness on the polar side in the exit area is often lacking. In this case an explanation may be postulated in terms of inadequate moisture to reveal the upward motion theoretically present there. These upper tropospheric systems are significant even to climatologists interested in weather patterns and sequences at the earth surface itself on account of jet stream influences on cloudiness, rainfall and temperature.

Meridional teleconnections

No review of middle latitudes would be complete without a mention of those interactions that are thought to relate certain subtropical and mid-latitudinal features of weather and climate. Two such interactions may be suggested by way of examples, the first involving the intertropical cloud band, the second tropical cyclones:

1 Periodically, portions of the main upper level westerly current of mid-latitudes shift from a dominantly zonal to a meridional flow regime. When this occurs large amplitude troughs develop, often extending to within 10° or 20° of the Equator. Satellite pictures show that amplification of the upper level troughs in mid-latitudes is accompanied by an extensive northward intrusion of clouds and moisture from the Tropics. Often these cloud surges emanate from within or near the intertropical cloud band. They are important weather producers in the subtropics and further north along, for example, the western coasts of the U.S.A. (Oliver, personal communication, 1972).

2 On other occasions, the tropical sources of broad and extensive cloud bands reaching into the zone of mid-latitude westerlies have been seen to be non-linear rather than linear, comprising tropical disturbances, especially hurricanes (Erickson and Winston 1972). Satellite photographs have revealed quite numerous cases of such cloud connections from western north Pacific tropical storms in autumn, the season of greatest hurricane activity. It has been established that these major cloud bands tend to be associated with upper tropospheric warmings in and immediately to the south of the bands. To a lesser degree cooling is noted on their northern flanks. Consistent with these tendencies, the zonal westerlies at 300 mb tend to increase both during and immediately after cloud band formation, not only in the vicinity of the bands themselves, but also over the entire north Pacific between 20° and 60° N. Corresponding energy cycles involve an increase in northern hemispheric kinetic energy.

Erickson and Winston concluded that, taken together, these and other features identify the cloud band connections between tropical and middle latitudes as visible manifestations of polewards energy ejections from tropical storms. Such occurrences have long been suspected but generally eluded confirmation in pre-satellite days. As early as 1957, Palmén and Riehl suggested that the energy export from tropical storms is likely to be an important factor in the maintenance of the westerly circulation of middle latitudes, and that, because of the highly non-uniform distribution of tropical storms, such transfer is likely to be sporadic in time and restricted in region. Satellite evidence at last enables reasonable calculations of such localized energy exchanges to be made. In general, the initial estimates indicate that the oblique cloud bands do indeed represent substantial transports of both heat and moisture signifying conversions to eddy forms of available potential and kinetic energy. The scales of the processes are sufficiently large to affect not only the north Pacific tropospheric circulation, but also that of the rest of the northern hemisphere. In short, these injections of energy into middle latitudes contribute significantly to the autumnal build-up of the hemispheric circulation towards its winter intensity maximum.

Future work will doubtless seek to establish further the extents of the contributions made by such interacting systems to the climatology of energy, moisture and momentum transports from low to higher latitudes. The most urgent need is for a more comprehensive and integrated approach than has been adopted hitherto. Big gaps exist in the present geographic coverage of infra-red and photographic studies, especially across Eurasia, and the coordination of satellite and conventional data must be improved if we are to obtain the maximum benefits from both. A promising beginning has been made by Winston (1967) through five day averaging of outgoing long-wave radiation data from Tiros IV over the Pacific sector and discussion of the results in terms of large-scale features of the northern hemisphere circulation. His conclusions may be summarized under two headings:

1 Latitudinally, well-marked radiation maxima and minima are associated with the principal weather belts, while being affected by the poleward radiation gradients from the equatorial zone. These maxima and minima fluctuated within 10° and 5° belts of latitude respectively during February–June 1962.
2 Longitudinally, radiation variations along 45° N were characterized chiefly by:

    (*a*) Persistent maxima and minima largely related to the 500 mb wave pattern.
    (*b*) Preferred positions of radiation minima between 0·2 and 0·6 of the distance from a 500 mb trough to the next ridge downstream, and also immediately to the rear of the trough.
    (*c*) Preferred positions of radiation maxima immediately to the rear of a 500 mb ridge, near the next trough, and halfway from the trough to the next ridge downstream.

Since the radiation fields are essentially functions of the major middle and high cloudiness and the vertical motion field, these relationships suggest broad-scale latitudinal assocations of vertical motion, and the zonal flow, i.e. stronger downward motion in the subtropics and stronger upward motion near the jet maximum when the westerlies and their anticyclonic shear are strong. Lesser-scale meridional variations are embroidered on the zonal pattern, associated with satellite-observed sequences of major frontal and cyclonic systems, and surface-observed sequences of weather.

We may conclude that a careful documentation of planetary scale characteristics of the radiative heat budget of the earth-atmosphere system from satellite radiation data should throw more light also upon essentially local features of a dynamic climatological nature while contributing to the ultimate solution of many problems in medium-range weather forecasting.

13 The polar regions

General considerations

The polar regions are among the most sparsely populated in the world. Hence, from the point of view of the direct influence that climate exerts on man, the polar regions are relatively unimportant climatologically. Viewed differently, however, it can be argued that climate is of paramount importance in causing these to be regions of extreme environmental difficulty, in which case their mean weather characteristics deserve specially careful consideration as the basic factors prohibiting fuller usages of such areas by man. Meteorologically the polar regions play their own special roles in daily weather development well beyond the Arctic and Antarctic Circles, and climatologically very high latitudes cannot be omitted from modern general circulation studies or from many studies in synoptic, dynamic and physical climatology. Economically, knowledge of the Arctic in particular has become more vital than before on account of its mineral resources, its increasing overflight by airliners following great circle routes from Europe to the Pacific, and its growing significance in contexts of global military strategy. In contrast, Antarctica has remained an area primarily of scientific interest, and, in the absence of strong economic or political stimuli, observational research in the south polar region has remained at a comparatively low level. So the polar regions may be sparsely populated and still relatively data-remote, but they cannot be omitted from a review such as the present one.

In the global circulation of the atmosphere the polar regions have particular significance as the final sinks for energy continually being transferred from equatorial regions. We are only just beginning to appreciate the precise natures of local radiation budgets near the poles, involving their levels, variations through time and differences from place to place. One of the most interesting recent conclusions regarding the Arctic region has been that its general energy budget is very sensitive to changes of albedo and ice cover. So the impression is deepening that the north polar basin can be regarded by no means as a stable climatic regime; a good deal of variability results from seemingly quite small changes in the terms that make up the energy budget as a whole (Vowinckel and Orvig 1967). Thus modern interest is focused roundly on cause-and-effect relationships rather than on the admittedly fascinating climatic extremes that are well known to occur in these, the most ultimate of all the 'ultimate corners' of the earth. Put differently, present climatological preoccupations in high latitudes have become much like those elsewhere: most emphasis is being placed on the processes and mechanisms contributing to polar climatology.

The patterns of early satellite orbits

Several of the earliest satellites were placed into orbits at angles of approximately 48° or 58° to the plane of the Equator (see table 2.1), and were thereby effectively prevented from making reasonable data collections from high latitudes. The first weather satellite to occupy a near polar orbit was Nimbus I, launched on 28 August 1964, almost four and half years after the launching of Tiros I. Only four of the first twelve American weather satellites orbited before the inauguration of the fully operational Essa system in February 1966 were capable of observing the poles. Hence relatively few polar observations have become available for climatological processing. Tiros radiation studies, for example, have been very definitely 'quasi-global' in that they could not be extended beyond about 60° N and S of the Equator (see for example, figs. 8.6–8.8).

The performances of satellite instrument systems

While measurements of polar regions have been made by now through most of the wavebands investigated by satellites sensors, photographic observations in particular have suffered from the simple solar climatic facts of life in very high latitudes. During the first decade of satellite meteorology no camera was able to function in the absence of direct solar illumination of its target. Consequently 'ultimate blanks' were obvious features of even the computer-mapped polar stereographic mosaics of satellite photographs, being caused by the local spread of 'polar night'. For the standard vidicon camera, for example, a broad area concentric about the North Pole remained data-remote during January while the South Pole was fully in view. The situation was reversed in July. In intervening months – for example, the equinoctial months of March and September – some losses of picture information were unavoidable in both regions at once.

When it is borne in mind additionally that the interpretation of satellite photographs from very high latitudes has been fraught with particularly acute difficulties stemming from basic similarities of appearance between clouds and frozen surfaces, it no longer seems surprising that few worthwhile photographic studies have been completed for either the Arctic or the Antarctic. It is within these regions above all others that infra-red evidence has had to be invoked from the early days of climatology from satellites.

Arctic and Antarctic regional climates

Before we proceed to review and summarize briefly some of the most interesting polar studies based on satellite data, we will do well to look more systematically at the climates of these two regions in order to be able to appreciate more fully the nature and extent of the satellite contribution. It is most convenient to organize this introduction along comparative lines: the Arctic and Antarctic regions possess some interesting dissimilarities despite the underlying similarities of their places and roles in the world pattern of climates. Let us consider in turn the similarities and dissimilarities between the north and south polar regions under the headings of geography and climatology:

Geographical similarities

The chief and most obvious geographical similarity is locational with respect to the disposition of the planet earth on its polar axis. This apart, no other similarities of major importance are readily apparent.

Geographical dissimilarities

These are more numerous, and can be assigned to two general groups:

(*a*) The north and south polar regions differ in their intrinsic topographic characters (figs. 13.1 and 13.2). Whereas the Arctic is covered largely by a deep polar ocean, much of which is more than 700 m in depth, the Antarctic is dominated by a large continental land mass, whose mean elevation of over 2300 m distinguishes it as the most elevated continent in the world. Hence, frozen surfaces within the Arctic Circle consist for the most part of a relatively thin cap of ice across the surface of the sea, in sharp contrast with those embraced by the Antarctic Circle, which are composed largely of the thick continental ice cap over the central land mass.

(*b*) In terms of simple geometry, the shapes and distributions of land and sea in the north polar region are much more complex than their southern

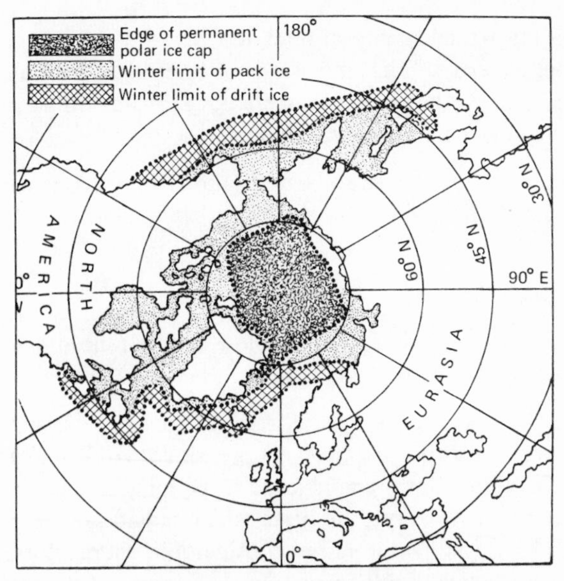

Fig. 13.1 The geography of the Arctic and surrounding regions, showing the seasonal differences in ice marginal positions. (From Barrett 1972)

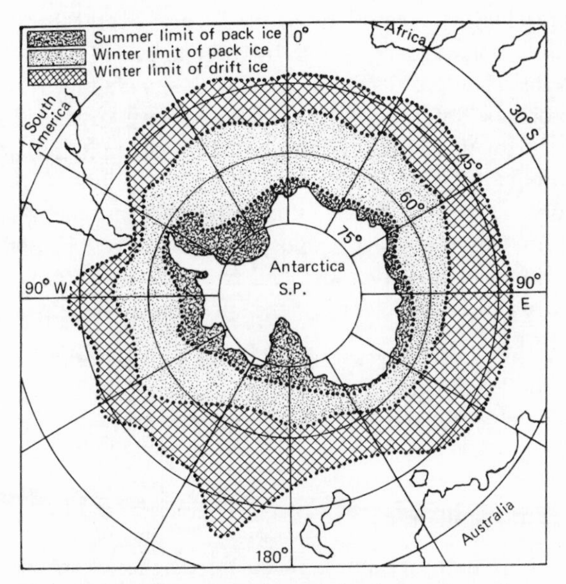

Fig. 13.2 The geography of the Antarctic and surrounding regions, showing the seasonal differences in ice marginal positions. (From Barrett 1972)

counterparts; arrangements around the South Pole are symmetrical in very general terms. The intrusion of the high ice-bound island of Greenland into the basin of the Arctic Ocean is a complication of major geographical and climatological significance.

Such quite profound geographical dissimilarities have very significant effects upon the climatologies of these two positionally similar regions.

Climatological similarities

These may be summarized under the now familiar headings of radiation, moisture and circulation.

(*a*) Radiation. As noted earlier, there is an obvious general similarity in that both regions constitute ultimate energy sinks within the radiation economy of the globe. Both regions experience long periods during the local winter during which very little direct solar radiation is received (see figs. 13.3 and 13.4), and both suffer considerable energy losses during the periods of more favourable insolation on account of the high albedoes of the frozen surfaces. Energy losses vary much less from season to season.

(*b*) Moisture. These regions both constitute 'cold deserts' across which mean annual precipitation, while being particularly difficult to evaluate, is generally very low. For the Antarctic, estimates of mean precipitation vary, representative suggestions being 27 mm overall by Loewe (1956), and 70 mm poleward of 70° S by Meinardus (1906). Snow drifting heightens the usual problem associated with snowfall measurement. Many of the notorious Antarctic blizzards consist very largely of surface snow whipped up by the wind. A representative precipitation figure for the Arctic is about 110 mm for the area from 80° to 90° N (Borisov 1965).

(*c*) Circulation. There is general similarity between the Arctic and the Antarctic in that the lower troposphere is marked in both regions by shallow centres of high pressure and anticyclonic outflow, above which low pressure and cyclonic convergence prevails in the upper troposphere.

It is below this level of gross generalities that the chief effects of geographical dissimilarities are most marked, prompting some quite striking contrasts at the synoptic and subsynoptic scale levels.

Climatological dissimilarities

(*a*) Radiation and energy regimes. In so far as these can be evidenced by surface temperatures, it is remarkable how much colder and less variable the climate at the South Pole seems to be. Reliable long-term mean temperature levels for the poles themselves are still hard to assess with confidence, but a highest daily value of −23 °C has been quoted for the Russian station, Vostok, recording since 1957 high on the interior plateau of Antarctica. This compares with fairly frequent winter temperatures of −65 °C and an absolute minimum of −88 °C (Rumney 1968). In contrast the variation at the North Pole is probably from something very close to freezing point occasionally in summer to as low as −80 °C in winter. Such contrasts in temperature ranges result largely from the greater geographical symmetry of Antarctic, whose continent effectively isolates the South Pole from incursions of relatively warm air such as those that may reach the North Pole in summer especially via the North Atlantic route. Similar contrasts of Q are indicated by figs 13.3 and 13.4.

On a more local scale, it is recognized now that the polar energy budgets are exceedingly complex (see e.g. Hare 1968). Both the type and

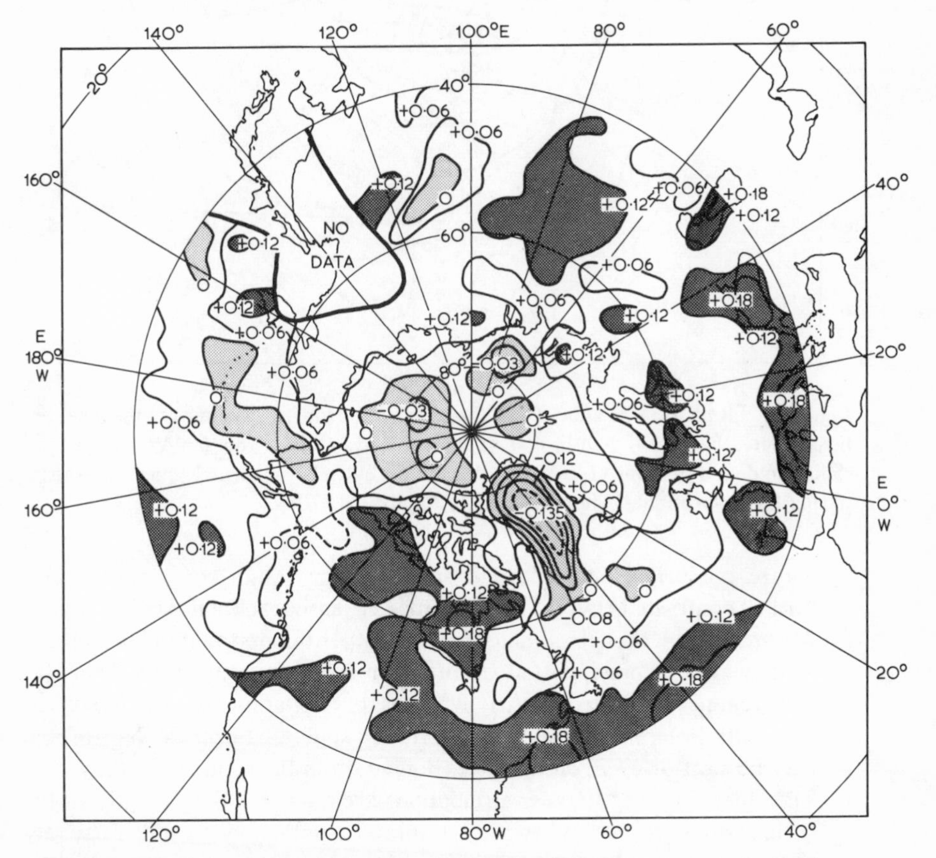

Fig. 13.3 Radiation balance, Q, of the earth–atmosphere system over the northern hemisphere, based on Nimbus II evidence, 1–15 July 1966 (cal/cm^2/min). (After Raschke & Bandeen 1970)

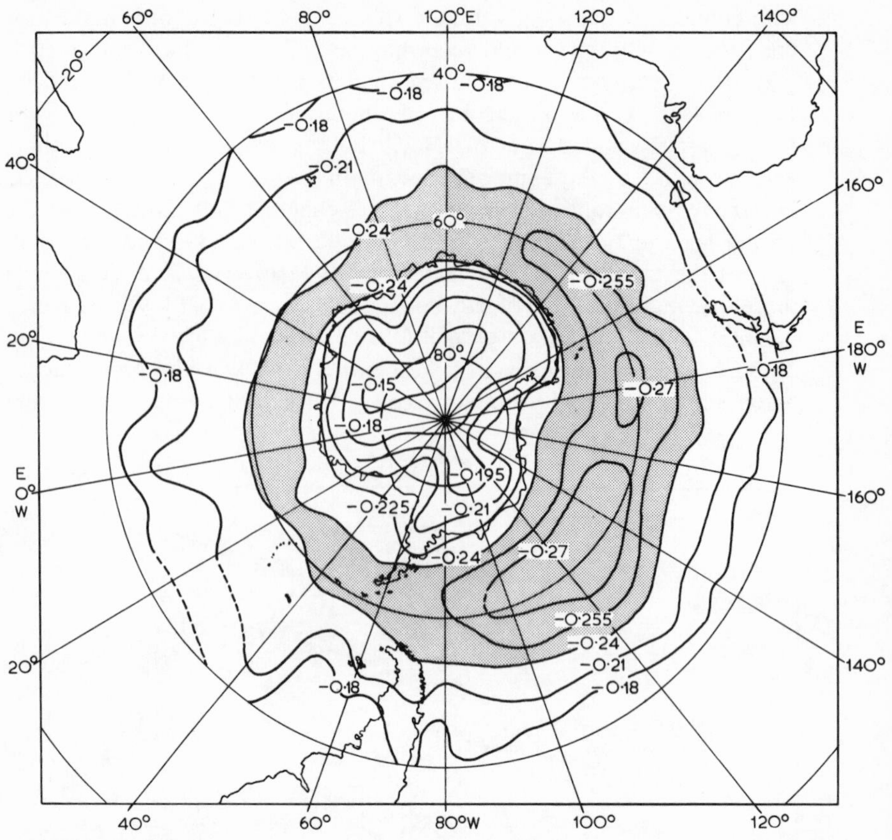

Fig. 13.4 Radiation balance, Q, of the earth–atmosphere system over the southern hemisphere, based on Nimbus II evidence, 1–15 July 1966 (cal/cm^2/min). The radiation deficit poleward of 68° S is due almost entirely to thermal emission to space during the polar night. (After Raschke & Bandeen 1970).

nature of surface ice influence such balances quite considerably, and further contrasts between the poles may be anticipated on these grounds.

(*b*) An interesting, though climatologically rather unimportant polar contrast with respect to precipitation results from a dissimilarity in the temperature regimes indicated in (*a*): whereas precipitation in the vicinity of the South Pole must always be in a frozen state, that near the North Pole may be sleet or even, on rare occasions, rain in the summer months.

(*c*) The chief contrasts in polar circulations are related to the strengths of the surface anticyclones. Whereas the Antarctic anticyclone varies relatively little in size and intensity, that over the Arctic is much more variable. Especially in summer, when the Arctic ice cap contracts from the northern shores of North American and Eurasia, the Arctic anticyclone shrinks markedly, the location of its centre often changing quite considerably from day to day. In this season warmer air masses may penetrate close to the pole itself. In winter the Arctic anticyclone is overshadowed by cells over North America and Siberia. Only in spring is the Arctic anticyclone a prominent feature on mean pressure charts. Fluctuations in the size and extent of the Antarctic anticyclone meanwhile accompany the waxing and waning of the frozen areas around the continent: from the climatological point of view, a frozen sea has the same influence as a cold continent. In winter, the extent of the 'land mass' is, in effect, supplemented by a broad fringe of ice shelves and pack ice (see figs. 13.1 and 13.2). Thus, in winter, 'Antarctica' may be said to expand beyond the Antarctic Circle to a latitude of about 60° S. Frontal weather sequences do, however, penetrate the edge of this continent from time to time, as chapter 12 suggested, and, despite the meridional differences outlined later, they do so more uniformly around its rim than is the case in the Arctic. Here, one strong cyclone penetration corridor crosses Iceland and Spitzbergen; in summer small, seemingly rather structureless lows develop and dissipate frequently around the shores of the northern continents, helping to prompt the observed variations in the ice cap anticyclone.

Satellite studies of the polar regions

Since it is not yet possible to compile a comprehensive picture of polar climatology from satellites, the satellite studies worthy of mention here will be organized systematically in two groups, depending on the types of data involved. As nephanalysis studies would be of little immediate value in such difficult regions from the interpretational point of view, the following remarks are under two headings, summarizing to infra-red and photographic investigations.

Infra-red studies

These were begun effectively by analyses of H.R.I.R. observations made by Nimbus I during its brief but useful operational lifetime in 1964. These demonstrated the particularly high potential value of such measurements through the narrow 3·4–4·2 μm atmospheric window waveband. Nordberg (1965) explained some of the physical variables that influenced H.R.I.R. observations from Antarctica on the one hand and Greenland on the other.

Across the interior ice cap of the Antarctic continent, Nimbus I recorded extremely uniform temperatures of 210–215 °K, whereas around the periphery of the continent surface temperatures averaged 240 °K, before becoming less uniform, seemingly in association with different kinds of ice. For example, the shelf of sea ice extending from the Weddell Sea to a latitude of only 57° S was about 12 °K less cold than the coastal continental ice. The temperature of the open sea around the continent was 275 °K, a little above freezing point. Meanwhile, a number of very narrow but quite distinct lines of higher temperatures crisscrossing the shelf of sea ice were apparently indicative of cracks and fissures in the ice, some of which must have been as much as 200 km in length. Along the continental coastline in the neighbourhood of Queen Maud Land the maximum temperatures in a belt about 100 km wide were 250 °K, higher than the usual sea shelf temperatures, but lower than those of free ocean water. Nordberg suggested that these intermediate temperatures might indicate a zone of ice floes. If the floes were fairly small and well distributed, the resolution of the H.R.I.R. radiometer would not have been sufficiently good to permit the individual masses of ice to be distinguished from sea water. As August 1964 gave way to September, some local changes in the radiation patterns were attributed to the progress of the spring thaw. More detailed studies along similar lines might well throw new light on the nature and ever-changing pattern of mesoscale energy budgets especially around the edges of the Antarctic continent.

Simultaneously Nimbus I was able to furnish radiation data from the north polar region, and Nordberg (1965) described some analytical results for Greenland, whose climatology is, if anything, more difficult to examine on the surface than that of the neighbouring polar ice cap on account of height and topographic roughness. It was possible, for example, to identify areas of surface temperature inversion (common over ice-bound surfaces in high latitudes) through the contrasts between their infra-red signatures and those from nearby cloud-covered regions. In the infra-red, some clouded areas showed 20 °K temperature advantages compared with non-cloudy areas, indicating that temperatures at some level within the atmosphere were higher than those elsewhere at the ground surface. It should become possible to assess the extents, intensities and persistences of such inversions much more comprehensively from meteorological satellite data than from the conventional station data which are still gathered all too thinly from across the major ice sheets of the world.

Later infra-red data analyses involving Nimbus II observations lent breadth to the range of features that seemed likely to benefit from measurement by satellites (Nordberg *et al.* 1966). Some such analyses were concerned with atmospheric moisture patterns established through M.R.I.R. water vapour channel data, interpreted alongside appropriate photographs and H.R.I.R. displays. In particular, immensely long, twisting cloud bands separated by much drier air were observed to extend frequently in winter from the tropics into south polar regions. Similar features have been noted and described by various groups of workers, including Allison *et al.* (1966), Widger *et al.* (1966) and Allison and Warnecke (1966). It is not uncommon for five or six such cloud bands to occur around the southern hemisphere on a typical winter's day. Some may attain unbroken lengths of more than 8000 km. Nimbus II data indicate that small tightly coiled secondary vortices occur quite commonly near the edges of the Antarctic ice in winter, between the giant quasi-frontal systems stretching from Antarctia into much lower latitudes.

Another fascinating fact to emerge from the analysis of Nimbus II radiation data concerned temperature variations aloft, in the upper troposphere and lower stratosphere as indicated through the CO_2 channel of the M.R.I.R. radiometer. As chapter 8 explained, CO_2 channel observations can be interpreted (in the absence of major high altitude cloud systems) in terms of the global scale stratospheric temperature field. This, in turn, can be related to stratospheric circulation. It emerged that, across the Antarctic regions, which had been beyond the scope of earlier studies, summer temperature patterns in the lower stratosphere were unexpectedly far from symmetrical. In particular, pools of warm air were observed in latitudes south of about 50° S. These were not stationary, but mobile, moving slowly in an eastward direction from the eastern Pacific sector in June across the south Atlantic to the Indian Ocean by July, and thence into a more permanent position over the south Pacific by the beginning of August. Meanwhile over the Arctic, warm air was distributed around the North Pole in nearly perfect symmetry during the entire summer period.

Clearly the distributions and exchanges of energy aloft, like those at the earth/atmosphere interface, are much more complex and variable than earlier climatological statements tended to suggest. So we await with keen interest and anticipation the results of currently continuing, more detailed, investigations of the polar atmospheres involving more recent Nimbus, Essa and Noaa satellite data.

Photographic studies

Dealing first with geographical interpretations of satellite photographic data, it is clear that polar orbiting weather satellites provide unique opportunities for the study of sea ice at least in the economically important Arctic region. The ice reconnaissance capabilities of satellite systems, particularly the near real-time A.P.T. facility, have been used for several years by E.S.S.A. and its successor N.O.A.A.

In association with the U.S. Naval Oceanographic Office and the Canadian Department of Transport, N.E.S.C. conducted an extensive operation in the Gulf of St Lawrence and east Newfoundland waters in 1962 to develop ice reconnaissance techniques (see Wark and Popham 1962). A resulting methodology has been applied since in the Great Lakes region, involving the preparation of ice reconnaissance charts at N.E.S.C. (and, more recently, at N.E.S.S.) for transmission on the national facsimile circuit as an adjunct to conventional ice observations (see plates 28*a* and *b*). Northern hemisphere ice and snow cover charts are also prepared weekly, for use primarily by scientists working on computer programmes involving extended weather prediction equations. Such studies, combined with sea surface temperature data from satellite infra-red sensors (McClain 1970) may help researchers to unravel something of the exceedingly long and complex chain of causes and effects involving the nature of the earth's surface and characteristics of its weather and climate. Early work related to such problems was begun by scientists such as Meinardus (1906), von Wiese (1924) and Brooks and Quennell (1928) in the early twentieth century. More recently, others like Schell (1970) have taken up the challenge anew armed with today's greater reserves of observational data. Satellites afford a new angle of view, and a more homogeneous and comprehensive one at that. So the complex interrelations between ice, sea surface temperatures, atmospheric pressure and pressure differences, the circulation of the atmosphere summarized by the zonal index, weather variables such as cloudiness, temperature and precipitation, and many others may become better understood.

In the meantime, the chief economically important contributions of satellite imagery to ice studies are found in ship routing by the International Ice Patrol (see plates 29*a* and *b*) in support of various scientific missions on the ground, and in petroleum exploration; satellite photographs provide an economical means of surveying ice conditions in areas especially along the Alaskan coast where exploratory offshore rigs might be located (N.E.S.C. 1968).

For the future, it seems that the mean brightness chart (see pp. 50–1) will become the most useful form of satellite photographic data for ice and snow studies especially the minimum brightness presentation. Five-day and seven-day composite minimum brightness charts have been found useful already for large-scale snow and ice mapping (McClain and Baker 1969). Such charts have been prepared with a resolution of approximately 45 km on a hemispheric basis, and about 3 km for the greater part of North America (see plate 30). Clouds, which are often comparable in brightness to ice and snow, are retained only when they are present in a given area every day of the compositing period. In a typical series of minimum brightness charts (3 km resolution), covering arctic North America at ten day intervals, Greenland's ice massif, as well as seasonal changes of sea ice in Hudson Bay and elsewhere in the Canadian archipelago, are readily seen even though occasional persistent cloud patches are not filtered off. Newer data forms of interest in ice and snow studies, and in other branches of oceanography and hydrology, include some from various new radiometers and spectrometers in Nimbus, Erts and Goes spacecraft, but their full impact on climatological understanding has yet to be felt.

Lastly, we may refer again to the photographic interpretational work undertaken by Streten (1968) and already discussed in part in the context of mid-latitude climatic features in chapter 12. Unfortunately, of course, it is almost impossible to identify, classify or track cloud organizations across broad frozen surfaces such as the Antarctic continent and its variable fringe of frozen sea. Thus the furthest Streten was able to take his study of cyclones in the southern hemisphere was the circumference of the ice-covered area. Fig. 13.5 indicates that in general vortex frequency apparently declined quite quickly into the latitudes of the Antarctic coasts in the local summer of 1966–7. Of equal interest is the fact that coastal vortices seem to suffer decay or passage out of sight inland most frequently in four identifiable regions:

1 Between 65° and 85° E, which is, geographically, the region of the Amery Ice Shelf fronting Prydz and Mackenzie Bays. This area of apparent cyclolysis was noted earlier as such by Streten (1961).

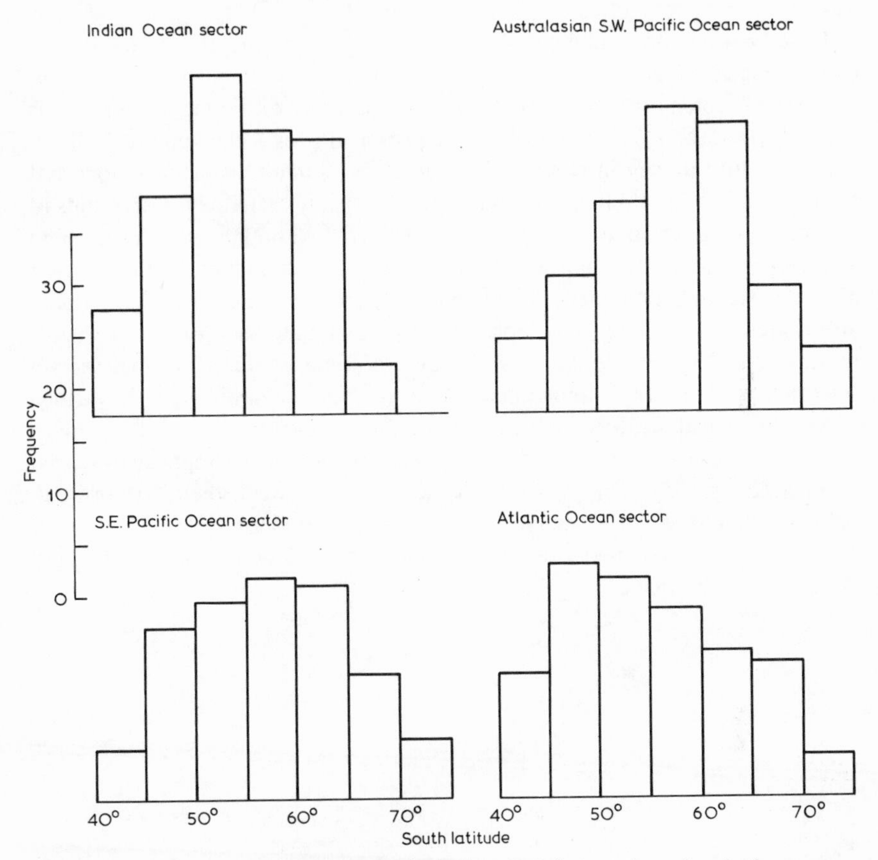

Fig. 13.5 Total vortex frequency in latitude zones for southern hemisphere ocean sectors, based on daily Essa satellite mosaics for 80 days in the local summer of 1966–7. (See figs. 12.2, 12.3 and 12.4) (From Streten 1968)

2 Between 150° and 180° W, to the west of the central and eastern parts of the Ross Sea. The Ross Sea generally was inferred to be a frequent region of cyclone decay by van Loon (1962).
3 Between 85° and 105° W, that is to say across the Bellingshausen Sea. This was noted previously as a cyclone-filling region by Astapenko (1964).
4 Between 5° and 25° W, to the north-east of the Weddell Sea.

Interspersed between these regions, Streten noted some coastal areas which, conversely, were visible more frequently than the average, under relatively clear skies. These included the Caird Coast (20–30° W), the Amery Shelf (70–80° E) and Adelie Coast (140–150° E). It is possible that these may have been related to relatively frequent offshore winds of more than average strength, blowing outward from the Antarctic anticyclone especially in regions of katabatic reinforcement.

Exactly what the implications of such distributions may be in the longer-term climatological sense is very much a matter for conjecture at present. However, Streten's studies, in keeping with other studies cited earlier in this chapter, appear to be among the most promising for follow-up work. They help to point the way in which the particularly difficult problem of how best to abstract facts of climatic significance from weather satellite observations of the polar regions may eventually be solved. Although the Arctic and Antarctic regions are admittedly remote from most human activities, their atmospheric variations are integral parts of those characterizing the global atmosphere as a whole. It is most desirable that they should become better understood. Whereas Walker (1947) could assume, at that time, that certain statistical relationships apparently linking Arctic conditions with various aspects of world weather had, most probably, occurred by chance, modern thinking on teleconnections and lag and feedback effects prompts us to search more actively today for possible physical links between them. At present, polar climatology from satellites is particularly underdeveloped. Hence it might be anticipated that this will soon become a key area in which rapid advances will be made.

PART SIX
Satellite data and climatic classification

14 The classification of climates

Introduction

For almost every topical subdivision of climatology outlined in chapter 1, an ultimate end product is a systematic arrangement of the spatial distribution of appropriate atmospheric parameters, or their effects upon other classes of phenomena. In each case the aim of the classification process should be a reasonable representation of climatic reality in map and/or tabular form, bearing in mind:

1 The purpose for which the scheme is required.
2 The desired degree of detail.
3 The types and quantities of available base data.
4 The contemporary level of relevant atmospheric concepts and theories.

Consequently, the features of an acceptable classification must include a logical, meaningful structure, sufficient detail to meet the needs of the average user and an appropriate display arrangement to permit ready and rapid extractions of required information as well as quick comparisons between individuals in different categories.

It is unfortunate, though scarcely surprising, that most past attempts to classify climates by conventional means have been descriptive rather than explanatory in nature, built of categorizations of many different meteorological elements and their periodic variations at the surface of the earth. By comparison very few schemes have been proposed that take account of the atmospheric processes and/or systems that combine to prompt the surface patterns of weather elements which are observed through time. In other words, simple statistical or descriptive classifications have been much more numerous than those of physical, dynamic, synoptic or meteorological essence. Three decades have elapsed since Haurwitz and Austin (1944) observed that a most desirable form of climatic classification would be 'a genetical classification based on the physical causes of climate', but no such scheme has yet received wide acceptance or recognition. Among the most widely known classifications is that proposed originally by Köppen (1918), and frequently modified thereafter by himself and others. Basically its purpose is to relate the distributions and types of natural vegetation over the earth to those weather elements that most significantly influence the growth of plants and are, at the same time, most universally measured. In a sense, therefore, Köppen's scheme is almost a botanical classification. Its climatic boundaries are expressed in a variety of largely plant-orientated temperature, precipitation and evaporation terms rather than as organizations of atmospheric facts alone. Despite – or perhaps because of – the wide use to which Köppen's classification has been put, countless arguments have been staged in the literature to suggest 'better' criteria for many of its suggested subdivisions. Hence, today, the scheme as a whole lacks uniformity and is extremely complex. Wilcock (1968) has helped to reduce the hypnotic effect that it still seems to exert on parts of the climatological community by a critical review of its scientific value after fifty years of 'development'.

Another, but more recent, classification of climates that has received wide publicity is that of Thornthwaite (1948), based upon the presumed 'effectiveness' of precipitation upon plant growth. Even more than Köppen's system, this classification employs the plant as an indicator of atmospheric conditions, thereby reducing still further the dependence of the classification directly upon the atmosphere itself. Thornthwaite defined 'effective precipitation' in terms of the ratio of precipitation to evaporation at a given locality, but, since appropriate records were inadequate, he developed a formula to provide an approximate relationship between temperature and precipitation measurements and the precipitation : evaporation ratio. Later schemes prepared by Miller (1951) and Trewartha (1954) are further descriptive or generic classifications widely used in encyclopaedic accounts of global climatology. Still other classifications have been proposed to account for such things as the 'cooling power' of temperature, moisture and wind on the human body, etc., but these have more limited applications. The 'Standard Atmospheres' (e.g. those of I.C.A.O. [the International Civil Aviation Authority 1954], and C.O.E.S.A. [U.S. Committee on Extension to the Standard Atmosphere 1962]) are in a real sense modern climatic classifications of a sort reflecting present-day interest in and demand for mean statements concerning the atmosphere in depth. These consist of hypothetical vertical distributions of atmospheric temperature, pressure and density, which, by international agreement are taken to be reasonable representations of the real atmosphere for various practical purposes, e.g. aircraft and missile design.

Meanwhile, what progress has been made towards that apparently worthy but neglected object of classifying climates in genetic terms? A number of interesting and quite diverse beginnings have been made towards such an end. Most are based on assessments of the atmospheric circulation, which undeniably forms the most immediate genetic basis for large-scale (or macro-) climates:

1 Hettner (1931) invoked wind systems, continentality, rainfall amount and duration, position relative to the sea, and elevation, as the contrast-causing variables within his global scale genetic scheme.
2 Alissov (1936) invoked the seasonal dominance of air masses as the major differentiating factor in his classification. This was an obvious choice for one working in the 'golden age' of air mass and frontal theories.
3 Flohn (1950) proposed that global wind belts and precipitation characteristics might afford even more suitable bases for a genetic classification of climates. His proposals were adopted in practice by several workers, including Neef, whose world map is presented and discussed by Flohn (1957).

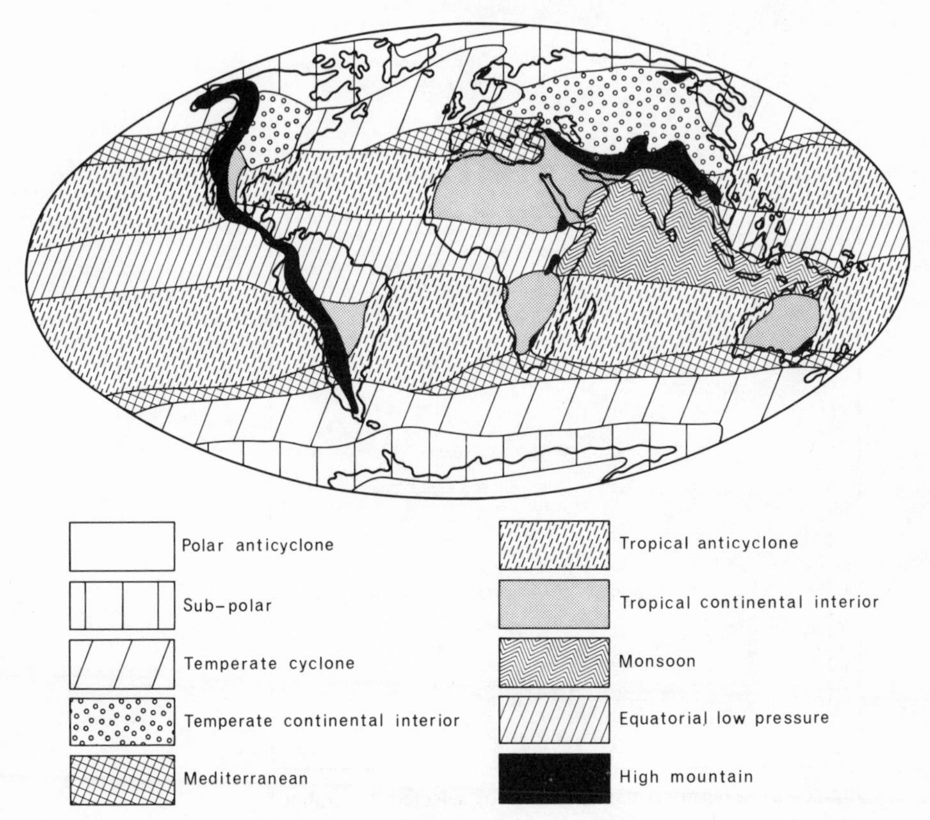

Fig. 14.1 A simple classification of world climates based on pressure and wind patterns and their local seasonality. (From Barrett 1971)

4 Hendl (1963) considered the global annual distribution and seasonal variations of the main lower tropospheric pressure belts, identifying three distinct types of zones – zones of permanent high pressure, zones of permanent low pressure, and intermediate zones of seasonal pressure reversal. In general terms a similar approach was followed and extended by
5 Barrett (1971) who suggested precipitation as a useful refining factor within a series of climatic regions established generally along Hendl's lines. He proposed further that high mountain and plateau regions should be viewed separately. This scheme, like Hendl's before it, but unlike most other classifications, was globally comprehensive: it covered the world's oceans as well as its continents (fig. 14.1).

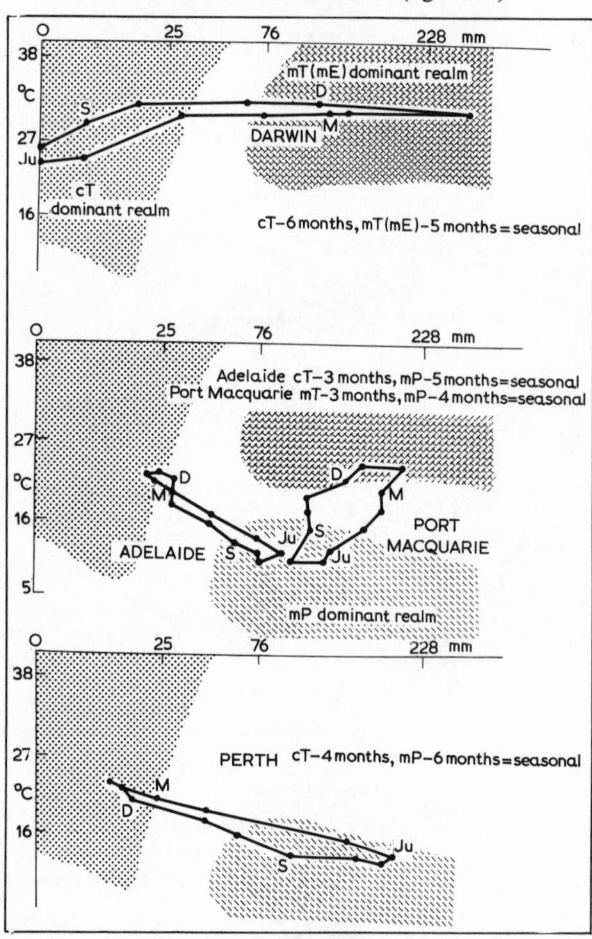

Fig. 14.2 Thermohyetal diagrams for selected Australian stations, illustrating different air mass regimes. (From Oliver 1970)

Recent approaches to the classification of climates

Before we look finally at satellite-based schemes we may note more carefully two further classifications as representative of modern 'conventional' thinking on this subject. These may be considered separately from those outlined above on the grounds that, to the time of writing, they have been tested only for a single continent (Australia in both cases). Logically there seem to be no powerful reasons why similar approaches should not be adopted more widely, but as yet no global statements have been published in those terms. They are introduced in order of appearance in the literature.

A genetic classification of climate based on an air frequency model (Oliver 1970)

This employs an air mass frequency model to facilitate the grouping of climates through quantitative means. The model uses Venn diagrams to dis-

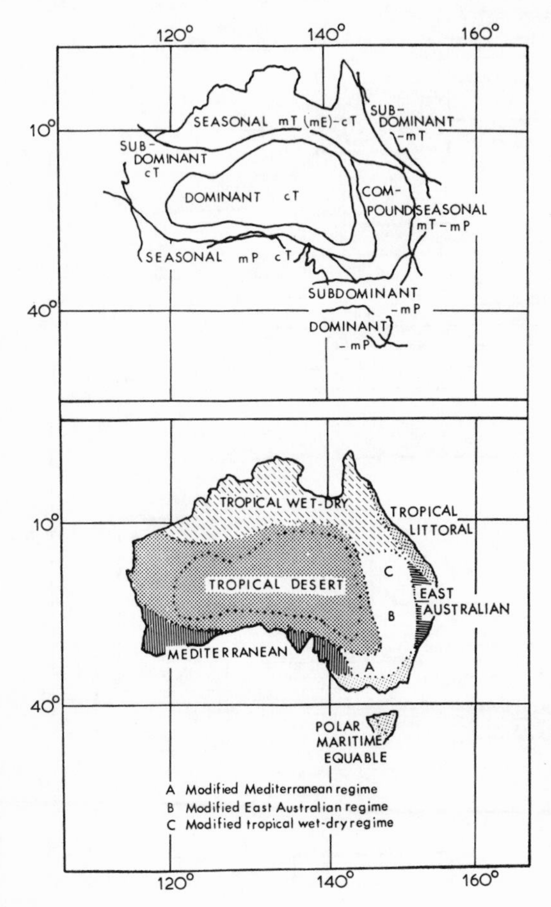

Fig. 14.3 Climatic regimes of Australia established by an air-mass frequency model. (From Oliver 1970)

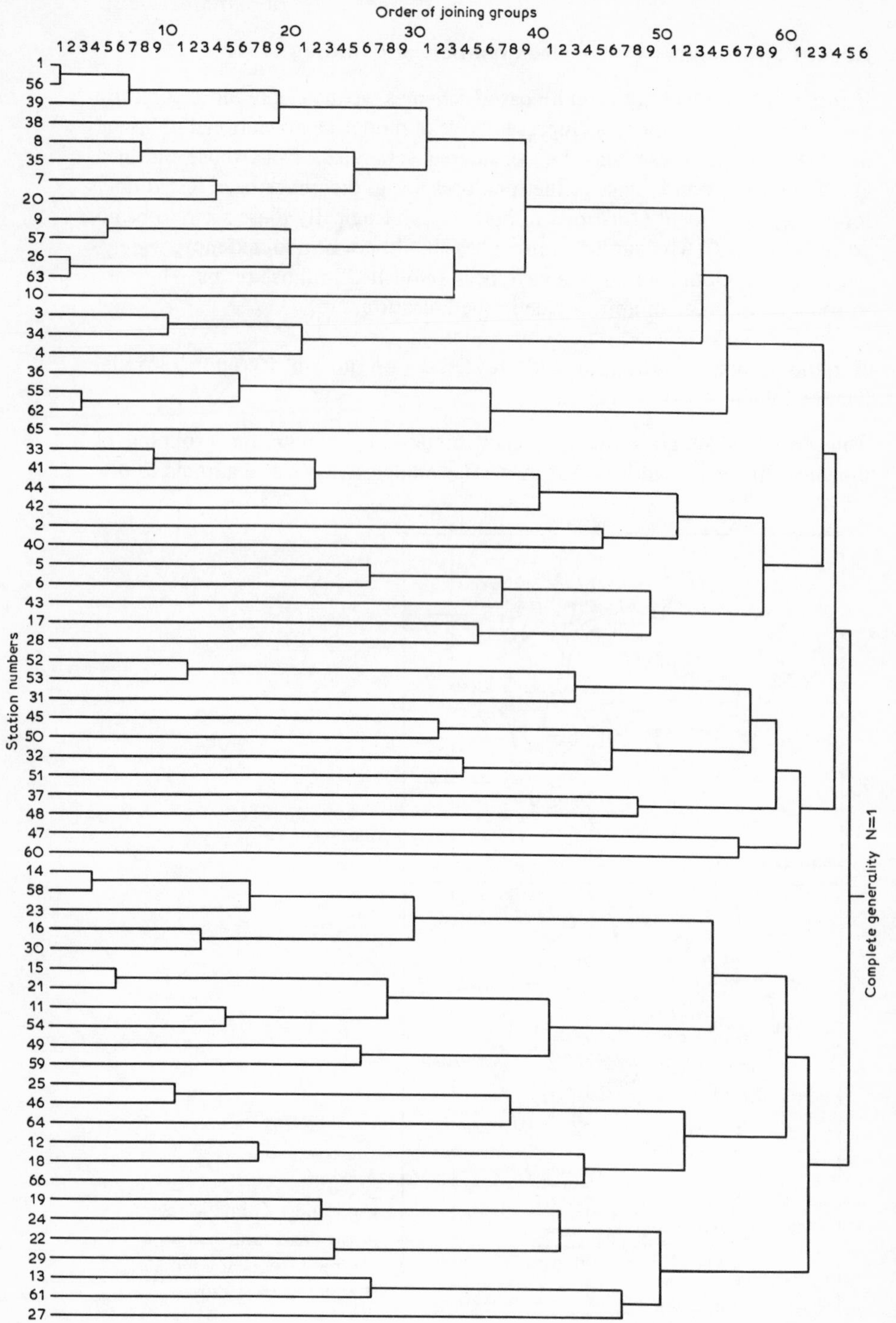

Fig. 14.4 Linkage tree of climatic stations. (From McBoyle 1971)

tinguish three basic climatic regions, each determined by the relative dominance of air masses. Subsequently subregions are distinguished on the basis of air mass type. The monthly air mass dominance is determined from actual station surface data by comparison with a specially prepared plan of realms of dominant air mass regimes. Oliver's plan was constructed on a thermohyet grid (see fig. 14.2). Using this diagram, a station dominated by a particular air mass falls almost entirely within a single shaded realm. A station marked by a seasonal regime would extend from one shaded area to another, whereas a plot of monthly data for a compound regime station might fall between all of the shaded realms. The precise classification of an individual station would entail additionally the specific air mass regimes observed. Fig. 14.3 displays climatic regimes of Australia according to Oliver's air mass frequency model.

A generic classification of climate by factor analysis (McBoyle 1971)

Starting with twenty weather variables from each of a number of initial meteorological stations, McBoyle compiled a matrix of simple correlations to be used as a starting point for a sequence of several statistical manipulations. This sequence entails the evolution of a smaller set of 'composite variables', i.e. a set of indices summarizing the original data matrix through factor analytical procedures. The end result was a computer grouping of climatic stations into twenty-eight distinct regions (see fig. 14.4). Three particular advantages were claimed for such an approach:

(*a*) It is dependent solely on climatic elements.
(*b*) The boundaries that result indicate where the most rapid change of climate occurs (the first order boundary indicating the most rapid climatic change, the second order the second most rapid and so on).
(*c*) The computer operation of the method readily permits the groupings to be changed as the context is changed from the local to the global. In other words, the more detailed the subdivisions are required to be, the more groups may be formulated by the factor analysis. An approach of this kind seems particularly amenable to the whole range from large- to small-scale studies.

Unfortunately both these methods are confronted in turn by problems similar to those that confronted the earlier approaches:

(1) *The atmosphere behaves like a fluid continuum*, being subject to kaleidoscopic variations in its form, detailed composition, and its effects upon the earth/atmosphere interface. It is difficult to decide through which terms it may be examined and represented best, even when specific classification 'consumers' require fairly specific types of information concerning its patterns through given periods of time.

(2) *It is difficult to define precise and widely accepted boundary criteria*

relating to the substructure of the atmosphere or the pattern of its effects on the surface of the earth. While it is not often hard to recognize at any scale 'core areas' that are clearly distinct from one another, it is almost impossible to avoid some degree of subjectivity or arbitrariness in regional or subregional boundary drawing.

At least partial solutions to these problems may be anticipated as our knowledge and understanding of the working of the atmosphere expands, and as more appropriate classification criteria are identified. Satellites may aid in both respects.

Rethinking the classification of climates in the satellite context

Theoretical considerations

In the continuing search for new, and perhaps even more acceptable modes of climatic classification, Barrett (1970) suggested that the time is ripe to rethink the whole matter from philosophical first principles – not because earlier classifications were necessarily 'wrong', or even 'inadequate' conceptually or in view of their intended applications – but because radically new data types give us an opportunity to begin the game again from scratch. More specifically, he observed that 'although in the past pressure, temperature, rainfall, etc., have been the chief elements invoked in classification procedures, it might prove more informative and helpful ultimately to consider qualities of the atmosphere that more readily classify themselves.' The argument was that it has become possible to envisage an alternative approach to the classification of climates resting not so much upon the intelligent selection of seemingly key boundary values by any climatologist, but rather upon simple analyses of patterns that are responsive to the posing of 'Yes/No' types of questions. In this way a scheme may be sought which would lend itself to operational completion by meteorologically unskilled data analysts; it is obviously desirable that any classification should yield the same results from the same base data when processed by any independent individual. But it is even more desirable basically that logical and physically meaningful boundary criteria should be identified in the initial planning stage. Three parameters in particular would seem to be especially strong candidates for consideration in new genetic classifications requiring simple and obvious natural groupings of observations into two categories alone, i.e. numerically positive and negative classes. The three most obvious starters are:

(1) Net radiation. The net radiation at the top of the earth's atmosphere has been described earlier in this book as the 'prime forcing function of the circulation of the earth's atmosphere'. As such, it is of extreme climatological and meteorological significance. This is the appropriate Yes/No question that can be asked in this context: 'Is the local net radiation balance at the top of the earth's atmosphere in surplus?' In answering this question, areas of surplus

net radiation (comprising net radiation sources), and areas of net radiation deficiency (net radiation sinks) become differentiated out.

(2) Vorticity. Mean circulation maps may be processed readily to yield spatial distributions of relative vorticity. Then the question may be asked: 'Is the local sign of relative vorticity positive or negative?' Once again the local answer must be 'yes', 'no' or 'neither' (in which case a regional boundary line is indicated). It can be argued that distributions of positive and negative vorticity afford more meaningful statements concerning atmospheric circulation and its related weather than maps of high and low pressure, which have been used more widely in the past. After all, mean sea level pressure (about 1013 mb) may be involved locally in either a cyclonic or an anticyclonic circulation of air at the surface of the earth. Weatherwise the sense of the rotation of air is more significant than the pressure exerted on the earth's surface by any column of air; the sense of vorticity is much more directly bound up with other significant weather-determining factors such as divergence and convection.

(3) Atmospheric moisture. In searching for an appropriate Yes/No type question relating to atmospheric moisture, the most obvious candidate would seem to involve the precipitation evaporation ratio, i.e. 'Is the local precipitation : evaporation ratio greater than unity?' In a sense this is not the best measure of moisture in the atmosphere on account of its emphasis on exchange processes affecting inputs and outputs of atmospheric water, but, once more, this question at least possesses the merit of reference to a single readily identifiable break point in the middle of its particular scale.

On a mean annual basis alone, the separate application of each question to globally complete measurements of radiation, circulation and moisture could only yield two types of climatic regions, those marked by positive values of each variable, and those marked by negative values. Taking all three together, however, a possible total of 2^3 combinations results. On a half-yearly basis the range of mathematical possibility expands to $(2^3)^2$, that is, to sixty-four in all. Of these sixty-four 'climatic regions' about one half appear to be meteorologically, as opposed to mathematically, feasible within the real world. On a quarterly or monthly basis further regional distinctions would become apparent through different sequences of combinations at different places. Even were each of the three questions to be asked individually in attempts to classify atmospheric energy regions, or regions of particular circulation or moisture characteristics, classifications of widely different degrees of complexity could be obtained as required from these simple either/or questions. It is not outside the realm of possibility that suitable data might be most readily forthcoming from satellite sources themselves. We have seen that radiation questions can already be answered well by such data while geosynchronous satellite procedures are currently under development to elucidate wide-scale circulation characteristics. Still other methods are being developed to give indications of precipitation, evaporation and other moisture-related parameters from data pools of satellite infra-red measurements.

Applied research

In the meantime, however, until a comprehensive satellite-based climatic classification becomes possible, most efforts are being concentrated upon self-standing classifications of individual satellite-observed atmospheric variables with special applications in mind. An example of such an effort is the 'global cloud model' prepared recently by Greaves (1971):

The development of a data base

This involved both conventional and satellite cloud statistics, special checks being devised to introduce internal consistency between the major 'unconditional' statistics (relating to non-random influences on cloudiness, e.g. geographic position, month of the year, and time of the day) and the 'conditional' statistics (involving relatively random day-to-day cloud variations).

The development of statistical adjustment techniques

These were required to reduce the full range of combinations and permutations of time and distance scales, as well as area sizes, to a manageable total. They were required also for handling special situations such as diurnal variations in temporal conditional statistics and the passage of the field of view of a satellite from one homogeneous cloud region to another.

The application of the model to specific mission problems

Simulation procedures were desirable to indicate the probable cloud cover conditions that might be anticipated in connection with a particular proposed mission.

Fig. 14.5 displays the global distribution of the twenty-nine cloud cover regions deemed adequate for immediate purposes by Greaves. Further details of their classification cannot and need not be given here. However, the broad outline of their scheme affords an interesting example of the current mode of thinking in climatology with particular respect to the possibilities, practicalities, and utilities of classifications of climatic data based jointly on the enormous viewing abilities of meteorological satellites and the large capacities of modern computers available for rapid and repetitive data manipulations.

Radiation observations

Lastly, it is well worth reviewing a proposed scheme for climatic classification based on net radiation. This is a good example of an approach developed initially for the analysis of conventional data, which would seem to be quite

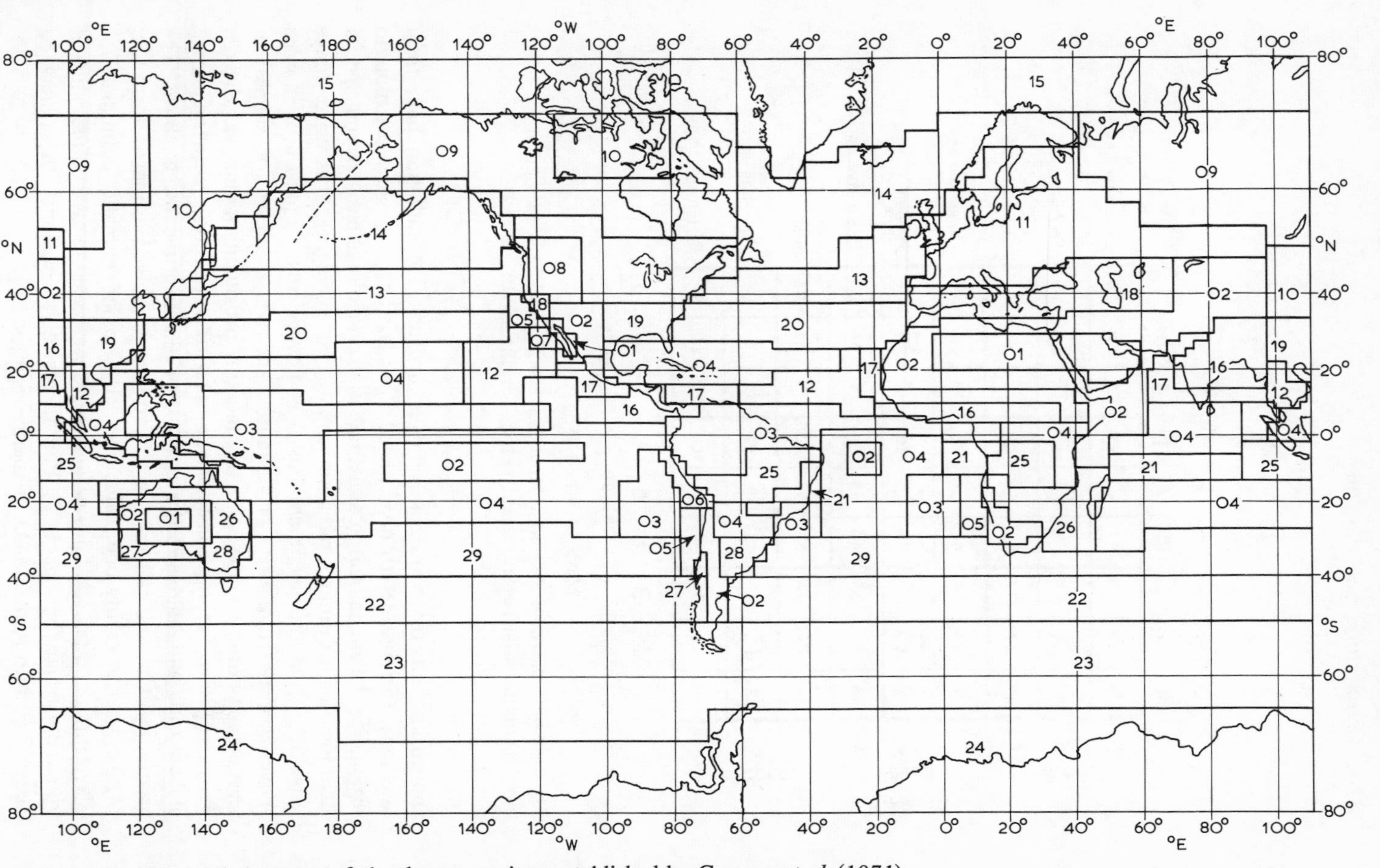

Fig. 14.5 The global pattern of cloud cover regions established by Greaves *et al.* (1971).

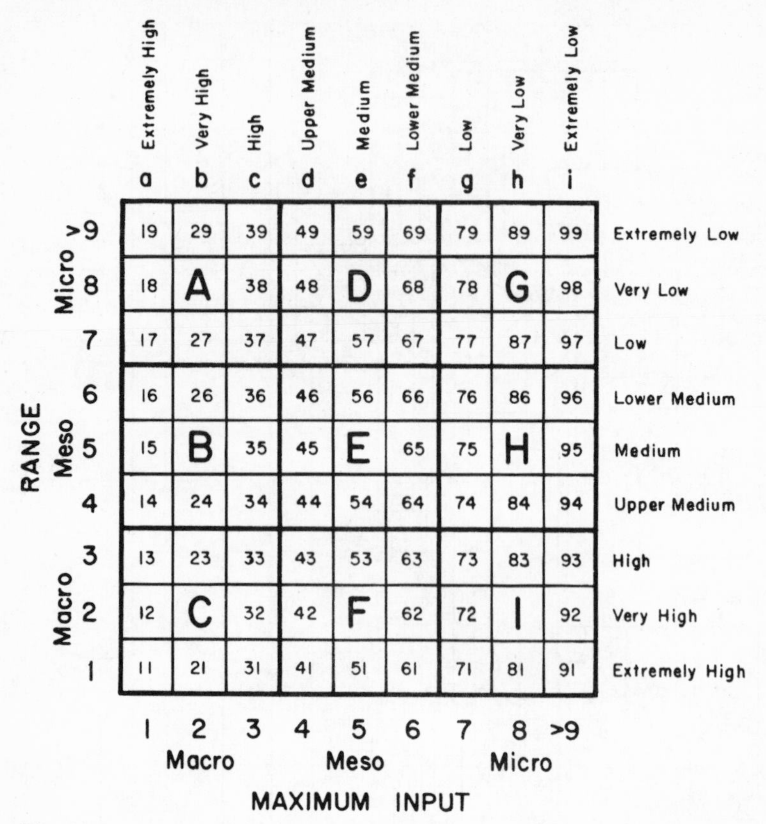

Fig. 14.6 A matrix of maximum input (waveheight) and range (deviation from maximum input) of insolation. (From Terjung 1970)

appropriate for the processing and summarizing of radiation data from satellites. Terjung (1970) constructed this classification, which attempts to regionalize the annual march of net radiation on a global basis, paying special attention to maximum radiation input (wave height), the wave range (deviation from the maximum input) and wavelength (number of months above zero net radiation) of radiation, and phase angles (or shapes) of seasonal net radiation curves. Taking monthly net radiation input values from Budyko (1963), Terjung constructed graphs for intersections across a chosen global grid, and analysed the graphs for the features listed above. The maximum input and range indices were then combined in matrix form (fig. 14.6) to give eighty-one cells and nine major climatic combinations. Wavelength and phase angle characteristics were seen to take on the range of forms illustrated by fig. 14.7. For comparative simplicity these may be categorized into four families of energy regimes:

1 Curve *e* (zenithal or summer solstice regime), subtype 1.
2 Curves *a* and *b* (equatorial regime), subtype 2.
3 Curves *c*, *h* and *f* (February–March–April maximum input), subtype 3.
4 Curves *d*, *i* and *g* (August–September–October maximum input), subtype 4.

Fig. 14.8 displays Terjung's tentative map of energy input climates. As the author pointed out, 'the system essentially portrays the input or cause, if you will, of climates'. This input is mainly astronomically determined by the solar radiation at the top of the atmosphere, the condition of the surface, and the resultant dynamic circulation, of which only the first two are primary causes. Chapters 3 and 6 indicate some of the complexities involved.

It is not difficult to see that satellite radiation observations could be employed well in global scale analyses of this kind, and the fairly immediate future must almost certainly see this come about. All we need are data of sufficient uniformity, areal breadth and extension through time. Further into the future, satellite-based classifications of dependent types might then emerge, perhaps depicting the horizontal divergence of non-latent energy through the atmosphere. Thus statements of 'radiation' or 'input climatology' could be supplemented by others of energy or moisture 'output climatology', especially relating to effective moisture provinces in the lower atmosphere.

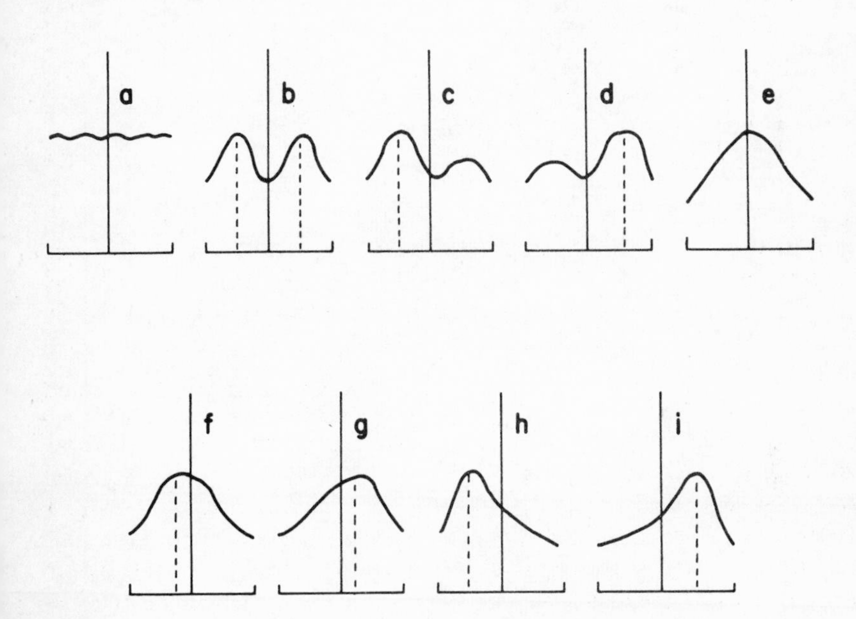

Fig. 14.7 Types of energy regimes: energy inputs at selected locations plotted month by month. (From Terjung 1970)

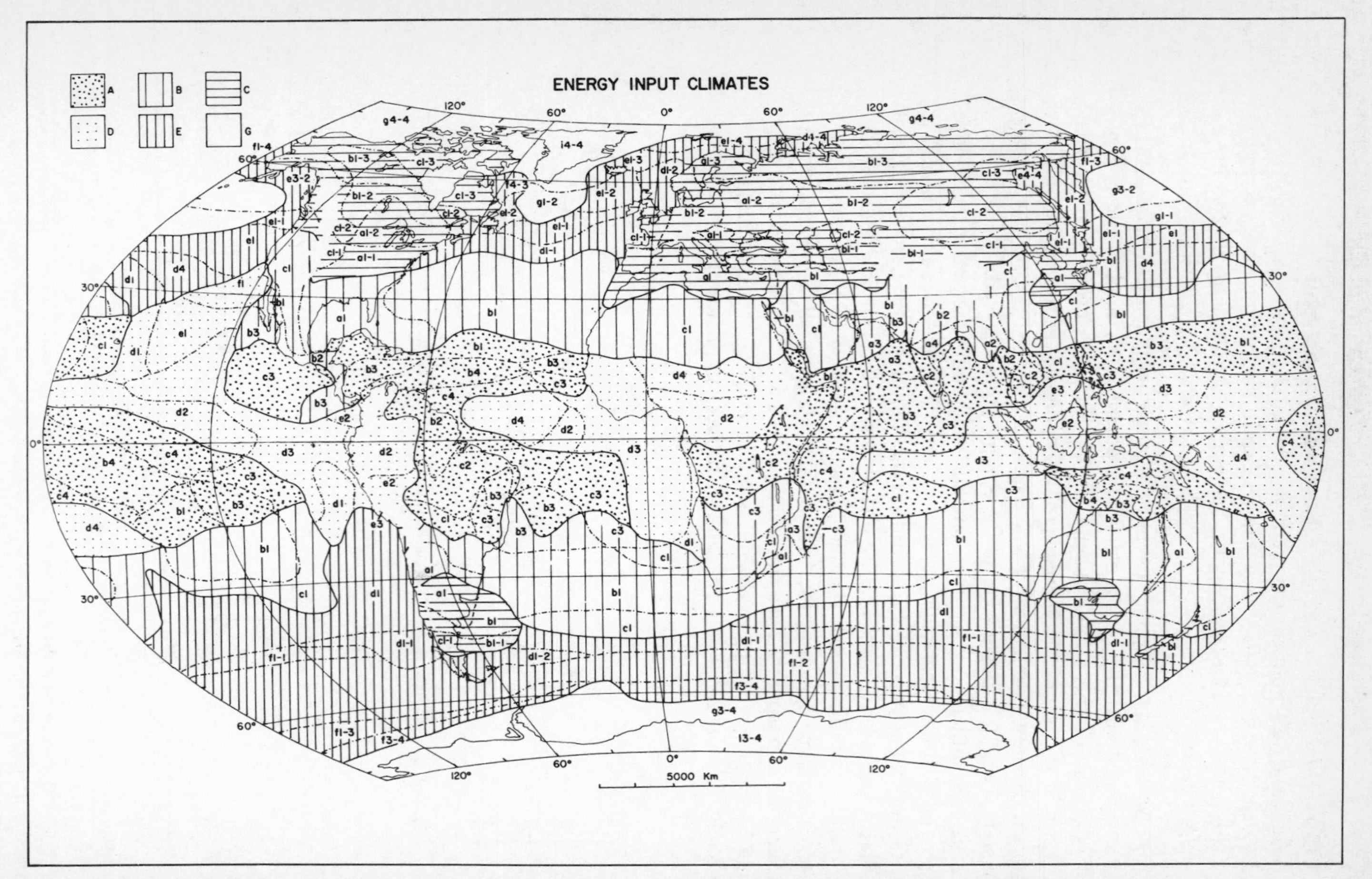

Fig. 14.8 A global classification of climates on the basis of energy input. Key letters refer to matrix subdivisions in fig. 14.6. (From Terjung 1970)

One further substantial benefit that must eventually accrue from large-scale satellite-based classifications of atmospheric parameters, however selected or organized, will be a new ability to read the factor of yearly variation into such schemes. Classifications in the past have all appeared unduly 'frozen' and 'final'. In the same way that modern climatology recognizes more fully than its forebears the many significances of seasonal, monthly or even shorter-term fluctuations in the mechanisms of climate, so future climatology will be able to identify and gauge with new-found accuracy the inconstancy of large-scale, yearly and longer-term climatic variations. Future computer-based classifications of satellite parameters, taken individually or in groups, should be included among routine climatological outputs from the chief data pools. Each should become available for expert analysis shortly after the end of each climatic unit period. Careful pattern comparisons should then help to indicate more positively where and why the chief variations in climate occur. Armed with such knowledge, considerable progress might then be made by workers seeking to extrapolate past trends into the future, to the enormous benefit of mankind.

References

Chapter 1

E. C. BARRETT (1967) *Viewing weather from space*, Longmans, London & Praeger, New York, 154 pp.

E. C. BARRETT (1970) Rethinking climatology, chapter 4 in *Progress in Geography*, Volume 2, C. Board, R. J. Chorley, P. Haggett & D. R. Stoddart (eds.), Edward Arnold, London, pp. 154–205.

R. G. BARRY (1963) Aspects of the synoptic climatology of central south England, *Meteorological Magazine* 92, pp. 300–8.

R. G. BARRY (1967) Models in meteorology and climatology, chapter 4 in *Models in geography*, R. J. Chorley & P. Haggett (eds.), Methuen, London, 801 pp.

R. G. BARRY (1970) A framework for climatological research with particular reference to scale concepts, *Transactions, Institute of British Geographers* 49, pp. 61–70.

R. G. BARRY & R. J. CHORLEY (1968) *Atmosphere, weather and climate*, Methuen, London, 319 pp. 2nd edition 1971, 380 pp.

J. E. BELASCO (1948) The incidence of anticyclonic days and spells over the British Isles, *Weather* 3, pp. 233–42.

T. T. FUJITA, K. WATANABE & T. IZAWA (1969) Formation and structure of equatorial anticyclones caused by large-scale cross-equatorial flows determined by ATS-1 photographs, *Journal of Applied Meteorology* 8, pp. 649–67.

R. GEIGER (1957) *The climate near the ground*, Harvard University Press, 611 pp.

R. E. HUSCHKE (ed.) (1959) *Glossary of meteorology*, American Meteorological Society, 525 pp.

W. G. KENDREW (1957) *Climatology*, Oxford, 400 pp.

H. H. LAMB (1966) Climate in the 1960's: World circulation reflected in prevailing temperatures, rainfall patterns and the levels of African lakes, *Geographical Journal* 132, pp. 183–212.

H. H. LAMB (1967) Britain's changing climate, *Geographical Journal* 133, pp. 445–68.

H. E. LANDSBERG (1958) *Physical climatology*, Craig, Pennsylvania, 446 pp.

R. LEE (1966) An organizational solution of the weather forecasting problem, *Bulletin of the American Meteorological Society* 47, pp. 438–44.

J. A. LEESE, A. L. BOOTH & F. A. GODSHALL (1970) *Archiving and climatological applications of meteorological satellite data*, ESSA Technical Report, N.E.S.C. 53, Washington, D.C., 118 pp.

P. E. LEHR (1962) Methods of archiving, retrieving and utilizing data acquired by TIROS meteorological satellites, *Bulletin of the American Meteorological Society* 43, pp. 539–48.

B. B. LUSIGNAN & J. R. KIELY (1970) *Global weather prediction: the coming revolution*, Holt, Rinehart & Winston, New York, 307 pp.

W. J. MAUNDER (1970) *The value of the weather*, Methuen, London, 388 pp.

A. A. MILLER (1957) *Climatology*, Methuen, London, (8th edition), 318 pp.

M. NEIBERGER & H. WEXLER (1961) Weather satellites, *Scientific American* 41, pp. 1–15.

G. R. RUMNEY (1968) *Climatology and the world's climates*, Macmillan, New York, & Collier-Macmillan, London, 656 pp.

J. S. SAWYER (1971) Possible effects of human activity on world climate, *Weather* 26 (6), pp. 251–62.

W. D. SELLERS (1969) *Physical climatology*, Chicago University Press, 272 pp.

SIR NAPIER SHAW (1926) Meteorology in History, *Manual of meteorology*, Volume 1, Cambridge University Press, 339 pp.

L. P. SMITH (1958) *Farming weather*, Nelson, 208 pp.

V. E. SUOMI (1970) Recent developments in satellite techniques for observing and sensing the atmosphere, *The global circulation of the atmosphere*, Royal Meteorological Society, London, pp. 222–34.

O. G. SUTTON (1965) The resurgence of interest in the observational sciences, *Weather* 20 (6), pp. 174–82.

C. W. THORNTHWAITE (1948) An approach towards a rational classification of climate, *Geographical Review* 38, pp. 55–94.

G. T. TREWARTHA (1954) *An introduction to climate*, MacGraw-Hill, New York, 402 pp.

I. P. VETLOV (1966) Role of satellites in meteorology, in *Interpretation and use of meteorological satellite data*, W.M.O. Training Seminar Report, Moscow, 20 pp.

A. A. WILCOCK (1968) Koppen after fifty years, *Annals of the Association of American Geographers* 58 (1), pp. 12–28.

J. S. WINSTON (1969) Global distribution of cloudiness and radiation as measured from weather satellites, chapter 6 in *Climate of the free atmosphere*, D. F. Rex (ed.), Elsevier, Amsterdam, London, New York, pp. 247–80.

W.M.O. (1967) *World Weather Watch: the plan and implementation programme*, W.M.O., Geneva, 56 pp.

Chapter 2

P. G. ABEL *et al.* (1970) The selective chopper radiometer for Nimbus D, *Proceedings, Royal Society London* A, 320, pp. 35–55.

ARACON (1966) *Nimbus II user's guide*, ARACON Geophysics Company, Concord, Massachusetts, 229 pp.

J. BARNETT *et al.* (1972) The first year of the selective chopper radiometer on Nimbus IV, *Quarterly Journal of the Royal Meteorological Society* 98, pp. 17–37.

E. C. BARRETT (1967) *Viewing weather from space*, Longmans, London, & Praeger, New York, 154 pp.

R. G. BARRY & R. J. CHORLEY (1968) *Atmosphere, weather and climate*, Methuen, London, 319 pp. 2nd edition 1971, 380 pp.

A. L. BOOTH & V. R. TAYLOR (1969) Mesoscale archive and computer products of digitized video data from Essa satellites, *Bulletin of the American Meteorological Society* 50, pp. 431–8.

C. L. BRISTOR (1966) *A summary of experiences in computer processing of Nimbus I data*, Meteorological Satellite Laboratory Report No. 27, Washington, D.C., 20 pp.

C. L. BRISTOR, W. M. CALLICOTT & R. E. BRADFORD (1966) Operation processing of satellite cloud pictures by computer, *Monthly Weather Review* 94, pp. 515–27.

COMMITTEE FOR SPACE RESEARCH (COSPAR) (1967) *Status report on the application of space technology to the World Weather Watch*, COSPAR Working Group, London, 144 pp.

R. HANEL & B. CONRATH (1969) Interferometer experiment on Nimbus III: Preliminary results, *Science* 165, pp. 1258–60.

L. F. HUBERT & P. E. LEHR (1967) *Weather satellites*, Blaisdell Publishing Co., Waltham, Massachusetts, 120 pp.

J. A. LEESE *et al.* (1970) *Archiving and climatological applications of meteorological satellite data*, ESSA Technical Report, N.E.S.C. 53, Washington, D.C., 118 pp.

R. W. LONGLEY (1970) *Elements of Meteorology*, Wiley, New York, 317 pp.

N.A.S.A. (1966) *Significant achievements in satellite meteorology, 1958–64*, Scientific and Technical Information Division, N.A.S.A., S.P. 96, Washington, D.C., 141 pp.

A. SCHWALB & J. GROSS (1969) *Vidicon data limitations*, ESSA Technical Memorandum, N.E.S.C.T.M. 17, Washington, D.C., 22 pp.

W. L. SMITH (1969) *Statistical estimation of the atmosphere's geopotential height distribution from satellite radiation measurements*, ESSA Technical Report, N.E.S.C. 48, Washington, D.C., 29 pp.

V. R. TAYLOR & J. S. WINSTON (1968) *Monthly and seasonal mean global charts of brightness from ESSA 3 and ESSA 5 digitized pictures, February 1967–February 1968*, ESSA Technical Report, N.E.S.C. 46, Washington, D.C., 9 pp.

M. TEPPER (1967) Space technology developments for the World Weather Watch, *Bulletin of the American Meteorological Society* 48, pp. 94–101.

M. TEPPER & D. S. JOHNSON (1965) Towards operational weather satellite systems, *Astronauts and Aeronautics* 2, pp. 16–26.

U.S. DEPARTMENT OF COMMERCE (1966) *Catalogs of Meteorological Satellite Data*, Washington, D.C., 5.311, ff.

D. Q. WARK (1970) Soundings from space platforms – a new era in global meteorological measurements, in *A century of weather progress*, American Meteorological Society, pp. 50–4.

D. Q. WARK & H. E. FLEMING (1966) Indirect measurements of atmospheric temperature profiles from satellites: I: Introduction, *Monthly Weather Review* 94, pp. 351–62.

W. K. WIDGER & C. P. WOOD (1961) An explanation of the limitations of the coverage provided by Tiros, *Weatherwise* 14, pp. 230–7.

WORLD METEOROLOGICAL ORGANIZATION (1967) *The role of meteorological satellites in the World Weather Watch*, World Weather Watch Planning Report, No. 18, Geneva, 38 pp.

WORLD METEOROLOGICAL ORGANIZATION (1970) *Scope of the 1972–75 plan*, World Weather Watch Planning Report, No. 30, Geneva, 141 pp.

G. E. WUKELIC (ed.) (1968) *Handbook of Soviet space science research*, Gordon & Breach, New York, 505 pp.

Chapter 3

ARACON (1970) *Nimbus IV user's guide*, ARACON Geophysics Company, Concord, Massachusetts, 214 pp.

J. BARNETT *et al.* (1972) The first year of the selective chopper radiometer on Nimbus IV, *Quarterly Journal of the Royal Meteorological Society* 98, pp. 17–37.

E. C. BARRETT (1967) *Viewing weather from space*, Longmans, London, & Praeger, New York, 154 pp.

E. C. BARRETT (1972) *Geography from space*, Pergamon, Oxford, 98 pp.

M. I. BUDYKO (1963) *Atlas of the heat balance of the globe*, Hydrometeorological Service of the U.S.S.R., Moscow, 69 pp.

M. I. BUDYKO & K. Y. KONDRATIEV (1964) The heat balance of the Earth, in *Research in geophysics: Volume 2; Solid earth and interface phenomena*, Massachusetts Institute of Technology, pp. 529–54.

P. A. DAVIS (1963) An analysis of the atmospheric heat budget, *Journal of the Atmospheric Sciences* 20 (1), pp. 5–22.

A. DRUMMOND (1970) Precision radiometry & its significance in atmospheric and space physics, *Advances in Geophysics*, Volume 14, Academic Press, New York, pp. 1–52.

G. M. HIDY (1967) *The winds*, d. van Nostrand, Princeton N.J., 174 pp.

H. G. HOUGHTON (1954) Annual heat balance of the northern hemisphere, *Journal of Meteorology* 11 (1), pp. 1–9.

J. T. HOUGHTON & S. D. SMITH (1970) Remote sounding of atmospheric temperature from satellites: I: Introduction, *Proceedings, Royal Society London* A, 320, pp. 23–33.

V. G. KUNDE (1965) *Theoretical relationships between equivalent blackbody temperatures & surface temperatures measured by the Nimbus HRIR radiometer*, N.A.S.A. SP-89, Goddard Space Flight Center, Greenbelt, Maryland, pp. 23–37.

J. LONDON (1957) *A study of the atmospheric heat balance*, Final Report, Contract AF19 (122)–165, Dept. of Meteorology and Oceanography, New York University, 99 pp.

E. N. LORENZ (1967) *The nature and theory of the general circulation of the atmosphere*, World Meteorological Organization, Technical Note No. 218, TP115, Geneva, 161 pp.

M. MARGULES (1903) *Uber die Energie der Sturme*, Jahrb Zentrahaust, Vienna, pp. 1–26.

D. H. MCINTOSH & A. S. THOM (1969) *Essentials of meteorology*, Wykeham, London & Winchester, 240 pp.

R. E. NEWELL, D. G. VINCENT, T. G. DOPPLICK, D. FERRUZA & J. W. KIDSON (1970) The energy balance of the global atmosphere, in *The global circulation of the atmosphere*, G. A. Corby (ed.), Royal Meteorological Society, London, pp. 42–90.

W. NORDBERG (1967) *Satellite studies of the lower atmosphere*, N.A.S.A. Report No. X-620-67-332, Goddard Space Flight Center, Greenbelt, Maryland, 50 pp.

J. POUQUET (1969) *Geopedological features derived from satellite measurements in the 3·4–4·2 micron and 0·7–1·3 micron spectral regions*, N.A.S.A. Report No. X-622-69-437, Goddard Space Flight Center, Greenbelt, Maryland, 29 pp.

E. RASCHKE & M. PASTERNAK (1967) *The global radiation balance of the*

Earth–atmosphere system obtained from the radiation data of the meteorological satellite Nimbus II, N.A.S.A. Report No. X-622-67-383, Goddard Space Flight Center, Greenbelt, Maryland, 19 pp.

W. L. SMITH (1969) *Statistical estimation of the atmosphere's geopotential height distribution from satellite radiation measurements*, ESSA Technical Report, N.E.S.C. 48, Washington, D.C., 29 pp.

W. L. SMITH, H. M. WOOLF & W. C. JACOB (1970) A regression method for obtaining global real-time temperature and geopotential height profiles from satellite spectometer measurements and its application to Nimbus III 'SIRS' observations, *Monthly Weather Review* 98, pp. 582–603.

T. H. VONDER HAAR & V. E. SUOMI (1971) Measurements of the Earth's radiation budget from satellites during a five year period. Part I: Extended time and space means, *Journal of the Atmospheric Sciences* 28, pp. 305–14.

D. Q. WARK, G. YAMAMOTO & J. H. LIENESCH (1962) Methods of estimating infra-red flux and surface temperature from meteorological satellites, *Journal of the Atmospheric Sciences* 19, pp. 369–84.

Chapter 4

L. J. ALLISON, E. R. KREINS, F. A. GODSHALL & G. WARNECKE (1969) *Examples of the usefulness of satellite data in general atmospheric circulation research, Part I – Monthly global atmospheric circulation characteristics as reflected in Tiros VII radiometric measurements*, N.A.S.A. Technical Note, N.A.S.A. TND-5630, Greenbelt, Maryland, 35 pp.

R. K. ANDERSON (1969) Introduction to data interpretation, in *Satellite Meteorology*, Bureau of Meteorology, Melbourne, Australia, pp. 15–16.

ARACON (1966) *Nimbus II user's guide*, ARACON Geophysics Company, Concord, Massachusetts, 229 pp.

A. ARKING (1964) The latitudinal distribution of cloud cover from Tiros photographs, *Science* 143, pp. 569–72.

W. R. BANDEEN, V. KUNDE, W. NORDBERG & H. P. THOMPSON (1964) Tiros III meteorological satellite radiation observations of a tropical hurricane, *Tellus* 16, pp. 481–502.

J. C. BARNES & D. CHANG (1968) *Accurate cloud cover determination and its effects on albedo computations*, Final Report, Contract No. NAS5-10478, Allied Research Associates, Concord, Massachusetts, 82 pp.

E. C. BARRETT (1968) Notes on the evolution and interpretation of satellite global-scale mosaics, *Weather* 23 (5), pp. 198–205.

E. C. BARRETT (1970b) Rethinking climatology, chapter 4 in *Progress in* data, *Monthly Weather Review* 98, pp. 322–7.

E. C. BARRETT (1970b) Rethinking climatology, Chapter 4 in *Progress in Geography*, Volume 2, C. Board, R. J. Chorley, P. Haggett & D. R. Stoddart (eds.), Edward Arnold, London, pp. 154–205.

E. C. BARRETT (1971) The tropical Far East; ESSA satellite evaluations of high season climatic patterns, *Geographical Journal* 137, pp. 535–55.

E. C. BARRETT (1972) *Geography from space*, Pergamon, Oxford, 98 pp.

E. C. BARRETT (1973) Daily and monthly rainfall estimates from weather satellite data, *Monthly Weather Review*, 101, pp. 215–22.

R. G. BARRY & R. J. CHORLEY (1968) *Atmosphere, weather and climate*, Methuen, London, 319 pp. 2nd edition, 1971, 380 pp.

A. L. BOOTH & V. R. TAYLOR (1969) Meso-scale archive and computer products of digitized video data from ESSA satellites, *Bulletin of the American Meteorological Society* 50, pp. 431–8.

P. F. CLAPP (1964) Global cloud cover for seasons using Tiros nephanalyses, *Monthly Weather Review* 92, pp. 495–507.

P. F. CLAPP & J. POSEY (1970) *Estimating environmental parameters from macro-scale video brightness data of ESSA satellites*, Extended Forecast Division, N.M.C., Washington, D.C., 27 pp.

COMMITTEE FOR SPACE RESEARCH (COSPAR) (1967) *Status Report on the application of space technology to the World Weather Watch*, COSPAR Working Group, London, 144 pp.

J. H. CONOVER (1962) *Cloud interpretation from satellite altitudes*, Research Note 81, Air Force Cambridge Research Laboratories, Cambridge, Massachusetts, 55 pp.

J. H. CONOVER (1963) *Cloud interpretation from satellite altitudes*, Research Note 81, Supplement 1, Air Force Cambridge Research Laboratories, Cambridge, Massachusetts, 18 pp.

A. S. DENNIS (1963) *Rainfall determination by meteorological satellite radar*, Standard Research Institute.

W. FOLLANSBEE (1973) *Estimation of average daily rainfall from satellite cloud photographs*, NOAA Technical Memorandum, N.E.S.S. 44, Washington, D.C., 39 pp.

S. FRITZ & P. K. RAO (1967) On the infra-red transmission through cirrus clouds and the estimation of relative humidity from satellites, *Journal of Applied Meteorology* 6, pp. 1088–96.

S. FRITZ & J. S. WINSTON (1962) Synoptic use of radiation measurements from satellite Tiros II, *Monthly Weather Review* 90, pp. 1–9.

W. L. GODSON (1958) Meteorological applications of earth satellites, *Royal Astronomical Society of Canada* 52, pp. 49–56.

J. R. HOPE (1966) Path of heavy rainfall photographed from space, *Bulletin of the American Meteorological Society* 47, pp. 371–3.

M. M. HOPKINS (1967) An approach to the classification of meteorological satellite data, *Journal of Applied Meteorology* 6, pp. 164–78.

M. J-C. HU (1963) *A trainable weather-forecasting system*, Technical Report 6759-1, Stanford Electronics Laboratories, California, 44 pp.

L. F. HUBERT (1963) Middle latitudes of the northern hemisphere, Tiros data as an analysis aid, in *Rocket and satellite meteorology*, H. Wexler & J. E. Caskey (eds.), North Holland, Amsterdam, pp. 312–16.

W. C. JACOBS (1951) The energy exchange between sea and atmosphere, and some of its consequences, *Bulletin of the Scripps Institute of Oceanography* 6, University of California, La Jolla, California, pp. 27–122.

D. H. JOHNSON, D. W. DENT & B. H. PREEDY (1969) Unpublished notes.

J. B. JONES & L. W. MACE (1963) Tiros meteorological satellite operational assists, *Weatherwise* 15, pp. 97–104.

J. KORNFIELD, A. F. HASLER, K. J. HANSON & V. E. SUOMI (1967) Photographic cloud climatology from ESSA III and V computer-produced mosaics, *Bulletin of the American Meteorological Society* 48, pp. 878–83.

R. LEE & C. I. TAGGART (1969) A procedure for satellite cloud photo-interpretation, and Appearance of clouds from satellite altitudes, in *Satellite Meteorology*, Bureau of Meteorology, Melbourne, Australia, pp. 17 (e)–(f).

M. LETHBRIDGE (1967) Precipitation probability and satellite radiation data, *Monthly Weather Review* 95, pp. 487–90.

M. LETHBRIDGE & H. A. PANOFSKY (1969) *Satellite radiation measurements and synoptic data*, Final Report, ESSA grant WBG-48, Department of Meteorology, Pennsylvania State University, University Park, Pennsylvania, 35 pp.

R. C. LO & D. R. JOHNSON (1969) *An investigation of the cloud distribution from satellite infra-red data*, Final Report, ESSA grants WBG-52 and E-8-69(G), Department of Meteorology, University of Wisconsin, Madison, Wisconsin, 47 pp.

E. P. MCCLAIN (1966) On the relation of satellite-viewed cloud conditions to vertically-integrated moisture fields, *Monthly Weather Review* 94, pp. 509–14.

D. B. MILLER (1971a) *Automated production of global cloud-climatology based on satellite data*, Proceedings of the 6th Automated Weather Support Technical Exchange Conference, U.S. Naval Academy, Air Weather Service, U.S.A.F., Technical Report 242, pp. 291–306.

D. B. MILLER (1971b) *Global atlas of relative cloud cover, 1967–70, based on data from meteorological satellites*, U.S. Dept. of Commerce, U.S. Air Force, Washington, D.C., 237 pp.

F. MÖLLER (1961) Atmospheric water vapour measurements at 6–7 microns from a satellite, *Planetary and Space Science* 5, pp. 202–6.

F. MÖLLER (1962) Einige vorlaüfige Auswertungen der Strahlungs-messungen von Tiros II, *Archiv für Meteorologie, Geophysik und Bioklimatologie*, Series B, Vienna, 12, pp. 78–93.

W. NORDBERG (1967) *Satellite studies of the lower atmosphere*, N.A.S.A. Report No. X-620-67-332, Goddard Space Flight Center, Greenbelt, Maryland, 50 pp.

F. J. ONDREJKA & J. H. CONOVER (1966) Notes on the stereo-interpretation of Nimbus II A.P.T. photography, *Monthly Weather Review* 94, pp. 611–14.

H. A. PANOFSKY & G. W. BRIER (1958) *Some applications of statistics to meteorology*, Pennsylvania State University Press, 244 pp.

P. K. RAO (1970) *Estimating cloud amount and height from satellite infra-red radiation data*, ESSA Technical Report, N.E.S.C. 54, Washington, D.C., 11 pp.

E. RASCHKE (1965) Auswertungen von infraroten Strahlungsmessungen des meteorologischen Satelliten Tiros III, *Beitr. zur Physik d. Atm.*, 38, pp. 97–120 & 153–87.

E. RASCHKE & W. R. BANDEEN (1967) A quasi-global analysis of tropospheric water vapour content from Tiros IV radiation data, *Journal of Applied Meteorology* 6, pp. 468–81.

C. A. ROSEN (1967) Pattern classification by adaptive machines, *Science* 156, pp. 38–44.

R. SABATINI, G. A. RABCHEVSKY & J. E. SISSALA (1971) *Nimbus earth resources observations*, Technical Report No. 2, Contract No. NAS-5-21617, Goddard Space Flight Center, Greenbelt, Maryland, 256 pp.

J. C. SADLER (1967) Average monthly cloud coverage of the global tropics determined from satellite observations, in *Proceedings of the Working Panel on Tropical Dynamic Climatology*, National Weather Research Facility, Norfolk, Virginia, pp. 154–63.

J. C. SADLER (1969) *Average cloudiness in the tropics from satellite observations*, East–West Center Press, Honolulu, 22 pp.

V. V. SALOMONSON (1969) *Cloud statistics in Earth resources technology satellite (ERTS) mission planning*, N.A.S.A. Report No. X-622-69-386, Goddard Space Flight Center, Greenbelt, Maryland, 19 pp.

J. S. SAWYER (1970) Large-scale disturbance of the equatorial atmosphere, *Meteorological Magazine* 99, pp. 1–9.

W. E. SHENK (1963) Tiros II window radiation and large-scale vertical motion, *Journal of Applied Meteorology* 2, pp. 770–5.

W. E. SHENK (1971) Stereo cloud plots from Apollo photos, *Bulletin of the American Meteorological Society* 52 (4), p. 238.

F. J. SMIGIELSKI & L. M. MACE (1970) *Estimating mean relative humidity from the surface to 500 mb. by use of satellite pictures*, ESSA Technical Memorandum, N.E.S.C.T.M., 23, Washington, D.C., 12 pp.

V. R. TAYLOR & J. S. WINSTON (1968) *Monthly and seasonal mean global charts of brightness from ESSA 3 and ESSA 5 digitized pictures, February 1967–February 1968*, ESSA Technical Report, N.E.S.C., 46, Washington, D.C., 9 pp.

P. THADDEUS (1966) *A micro-wave radiometer for the Nimbus D meteorological satellite*, Proposal, N.A.S.A., Goddard Space Flight Center, G.I.S.S.

A. THOMPSON & P. W. WEST (1967) Use of satellite cloud pictures to estimate average relative humidity below 500 mb., with application to the Gulf of Mexico area, *Monthly Weather Review* 95, pp. 791–8.

G. B. TUCKER (1961) Precipitation over the North Atlantic Ocean, *Quarterly Journal of the Royal Meteorological Society* 87, pp. 147–58.

L. F. WHITNEY (1966) On locating jet streams from Tiros photographs, *Monthly Weather Review* 94, pp. 127–38.

W. K. WIDGER, J. C. BARNES, E. S. MERRITT & R. B. SMITH (1965) *Meteorological interpretation of Nimbus High Resolution Infra-red data*, Final Report, Contract No. NAS5-9554, ARACON Geophysics Company, Concord, Massachusetts.

W. L. WOODLEY & B. SANCHO (1971) A first step towards rainfall estimation from satellite photographs, *Weather* 26, pp. 279–89.

WORLD METEOROLOGICAL ORGANIZATION (1956) *International cloud atlas*, Volumes I and II, World Meteorological Organization, Geneva.

WORLD METEOROLOGICAL ORGANIZATION (1966) *The use of satellite pictures in weather analysis and forecasting*, W.M.O. Technical Note, No. 75, Geneva, 150 pp.

R. J. YOUNKIN, J. A. LA RUE & F. SANDERS (1965) The objective prediction of clouds and precipitation using vertically-integrated moisture and adiabatic vertical motions, *Journal of Applied Meteorology* 4, pp. 3–17.

Chapter 5

R. K. ANDERSON (1969) Exercise in estimating high-level winds, in *Satellite Meteorology*, Bureau of Meteorology, Melbourne, Australia, pp. 35–6.

E. C. BARRETT (1967) *Viewing weather from space*, Longmans, London, & Praeger, New York, 154 pp.

E. C. BARRETT (1970a) A contribution to the dynamic climatology of the equatorial eastern Pacific and Central America, based on meteorological satellite data, *Transactions, Institute of British Geographers* 50, pp. 25–53.

E. C. BARRETT (1970b) Rethinking climatology, chapter 4 in *Progress in Geography*, Volume 2, C. Board, R. J. Chorley, P. Haggett & D. R. Stoddart (eds.), Edward Arnold, London, pp. 154–205.

H. A. BEDIENT, W. G. COLLINS & G. DENT (1967) An operational tropical analysis system, *Monthly Weather Review* 95, pp. 942–9.

K. P. CHOPRA & L. F. HUBERT (1965) Mesoscale eddies in wake of islands, *Journal of Atmospheric Sciences* 22, pp. 652–5.

J. H. CONOVER (1964) The identification and significance of orographically-induced clouds observed by Tiros satellites, *Journal of Applied Meteorology* 3, pp. 226–34.

J. W. COOLEY & J. W. TUKEY (1965) An algorithm for the machine computation of complex Fourier series, *Mathematical Computation* 19, pp. 297–301.

R. M. ENDLICH, D. E. WOLF, D. J. HALL & A. E. BRAIN (1971) Use of a pattern recognition technique for determining cloud motions from sequences of satellite photographs, *Journal of Applied Meteorology* 10, pp. 105–17.

C. O. ERICKSON (1964) Satellite photographs of convective clouds and their relation to vertical wind shear, *Monthly Weather Review* 92, pp. 283–96.

W. E. EVANS & S. M. SEREBRENY (1969) *Construction of A.T.S. cloud console*, Final Report, Contract NAS 5-11652, Stanford Research Institute, Menlo Park, California, 40 pp.

A. J. FALLER (1965) Large eddies in the atmospheric boundary layer and their probable role in the formation of cloud rows, *Journal of Atmospheric Sciences* 22, pp. 176–84.

S. FRITZ (1965) The significance of mountain lee waves as seen from satellite pictures, *Journal of Applied Meteorology* 4, pp. 31–7.

T. T. FUJITA, D. L. BRADBURY, C. MURINO & L. HULL (1968) *A study of meso-scale cloud motions computed from A.T.S.-1 and terrestrial photographs*, S.M.R.P. Research Paper 71, University of Chicago, 25 pp.

T. T. FUJITA, K. WATANABE & T. IZAWA (1969) *Formation and structure of equatorial anticyclones caused by large-scale cross-equatorial flows determined by ATS-I photographs*, S.M.R.P. Research Paper No. 78, University of Chicago, 37 pp.

D. C. GABY (1967) Cumulus cloud lines versus surface winds in equatorial latitudes, *Monthly Weather Review* 95, pp. 203–8.

L. F. HUBERT & A. TIMCHALK (1969) Estimating hurricane wind speeds from satellite pictures, *Monthly Weather Review* 97, pp. 382–3.

L. F. HUBERT, A. TIMCHALK & S. FRITZ (1969) *Estimating maximum wind speed of tropical storms from high resolution infra-red data*, ESSA Technical Report, N.E.S.C. 50, 33 pp.

L. F. HUBERT & L. F. WHITNEY (1971) Wind estimation from geostationary satellite pictures, *Monthly Weather Review* 99, pp. 665–72.

G. JAGER (1968) *Operational utilization of upper tropospheric wind estimates based on meteorological photographs*, ESSA Technical Memorandum, N.E.S.C.T.M.-8, 23 pp.

G. JAGER, W. A. FOLLANSBEE & V. J. OLIVER (1968) *Operational utilization of upper tropospheric wind estimates based on meteorological satellite photographs*, ESSA Technical Memorandum, N.E.S.C.T.M.-8, Washington, D.C., 23 pp.

J. P. KUETTNER (1959) The band structure of the atmosphere, *Tellus* 11, pp. 267–72.

J. A. LEESE & C. S. NOVAK (1971) An automated technique for obtaining cloud motion from geosynchronous satellite data using cross-correlation, *Journal of Applied Meteorology* 10, pp. 118–32.

E. N. LORENZ (1967) *The nature and theory of the general circulation of the atmosphere*, World Meteorological Organization, Technical Note No. 218, TP115, Geneva, 161 pp.

W. A. LYONS & T. T. FUJITA (1968) Mesoscale motions in oceanic stratus as revealed by satellite data, *Monthly Weather Review* 96, pp. 304–14.

J. S. MALKUS & H. RIEHL (1964) *Cloud structure and distributions over the tropical Pacific Ocean*, University of California Press, Berkeley & Los Angeles, 230 pp.

E. S. MERRITT & C. W. C. ROGERS (1965) *Meteorological satellite studies of mid-latitude atmospheric circulation*, Final Report, Contract No. N62306-1584, ARACON Geophysics Co., Concord, Massachusetts.

V. J. OLIVER (1969a) Tropical storm classification system, in *Satellite Meteorology*, Bureau of Meteorology, Melbourne, Australia, pp. 27–9 (d).

V. J. OLIVER (1969b) Estimating high-level winds from picture data, in *Satellite Meteorology*, Bureau of Meteorology, Melbourne, Australia, pp. 33–4.

V. J. OLIVER & R. K. ANDERSON (1969) Circulation in the tropics as revealed by satellite data, *Bulletin of the American Meteorological Society* 50, pp. 702–7.

H. A. PANOFSKY & G. W. BRIER (1958) *Some applications of statistics to meteorology*, Pennsylvania State University Press, 224 pp.

A. TIMCHALK, L. F. HUBERT & S. FRITZ (1965) *Wind speeds from Tiros pictures of storms in the Tropics*, M.S.L. Report No. 33, U.S. Weather Bureau, Washington, D.C., 33 pp.

L. F. WHITNEY (1966) On locating jet streams from Tiros photographs, *Monthly Weather Review* 94, pp. 127–38.

J. S. WINSTON & V. R. TAYLOR (1967) *Atlas of world maps of long-wave radiation and albedo for seasons and months based on measurements from Tiros IV & VII*, ESSA Technical Report, N.E.S.C. 43, Washington, D.C., 32 pp.

Chapter 6

W. R. BANDEEN, M. HALEV & I. STRANGE (1965) *A radiation climatology in the visible and infra-red from the Tiros meteorological satellites*, N.A.S.A. Technical Note No. TN D-2534, Goddard Space Flight Center, Greenbelt, Maryland, 30 pp.

M. I. BUDYKO (1963) *Atlas of the heat balance of the Earth*, Moscow, 69 pp.

A. J. DRUMMOND, J. A. HICKEY, W. J. SCHOLES & E. G. LANE (1968) New value for the solar constant of radiation, *Nature* 218, pp. 258–61.

H. FLOHN (1969) *Climate and weather*, Weidenfeld & Nicolson, London, 253 pp.

W. L. GODSON (1958) Meteorological applications of Earth satellites, *Royal Astronomical Society of Canada* 52, pp. 49–56.

F. S. JOHNSON (1954) The solar constant, *Journal of Meteorology* 11, pp. 431–9.

K. Y. KONDRATIEV & G. A. NIKOLSKY (1968) Direct solar radiation & areas of structure of the atmosphere from balloon measurements in the period of I.Q.S.Y., *Scientific Papers dedicated to A. Angström*, Stockholm, Sweden, pp. 14–23.

H. H. LETTAU (1954) A study of the mass, momentum and energy budget of the atmosphere, *Archiv für Meteorologie, Geophysik und Bioklimatologie* Series A, Band 7, pp. 133–57.

J. LONDON (1957) *A study of the atmospheric heat balance*, Final Report, Contract AF19(122)-165, Dept. of Meteorology and Oceanography, New York University, 99 pp.

T. H. MACDONALD (1957) Personal communication.

E. RASCHKE & W. R. BANDEEN (1970) The radiation balance of the planet Earth from radiation measurements of the satellite Nimbus II, *Journal of Applied Meteorology* 9, pp. 215–38.

E. RASCHKE & M. PASTERNAK (1967) *The global radiation balance of the Earth-atmosphere system obtained from the radiation data of the meteorological satellite Nimbus II*, N.A.S.A. Report No. X-622-67-383, Goddard Space Flight Center, Greenbelt, Maryland, 19 pp.

S. I. RASOOL (1964) Global distribution of the net energy balance of the atmosphere from Tiros radiation data, *Science* 143, pp. 567–9.

S. I. RASOOL & C. PRABHAKARA (1965) *Radiation studies from meteorological satellites*, Report No. 65-1, Geophysical Sciences Laboratory, New York University, 32 pp.

S. I. RASOOL & C. PRABHAKARA (1966) Heat budget of the southern hemisphere, in *Problems of atmospheric circulations*, Spartan Books, Washington, D.C., pp. 75–92.

W. D. SELLERS (1965) *Physical climatology*, University of Chicago Press, 272 pp.

O. G. SUTTON (1965) The energy of the atmosphere, *Science Journal*, I, pp. 78–81.

R. STAIR & R. G. JOHNSTON (1956) Prelimary spectroradiometric measurements of the solar constant, *Journal of Research of the National Bureau of Studies* 57, pp. 205–11.

M. P. THEKAEKARA *et al.* (1968) *The solar constant*, G.S.F.C. Document X-322-68-304, pp. 60–3.

T. H. VONDER HAAR (1968) Variations of the Earth's radiation budget, in *Meteorological Satellite Instrumentation and Data Processing*, Final Report,

Contract NASw-65, 1958–68, Dept. of Meteorology, University of Wisconsin, 179 pp.

T. H. VONDER HAAR & K. J. HANSON (1969) Absorption of solar radiation in the tropics, *Journal of the Atmospheric Sciences* 26, pp. 652–5.

T. H. VONDER HAAR & V. E. SUOMI (1969) Satellite observations of the Earth's radiation budget, *Science* 163, pp. 667–9.

T. H. VONDER HAAR & V. E. SUOMI (1971) Measurements of the Earth's radiation budget from satellites during a five-year period. Part I: Extended time and space means, *Journal of the Atmospheric Sciences* 28, pp. 305–14.

W. WASHINGTON (1968) Computer simulation of the Earth's atmosphere, *Science Journal* 4, pp. 36–41.

J. S. WINSTON (1967) Planetary-scale characteristics of monthly mean long-wave radiation and albedo and some year-to-year variation, *Monthly Weather Review* 95, pp. 235–56.

J. S. WINSTON (1969) Global distribution of cloudiness and radiation as measured from weather satellites, in *World Survey of Climatology*, Volume 4, *Climate of the free atmosphere*, D. F. Rex (ed.), Elsevier, Amsterdam – London – New York, pp. 247–80.

J. S. WINSTON & V. R. TAYLOR (1967) *Atlas of world maps of long-wave radiation and albedo for seasons and months based on measurements from Tiros IV & VII*, ESSA Technical Report, N.E.S.C. 43, Washington, D.C., 32 pp.

Chapter 7

J. ADEM (1964a) On the physical basis for the numerical prediction of monthly and seasonal temperatures in the troposphere-ocean-continent system, *Monthly Weather Review* 92, pp. 91–104.

J. ADEM (1964b) On the normal state of the troposphere-ocean-continent system in the northern hemisphere, *Geofisica Internacionale* 4, pp. 3–16.

J. ADEM (1965) Experiments aimed at monthly and seasonal numerical weather prediction, *Monthly Weather Review* 93, pp. 495–503.

J. ADEM (1967) Relations among winds, temperature, pressure and density with particular reference to monthly averages, *Monthly Weather Review* 95, pp. 531–9.

J. F. ANDREWS (1964) The circulation and weather of 1963, *Weatherwise* 71, pp. 9–15.

W. R. BANDEEN, M. HALEV & I. STRANGE (1965) *A radiation climatology in the visible and infra-red from the Tiros meteorological satellites*, N.A.S.A. Technical Note No. TN D-2534, Goddard Space Flight Center, Greenbelt, Maryland, 30 pp.

J. C. BARNES (1966) Note on the use of satellite observations to determine average cloudiness over a region, *Journal of Geophysical Research* 75, pp. 6137–40.

E. C. BARRETT (1971) The tropical Far East: ESSA satellite evaluations of high season climatic patterns, *Geographical Journal* 137, pp. 535–55.

P. F. CLAPP (1964) Global cloud cover for seasons using Tiros nephanalyses, *Monthly Weather Review* 92, pp. 495–507.

P. F. CLAPP (1965) *Parameterization of certain atmospheric heat sources and sinks for use in a numerical model for monthly and seasonal forecasting*, Mimeo Report, U.S. Weather Bureau, Extended Forecast Division, 55 pp.

P. F. CLAPP (1968) *Northern hemispheric cloud cover for selected late fall seasons using Tiros nephanalyses*, ESSA Technical Memorandum, W.B.T.M. N.M.C. 44, Washington, D.C., 12 pp.

P. R. CROWE (1949) The trade-wind circulation of the world, *Transactions, Institute of British Geographers* 15, pp. 37–56.

P. R. CROWE (1950) The seasonal variation in the strength of the trades, *Transactions, Institute of British Geographers* 17, pp. 21–76.

H. E. LANDSBERG (1945) Climatology, in *Handbook of meteorology*, F. A. Berry, E. Bollay & N. R. Beers (eds.), American Meteorological Society, Boston, Massachusetts, pp. 927–98.

J. NAMIAS (1960) Influence of abnormal surface heat sources and sinks on atmospheric behaviour, *Proceedings of the International Symposium on National Weather Prediction*, Tokyo, pp. 42–65.

J. W. POSEY (1962) The weather and circulation of March, 1962 – A month with an unusually strong high-latitude block, *Monthly Weather Review* 90, pp. 252–8.

H. RIEHL (1954) *Tropical meteorology*, McGraw-Hill, New York & London, 392 pp.

G. R. RUMNEY (1968) *Climatology and the world's climates*, Macmillan, New York, & Collier-Macmillan, London, 656 pp.

V. V. SALOMONSON (1969), *Cloud statistics in Earth Resources Technology Satellite (ERTS) mission planning*, N.A.S.A. Report No. X-622-69-386, Goddard Space Flight Center, Greenbelt, Maryland, 19 pp.

J. M. YOUNG (1967) Variability in estimating total cloud cover from satellite pictures, *Journal of Applied Meteorology* 6, pp. 573–9.

Chapter 8

L. J. ALLISON, T. L. GRAY & G. WARNECKE (1964) *A quasi-global presentation of Tiros III radiation data*, N.A.S.A. Special Publication SP-53, Greenbelt, Maryland, 35 pp.

L. J. ALLISON, E. R. KREINS, F. A. GODSHALL & G. WARNECKE (1969) *Examples of the usefulness of satellite data in general atmospheric circulation research, Part I – Monthly global atmospheric circulation character-*

istics as reflected in Tiros VII radiometric measurements, N.A.S.A. Technical Note, N.A.S.A. TN D-5630, Greenbelt, Maryland, 35 pp.

E. C. BARRETT (1967) *Viewing weather from space*, Longmans, London, & Praeger, New York, 154 pp.

R. G. BARRY & R. J. CHORLEY (1968) *Atmosphere, weather and climate*, Methuen, London, 319 pp. 2nd edition 1971, 380 pp.

B. W. BOVILLE (1963) What are the causes of the Aleutian anticyclone?, *Proceedings of the International Symposium on the Stratospheric Mesospheric Circulation*, Berlin, pp. 107–20.

S. FRITZ (1969) On the question of measuring the vertical temperature distribution of the atmosphere from satellites, *Monthly Weather Review* 97, pp. 712–15.

S. FRITZ (1970) Earth's radiation to space at 15 microns – stratospheric temperature variations, *Journal of Applied Meteorology* 9, pp. 815–24.

S. FRITZ & S. D. SOULES (1970) Large-scale temperature changes in the stratosphere observed from Nimbus III, *Journal of the Atmospheric Sciences* 27, pp. 1091–7.

F. K. HARE (1962) The stratosphere, *Geographical Review* 52, pp. 525–47.

H. W. HILL (1963) The weather in lower latitudes of the south-west Pacific associated with the passage of disturbances in the middle latitude westerlies, *Proceedings, Symposium of Tropical Meteorology*, New Zealand, pp. 352–65.

H.M.S.O. (1971) *Handbook of aviation*, H.M.S.O., Met. Office 818, (A.P.3340), 404 pp.

G. JAGER, W. FOLLANSBEE & V. J. OLIVER (1968) *Operational utilization of upper tropospheric wind estimates based on meteorological satellite photographs*, ESSA Technical Memorandum, N.E.S.C.T.M.-8, Washington, D.C., 23 pp.

J. KORNFIELD, A. F. HASLER, K. J. HANSON & V. E. SUOMI (1967) Photographic cloud climatology from ESSA 3 and 5 computer-produced mosaics, *Bulletin of the American Meteorological Society* 48, pp. 878–83.

V. E. LALLY (1967) *Superpressure balloons for horizontal soundings of the atmosphere*, N.C.A.R. Technical Notes, N.C.A.R.-TN-28, National Center for Atmospheric Research, Boulder, Colorado, 167 pp.

V. E. LALLY, E. W. LICHFIELD & S. B. SOLOT (1966) The southern hemisphere *GHOST* experiment, *W.M.O. Bulletin* 15, pp. 124–8.

E. N. LORENZ (1967) *The nature and theory of the general circulation of the atmosphere*, W.M.O., Geneva, No. 218, T.P. 115.

S. MANABE & G. B. HUNT (1968) Experiments with a stratospheric general circulation model: I: Radiation and dynamic aspects, *Monthly Weather Review* 96, pp. 477–502.

B. J. MASON (1971) Global atmospheric research programme, *Nature* 233, pp. 382–8.

Y. MINTZ & G. DEAN (1952) *The observed mean field of motion of the atmosphere*, Geophysical Research Paper, 17, Air Force Cambridge Research Laboratories, Cambridge, Massachusetts, 37 pp.

W. NORDBERG (1965) Geophysical observations from Nimbus I, *Science* 150, pp. 559–72.

E. PALMEN (1951) The role of atmospheric disturbances in the general circulation, *Quarterly Journal of the Royal Meteorological Society* 77, pp. 337–54.

S. PETTERSSEN (1969) *Introduction to meteorology*, 3rd edition, McGraw-Hill, New York, 333 pp.

E. REITER (1969) Tropospheric circulation and jet streams, in *World Survey of Climatology*, Volume 4, *Climate of the free atmosphere*, D. F. Rex (ed.), Elsevier, Amsterdam, London & New York, pp. 85–204.

H. RIEHL (1954) *Tropical meteorology*, McGraw-Hill, New York & London, 392 pp.

C. G. ROSSBY (1949) On the nature of the general circulation of the lower atmosphere, in *The atmosphere of the earth and planets*, G. P. Kuiper (ed.), Chicago, pp. 16–48.

K. WEGE (1957) Druck-, Temperatur-, und Stromungsverhältnisse in der Stratosphäre über der Nordhalbkugel, *Meteorol. Abhandl. di. Feien Universität Berlin*, Band V, Heft 4.

H. C. WILLETT & F. SANDERS (1959) *Descriptive Meteorology*, Academic Press, New York, (2nd Edn.), 355 pp.

Chapter 9

M. ALAKA (1964) *Problems of tropical meteorology: A survey*, Technical Note No. 62, World Meteorological Organization, Geneva, 36 pp.

L. J. ALLISON (1972) *Air-sea interaction in the tropical Pacific Ocean*, National Aeronautics and Space Administration, Technical Note. TN D-6684, Washington D.C., 84 pp.

AMERICAN METEOROLOGICAL SOCIETY (1959) *Glossary of meteorology*, R. E. Huschke (ed.), American Meteorological Society, Boston, Massachusetts.

G. C. ASNANI (1968) The equatorial cell in the general circulation, *Journal of the Atmospheric Sciences* 25, pp. 133–4.

E. C. BARRETT (1970a) A contribution to the dynamic climatology of the equatorial eastern Pacific and Central America based on meteorological satellite data, *Transactions, Institute of British Geographers* 50, pp. 25–53.

E. C. BARRETT (1970b) Rethinking climatology, in *Progress in Geography*, Volume II, C. Board, R. J. Chorley, P. Haggett & D. R. Stoddart (eds.), Edward Arnold, London, pp. 154–205.

R. G. BARRY & R. J. CHORLEY (1968) *Atmosphere, weather and climate*, Methuen, London, 319 pp. 2nd Edition 1971, 380 pp.

J. BJERKNES (1969) Atmospheric teleconnections from the equatorial Pacific, *Monthly Weather Review* 97, pp. 163–72.

T. N. CARLSON (1969) Synoptic histories of three African disturbances that developed into Atlantic hurricanes, *Monthly Weather Review* 97, pp. 256–76.

C-P. CHANG (1970) Westward propagating cloud patterns in the tropical Pacific as seen from time-composite satellite photographs, *Journal of the Atmospheric Sciences* 27, pp. 133–8.

P. R. CROWE (1951) Wind and weather in the equatorial zone, *Transactions, Institute of British Geographers* 17, pp. 21–48.

P. R. CROWE (1971) *Concepts in climatology*, Longmans, London, 589 pp.

R. W. FETT (1966) Upper-level structure of the formative tropical cyclone, *Monthly Weather Review* 94, pp. 9–18.

H. FLOHN (1969) *Climate and weather*, Weidenfeld & Nicolson, London, 253 pp.

N. FRANK (1969) The 'inverted-V' cloud pattern – an easterly wave?, *Monthly Weather Review* 97, pp. 130–40.

S. FRITZ (1964) Pictures from meteorological satellites and their interpretation, *Space Science Reviews* 3, pp. 541–80.

T. T. FUJITA, K. WATANABE & T. IZAWA (1969) Formation and structure of equatorial anticyclones caused by large-scale cross-equatorial flows determined by ATS-1 photographs, *Journal of Applied Meteorology* 8, pp. 649–67.

M. GARBELL (1947) *Tropical and equatorial meteorology*, Pitman, London, 237 pp.

GLOBAL ATMOSPHERIC RESEARCH PROGRAM (1968) Joint Organising Committee of GARP Report on the first session of the Study Group on tropical disturbances, Madison, October 1968.

F. A. GODSHALL (1968) Intertropical convergence zone and mean cloud amount in the tropical Pacific Ocean, *Monthly Weather Review* 96, pp. 172–5.

A. GRUBER (1972) Fluctuations in the position of the ITCZ in the Atlantic and Pacific Oceans, *Journal of the Atmospheric Sciences* 29, pp. 193–7.

S. HASTENRATH (1968) On mean meridional circulation in the tropics, *Journal of the Atmospheric Sciences* 25, pp. 979–83.

W. HEWES (1968) Unpublished Ph.D. thesis, Oregon State University.

J. J. HIDORE (1969) *A geography of the atmosphere*, William C. Brown, Dubuque, Iowa, 106 pp.

J. R. HOLTON, J. M. WALLACE & J. A. YOUNG (1971) On boundary layer dynamics and the I.T.C.Z., *Journal of the Atmospheric Sciences* 28, pp. 275–80.

L. F. HUBERT, A. F. KRUEGER & J. S. WINSTON (1969) The double intertropical convergence zone – fact or fiction?, *Journal of the Atmospheric Sciences* 26, pp. 771–3.

D. H. JOHNSON (1970) The role of the tropics in the global circulation, in *The Global Circulation of the Atmosphere*, Royal Meteorological Society, London, pp. 113–36.

J. KORNFIELD & A. F. HASLER (1970) A photographical summary of the Earth's cloud cover for the year 1967, *Journal of Applied Meteorology* 8, pp. 687–700.

A. F. KRUEGER & T. I. GRAY (1969) Long-term variations in equatorial circulation and rainfall, *Monthly Weather Review* 97, pp. 700–11.

V. J. OLIVER & R. K. RICHARDSON (1971) Seasonal changes in tropical cloud distribution, in *Application of meteorological satellite data in analysis and forecasting*, ESSA Technical Report, N.E.S.C.-51, pp. 4-B-1 to 4-B-5.

C. E. PALMER (1951) Tropical Meteorology, in *Compendium of Meteorology*, T. F. Malone (ed.), American Meteorological Society, Boston, Massachusetts, pp. 859–80.

A. C. PIKE (1971) Intertropical convergence zone studied with an interacting atmosphere and ocean model, *Monthly Weather Review* 99, pp. 469–77.

H. RIEHL (1945) *Waves in the easterlies & the polar front in the tropics*, Miscellaneous Report No. 17, Department of Meteorology, University of Chicago.

H. RIEHL (1954) *Tropical meteorology*, McGraw-Hill, New York & London, 392 pp.

H. RIEHL (1969) On the role of the tropics in the general circulation of the atmosphere, *Weather* 24, pp. 288–308.

H. RIEHL & J. S. MALKUS (1958) On the heat balance in the equatorial trough zone, *Geophysica* 6, pp. 503–38.

J. C. SADLER (1959) Wind regimes of the troposphere and stratosphere over the equatorial and sub-equatorial central Pacific, in Volume 13, Secretariat, North Pacific Congress, Department of Science, Bangkok, 1959.

J. C. SADLER (1966) *The easterly wave – the biggest hoax in tropical meteorology*, Paper presented at the National Center for Atmospheric Research, Boulder, Colorado, 1966.

K. SAHA (1971) Mean cloud distributions over tropical oceans, *Tellus* 23, pp. 183–95.

D. N. SIKDAR & V. E. SUOMI (1971) Time variations of tropical energetics as viewed from a geostationary altitude, *Journal of the Atmospheric Sciences* 28, pp. 170–80.

R. H. SIMPSON, N. L. FRANK, D. SHIDELER & H. M. JOHNSON (1968) Atlantic tropical disturbances, 1967, *Monthly Weather Review* 96, pp. 251–9.

V. R. TAYLOR & J. S. WINSTON (1968) *Monthly and seasonal mean global charts of brightness from ESSA 3 and ESSA 5 digitized pictures, February 1967–February 1968*, ESSA Technical Report, N.E.S.C. 46, Washington, D.C., 9 pp.

B. W. THOMPSON (1965) *The climate of Africa*, Oxford University Press, New York.

G. T. TREWARTHA (1968) *An introduction to climate*, McGraw-Hill, New York & London, 395 pp.

T. H. VONDER HAAR & K. J. HANSON (1969) Absorption of solar radiation in the tropics, *Journal of the Atmospheric Sciences* 26, pp. 652–5.

SIR GILBERT WALKER (1923) Correlations of seasonal variations of weather, VIII, A preliminary survey of world weather, *Memoirs of the Indian Meteorological Department* 24, pp. 75–131.

SIR GILBERT WALKER (1924) Correlations of seasonal variations of weather, IX, A further study of world weather, *Memoirs of the Indian Meteorological Department* 25, pp. 275–332.

J. M. WALLACE (1970) *Time-longitude sections of tropical cloudiness*, ESSA Technical Report, N.E.S.C.-56, Washington, D.C., 37 pp.

J. S. WINSTON (1967) Planetary-scale characteristics of monthly mean long-wave radiation and albedo and some year-to-year variations, *Monthly Weather Review* 95, pp. 235–56.

M. YANAI, T. MARUYAMA, T. NITTA & Y. HAYASHI (1968) Power spectra of large-scale disturbances over the Tropical Pacific, *Journal of the Meteorological Society of Japan* 46, pp. 308–23.

Chapter 10

M. ALAKA (1964) *Problems of tropical meteorology: A survey*, Technical Note No. 62, World Meteorological Organization, Geneva, 36 pp.

E. C. BARRETT (1970) Rethinking climatology, chapter 4 in *Progress in Geography*, 2, C. Board, R. J. Chorley, P. Haggett & D. R. Stoddart (eds.), Edward Arnold, London, pp. 154–205.

G. W. BRIER & J. SIMPSON (1969) Tropical cloudiness and rainfall related to pressure and tidal variations, *Quarterly Journal of the Royal Meteorological Society* 95, pp. 120–47.

A. T. BRUNT (1966) Rainfall associated with tropical cyclones in the north-east Australian region, *Australian Meteorological Magazine* 14, pp. 85–109.

A. T. BRUNT (1968) Space-time relations of cyclonic rainfall in the north-east Australian region, *Civil Engineering Transactions*, Institute of Engineers, Australia, pp. 40–6.

T. N. CARLSON (1969) Some remarks on African disturbances and their progress over the tropical Atlantic, *Monthly Weather Review* 97, pp. 716–26.

T. N. CARLSON (1971) Weather note – An apparent relationship between the sea-surface temperature of the tropical Atlantic and the development of African disturbances into tropical storms, *Monthly Weather Review* 99, pp. 309–10.

M. COX & G. JAGER (1969) *A satellite analysis of twin tropical cyclones in the western Pacific*, ESSA Technical Memorandum, WBTM SOS 5, Silver Spring, Maryland, 24 pp.

W. J. DENNEY (1969) Eastern Pacific hurricane season of 1968, *Monthly Weather Review* 97, pp. 207–18.

W. J. DENNEY (1971) Eastern Pacific hurricane season of 1970, *Monthly Weather Review* 99, pp. 286–301.

G. E. DUNN & B. I. MILLER (1964) *Atlantic Hurricanes*, Louisiana State University Press, Baton Rouge, 326 pp.

V. F. DVORAK (1972) *A technique for the analysis and forecasting of tropical cyclone intensities from satellite pictures*, NOAA Technical Memorandum, NESS 36, Washington, D.C., 15 pp.

C. O. ERICKSON (1967) Some aspects of the development of hurricane Dorothy, *Monthly Weather Review* 95, pp. 121–30.

R. W. FETT (1964) Aspects of hurricane structure – new model considerations, *Monthly Weather Review* 92, pp. 43–52.

R. W. FETT (1966) Upper-level structure of the formative tropical cyclone, *Monthly Weather Review* 94, pp. 9–18.

R. W. FETT (1968a) Typhoon formation within the zone of the inter-tropical convergence, *Monthly Weather Review* 96, pp. 106–17.

R. W. FETT (1968b) Some unusual aspects concerning the development & structure of typhoon Billie, July 1967, *Monthly Weather Review* 96, pp. 637–48.

N. L. FRANK (1971) Atlantic tropical systems of 1970, *Monthly Weather Review* 99, pp. 281–5.

N. L. FRANK & S. A. HUSAIN (1971) The deadliest tropical cyclone in history?, *Bulletin of the American Meteorological Society* 52, pp. 438–45.

S. FRITZ, L. F. HUBERT & A. TIMCHALK (1966) Some inferences from satellite pictures of tropical disturbances, *Monthly Weather Review* 94, pp. 231–6.

M. GARSTANG & T. R. VISVANATHAN (1967) *Solar and lunar influences on rainfall*, Final Report, Part I, ESSA Grant No. E-18-67 (G), Washington, D.C.

D. A. GOUDEAU & W. C. CONNOR (1968) Storm surge over the Mississippi river delta accompanying hurricane Betsy 1965, *Monthly Weather Review* 96, pp. 118–24.

W. M. GRAY (1968) A global view of the origin of tropical disturbances & storms, *Monthly Weather Review* 96, pp. 669–700.

R. L. HOLLE (1969) Some aspects of tropical oceanic cloud populations, *Journal of Applied Meteorology* 7, pp. 173–83.

J. W. HUTCHINGS (ed.) (1964) *Proceedings of the symposium on tropical meteorology*, New Zealand Meteorological Service, Wellington, 737 pp.

D. H. JOHNSON (1970) The role of the tropics in the global circulation, in *The global circulation of the atmosphere*, Royal Meteorological Society, London, pp. 113–36.

W. L. KISER, T. H. CARPENTER & G. W. BRIER (1963) The atmospheric tides at Wake Island, *Monthly Weather Review* 91, pp. 556–72.

R. L. LAVOIE (1963) *Some aspects of the meteorology of the tropical Pacific viewed from an atoll*, Report No. 27, Hawaii Institute of Geophysics, 25 pp.

R. M. LEIGH (1969) A meteorological satellite study of a double vortex system over the western Pacific Ocean, *Australian Meteorological Magazine* 17, pp. 48–62.

J. MACDONALD (1968) Weather modifications, *Science Journal* 5, pp. 39–44.

J. S. MALKUS (1963) Convective processes in the tropics, *Proceedings of the symposium on Tropical Meteorology*, W.M.O., Rotorua, New Zealand, pp. 247–77.

J. S. MALKUS & R. H. SIMPSON (1964) Note on the potentialities of cumulonimbus & hurricane seeding experiments, *Journal of Applied Meteorology* 3, pp. 470–5.

D. W. MARTIN & O. KARST (1968) *A census of cloud systems over the tropical Pacific, Studies in Atmospheric energetics based on aerospace probing*, Space Science and Engineering Center, University of Wisconsin, pp. 37–50.

E. S. MERRITT & C. J. BOWLEY (1966) *Analyses of diurnal variations in Tiros VII 8–12 micron window radiation over Indonesia and Malayasia*, Second Quarterly Report, Contract No. NAS-10151 Goddard Space Flight Center, Greenbelt, Maryland.

B. I. MILLER (1967) Characteristics of hurricanes, *Science* 157, pp. 1389–99.

J. NAMIAS (1969) On the causes of the small number of Atlantic hurricanes in 1968, *Monthly Weather Review* 97, pp. 346–8.

V. J. OLIVER (1969) Tropical storm classification system, in *Satellite Meteorology*, Bureau of Meteorology, Melbourne, Australia, pp. 27–9.

D. E. PEDGLEY (1969) Cyclones along the Arabian coast, *Weather* 24, pp. 456–68.

H. RIEHL (1954) *Tropical meteorology*, McGraw-Hill, New York & London, 392 pp.

H. RIEHL & J. S. MALKUS (1961) Some aspects of hurricane Daisy, 1958, *Tellus* 13, pp. 181–213.

G. R. RUMNEY (1968) *Climatology & the world's climates*, Macmillan, New York, & Collier-Macmillan, London, 656 pp.

J. C. SADLER (1963) Utilization of meteorological satellite cloud data in tropical meteorology, in *Rocket and Satellite Meteorology*, H. Wexler & J. E. Caskey (eds.), North Holland, Amsterdam, pp. 333–57.

J. C. SADLER (1964) Tropical cyclones of the eastern North Pacific as revealed by Tiros observations, *Journal of Applied Meteorology* 3, pp. 347–66.

S. C. SERRA (1971) Hurricanes and tropical storms of the west of Mexico, *Monthly Weather Review* 99, pp. 302–8.

D. N. SIKDAR & V. E. SUOMI (1971) Time variations of tropical energetics as viewed from a geostationary altitude, *Journal of the Atmospheric Sciences* 28, pp. 170–80.

J. SIMPSON, M. GARSTANG, E. J. ZIPSER & G. A. DEAN (1967) A study of a non-deepening tropical disturbance, *Journal of Applied Meteorology* 6, pp. 237–54.

A. L. SUGG & P. J. HEBERT (1969) The Atlantic hurricane season of 1968, *Monthly Weather Review* 97, pp. 225–39.

B. W. THOMPSON (1951) An essay on the general circulation over south-east Asia and the west Pacific, *Quaterly Journal of the Royal Meteorological Society* 77, pp. 569–97.

D. Q. WARK, G. YAMAMOTO & J. LIENESCH (1962) Methods of estimating infra-red flux and surface temperature from meteorological satellites, *Journal of the Atmospheric Sciences* 19, pp. 369–84.

G. WARNECKE, L. J. ALLISON, E. R. KREINS & L. M. MCMILLIN (1968) *A satellite view of typhoon Marie 1966 development*, National Aeronautics and Space Administration, Technical Note, NASA-TN D-4757, Washington, D.C., 94 pp.

J. S. WINSTON (1969) Global distribution of cloudiness and radiation as measured from weather satellites, in *World Survey of Climatology*, Volume 4, *Climate of the Free Atmosphere*, D. F. Rex (ed.), Elsevier, Amsterdam – London – New York, pp. 247–80.

WORLD METEOROLOGICAL ORGANIZATION (1959) *Tropical circulation patterns*, W.M.O. Technical Note No. 9, Geneva.

E. J. ZIPSER (1970) The Line Islands experiment: its place in tropical meteorology, *Bulletin of the American Meteorological Society* 51, pp. 1130–46.

Chapter 11

B. W. ATKINSON (1968) *The weather business*, Aldus Books, London, 192 pp.

E. C. BARRETT (1971) The tropical Far East: ESSA satellite evaluations of high season climatic patterns, *Geographical Journal* 137, pp. 535–55.

R. G. BARRY & R. J. CHORLEY (1968) *Atmosphere, weather and climate*, Methuen, London, 319 pp. 2nd edition 1971, 380 pp.

A. F. BUNKER (1967) Cloud formations leeward of India during the north-east monsoon, *Journal of the Atmospheric Sciences* 24, pp. 497–507.

J. FINDLATER (1969) A major low-level air current near the Indian Ocean during the northern summer, *Quarterly Journal of the Royal Meteorological Society* 95, pp. 362–80.

M. G. HAMILTON (1973) *Satellite studies of the south Asian Summer monsoon*, Unpublished Ph.D. thesis, University of Bristol, 351 pp.

P. KOTESWARAM (1958) The easterly jet stream in the tropics, *Tellus* 10, pp. 43–57.

E. R. KREINS & L. J. ALLISON (1969) *An atlas of Tiros VII monthly maps of emitted radiation in the 8–12 micron atmospheric window over the Indian Ocean area*, N.A.S.A. Technical Note, TN D-5101, Goddard Space Flight Center, Greenbelt, Maryland, 25 pp.

J. G. LOCKWOOD (1965) The Indian monsoon – a review, *Weather* 20, pp. 2–8.

A. MILLER (1966) *Meteorology*, Charles E. Merrill, Columbus, Ohio, 128 pp.

C. S. RAMAGE (1971) *Monsoon meteorology*, Academic Press, New York & London, 296 pp.

K. M. RAMAMURTHI & F. JAMBUNATHAN (1965) On the onset of the south-west monsoon rains along extreme south-west coast of peninsula India, *Proceedings of the Symposium on the Meteorological Results of the International Indian Ocean Expedition*, Bombay, pp. 374–9.

G. R. RUMNEY (1968) *Climatology and the world's climates*, Macmillan, New York, & Collier-Macmillan, London, 656 pp.

J. S. SAWYER (1947) The structure of the intertropical front over north-west India during the south-west monsoon, *Quarterly Journal of the Royal Meteorological Society* 73, pp. 346–69.

O. G. SUTTON (1964) *The challenge of the atmosphere*, Hutchinson, London, 228 pp.

B. W. THOMPSON (1951) An essay on the general circulation over south-east Asia and the West Pacific, *Quaterly Journal of the Royal Meteorological Society* 77, pp. 569–97.

D. Q. WARK, G. YAMAMOTO & J. H. LIENESCH (1962) Methods of estimating infra-red flux and surface temperature from meteorological satellites, *Journal of the Atmospheric Sciences* 19, pp. 369–84.

J. S. WINSTON (1969) Global distribution of cloudiness and radiation as measured from weather satellites, in *World Survey of Climatology*, Volume 4, *Climate of the Free Atmosphere*, D. F. Rex (ed.), Elsevier, Amsterdam – London – New York, pp. 247–80.

Chapter 12

R. K. ANDERSON, J. P. ASHMAN, F. BITTNER, G. R. FARR, E. W. FERGUSON, V. J. OLIVER & A. H. SMITH (1969) *Application of meteorological satellite data in analysis and forecasting*, Air Weather Service, U.S.A.F., Technical Report 212, 223 pp.

R. K. ANDERSON, E. W. FERGUSON & V. J. OLIVER (1966) *The use of satellite pictures in weather analysis and forecasting*, W.M.O. Technical Note No. 75, World Meteorological Organization, Geneva, Switzerland, 184 pp.

P. D. ASTAPENKO (1964) *Atmospheric processes in the high latitudes of the southern hemisphere*, Israel Program for Scientific Translations, Jerusalem, 286 pp.

E. C. BARRETT (1970) Rethinking climatology, in *Progress in Geography*, Volume 2, C. Board, R. J. Chorley, P. Haggett & D. R. Stoddart (eds.), Edward Arnold, London, pp. 154–205.

H. BERG (1968) In *Atmosphere, weather and climate*, R. G. Barry & R. J. Chorley, Methuen, London, p. 173, 2nd edition, 1971, p. 207.

R. J. BOUCHER (1963) *Synoptic interpretations of cloud vortex patterns as observed by meteorological satellites*, Final Report, Contract No. Czb.-10630, ARACON Geophysics Company, Concord, Massachusetts, pp. 176–80.

R. J. BOUCHER & H. J. NEWCOMB (1962) Synoptic interpretations of some Tiros vortex patterns; a preliminary cyclone model, *Journal of Applied Meteorology* 1, pp. 127–36.

J. H. CONOVER (1962) *Cloud interpretation from satellite altitudes*, Research Note 81, Air Force Cambridge Research Laboratories, Cambridge, Massachusetts, 55 pp.

J. H. CONOVER (1963) *Cloud interpretation from satellite altitudes*, Research Note 81, Supplement 1, Air Force Cambridge Research Laboratories, Cambridge, Massachusetts, 18 pp.

J. H. CONOVER (1964) The identification and significance of orographically-induced clouds observed by Tiros satellites, *Journal of Applied Meteorology* 3, pp. 226–32.

C. O. ERICKSON & J. S. WINSTON (1972) Tropical storm, mid-latitude, cloud-band connections and the autumnal build-up of the planetary circulation, *Journal of Applied Meteorology* 11, pp. 23–36.

S. FRITZ (1965) The significance of mountain lee waves as seen from satellite pictures, *Journal of Applied Meteorology* 4, pp. 31–7.

J. GENTILLI (1971) *Climates of Australia and New Zealand*, Volume 13 in *World Survey of Climatology*, H. E. Landsberg (ed.), Elsevier, Amsterdam, London & New York, 405 pp.

W. J. GIBBS (1949) A period of analysis for the Southern Ocean and implications in southern hemisphere circulation, *Weather Development Research Bulletin, (Australian)* 12, pp. 5–17.

W. J. GIBBS (1960) Antarctic synoptic analysis, in *Antarctic Meteorology*, Pergamon Press, Oxford, 483 pp.

W. GORCZYNSKI (1920) Sur le calcul du degre du continentalisme et son application dans la climatologie, *Geografiska Annaler* 2, pp. 324–31.

L. B. GUYMER (1969) Estimation of 1000–500 mb thickness patterns from satellite of convective areas, in *Satellite Meteorology*, Bureau of Meteorology, Melbourne, Australia, pp. 51–5.

J. C. LANGFORD (1957) Southern Ocean analysis with special reference to the period December 1954 to March 1955, *Australian Meteorological Magazine* 16, pp. 1–22.

H. VAN LOON (1965) A climatological study of the atmospheric circulation in the southern hemisphere during the I.G.Y., Part I: 1 July 1957 to 31 March 1958, *Journal of Applied Meteorology* 4, pp. 479–91.

H. VAN LOON (1967) A climatological study of the atmospheric circulation in the southern hemisphere during the I.G.Y., Part II, *Journal of Applied Meteorology* 6, pp. 803–15.

D. W. MARTIN (1968a) A re-analysis of synoptic conditions over and south of Australia from 10 September to 14 September 1964, Unpublished notes, Bureau of Meteorology, Melbourne, Australia.

D. W. MARTIN (1968b) *Satellite studies of cyclonic development over the Southern Ocean*, I.A.M.R.C. Technical Report No. 9, Bureau of Meteorology, Melbourne, Australia.

D. W. MARTIN (1968c) Unpublished Ph.D. thesis, University of Wisconsin.

J. NAMIAS (1964) Seasonal persistence and recurrence of European blocking during 1958–60, *Tellus* 3, pp. 394–407.

C. W. NEWTON (1970) The role of extra-tropical disturbances in the global atmosphere, in *The global circulation of the atmosphere*, G. A. Corby (ed.), Royal Meteorological Society, London, pp. 137–58.

V. J. OLIVER, R. K. ANDERSON & E. W. FERGUSON (1964) Some examples of the detection of jet streams from Tiros photographs, *Monthly Weather Review* 92, pp. 441–8.

E. PALMEN & H. RIEHL (1957) Budget of angular momentum and energy in tropical cyclones, *Journal of Meteorology* 14, pp. 150–9.

F. C. PARMENTER (1967) Picture of the month, *Monthly Weather Review* 95, pp. 153–4.

E. REITER (1969), Tropospheric circulation and jet streams, in *World Survey of Climatology*, Volume 4, *Climate of the free atmosphere*, D. F. Rex (ed.), Elsevier, Amsterdam, London & New York, pp. 85–204.

H. RIEHL (1969) On the role of the tropics in the general circulation of the atmosphere, *Weather* 24, pp. 288–308.

H. RIEHL *et al.* (1952) *Forecasting in middle latitudes*, Meteorological Monographs 1 (5), American Meteorological Society, Boston, Massachusetts, 80 pp.

G. RUMNEY (1971) *Climatology and the world's climates*, Macmillan, New York, & Collier-Macmillan, London, 656 pp.

A. H. SMITH (1968) Unpublished notes regarding the correlation of sun glints and the location and orientation of surface ridgelines, E.T.A.C. Satellite Section, U.S. Air Force.

N. A. STRETEN (1968a) A note on multiple image photo-mosaics for the southern hemisphere, *Australian Meteorological Magazine* 16, pp. 127–36.

N. A. STRETEN (1968b) Some aspects of high latitude southern hemisphere summer circulation as viewed by ESSA 3, *Journal of Applied Meteorology* 7, pp. 324–32.

N. A. STRETEN & A. J. TROUP (1973) A synoptic climatology of satellite observed cloud vortices over the Southern Hemisphere, *Quarterly Journal of the Royal Meteorological Society* 99, pp. 56–72.

O. G. SUTTON (1965) The energy of the atmosphere, *Science Journal* 1, pp. 76–81.

J. J. TALJAARD (1967) Development, distribution and movement of cyclones and anticyclones in the southern hemisphere during the I.G.Y., *Journal of Applied Meteorology* 6, pp. 324–32.

A. J. TROUP & N. A. STRETEN (1972) Satellite observed Southern Hemisphere cloud vortices in relation to conventional observations, *Journal of Applied Meteorology* 11, pp. 909–17.

I. P. VETLOV (1966) Role of satellites in meteorology, in *Interpretation and use of meteorological satellite data*, W.M.O. Training Seminar Report, 20 pp.

W. VIEZEE, S. M. SEREBRENY, R. M. ENDLICH & R. M. TRUEDEAN (1966) *Tiros-viewed jet stream cloud patterns in relation to wind, temperature and turbulence*, Contract Cwb. 11129, Stanford Research Institute, Menlo Park, California, 83 pp.

L. F. WHITNEY, A. TIMCHALK & T. I. GRAY (1966) On locating jet streams from Tiros photographs, *Monthly Weather Review* 94, pp. 127–38.

W. K. WIDGER (1964) A synthesis of interpretations of extra-tropical vortex patterns as seen by Tiros, *Monthly Weather Review* 92, pp. 263–82.

J. S. WINSTON (1967) Zonal and meridional analysis of 5-day averaged outgoing long-wave radiation data from Tiros IV over the Pacific sector in relation to the northern hemisphere circulation, *Journal of Applied Meteorology* 6, pp. 453–63.

J. W. ZILLMAN (1969) Interpretations of satellite data over the Southern Ocean using the technique of Martin (1968), in *Satellite Meteorology*, Bureau of Meteorology, Melbourne, Australia, pp. 43–7.

Chapter 13

L. J. ALLISON, J. S. KENNEDY & G. W. NICHOLAS (1966) Examples of the meteorological capability of the Nimbus satellite, *Journal of Applied Meteorology* 5, pp. 314–33.

L. J. ALLISON & G. WARNECKE (1966) The synoptic interpretation of Tiros III radiation data recorded on 16 July 1961, *Bulletin of the American Meteorological Society* 47, pp. 374–83.

P. D. ASTAPENKO (1964) *Atmospheric processes in the high latitudes of the southern hemisphere*, Israel Program for Scientific Translations, Jerusalem, 286 pp.

E. C. BARRETT (1972) *Geography from space*, Pergamon, Oxford, 98 pp.

A. A. BORISOV (1965) *Climates of the U.S.S.R.* (edited by C. A. Halstead), Oliver & Boyd, 255 pp.

C. E. P. BROOKS & W. A. QUENNELL (1928) *The influence of the Arctic Ice on the subsequent distribution of pressure over the eastern North Atlantic and Western Europe*, Geophysical Memoirs, Meteorological Office, London, 5, No. 41, 36 pp.

F. K. HARE (1968) The Arctic, *Quarterly Journal of the Royal Meteorological Society* 94, pp. 439–59.

F. LOEWE (1956) Precipitation and evaporation in the Antarctic, in *Meteorology of the Antarctic*, M. P. van Rooy (ed.), Weather Bureau, Department of Transport, Pretoria, South Africa, 240 pp.

H. VAN LOON (1962) On the movement of lows in the Ross and Weddell Sea sectors in summer, *Notos* 11, pp. 47–50.

E. P. MCCLAIN (1970) *Applications of environmental satellite data to oceanography and hydrology*, ESSA Technical Memorandum, N.E.S.C.T.M.-19, Washington, D.C., 12 pp.

E. P. MCCLAIN & D. R. BAKER (1969) *Experimental large-scale snow & ice mapping with composite minimum brightness charts*, ESSA Technical Memorandum, N.E.S.C.T.M.-12, Washington, D.C., 19 pp.

W. MEINARDUS (1906) Periodische Schwankungen der Eistrift bei Island, *Annalen der Hydrographie und Maritimen Meteorologie* 34, pp. 148–62, 227–39 and 278–85.

W. MEINARDUS (1938) Klimakunde der Antarktis, in *Handbuch der Klimatologie*, Band IV (U), W. Koppen & R. Geiger (eds.), Berlin.

N.E.S.C. (1968) *Arctic sea ice studies with the aid of polar-orbiting satellites*, in U.S. Senate Document 71, Government Printing Office, Washington, D.C., pp. 39–40.

W. NORDBERG (1965) Geophysical observations from Nimbus I, *Science* 150, No. 3696, pp. 559–72.

W. NORDBERG, A. W. MCCULLOCH, L. L. FOSHEE & W. R. BANDEEN (1966) Prelimary results from Nimbus II, *Bulletin of the American*

Meteorological Society 47, pp. 857–72.

E. RASCHKE & W. R. BANDEEN (1967) A quasi-global analysis of tropospheric water vapour content from Tiros IV radiation data, *Journal of Applied Meteorology* 6, pp. 468–81.

G. R. RUMNEY (1968) *Climatology and the world's climates*, Macmillan, New York, & Collier-Macmillan, London, 656 pp.

I. I. SCHELL (1970) Arctic ice and sea temperature anomalies in the northeast Atlantic and their significance for seasonal fore-shadowing locally and to the eastward, *Monthly Weather Review* 98, pp. 833–50.

N. A. STRETEN (1961) A note on observations at Mawson and field stations on MacRobertson land, in 1960, *Australian Meteorological Magazine* 34, pp. 45–62.

N. A. STRETEN (1968) Some aspects of high latitude southern hemisphere summer circulation as viewed by ESSA 3, *Journal of Applied Meteorology* 7, pp. 324–32.

E. VOWINCKEL & S. ORVIG (1967) Climate change over the Polar Ocean, I, The radiation budget, *Arch. Met. Geophys. Biokl.*, B.15, pp. 1–23.

G. T. WALKER (1947) Arctic conditions and world weather, *Quarterly Journal of the Royal Meteorological Society* 73, pp. 226–56.

D. Q. WARK & R. W. POPHAM (1962) *Ice photography from meteorological satellites Tiros I and Tiros II*, Meteorological Satellite Laboratory Report No. 8, U.S. Department of Commerce, Washington, D.C., 68 pp.

W. K. WIDGER, J. C. BARNES, E. S. MERRITT & R. B. SMITH (1966) *Meteorological interpretation of Nimbus H.R.I.R. data*, Allied Research Associates Inc., Concord, Massachusetts, Contract Report NAS5-9554, 150 pp.

W. VON WIESE (1924) Polareis und atmosphärische Schwankungen, *Geografiska Annaler* 6, pp. 273–99.

Chapter 14

B. P. ALISSOV (1968) In *Atmosphere, weather and climate*, R. G. Barry & R. J. Chorley, Methuen, London, p. 173, 2nd ed. 1971, pp. 380.

E. C. BARRETT (1967) *Viewing weather from space*, Longmans, London, & Praeger, New York, 154 pp.

E. C. BARRETT (1970) Rethinking climatology, in *Progress in Geography*, Volume II, C. Board, R. J. Chorley, P. Haggett & D. R. Stoddart (eds.), Edward Arnold, London, pp. 154–205.

E. C. BARRETT (with J. O. BAILEY) (1971) *Weather and climate*, Collins, Glasgow, 96 pp.

M. I. BUDYKO (1963) *Atlas of the heat balance of the globe*, Hydrometeorological Service of the U.S.S.R., Moscow, 69 pp.

C.O.E.S.A. (1962) *U.S. Standard Atmosphere*, U.S. Committee for the

Extension of the Standard Atmosphere, Washington, D.C., 278 pp.

H. FLOHN (1950) Neue Ausschauungen über die allgemeine Zirkulation der Atmosphäre unde ihre Klimatische Bedeutung, *Erdkunde* 4, pp. 141–62.

H. FLOHN (1957) Zur Frage der Einteilung der Klimazoner, *Erdkunde* 11, pp. 161–75.

J. F. GREAVES (1971) *Development of a global cloud model for simulating Earth-viewing space missions*, ARACON Geophysics Research Company, Concord, Massachusetts, 88 pp.

B. HAURWITZ & J. M. AUSTIN (1944) *Climatology*, McGraw-Hill, New York, 410 pp.

M. HENDL (1963) *Einführung in die Physikalische Klimatologie*, Band II, *Systemische Klimatologie*, V.E.B. Deutschen Verlag der Wissenschaften, Berlin.

A. HETTNER (1931) Die Klimate der Erde, *Geographische Schriften* 5.

I.C.A.O. (1954) *Manual of the I.C.A.O. Standard Atmosphere*, International Civil Aviation Authority, Document 7488, 132 pp.

W. KÖPPEN (1918) Klassification der Klimate, nach Temperatur, Niederschlag, und Jahreslauf, *Petermanns Geogr. Mitt.*, pp. 193–203 and pp. 243–8.

G. R. MCBOYLE (1971) Climatic classification of Australia by computer, *Australian Geographical Studies*, pp. 1–14.

A. A. MILLER (1951) Three new climatic maps, *Transactions, Institute of British Geographers* 17, pp. 13–20.

J. E. OLIVER (1970) A genetic approach to climatic classification, *Annals of the Association of American Geographers* 60, pp. 615–37.

W. H. TERJUNG (1970) An approach to climatic classification based on net radiation, *Proceedings of the Association of American Geographers* 2, pp. 140–4.

C. W. THORNTHWAITE (1948) An approach towards a rational classification of climate, *Geographical Review* 38, pp. 55–94.

G. T. TREWARTHA (1954) *An introduction to climate*, McGraw-Hill, New York, 402 pp.

A. A. WILCOCK (1968) Köppen after 50 years, *Annals of the Association of American Geographers* 58, pp. 12–28.

Indexes

Author index (references excluded)

Geographical index

Subject index